Flavonoids in Health and Disease

OXIDATIVE STRESS AND DISEASE

Series Editors

LESTER PACKER, PH.D.
ENRIQUE CADENAS, M.D., PH.D.
University of Southern California School of Pharmacy
Los Angeles, California

1. Oxidative Stress in Cancer, AIDS, and Neurodegenerative Diseases, *edited by Luc Montagnier, René Olivier, and Catherine Pasquier*
2. Understanding the Process of Aging: The Roles of Mitochondria, Free Radicals, and Antioxidants, *edited by Enrique Cadenas and Lester Packer*
3. Redox Regulation of Cell Signaling and Its Clinical Application, *edited by Lester Packer and Junji Yodoi*
4. Antioxidants in Diabetes Management, *edited by Lester Packer, Peter Rösen, Hans J. Tritschler, George L. King, and Angelo Azzi*
5. Free Radicals in Brain Pathophysiology, *edited by Giuseppe Poli, Enrique Cadenas, and Lester Packer*
6. Nutraceuticals in Health and Disease Prevention, *edited by Klaus Krämer, Peter-Paul Hoppe, and Lester Packer*
7. Environmental Stressors in Health and Disease, *edited by Jürgen Fuchs and Lester Packer*
8. Handbook of Antioxidants: Second Edition, Revised and Expanded, *edited by Enrique Cadenas and Lester Packer*
9. Flavonoids in Health and Disease: Second Edition, Revised and Expanded, *edited by Catherine A. Rice-Evans and Lester Packer*

Related Volumes

Vitamin E in Health and Disease: Biochemistry and Clinical Applications, *edited by Lester Packer and Jürgen Fuchs*

Vitamin A in Health and Disease, *edited by Rune Blomhoff*

Free Radicals and Oxidation Phenomena in Biological Systems, *edited by Marcel Roberfroid and Pedro Buc Calderon*

Biothiols in Health and Disease, *edited by Lester Packer and Enrique Cadenas*

Handbook of Antioxidants, *edited by Enrique Cadenas and Lester Packer*

Handbook of Synthetic Antioxidants, *edited by Lester Packer and Enrique Cadenas*

Vitamin C in Health and Disease, *edited by Lester Packer and Jürgen Fuchs*

Lipoic Acid in Health and Disease, *edited by Jürgen Fuchs, Lester Packer, and Guido Zimmer*

Additional Volumes in Preparation

Flavonoids in Health and Disease

Second Edition
Revised and Expanded

edited by

Catherine A. Rice-Evans
King's College London
London, England

Lester Packer
University of Southern California School of Pharmacy
Los Angeles, California

CRC Press
Taylor & Francis Group
Boca Raton London New York

CRC Press is an imprint of the
Taylor & Francis Group, an **informa** business

First published 2003 by Marcel Dekker, Inc.

Published 2019 by CRC Press
Taylor & Francis Group
6000 Broken Sound Parkway NW, Suite 300
Boca Raton, FL 33487-2742

© 2003 by Taylor & Francis Group, LLC
CRC Press is an imprint of Taylor & Francis Group, an Informa business

First issued in paperback 2019

No claim to original U.S. Government works

ISBN-13: 978-0-367-44672-7 (pbk)
ISBN-13: 978-0-8247-4234-8 (hbk)

Visit the Taylor & Francis Web site at
http://www.taylorandfrancis.com

and the CRC Press Web site at
http://www.crcpress.com

Library of Congress Cataloging-in-Publication Data
A catalog record for this book is available from the Library of Congress.

Series Introduction

Oxygen is a dangerous friend. Overwhelming evidence indicates that oxidative stress can lead to cell and tissue injury. However, the same free radicals that are generated during oxidative stress are produced during normal metabolism and thus are involved in both human health and disease.

> Free radicals are molecules with an odd number of electrons. The odd, or unpaired, electron is highly reactive as it seeks to pair with another free electron.
>
> Free radicals are generated during oxidative metabolism and energy production in the body.
>
> Free radicals are involved in:
>> Enzyme-catalyzed reactions
>> Electron transport in mitochondria
>> Signal transduction and gene expression
>> Activation of nuclear transcription factors
>> Oxidative damage to molecules, cells, and tissues
>> Antimicrobial action of neutrophils and macrophages
>> Aging and disease

Normal metabolism is dependent on oxygen, a free radical. Through evolution, oxygen was chosen as the terminal electron acceptor for respiration. The two unpaired electrons of oxygen spin in the same direction; thus, oxygen is a biradical, but is not a very dangerous free radical. Other oxygen-derived free radical species, such as superoxide or hydroxyl radicals, formed during metabolism or by ionizing radiation are stronger oxidants and are therefore more dangerous.

In addition to research on the biological effects of these reactive oxygen species, research on reactive nitrogen species has been gathering momentum. NO, or nitrogen monoxide (nitric oxide), is a free radical generated by NO synthase (NOS). This enzyme modulates physiological responses such as vasodilation or signaling in the brain. However, during inflammation, synthesis of NOS (iNOS) is induced. This iNOS can result in the overproduction of NO, causing damage. More worrisome, however, is the fact that excess NO can react with superoxide to produce the very toxic product peroxynitrite. Oxidation of lipids, proteins, and DNA can result, thereby increasing the likelihood of tissue injury.

Both reactive oxygen and nitrogen species are involved in normal cell regulation in which oxidants and redox status are important in signal transduction. Oxidative stress is increasingly seen as a major upstream component in the signaling cascade involved in inflammatory responses, stimulating adhesion molecule and chemoattractant production. Hydrogen peroxide, which breaks down to produce hydroxyl radicals, can also activate NF-κB, a transcription factor involved in stimulating inflammatory responses. Excess production of these reactive species is toxic, exerting cytostatic effects, causing membrane damage, and activating pathways of cell death (apoptosis and/or necrosis).

Virtually all diseases thus far examined involve free radicals. In most cases, free radicals are secondary to the disease process, but in some instances free radicals are causal. Thus, there is a delicate balance between oxidants and antioxidants in health and disease. Their proper balance is essential for ensuring healthy aging.

The term *oxidative stress* indicates that the antioxidant status of cells and tissues is altered by exposure to oxidants. The redox status is thus dependent on the degree to which a cell's components are in the oxidized state. In general, the reducing environment inside cells helps to prevent oxidative damage. In this reducing environment, disulfide bonds (S–S) do not spontaneously form, because sulfhydryl (SH) groups kept in the reduced state prevent protein misfolding or aggregation. This reducing environment is maintained by oxidative metabolism and by the action of antioxidant enzymes and substances such as glutathione, thioredoxin, vitamins E and C, and enzymes such as superoxide dismutase (SOD), catalase, and the selenium-dependent glutathione and thioredoxin hydroperoxidases, which serve to remove reactive oxygen species.

Changes in the redox status and depletion of antioxidants occur during oxidative stress. The thiol redox status is a useful index of oxidative stress mainly because metabolism and NADPH-dependent enzymes maintain cell glutathione (GSH) almost completely in its reduced state. Oxidized glutathione (glutathione disulfide, GSSG) accumulates under conditions of oxidant exposure, and this changes the ratio of oxidized to reduced glutathione; an increased ratio indicates oxidative stress. Many tissues contain large amounts of glutathione, 2–4 mM in erythrocytes or neural tissues and up to 8 mM in hepatic tissues. Reactive oxygen and nitrogen species can directly react with glutathione to lower the levels of this substance, the cell's primary preventative antioxidant.

Current hypotheses favor the idea that lowering oxidative stress can have a clinical benefit. Free radicals can be overproduced or the natural antioxidant system defenses weakened, first resulting in oxidative stress, and then leading to oxidative injury and disease. Examples of this process include heart disease and cancer. Oxidation of human low-density lipoproteins is considered the first step in the progression and eventual development of atherosclerosis, leading to cardiovascular disease. Oxidative DNA damage initiates carcinogenesis.

Compelling support for the involvement of free radicals in disease development comes from epidemiological studies showing that an enhanced antioxidant status is associated with reduced risk of several diseases. Vitamin E and prevention of cardiovascular disease is a notable example. Elevated antioxidant status is also associated with decreased incidence of cataracts and cancer, and some recent reports have suggested an inverse correlation between antioxidant status and occurrence of rheumatoid arthritis and diabetes mellitus. Indeed, the number of indications in which antioxidants may be useful in the prevention and/or the treatment of disease is increasing.

Oxidative stress, rather than being the primary cause of disease, is more often a secondary complication in many disorders. Oxidative stress diseases include inflammatory bowel diseases, retinal ischemia, cardiovascular disease and restenosis, AIDS, ARDS, and neurodegenerative diseases such as stroke, Parkinson's disease, and Alzheimer's disease. Such indications may prove amenable to antioxidant treatment because there is a clear involvement of oxidative injury in these disorders.

In this new series of books, the importance of oxidative stress in diseases associated with organ systems of the body will be highlighted by exploring the scientific evidence and the medical applications of this knowledge. The series will also highlight the major natural antioxidant enzymes and antioxidant substances such as vitamins E, A, and C, flavonoids, polyphenols, carotenoids, lipoic acid, and other nutrients present in food and beverages.

Oxidative stress is an underlying factor in health and disease. More and more evidence indicates that a proper balance between oxidants and antioxidants is involved in maintaining health and longevity and that altering this balance in favor of oxidants may result in pathological responses causing functional disorders and disease. This series is intended for researchers in the basic biomedical sciences and clinicians. The potential for healthy aging and disease prevention necessitates gaining further knowledge about how oxidants and antioxidants affect biological systems.

Flavonoids and flavonoid-rich botanical extracts increasingly have been a subject of interest for scientific research, conferences in the scientific community, the herbal products and dietary supplement industry, and the consumer. This interest has led to research exploring the molecular basis of their action in health. *Flavonoids in Health and Disease, Second Edition, Revised and Expanded* highlights the recent advances in this rapidly developing field of study.

Lester Packer
Enrique Cadenas

Preface

The evidence for flavonoid-rich food components as cardioprotective, neuro-protective, and chemopreventive agents is steadily accumulating. However, their mechanisms of action still remain to be elucidated. Although there are hundreds of flavonoid molecules in dietary plants, the major components of current interest for their beneficial health effects are members of the flavonol, flavanol, flavanone, anthocyanin, and hydroxycinnamate families, as well as specific stilbene and chalcone structures. Key molecules of common interest are the flavanol components of grapes, wine, and teas (epicatechin/catechin family with their procyanindin oligomers and gallate esters) in relation to cardioprotection; the berry fruit flavonoids, especially in the context of neuroprotection; and citrus flavanones and the chemopreventive properties of a variety of molecules including the green tea constituents resveratrol and quercetin. In addition, a number of more unusual sources of flavonoids are presented and discussed as potential lead molecules or pro-nutraceuticals for the future.

Flavonoids in Health and Disease, Second Edition, Revised and Expanded, reflects the major advances made in this exciting area of research over the last 5 years and the involvement of flavonoids and phenolic compounds in biochemical, physiological, and pharmacological processes that are implicated in disease prevention. In particular, the field has gained much momentum through the knowledge that flavonoids are modified on absorption by the small intestine, through conjugation and metabolism, and by the large intestine, through the actions of the colonic microflora, and by subsequent hepatic metabolism of the components that are absorbed. Thus, the flavonoid forms reaching the cells and tissues after leaving the gastrointestinal tract are chemically, biologically, and, in many instances, functionally distinct from the dietary forms, such features underlying their bioactivities. This new edition thoroughly scrutinizes and updates these aspects. In addition, the current thinking on and evidence for the modes of action of in vivo flavonoid metabolite forms, in particular under conditions of oxidative stress, are addressed. It is clear from the studies reported in the literature since the first edition that the vision of flavonoids functioning as hydrogen-donating antioxidants in vivo is receding and that their "antioxidant ability" to protect cells and tissues from oxidative stress is more likely to be through their influences on signal transduction pathways and gene expression, the key putative modes of action of these exciting physiological molecules. All these aspects are explored in this volume.

In addition, the book provides a thorough update on the flavonoid and phenolic profiles of dietary plants and herbs and the advances made in the analysis of these molecules in the environment of the plant, in chemical systems, on interaction with cells, and in vivo in biological fluids.

Catherine A. Rice-Evans
Lester Packer

Contents

ix

Contributors

Michael Aviram, D.Sc. Lipid Research Laboratory, Technion Faculty of Medicine, Rambam Medical Center, Haifa, Israel

Clinton S. Boyd Department of Cell and Molecular Pharmacology and Toxicology, University of Southern California School of Pharmacy, Los Angeles, California, U.S.A.

Enrique Cadenas, M.D., Ph.D. Department of Cell and Molecular Pharmacology and Toxicology, University of Southern California School of Pharmacy, Los Angeles, California, U.S.A.

Cécile Cren-Olivé Lille University of Science and Technology, Villeneuve d'Ascq, France

Andrea J. Day Procter Department of Food Science, University of Leeds, Leeds, England

Jennifer L. Donovan Laboratory of Drug Disposition and Pharmacogenetics, Medical University of South Carolina, Charleston, South Carolina, U.S.A.

Annie Fleuriet, Dr. Sci. Department of Plant Biochemistry and Molecular Biology, University of Montpellier II, Montpellier, France

Bianca Fuhrman, D. Sc. Lipid Research Laboratory, Technion Faculty of Medicine, Rambam Medical Center, Haifa, Israel

Alema Galijatovic, Ph.D. Department of Molecular Pharmacology and Experimental Therapeutics, Medical University of South Carolina, Charleston, South Carolina, U.S.A.

Claudio Gardana Institute of Biomedical Technologies–National Council of Research, Segrate, Italy

James A. Joseph, Ph.D.   Jean Mayer USDA Human Nutrition Center on Aging at Tufts University, Boston, Massachusetts, U.S.A.

Gunter G. C. Kuhnle, Dr. rer. nat.   Wolfson Center for Age-Related Diseases, King's College London, London, England

Jean-Jacques Macheix, Dr. Sci.   Department of Plant Biochemistry and Molecular Biology, University of Montpellier II, Montpellier, France

Kenneth R. Markham, Ph.D., F.N.Z.I.C., F.R.S.N.Z.   New Zealand Institute for Industrial Research and Development, Lower Hutt, New Zealand

Annamaria Pietta, Pharm. D.   Farmacia Dr. Carlo Bravi, Brescia, Italy

Piergiorgio Pietta, Ph.D.   Institute of Biomedical Technologies–National Council of Research, Segrate, Italy

Anna R. Proteggente, Pharm. D.   Wolfson Center for Age-Related Diseases, King's College London, London, England

Andreas R. Rechner   Wolfson Center for Age-Related Diseases, King's College London, London, England

Catherine A. Rice-Evans, Ph.D., D.Sc., F.R.C.Path.   Wolfson Center for Age-Related Diseases, King's College London, London, England

Bertrand Henri Rihn   Teaching Hospital of Brabois, Vandoeuvre, France

Christian Rolando   Lille University of Science and Technology, Villeneuve d'Ascq, France

Claude Saliou   Johnson & Johnson Consumer Products Worldwide, Skillman, New Jersey, U.S.A.

Hagen Schroeter, Ph.D.   Department of Cell and Molecular Pharmacology and Toxicology, University of Southern California School of Pharmacy, Los Angeles, California, U.S.A., and Wolfson Center for Age-Related Diseases, King's College London, London, England

Jeremy P. E. Spencer, Ph.D.   Wolfson Center for Age-Related Diseases, King's College London, London, England

Surjit Kaila Singh Srai, Ph.D.   Department of Biochemistry and Molecular Biology, Royal Free University College Medical School, London, England

Ewald E. Swinny, Ph.D.   Department of Horticulture, Viticulture, and Enology, University of Adelaide, Adelaide, South Australia

Jaya b. Vaidyanathan Department of Cell and Molecular Pharmacology and Experimental Therapeutics, Medical University of South Carolina, Charleston, South Carolina, U.S.A.

Frans H. M. M. van de Put, Ph.D. Unilever Health Institute, Vlaardingen, The Netherlands

Richard A. Walgren, M.D., Ph.D. Department of Cell and Molecular Pharmacology and Experimental Therapeutics, Medical University of South Carolina, Charleston, South Carolina, U.S.A.

Thomas Walle, Ph.D. Department of Cell and Molecular Pharmacology and Experimental Therapeutics, Medical University of South Carolina, Charleston, South Carolina, U.S.A.

U. Kristina Walle, R. Ph. Department of Cell and Molecular Pharmacology and Experimental Therapeutics, Medical University of South Carolina, Charleston, South Carolina, U.S.A.

Andrew L. Waterhouse, Ph.D. Department of Viticulture and Enology, University of California, Davis, Sacramento, California

Gary Williamson, Ph.D. Institute of Food Research, Norwich, England

Sheila Wiseman, Ph.D. Unilever Health Institute, Vlaardingen, The Netherlands

Kuresh A. Youdim, Ph.D. Wolfson Center for Age-Related Diseases, King's College London, London, England

Flavonoids in Health and Disease

1
Phenolic Acids in Fruits and Vegetables

Annie Fleuriet and Jean-Jacques Macheix
University of Montpellier II
Montpellier, France

I. INTRODUCTION

The several thousand polyphenols that have been described in plants can be grouped into distinct classes, most of which are found in fruits and vegetables [1,2]. Distinctions among these classes are drawn first on the basis of the number of constitutive carbon atoms and then in light of the structure of the basic skeleton. Phenolic acids belong to two different classes, hydroxybenzoic acids (HBA) and hydroxycinnamic acids (HCA), which derive from two nonphenolic molecules, benzoic and cinnamic acid, respectively. In contrast to other phenolic compounds, HBA and HCA present an acidic character because of the presence of one carboxylic group in the molecule. They are widely represented in plants, although their distribution may strongly vary with species, cultivar, and physiological stage. They clearly play a role both in the interactions between the plant and its biotic or abiotic environment and in the organoleptic and nutritional qualities of fruits, vegetables, and derived products, e.g., fruit juices, wines, and ciders. Furthermore, their antioxidant properties are essential in the stability of food products and in antioxidant defense mechanisms of biological systems. These last aspects are largely developed elsewhere in this volume.

 Plant organs consumed by humans as vegetables have various botanical origins, e.g., leaves, stems, shoots, flowers, roots, rhizomes, tubers, bulbs, seeds, pods, and even some fleshy fruits. In some cases, it is not easy to distinguish between fruits and vegetables, as there is no concordance between the botanical definitions and the common use of plant organs by the consumer. For instance, bean pods, tomatoes, eggplant fruits, and sweet peppers are fruits in a botanical sense, whereas they generally are commercially marketed as vegetables.

Qualitative and quantitative determinations of phenolic acids, especially the combined forms, have been significantly improved during the last two decades, allowing one to draw a general picture of their distribution in fruits and vegetables and their importance as food constituents. In the comprehensive reviews on these topics that have already been published [1–5] most of the oldest references may be found. In the present review, our attention is focused on the presence and content of phenolic acids in fruits (mainly fleshy fruits with their seeds) and vegetables, and on the main parameters that can modify them.

II. ANALYSIS

Soluble HBA or HCA derivatives are frequently extracted from fruits and vegetables with ethanol or methanol-water solutions (80/20, v/v), using low temperatures and adding an antioxidant to prevent oxidation during the extraction procedure. Chemical or enzymatic hydrolysis of the plant material is necessary when phenolic acids are linked to cell wall constituents to give insoluble forms [6]. Apolar solvents or supercritical carbon dioxide may be useful to extract phenolic lipids [7,8]. In the case of acylated flavonoids, solvents must be adapted to the characteristics of the flavonoid itself, e.g., acidic methanol for fruit anthocyanins, although some artefacts may appear under these conditions.

Purification of the raw extract is essential. This may be performed in a first stage by removing chlorophylls and carotenoids and in a second stage by extracting phenolic acids with ethyl acetate from the depigmented aqueous extract, using a method previously described for fruits [2]. A preliminary analysis on a polyamide column has the advantage of separating the two groups of HCA derivatives: glucose derivatives on the one hand and quinic, tartaric, malic, or galactaric derivatives on the other [7]. Paper chromatography, classical or high-performance thin-layer chromatography, and column chromatography have been used extensively since the 1960s to separate phenolic acids, both before and after hydrolysis of esters and glycosides. Furthermore, separation of phenolic acid conjugates has greatly progressed thanks to high-performance capillary electrophoresis [9,10] and high-performance liquid chromatography (HPLC), which also allows quantitative determinations. In particular, the development of reversed-phase columns has greatly improved the separation performance of HCA and HBA derivatives [7].

In addition to analytical separations, the identification of phenolic acids has greatly benefited from the development of modern techniques (infrared [IR] and nuclear magnetic resonance [NMR] spectroscopy, mass spectrometry, etc.), that have added to the accurate knowledge of the structure of natural phenolic molecules [7]. New analytical approaches, including Raman spectroscopy, also allow in situ detection of HCA covalenty linked to cell wall constituents [6]. Some early approximate identifications have now been rectified, but there may be others as yet unrecognized [4].

In some unusual cases, spectrophotometric estimation of a major phenolic acid may be performed directly in plant extracts, such as chlorogenic acid in apples, pears, or potatoes [2,11], but this gives approximative information. From a quantitative point of view, HPLC techniques appear to be the most suitable, and they have been widely developed for estimating individual plant phenolic acids in their native forms [7]. Numerous examples concerning fruits and vegetables have already been reported [1,2]. Nevertheless, given the diversity and complexity of the combined forms naturally present, it has often been easier to determine phenolic acids released after hydrolysis of the extract, although some molecules might then be degraded.

A rapid fluorometric determination of p-coumaric, protocatechuic, and gallic acids has also been proposed in persimmon [12], but interference with other phenolic compounds is likely. Moreover, the radical scavenging activities of HBA and HCA may be used for their quantitative determination by chemiluminescence in the presence of hydrogen peroxide [13].

III. OCCURRENCE IN FRUITS AND VEGETABLES

In most cases, phenolic acids are not found in a free state, except in trace levels, but as combined forms, either soluble and then accumulated in the vacuole or insoluble when linked to cell wall components. Nevertheless, some exceptional situations can cause phenolic acids to accumulate in the free form [2]: brutal extraction conditions, physiological disturbances, contamination by microorganisms, anaerobiosis, processing of fruit juices, and winemaking. As they also accumulate when plant extracts are submitted to hydrolysis, the free HCA profile may characterize the plant material, and it has been used to discriminate between blood and blond oranges [14]. In rare cases, for example, in *Capsicum* species, the balance between free and combined forms may serve as a chemotaxonomic criterion: free phenolic acids are present in fruits of *C. annum*, whereas only the glycosylated forms appear in *C. frutescens* [15].

A. Hydroxybenzoic Acids

HBAs have a general structure of the C6-C3 type derived directly from benzoic acid (Fig. 1), and variations in structure lie in the hydroxylations and methoxylations of the aromatic cycle. They are mainly present in fruits and vegetables in the form of *O*-glucosides, but glucose esters of p-hydroxybenzoic, vanillic, or syringic acids have also been reported, e.g., in garden cress (Table 1). In most of

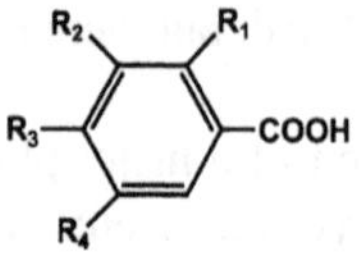

R₁=R₂=R₄=H, R₃=OH p-hydroxybenzoic acid
R₁=R₄=H, R₂=R₃=OH   protocatechuic acid
R₁=R₄=H, R₂= OCH₃, R₃=OH   vanillic acid
R₁=H, R₂=R₃=R₄=OH   gallic acid
R₁=H, R₂=R₄=OCH₃, R₃=OH   syringic acid
R₁=OH, R₂=R₃=R₄=H   salicylic acid
R₁=R₄=OH, R₂=R₃=H   gentisic acid

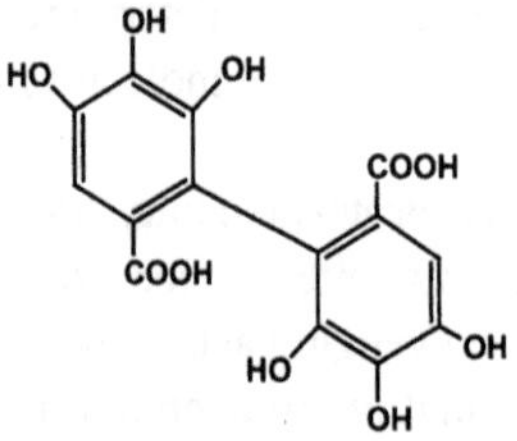

hexahydroxydiphenic acid

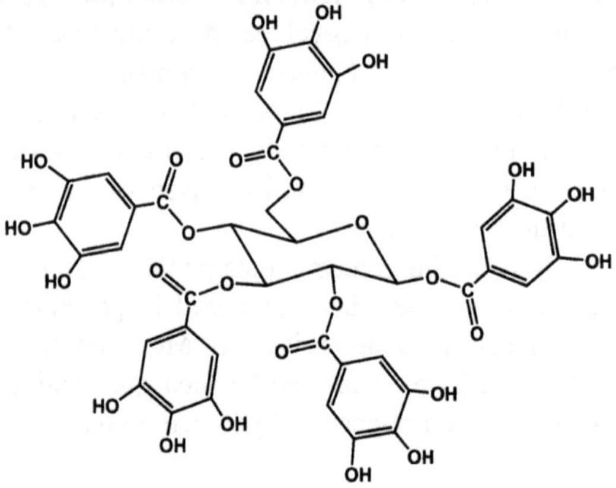

1,2,3,4,6-pentagalloylglucose

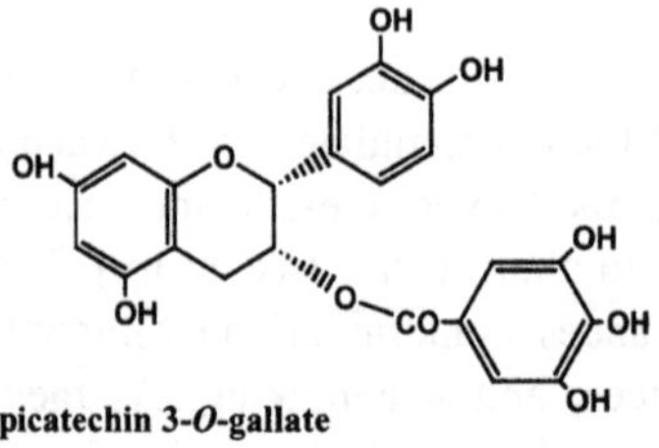

(-)-epicatechin 3-O-gallate

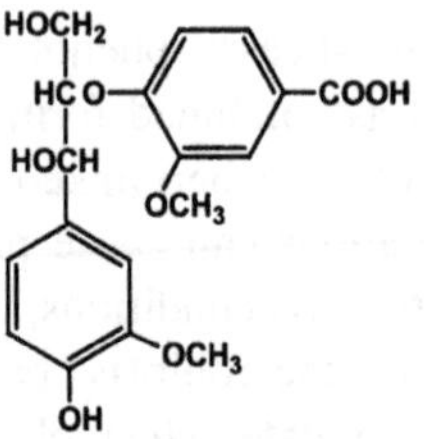

guaiacylglycerol-8'-vanillic acid ether

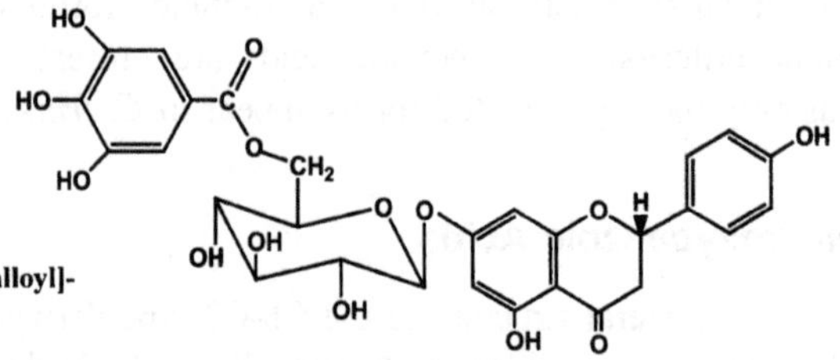

(-)-naringenin-7-O-β-D-[6''-O-galloyl]-
glucopyranoside

Figure 1 Chemical structure of hydroxybenzoic acids and some derivatives identified in fruits and vegetables.

Table 1 Contents of Hydroxybenzoic Acids in Vegetables and Ripe Fruits[a]

Compounds	Grape	Tomato	Cherry	Plum	Strawberry	Blackberry	Black currant	Blueberry	Horseradish leaf	Onion peel	Garden cress	Potato peel
Free acids[b]												
p-hydroxybenzoic (p-HB)	0–0.07		<0.5–5		10–36	6–16	0–16	—				
Protocatechuic (P)	t				<0.5–6	68–189	<10–52	—				37.7
Vanillic (V)	0.07–2.275	<0.5–1	2–11	2–12	<0.5–4							
Salicylic	0.04	<0.5–1										
Gallic (G)	t–0.46		11–44			8–67	30–62	1–2				129.6
Syringic (Sy)	t–11											
Derivatives												
p-HB glucoside			14	3	3–7	4–18	4–10	4–5	38	9	3	t
p-HB glucose									—	—	3	—
V glucoside			6	5					66	3	2	1
V glucose				3					6	—	45	—
Sy glucoside											2	—
Sy glucose											6	—
P glucoside			2	1		2–4	1	3–6		91	—	—
G glucoside						1	1	2–9			—	—
G glucose						3	4–7	—				—
G quinic acid					1	—	1	—				—
Ellagic acid					19.9							

[a] Extreme value for several cultivars (milligrams per kilogram fresh weight); t, traces;—, not detectable.
[b] Determined after hydrolysis in most cases.
Source: Refs. 1–3,19,123.

the important species of fruits and vegetables, because HBA conjugates are only found in trace concentrations, their identification is difficult. The presence of free HBA likely corresponds to degradation products from conjugates forms, during either extraction or subsequent hydrolysis. For example, salicylic, p-hydroxybenzoic, vanillic, gentisic, 3,4-dihydroxybenzoic, syringic, p-coumaric, and gallic acids were identified in the fruit of *Diospyros lotus*, whereas no information was reported about native forms [16].

Three HBAs (*p*-hydroxybenzoic, vanillic, and protocatechuic) are apparently universal in the angiosperms, and others (e.g., syringic, gallic, salicylic) are also frequently present in either complex structures, i.e., hydrolysable tannins, or as simple derivatives in combination with sugars or organic acids.

Gallic acid, hexahydroxydiphenic acid, and pentagalloylglucose (Fig. 1) are also constituents of hydrolyzable tannins. In addition, very low concentrations of gallic acid are found in fruits in the form of esters with quinic acid (theogallin) or glucose (glucogallin) and in the form of glucosides. Glucogallin has also been identified in persimmon and isolated only from astringent immature fruit, whereas free gallic acid was found in immature fruit of both astringent and nonastringent varieties [2]. Glucogallin was thus proposed as a good index for distinguishing between astringent and nonastringent varieties. Gallic acid is also found combined with naringenin in fruits of *Acacia farnesiana* or with (−)−epicatechin to form epicatechin 3-O-gallate, a constituent of unripe grapes [2] (Fig. 1). Ellagic acid, a dimer of gallic acid, is a component of ellagitannins, but it has also been reported in the free form and as arabinoside, acetylxyloside, or acetylarabinoside in raspberry and strawberry [17–19].

p-Hydroxybenzoic and vanillic acids are also present in numerous fruits and vegetables [1], and the native forms are frequently simple combinations with glucose (Table 1). Other derivatives have been detected in certain fruits [1,2]: the methyl ester of p-hydroxybenzoic acid in passion fruit, 3,4-dihydroxybenzoic aldehyde in banana, a phenylpropene benzoic acid derivative in fruits of Jamaican *Piper* species, and benzoyl esters and other derivatives in the fruits of *Aniba riparia*. Different new glycosides of HBA showing radical-scavenging activity [e.g., a new guaiacylglycerol-vanillic acid ether (Fig. 1)] have been identified in the fruits of *Boreava orientalis* [20].

Syringic acid or its glucoside has been reported in grape, plum, and some vegetables (Table 1), but its distribution appears to be very limited. It is very likely that p-hydroxybenzoic, vanillic, and syringic acids derive, at least partially, from the degradation of certain lignified zones of the fruit when these exist (stone, seed teguments, etc.).

Protocatechuic acid is found in a number of soft fruits and vegetables in the form of glucosides (Table 1), generally much less abundantly than those of p-hydroxybenzoic acid [1,2], except in onion peel, where it is prominent [21]. Salicylic and gentisic acids have been reported in very small quantities in the fruits of certain Solanaceae (tomato, eggplant, pepper), Cucurbitaceae (melon, cucumber) and other species (e.g., kiwi fruit, grapefruit, grape).

Very low concentrations of *p*-hydroxybenzoic, protocatechuic, and *t*-cinnamic acids have been reported in different species of mushrooms (*Agaricus* and *Lentinus* species), along with traces of caffeic acid [22].

B. Hydroxycinnamic Acids

Among fruit and vegetable phenolics, HCA derivatives play an important role that is due to their abundance and diversity. They all derive from cinnamic acid and are essentially present as combined forms of four basic molecules: coumaric, caffeic, ferulic, and sinapic acids (Fig. 2). Two main types of soluble derivatives have been identified (Fig. 2): (1) those involving an ester bond between the carboxylic function of phenolic acid and one of the alcoholic groups of an organic compound (e.g., quinic acid, glucose), for example, chlorogenic acid, which has been identified in numerous fruits and vegetables; and (2) those that involve a bond with one of the phenolic groups of the molecule, e.g., *p*-coumaric acid *O*-glucoside in tomato fruit. The diversity of HCA conjugates thus results from the nature of the bonds and that of the molecule(s) involved. In addition, for each of these compounds, the presence of a double bond in the lateral chain leads to the possible existence of two isomeric forms: *cis* (Z) and *trans* (E). Although native compounds are mainly of the *trans* form, isomerization occurs during extraction, purification, and processing under the effect of light or other chemical and physical factors.

1. Hydroxycinnamic Acids in Fruits

Quinic esters of HCA have been reported for a long time in fruits. The first were chlorogenic acid (5-*O*-caffeoylquinic acid*) and *p*-coumaroylquinic acid in apple [23]. Chlorogenic acid was subsequently found in many other fruits (Table 2), often accompanied by other caffeoylquinic isomers such as neochlorogenic acid (3-*O*-caffeoylquinic acid) and cryptochlorogenic acid (4-*O*-caffeoylquinic acid)/ isochlorogenic acid (a mixture of several di-*O*-caffeoylquinic acids) in coffee beans or coffee pulp, apple, avocado, pineapple, cherry, peach, eggplant [1,2], and in loquat fruit [24]. Red berries are particularly rich in caffeoylquinic esters, which confer on them, along with anthocyanins, high antioxidant activity [25,26]. The presence of tri-or tetra-*O*-caffeoylquinic acids in some fruits or leaves seems to be rather uncommon [4].

---

* The nomenclature of HCA quinic esters is in conformity with the International Union of Pure and Applied Chemistry (IUPAC) recommendations. Chlorogenic acid is thus 5-caffeoylquinic acid and not 3-caffeoyquinic acid as it was originally called.

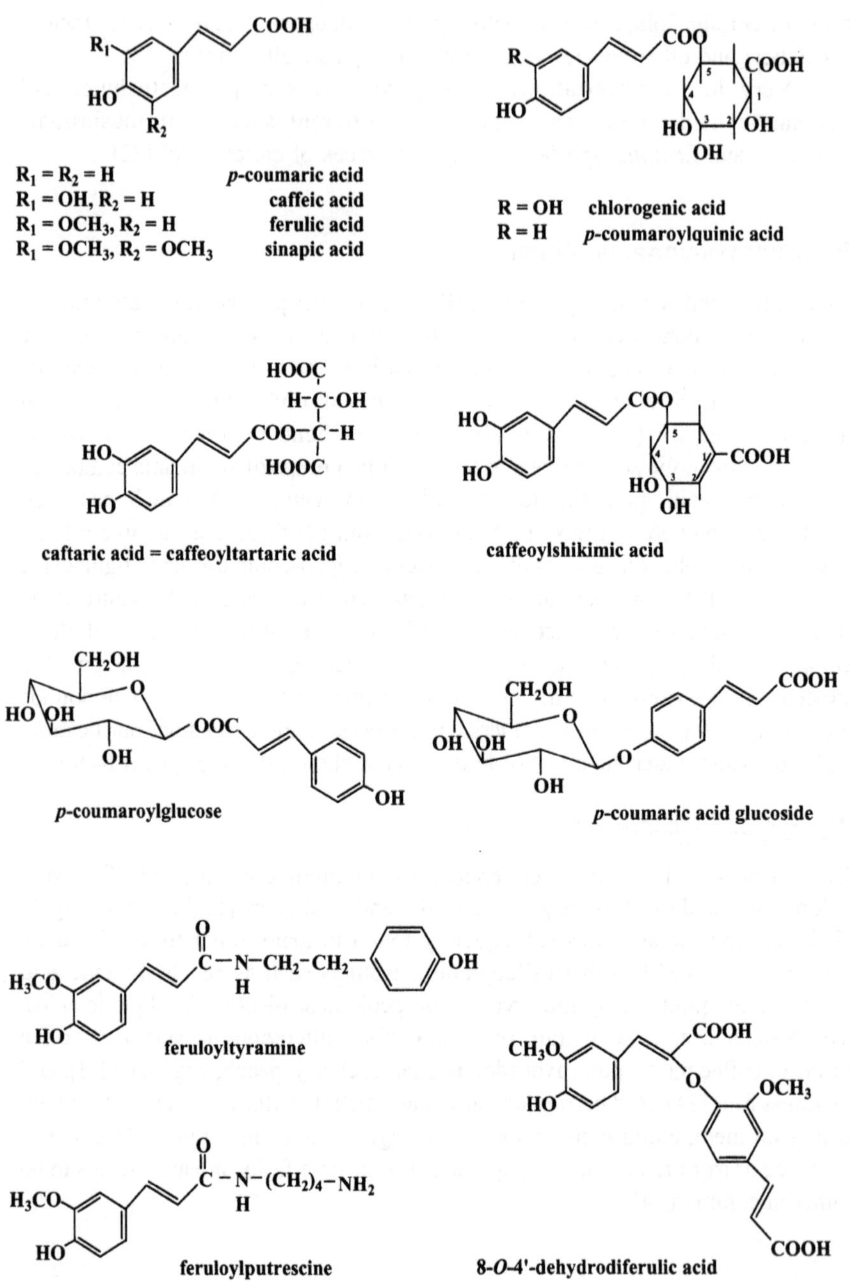

$R_1 = R_2 = H$ *p*-coumaric acid
$R_1 = OH, R_2 = H$ caffeic acid
$R_1 = OCH_3, R_2 = H$ ferulic acid
$R_1 = OCH_3, R_2 = OCH_3$ sinapic acid

R = OH   chlorogenic acid
R = H   *p*-coumaroylquinic acid

caftaric acid = caffeoyltartaric acid

caffeoylshikimic acid

p-coumaroylglucose

p-coumaric acid glucoside

feruloyltyramine

feruloylputrescine

8-*O*-4'-dehydrodiferulic acid

Figure 2 Chemical structure of hydroxycinnamic acids and some common derivatives identified in fruits and vegetables.

Table 2 Contents of Hydroxycinnamic Derivatives in Ripe Fruits[a]

Phenolic acid derivatives	Apple	Pear	Grape[b]	Tomato	Cherry	Plum	Peach	Apricot	Strawberry	Blackberry	Black currant	Blueberry	Loquat
Caffeic													
5-CQ	26–510	10–516	—	12–71	11–140	15–142	30–282	37–123	—	t–3	1–2	1851–2075	329–907
4-CQ	t–12	—	—	5–11	1–21	6–100	—	t	—	1	3–5	2–5	5–43
3-CQ	—	—	—	11.2	73–620	88–771	29–142	26–132	—	41–52	38–48	5–7	95–207
CG	t–6	t	—	t	—	2–7	—	—	t–2	3–6	19–30	t	—
C Gluc				15–48							2	3	
CT	—		6–621										
p-Coumaric													
5-p CQ	t–12	2–60		3.5	1–2	t–2	t	t–2				2–5	
4-p CQ	3–46	—			8–25	t–4	—	t–2			1–2	—	
3-p CQ	2–4	—			40–450	4–40	2–3	2–9		2–5	13–21		
p-CG	t–19	t–8		6–19	t	3–34	t		14–27	4–11	10–14	t	
p-C Gluc									t	2–5	3–10	3–15	
p-CT			1.8–484	19–68									
Ferulic													
FQ	2–4	—		2	t–13	1–34	2–9	5–22	t–2	2–4	1–3	8	28–145
FG	t–9	—		t	t	2–12		t		2–6	11–15	t	
F Gluc				8–15							2–4	5–10	
FT			0.98–65										
Sinapic													
SG	—	—		2.6		—			—			—	—

[a] Extreme values for several cultivars (milligrams per kilogram fresh weight); CQ, p CQ: FQ, caffeoyl, p-coumaroyl, feruloylquinic acids; CT, p-CT; FT, caffeoyl, p-coumaroyl, feruloyl tartaric acids; CG, p-CG, FG; SG, glucose esters; C Gluc, p-C Gluc, F Gluc, glucoside derivatives; t, traces; —, not detectable.

[b] Skin of white or red grapes.

Source: Refs. 1–3,24,66,68.

Quinic derivatives of other HCAs have also been identified in numerous fruits, e.g., several isomers of *p*-coumaroylquinic acid in apple and 5-*O*-feruloylquinic acid in tomato [2]. Although quinic derivatives are generally abundant in fruits, some contain none at all, e.g., grape, cranberry, and strawberry (Table 2). Mixed quinic di-esters of caffeic and ferulic acids are also present in robusta coffee beans [4].

Tartaric esters are limited to certain fruits of *Vitis* species and to some vegetables of the Asteraceae family (Tables 2 and 3). HPLC separations during the 1980s fully confirmed previous data by showing that the only combined form of caffeic acid in grape was in fact caffeoyltartaric acid (=caftaric acid) (Fig. 2). In addition, *p*-coumaroyl and feruloyl-tartaric acids (respectively named coutaric and fertaric acids) were found in varying proportions according to species and physiological stages [27]. Caffeoylshikimic esters (Fig. 2) are not widespread in plants, but they are very abundant in date fruit, where they participate in enzymic browning [2].

HCA derivatives with other hydroxyacids have rarely been identified in fruits, although *p*-coumaroylmalic acid is present in pear skin [28] and 2'-*O*-*p*-coumaroyl-, 2'-*O*-feruloylgalactaric acids, 2'-*O*-*p*-coumaroyl-, 2'-*O*-feruloyl-, and 2',4'-*O*-diferuloylglucaric acids in the peel of citrus fruits [29].

Since the identification of 1-*O*-*p*-coumaroylglucose (Fig. 2) and caffeic acid 3-*O*-glucoside in potato berry, numerous derivatives of HCA with simple sugars have been identified in various fruits [2] (Table 2), and cinnamoylglucose itself has been reported in blood orange [30]. Glucose esters and glucosides may be present simultaneously, for example, in tomato fruit, where *p*-coumaric and ferulic acids are present both as glucosides and as glucose esters (Fig. 2), whereas caffeic acid is only represented by caffeoylglucose. Glucose esters of sinapic acid have also been reported in tomato and in *Boreava orientalis*, where it is present along with a glucosinolate salt [31,32]. Different new phenyl-propanoid derivatives with simple sugars have been shown in the fresh fruit of *Piscrama quassioides* [33]. Verbascoside (Fig. 3) is an example of a rather more complex chemical combination that was identified in olives and in the fruit of different members of the Oleaceae family, along with several other caffeoyl glycosides [2].

Although HCA derivatives with sugars and hydroxyacids are simultane-ously present in numerous fruits [e.g., apple, tomato, cherry (Table 2)], several exceptions should be reported. Glucose derivatives of HCA are not present or are present only as traces in pear and in grape, whereas HCAs are only present in the form of conjugates with sugars in strawberry and cranberry [1,2].

The presence of hydroxycinnamoyl amides in fruits and vegetables has rarely been reported. Feruloyputrescine (Fig. 2) occurs in grapefruit and orange juice [29] but has not been found in tangerine or lemon. The *p*-coumaroyl or caffeoyl amides of hydro-or dihydroxyphenylalanine have been reported in cocoa [34]. Two new phenolic amides were isolated from the fruit of white pepper (*Piper nigrum* L.), *N-trans*-feruloyltyramine (Fig. 2) and *N-trans*-feruloylpiperidine, together with some other derivatives of piperidine and phenolics [2].

Acylation of anthocyanins with certain phenolic acids has been known for a long time [35]. Grape has been studied extensively, and it was shown that *p*-coumaric acid plays a major role in the acylation of malvidin (Fig. 3) and of all the other anthocyanins present, whereas caffeic acid combines only with malvidin 3-glucoside, a condition common in fruits and vegetables [35]. In eggplant, delphinidin is acylated with coumaric and caffeic acids; delphinidin 3-(*p*-coumaroylrutinoside)-5-glucoside is a major pigment in purple-skinned varieties. In the fruit of *Solanum guineese* (garden huckleberry), petunidin 3-(*p*-coumaroyl-rutinoside)-5-glucoside forms at least 70% of anthocyanins and is accompanied by very small quantities of several other acylated derivatives [36]. An extreme case concerns the blue berries of *Dianella* species, which contain delphinidin tetraglucosides bearing *p*-coumaroyl groups on two, three, or four of the sugars [37] (Fig. 3).

Flavonoid glycosides other than anthocyanins can also be acylated with HCA, but they have only rarely been reported in fruits, e.g., in the form of kaempferol *p*-coumaroylglycosides in *Tribulus terrestris*, 7-O-*p*-coumaroylglycoside-naringenin in *Mabea caudata*, or rhamnetin-3-*p*-coumaroylrhamninoside in *Rhamnus petiolaris* [38]. *p*-Coumaric and ferulic acids are also present in combination with betanidin monoglucoside (Fig. 3) in fruits of *Basella rubra* [39].

HCA may also be covalenty attached to aliphatic components of cutin and suberin. The amount of covalently bound phenolic compounds (*m*-, *p*-coumaric acids and flavonoids) in tomato fruit cutin increased during fruit development and accounted for as much as 6% of cutin membranes. Protoplasts isolated from immature tomato fruit secrete a wall that has been shown to contain suberin, in which phenolic compounds formed 25% of total monomers [3].

2. Hydroxycinnamic Acids in Vegetables and Cereals

Most of HCA conjugates previously described in fruits are also present in vegetables [1,4], but concentrations may be very different according to the botanical origin and the nature of the plant organ. An extensive review of the different HCA conjugates encountered in most vegetables consumed was published in 1999 [4], and here we only summarize some peculiar points.

Caffeoylquinic esters have been reported in most vegetables (Table 3): cabbages, endive, artichoke, potatoes, carrot, etc. In addition to the classical dicaffeoylquinic acids, diferuloylquinic acids are present in carrot root [40]. Several points must be underlined in Brassica vegetables: (1) 3-O-caffeoyl-

Table 3 Contents of Hydroxycinnamic Derivatives in Vegetables[a]

Phenolic acid derivatives	Cabbage	Broccoli	Radish	Lettuce	Endive	Artichoke	Spinach	Potato	Carrot	Rhubarb leaf	Faba bean pod	Corchurus olitorius leaves
Caffeic												
5-CQ	t-10	t	2	5-39	36-124	433	—	22-71[c]	5-541	—	—	3840
4-CQ	2-19	3	—	—	—	11	—	4-20	t	—	—	
3-CQ	6-120	58	t	—	—	12	—	3-9	t-9	—	—	
di-CQ						160-260[b]		3	1-58	—	—	1020
CG	t-2	2	2				16-21	—	—	8-13	—	
C Gluc											t	
CT				10-17	21-31							
di-CT				38-73	163-334							
CM			11-53	2-18	10-13						t-9	
p-Coumaric												
5-p CQ	—					10		—	t-6	t-8	—	
4-p CQ	2-17							—	2-3	3-23	—	
3-p CQ	2-104	7						—	—	6-48	—	
p-CG	t		t				16-21	—	—	20-22	t	
p-C Gluc							189-230					
p-C meso T							23-29					
p-CM	3-10										t-61	

Ferulic								
FQ	2–17	—	—	—	—	—	t–83	—
di-FQ							t–17	
FG	3–71	7	—	2	42–64	—	9	t–2
F Gluc					t			3–2
FT								
FM	7–33	14–16	—	2	—	—		t–31
Sinapic								
SG	4–273	10	t	—	2	—	3–21	3–4

[a] Extreme values for several cultivars (milligrams per kilogram fresh weight); t, traces;—, not detectable; CQ, p CQ; FQ, caffeoyl, p-coumaroyl, feruloylquinic acids; CT, p-CT; FT, caffeoyl, p-coumaroyl, feruloyl tartaric acids; CM, p-CM; FM, caffeoyl, p-coumaroyl, feruloyl malic acids; CG, p-CG, FG; SG, glucose esters; C Gluc, p-C Gluc; F Gluc, glucoside derivatives.
[b] Estimate for cynarine (1,3 di-CQ).
[c] Up to 229.1 in potato peel (Ref. 123).
Source: Refs. 1,2,4,40,43,44,89,123.

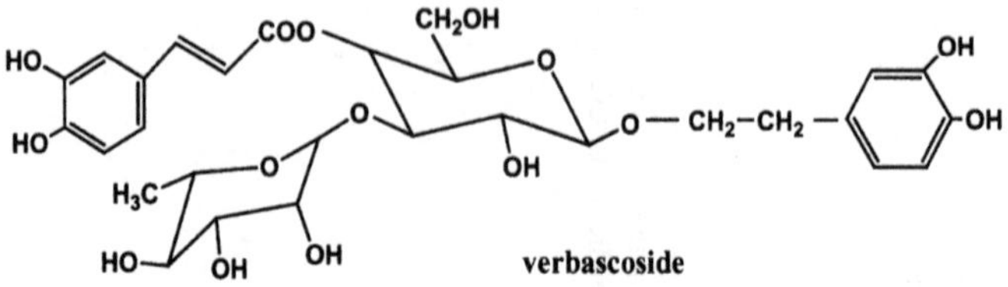

verbascoside

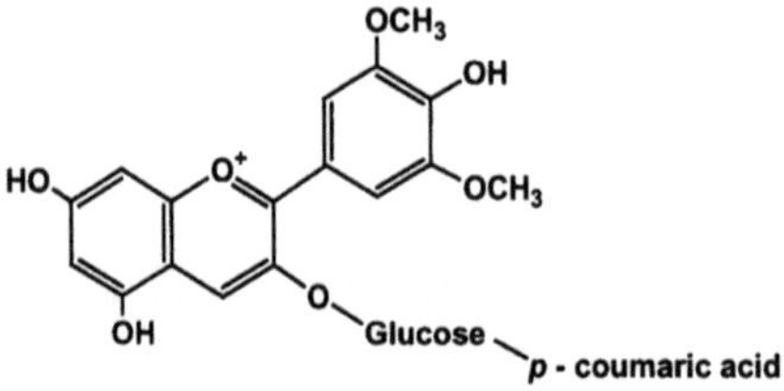

malvidin 3-(*p*-coumaroylglucoside)

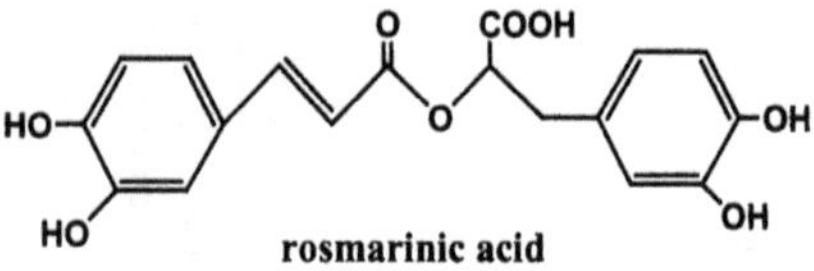

rosmarinic acid

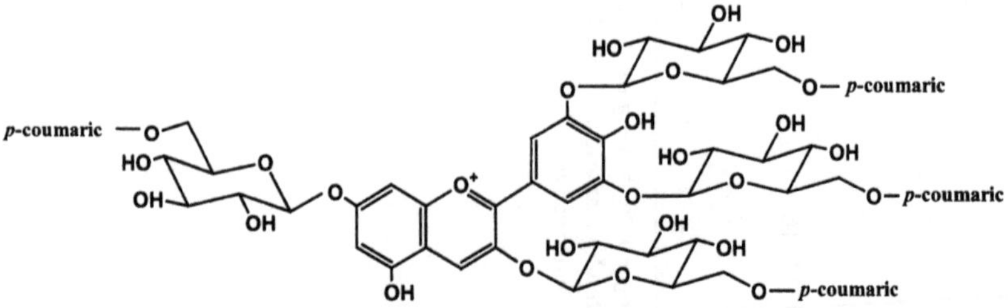

delphinidin 3,7,3',5'-tetra (6-*O*-*p*-coumaroyl-β-glucoside)

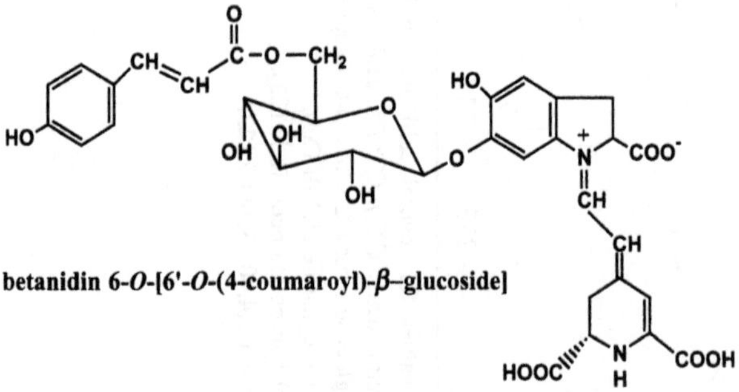

betanidin 6-*O*-[6'-*O*-(4-coumaroyl)-β-glucoside]

Figure 3 Chemical structure of some complex hydroxycinnamic derivatives identified in fruits and vegetables.

quinic acid is always isomers and a similar condition is found for p-coumaroyl and feruloyl quinic esters; (2) feruloyl and sinapoyl glucose esters are important in kale and red cabbage; (3) mixed feruloyl-sinapoyl esters of gentibiose are present in broccoli [41]; (4) malic esters are present in radish tuber and leaf, whereas quinic and glucose esters are present only as traces [1].

Chlorogenic acid is also detected in fennel teas prepared by infusion or decoction [42] and in the leaves of *Corchurus olitorius* used as a vegetable for soup [43]. Artichoke capitula is characterized by significant amounts of chlorogenic acid and various dicaffeoylquinic esters, especially 1,3-dicaffeoyl-quinic acid, known as cynarin [44]. In addition to the previous compounds, 3,5-dicaffeoyl-4-succinylquinic acid is present in garland [45] and several caffeoyl-methylquinic acids with a strong antioxidant activity were character-ized in bamboo shoots [46].

Along with quinic esters, caffeoyl and dicaffeoyltartaric acids are prom-inent in the leaves of some of the Asteraceae [1], e.g., lettuce, endive, and chicory. Although they are rarely present in fruits, malic esters of HCA are more frequently found in vegetables, e.g., in the leaves and pods of faba bean and in lettuce or spinach leaves (Table 3). Nevertheless, in the latter case, the prominent HCA conjugate is p-coumaroyl-*meso*-tartaric acid [1,47]. Tartaric acid occurs as p-coumaroyl, feruloyl, and caffeoyl-tartronic esters in the leaves of mung bean (*Vigna radiata*) [48]. Rosmarinic acid, a caffeic ester of 3,4-dihydroxyphenyllactic acid (Fig. 3), is found at a high level in extracts of various culinary and medicinal herbs (up to 1 g/kg fresh weight in thyme), where it shows remarkable antioxidant activity [49–51].

As previously shown in the case of fruits, sugar esters of HCA are also present in numerous vegetables, especially p-coumaroyl, caffeoyl, and sinapoyl glucose esters in Brassiceae, spinach leaves, and rhubarb stalk (Table 3). Root and/or derived cell cultures of red beet are rich in different HCA esters, e.g., several feruloylglucose conjugates, a feruloylsucrose monoester, a ferulic-aspartic acid amide, and a feruloylglycerol glucuronide [52,53]. Furthermore, red beet also contains low concentrations of two conjugates of HCA with betacyanins (the major coloring substances of red beet): lampranthin I (p-coumaroylbetanin) and lampranthin II (feruloylbetanin) [53]. In addition to the case of tomato fruit previously reported (Table 2), HCA glucosides have been identified in faba beans (leaves and pods) and are present as traces in carrot [1].

Although the presence of chlorogenic acid itself has rarely been reported in barley grains [54], HCAs, and ferulic acid in particular, are generally found as insoluble forms in various glucidic fractions of the cell wall. These compounds have not been reported in fleshy fruits but exist in Graminaeae and some other plants from which they are easily liberated by chemical or enzymatic hydrolysis. Several reviews on the subject were published in 1999 [6,55], and only a brief summary is given here. Ferulic and *p*-coumaric acids are bound through an ester linkage to the arabinoxylans or xyloglucans of Gramineae (wheat, maize, barley, rice, etc.), leaves, straw, and grain (bran and aleurone layer). A part of ferulic acid also exists as dehydrodimers (Fig. 2) (e.g., in grasses, cereals, Chinese water chestnut, sugar beet, carrot), which cross-link and strengthen the wall [56–58], and a small amount of ferulic acid is also found in the cell walls of the thick cuticle of fleshy scales of onions [21]. In some dicotyledons (e.g., sugar beet, spinach, beans) ferulic and *p*-coumaric acid are also bound to the galactose or arabinose residues of pectins [4,59].

Apolar esters of sterols and stanols with ferulic or *p*-coumaric acids have been reported in corn bran and other cereals [60]. Furthermore, oats contain numerous caffeic and ferulic esters of glycerol, long-chain alkanols, and ω-hydroxyacids, in addition to avenanthramides (esters of anthranilic acid with either *p*-coumaric, caffeic, or ferulic acids) [61].

Suberized potato includes long-chain fatty acids and phenolic derivatives [62]. Furthermore, in addition to the HCA esters of *p*-coumaric acid, a significant number of ligninlike monolignol structures exist within suberin [63].

IV. FACTORS AFFECTING THE PHENOLIC ACID CONTENT

Accumulation of phenolic acids in fruits and vegetables varies strongly in relation to different factors: (1) the genetic background, (2) the stage of development of the plant organ, and (3) the environmental and culture conditions. All these changes involve the regulation of phenolic metabolism (biosynthesis and degradation) and its integration in the program of cell and tissue differentiation, the control of gene expression, and the regulation of enzyme activities and of their compartmentation. The biosynthetic pathway of phenolic compounds is now well known and is not discussed here as it is not specific to fruits or vegetables. The deamination of phenylalanine, previously formed via the shikimate pathway, yields the nonphenolic cinnamic acid that is the direct precursor of the different HCAs and of their coenzyme A (CoA) esters through the general phenylpropanoid metabolism [64]. CoA esters of HCAs are common precursors of various other classes of phenolic compounds (e.g., HBA, anthocyanins, tannins, lignins). The structure and regulation of genes encoding the enzymes of the general phenylpropanoid metabolism from a number of plant species have been studied. Gene expression and enzyme activity are subject to large fluctuations in relation to endogenous and external factors (e.g., temperature, light, various types of stress) [64]. Furthermore, the enzymatic oxidation of phenolic compounds is of vital importance to the quality of fruits, vegetables, and their products, because of the formation of undesirable color and flavor and the loss of nutrients during processing (see Secs. V and VI).

A. Changes in the Phenolic Acid Patterns According to Species and Cultivars

Numerous factors may influence considerable qualitative and quantitative modifications in the phenolic acid patterns of fruits and vegetables from different species and cultivars. Although HBAs are present in most fruits and vegetables, the HBA content of fruits is generally low, except in certain fruits of the Rosaceae family and in particular blackberry, in which protocatechuic and gallic acid content may be very high: respectively, 189 and 67 mg/kg fresh weight in the richest cultivars (Table 1). Great interspecific differences in HBA exist in fruits and vegetables with regard to both quality and quantity, and such differences are also found among the varieties of the same species. In fact, qualitative and quantitative investigation of the native molecules of HBA derivatives is still inconclusive and it is difficult to draw general and final conclusions.

Comparing HCA content in numerous fruits and vegetables reveals enormous variations among species (Tables 2 and 3). Chlorogenic acid itself is especially abundant in coffee beans (6–10% on a dry matter basis), *C. olitorius* leaves (3.8 g/kg fresh weight [FW]), blueberries (2 g/kg FW), corn salad (approximately 1 g/kg FW), loquat fruit (329–907 mg/kg FW), eggplant (575–632 mg/kg FW), purple carrot (541 mg/kg FW), and artichoke (433 mg/kg FW), whereas it is present only as traces in Cucurbitaceae [1,2,4,24,40,43]. Similar variations are also frequently reported between cultivars of the same species, for example, 26 to 510 mg/kg chlorogenic acid in apples [2,65–68] or 6 to 621 mg/kg caftaric acid in grapes [2].

The relative proportions of each HCA are a characteristic of fruit in the mature stage. Caffeic acid is frequently the most abundant phenolic acid. It commonly exceeds 75% of total HCA in numerous fruits and vegetables (e.g., apple, plum, tomato, grape, red cabbage, endive, artichoke, potatoes) and may even form almost the entire HCA content in extreme cases, such as eggplant or certain blueberries. In some cases (e.g., pineapple, white currants, savoy cabbage, faba bean pod, spinach), *p*-coumaric acid is predominant, and in rare cases—in a few varieties of raspberry, for example—only traces of caffeic acid are found, whereas the other HCAs are dominant. Whereas ferulic acid usually forms only a small percentage of total HCA in fruits and vegetables, it can reach and even exceed 50% in peppers, some citrus, and some white grape cultivars [27]. Sinapic acid has been reported more rarely in fruits and is generally only observed as traces [31,32], whereas it may be abundant in various Brassica vegetables (Table 3).

The balance of the various HCA conjugates also characterizes fruit and vegetable species and cultivars. Thus, the HCA quinic ester patterns of stone and pome fruit differ considerably: the 3-isomers are major constituents in cherry and plum, whereas the 5-isomers are principally found in apple and pear

[[1,2,65,69,70] (Table 2)]. An identical difference is observed between Brassica (rich in 3-isomers) and Asteraceae (rich in 5-isomers) vegetables (Table 3). In most cases (e.g., apple, tomato, artichoke, carrot), glycosylated derivatives are distinctly less abundant than quinic esters, whereas the opposite proportion is more rarely observed (kale, raspberry). The relative proportions of glucose esters and glucosides of both HBAs and HCAs are also variable with the different species of fruits and vegetables (Tables 1, 2, and 3).

HCA ester content can be selected, among other parameters, to discriminate between grape species, but the most reliable criterion when comparing cultivars of the same species appears to be the percentage of each HCA, as shown in the case of *V. vinifera*, where the percentage of *p*-coumaroyl and caffeoyltartaric esters can be used to discriminate between varieties for taxonomic purposes [27].

B. Changes with Tissular Localization

The highest levels of HCA derivatives are frequently found in the external part of fleshy fruits, as shown for chlorogenic acid in pear and peach and 3-caffeoyl-quinic acid in cherry [2,70,71]. On the contrary, chlorogenic acid is often more abundant in the core than in the peel of apple, although this distribution depends on the cultivar [67,72]. Tomato is one of the better-known examples of HCA distribution: quantity of quinic esters in ripe fruit was found to be higher in the pulp than in the pericarp, whereas the opposite was found for glucose derivatives [2]. Tissue compartmentation of *p*-coumaroylglucose, caffeoyl, and 3,4-dime-thoxycinnamoyl glycosides makes it possible to discriminate clearly between the placenta and the pericarp of *Capsicum frutescens* [15]. In grape, although the level was always higher in skin than in pulp, the percentage of caffeoyltartaric acid was higher in the pulp, whereas the opposite was true for *p*-coumaroyltarta-ric acid [27]. *p*-Coumaroylgalactaric and feruloylgalactaric acids are also more abundant in the outer part of flavedo and albedo of citrus [2], and ferulic and sinapic acid concentrations are higher in sour orange peel [73]. Distribution of HCA derivatives is even more complex in certain cases, such as pineapple: in addition to the gradients between the inside and outside, there are very important longitudinal gradients, probably related to different stages of maturity of each of individual fruits that make up the pineapple [2].

The outer part of many plant organs consumed as vegetables also shows the highest concentrations of HBA and HCA conjugates. For example, puree from nonpeeled carrot roots contained 104 mg/kg chlorogenic acid, whereas only 28.3 mg/kg were found after removing approximatively 2 mm of periderm tissue [74]. In potato tuber, about 50% of the phenolic compounds (mostly chlorogenic acid and other mono- or di-caffeoylquinic esters) were located in the peel; the remainder decreased in concentration from the outside to the center of tubers [11]. In cereals, bran always presents the highest antioxidant activity, which is due to the localization of bound phenolic acids in the grain: the outer layers (husk, pericarp, testa, and aleurone layer) contain the greatest concentrations of total phenolics, whereas concentrations are considerably lower in the endosperm. About 80% of ferulic acid of both rye and wheat grain was found in the bran [25,75].

C. Changes with Physiological Stage

Concentrations of phenolic acids in a plant organ result from a balance between biosynthesis and further metabolism, including turnover and catabolism. Considerable variations are generally observed in the amount of phenolic acids according to the physiological stage when plant organs are picked up to be consumed or transformed by humans. This may concern each type of organ (leaves, flowers, stalks, tubers, roots, etc.), but the most spectacular cases are those of fruits, as considerable variations in phenolic compounds occur during maturation.

Concentrations of soluble forms of phenolic acid conjugates (expressed per unit of fresh or dry weight) are generally highest in young fruits, with a maximum during the early weeks after blossoming and a rapid decrease during fruit development [2,27]. In different apple cultivars, for example, maximal concentrations of chlorogenic acid, p-coumaroylquinic acid, and p-coumaroylglucose were found in very young fruits, followed by a constant decrease [2,72]. These changes make it possible to divide the life of a fruit into two main periods. During the first (approximately 2 months in apple), HCA derivatives accumulate in the fruit with a positive balance among in situ biosynthesis, migration, and possible reutilization. However, in the second period, this balance becomes negative and the overall HCA content in the fruit falls.

Loquat fruit represents an exception as the concentration of chlorogenic acid increased during maturation and became more prominent than neochlorogenic acid and all other phenolic compounds identified in this fruit [24]. The late accumulation of chlorogenic acid in this fruit results from the activation of its metabolism and especially from an increase in the enzyme activities of phenylalanine ammonia-lyase, 4-coumarate:CoA ligase, and hydroxycinnamoyl CoA:quinate hydroxycinnamoyltransferase. Consequently, the metabolism of chlorogenic acid may be considered to be a biochemical marker for maturation of loquat fruit.

In certain fruits the disappearance of phenolic acids may occur in relation to the biosynthesis of other phenolic compounds, for example, the decrease in hydroxycinnamic conjugates during growth and maturation of *Vitis vinifera* berries and the rapid increase in anthocyanin levels that occurs at the same time in the red cultivars [2]. In fruits of chili pepper *(Capsicum frutescens)* the onset of capsaicinoid accumulation and "ligninlike" material parallels the disappearance of the three cinnamoyl glycosides, which may be considered a source of precursors in capsaicinoid biosynthesis [15].

Quantitative changes are sometimes accompanied by qualitative ones. Thus, in tomato (cv. *cerasiforme*), most HCA conjugates appear during growth, and ripe fruit contains 11 different conjugates, whereas very young green fruit contains only chlorogenic acid [2]. Some compounds are characteristic of a physiological stage: chlorogenic acid forms 76% of total HCA derivatives in the unripe fruit, then falls to 15% in ripe fruit. By contrast, HCA glucosides form 23% and 84% during the same periods, a finding that may suggest certain metabolic relations between these compounds. Thus, growth and matutation of the tomato fruit are characterized by different expressions of the metabolism of HCA derivatives. In growing pulp it is mainly oriented toward the accumulation of quinic derivatives, whereas glucose derivatives (particularly glucosides) accumulate in the pericarp during maturation. A good correlation between variations in activities of enzymes of phenylpropanoid pathway and accumulation of phenolic compounds is observed in tomato. These data led to the notion that phenolic acids and their metabolism may be suitable markers of maturation.

Variations in the phenolic acid levels during growth and development of vegetables and cereals are not so homogeneous as those reported for fleshy fruits. It depends mainly on two parameters: the nature of the plant organ (leaves, tubers, grains, etc.) and the subcellular localization of the phenolic conjugates that accumulate either as soluble forms in the vacuole or linked to the cell wall.

During the development of soft or durum wheat grains, ferulic acid is mainly present as bound forms and accumulates during the milk stage, in correlation with high activities of phenylalanine and tyrosine-ammonia-lyases [76,77]. A decrease in ferulic acid level is then observed during grain ripening, but this may be due in part to the formation of alkali-resistant bonds in cross-linked polymers in cell walls, parallel to the progressive decrease in the grain water content. Peroxidases should play a role in the formation of these covalent linkages [56] as their activity occurs even in the last stages of grain ripening [77]. Changes in the ratios of different dehydrodiferulic acids were also observed in the cell walls during the growth period of sugar beet root along with a decrease of more than 50% [78]. These changes may be related to the major expansion of the storage root during the latter part of the growth period.

The integration of phenolic metabolism in the program of plant development raises the question of the possible role of these substances in physiological regulations. They have sometimes been implicated in the control of growth, maturation, and abscission [2,64], but these aspects are rather speculative and are not discussed here. HCA derivatives and coumarins may act as in situ inhibitors of seed germination in berries or other fleshy fruits, either directly or indirectly through the control of oxygen consumption [79]. It is also well known that various phenolic compounds play a role in the interactions between plants and microorganisms [64], and allelopathic effects of HBA and HCA have been reported in wheat root exudates [80]. Furthermore, they clearly participate in the resistance of plants to biological and environmental stresses [64] and play an important role in the browning capacity of plant organs [2].

D. Changes with Environmental Factors

Secondary metabolism, and in particular phenolic metabolism, largely depends on external factors such as light, temperature, and various stresses [64]. Although this has been mainly studied in relation to flavonoid production, we report here only information related to phenolic acids.

Caffeoyl quinic ester concentrations of potato tubers steadily increased after light exposure, whereas they decline during prolonged storage in the dark [81,82]. Rates of accumulation were influenced by cultivar, storage period, and light source. Furthermore, ratios of 5-:4-:3-caffeoylquinic esters were modified, but this had no effect on blackening of tubers. In *Vitis vinifera* berries, some phenolic components of aroma (e.g., methyl vanillate, methyl syringate) were significantly less abundant after bunch shading, whereas this treatment did not modify some other compounds (e.g., methyl salicylate, 4-vinylphenol) [83].

γ-Irradiation treatment degradates phenolic acids in a first step, but it later stimulates phenylalanine ammonia-lyase and the biosynthesis of *p*-coumaric acid, flavanones, and flavones in clementines [84]. *p*-Coumaric content was then particularly high in irradiated fruits after 49 days of storage at 3 °C and could be related to better resistance to pathogens. Similar data have been obtained in potato: irradiation of tubers to inhibit sprouting caused an initial reduction in phenolics, but an increased formation of chlorogenic acid and its isomers was later observed during storage [11].

Two major types of observations reveal relations between phenolic acids and temperature. First, various data link the overall effects of climate and local environmental conditions with the accumulation of phenols. Another group of results is taken from the frequent use of low temperatures in postharvest storage of fruits and vegetables. In this case, physiological disturbances may occur even when temperatures are maintained above 0 °C. Such effects are generally referred to as *chilling injury*, and they frequently take the form of discoloration, for which phenolic compounds may be directly responsible. Several examples show great variations in phenolic compounds or in the enzymes of their metabolism during cold, but the changes vary greatly, depending on the species and cultivar [2]. Cold storage of apples did not induce important variations in their chlorogenic acid content for up to 9 months, and the health benefits of phenolics should be maintained during long-term storage, although the response may slightly differ with cultivars [66,67]. On the contrary, numerous other examples indicate an increase in phenolic acids during cold storage: chlorogenic acid in Anjou pear [2], ellagitannins and p-coumaroylglucose in strawberries [19], a sucrose ester of ferulic acid in the peel of beetroot [53], chlorogenic acid in different cultivars of potato tubers stored at 0 °C [11,85], whereas a decrease was also observed at 5 °C in other cultivars [82]. p-Coumaric and caffeic derivatives also accumulate in pineapple during storage at 8 °C, and the content in phenolic acids was multiplied 10-fold 15 days after the fruits were returned at 20 °C [2]. In tomato, a fairly specific action of low-temperature storage was observed on chlorogenic acid metabolism. Among the enzymes tested, levels of phenylalanine ammonia-lyase and hydroxycinnamoyl-quinate transferase, two enzymes that allow synthesis of chlorogenic acid, increased considerably, in relation to chlorogenic acid accumulation [2].

Storage of bean seeds at high temperature (35 °C) and humidity causes textural defects along with an increase in free phenolic acids (caffeic, p-coumaric, ferulic and sinapic acids), a decrease in soluble esters, and a strong increase in ferulic acid bound to soluble pectins [59]. These modifications result in poor soaking imbibition of seeds and in prolonged cooking time.

Phenolic acids are directly implied in the response of plant organs to different kinds of stresses [64]: mechanical (wounding), chemical (various types of treatment), or microbiological (pathogen infection). Phenolic acids are involved in resistance in two ways: (1) by contributing to the healing of wounds by lignification of cell walls around wounded zones [86] and (2) through the antimicrobial properties demonstrated for many of them [87]. The compounds involved can be classified in three groups: (1) some are already present in the plant, and their level generally increases after stress; (2) others are formed only after injury but are derived from existing substances by hydrolysis or oxidation; (3) still others are biosynthesized de novo and can be classified as phytoalexins.

The effect of wounding has been particularly well studied in fruits [2] and in minimally processed fruits and vegetables [88]. The most immediate response to wounding is the oxidation of preexisting phenolic compounds and hence their degradation. Thus, the chlorogenic acid content of tomato fruit pericarp falls for 6 h after wounding, and activation of phenolic metabolism occurs later, with an increase in phenylalanine ammonia-lyase activity and consequent accumulation of chlorogenic acid, feruloyl-, p-coumaroyl, and sinapoyl-glucose [2]. Such an accumulation of caffeoylquinic acids or caffeoyl-and dicaffeoyl tartaric acids is also observed in shredded carrots and in minimally processed lettuce leaves [89–91], although its intensity may depend on storage conditions, either in air or in controlled atmospheres. Along with polyphenoloxidase and peroxidase activities, this increase in phenolic substrates is responsible of the browning of wounded tissues that shortens storage life of the product. Finally, the third aspect of response to wounding is the formation of healing tissues ("wound lignin" or suberin) that protect plant organs from water loss and also form a physiological barrier that prevents possible penetration of pathogens [2,11].

The increase in phenolic content with pathogen infection has been well documented in cell suspension cultures, especially at a molecular level [64]. It was postulated that as a defense mechanism, ferulic and *p*-coumaric acids are esterified to wall polysaccharides, possibly rendering the wall resistant to fungal enzymes either by masking the substrate or by altering the solubility properties of these wall polysaccharides [56,64]. Phenolic acids also possess antimicrobial properties, and chlorogenic acid and related compounds may function to arrest *Molinia fructicola*, in quiescent infections associated with immature and ripening peach fruits [92]. Chlorogenic and *p*-coumaroylquinic acids in apple are inhibitors of both *Botrytis* sp. spore germination and mycelial growth [2]. When their effects on growth of certain fungi are compared, *p*-coumaroylquinic is more inhibitory than chlorogenic at the same concentrations for *B. cinerea* and *Alternaria* sp., whereas *P. expansum* is less sensitive. Both quinic derivatives are stimulatory at low concentrations for *Botrytis* and *Penicillium* sp.

The acquisition of antimicrobial properties by phenolic compounds may derive from oxidation or hydrolysis. First, *o*-quinones are generally more active than *o*-diphenols and browning intensity is often greatest in highly resistant plants, suggesting that black and brown pigments contribute to resistance. Hydrolysis may be carried out by fungal pectic enzymes, as suggested by the appearance of free *p*-coumaric, caffeic, and ferulic acids in apple infected with *P. expansum* [2]. In this case, the damage does not cause much browning around the infection site because of the inhibition of the phenolase system by acids released after hydrolysis of chlorogenic and *p*-coumaroylquinic acids in the fruit. Again in apple, antifungal compounds, such as 4-hydroxybenzoic acid produced after infection with *Sclerotinia fructigena*, are thought to be derived from the transformation of chlorogenic acid by the fungus. Free phenolic acids are the best inhibitors of growth of the fungi appearing during the postharvest storage, and their structure/activity relationships have been studied in vitro [87]. An additional methoxy group caused increased activity of HBA and HCA derivatives. Thus, ferulic and 2,5-methoxybenzoic acids showed a strong inhibition against all fungi tested.

Certain phenolic compounds can be biosynthesized de novo after infection. These compounds, which do not exist before infection and which have antimicrobial properties, are called *phytoalexins*. They are produced by plants as defense mechanisms in response to microbial infection [64,93], but the accumulated compounds are often flavonoids or coumarins, although benzoic acid itself has been shown in apple after infection.

Both constitutive phenolic acids and phytoalexins may be involved in the resistance of potatoes to *Erwinia* sp. [94]. This resistance could result from the increase of caffeic, chlorogenic, and ferulic acids and the formation of suberized barriers after wound injury or pathogen attacks [11,63,94]. Both polyester-forming HCAs and ligninlike monolignols of potato suberin constitute a dense covalent network capable of repelling water and protecting the cell wall from pathogenic attack [63]. The involvement of phenolic acids in potato resistance is nevertheless selective as they protect against some but not all pathogens: although levels of chlorogenic acid increased after infection of potato tubers with the fungus *Phytophtora infestans*, no differences in the levels were observed in resistant or nonresistant cultivars [95].

V. ROLE OF PHENOLIC ACIDS IN THE ORGANOLEPTIC AND NUTRITIONAL QUALITY OF FRUITS AND VEGETABLES

Phenolic acids contribute to the sensory and nutritional qualities of fruits, vegetables, and derived foods. Directly or indirectly, they play a role in color, astringency, bitterness, and aroma, and they also are of great interest to humans, because of their antioxidant capacity [2,4].

A. Phenolic Acids and Food Flavor

As reported, acylation of anthocyanins with *p*-coumaric and caffeic acids is common in fruits, and it is responsible for better color stability in fruit products [35]. For example, it has been shown that the difference of stability to light and heat of different *Sambucus* species results from the degree of anthocyanin acylation [96,97]. Diacylated anthocyanins are stabilized by the sandwich-type stacking caused by hydrophobic interaction between the anthocyanidin ring and the two aromatic acyl groups [98]. Intramolecular copigmentation involving *p*-coumaroyl-glucose units at three or four positions of delphinidin is responsible for the exceptionally deep blue color of *Daniella* sp. berries [37]. Furthermore, numerous flavonoids and HCA derivatives play a role in the intermolecular copigmentation by stabilizing the pigment in its colored form and being the cause of a bathochromic shift and of an increase in the absorbance in the visible band [98].

The color of plant organs may also be strongly modified by the appearance of brown compounds, which generally result from the enzymatic oxidation of phenolic compounds including caffeic esters [2,99]. These melanin-type pigments may appear naturally during maturation of certain fruits, but they generally occur after wounding and crushing of plant organs. The resultant discoloration affects both commercial quality and nutritional parameters. These aspects are discussed late in relation to the processing of fruits and vegetables.

The astringency of fruits results from the interaction of salivary proteins with tannins or other phenolics. Although chlorogenic itself has sometimes been reported to be astringent, HBAs play a major role as they participate in the formation of hydrolysable tannins [100]. For example, ellagitannins (ellagic acid esters of glucose) are responsible for the strong astringency of various fruits, e.g., pomegranate, persimmon, chestnuts, and fruits of Rosaceae. Furthermore, in rather rare cases, HBAs are also present in condensed tannins in the form of epicatechin-gallate [27].

The role of phenolic acids in the bitterness of fruit and fruit products is still a matter of discussion, but it was concluded that HCAs do not play any role in the taste of wines, even at high concentrations of caftaric acid and glutathionylcaftaric acid [27]. Verbascoside may contribute to bitterness in olives, but its concentration is always low in comparison to oleuropein concentration [2]. Phenylpropanoid sucrose esters with several acetyl groups are also responsible for the bitter taste of stone fruits of *Prunus* sp. [101].

The importance of phenolic acids in fruit aroma is low, though many simple aromatic phenols may be released by enzymatic or chemical reactions from glycosylated precursors during maturation or processing, e.g., in vanilla, passion fruit, mango, and apricot [102]. Such transformations are also at the origin of some aroma constituents in wines, ciders, and fruit juices, through the degradation of HCA conjugates. Vanillic acid participates, in addition to vanillin, in vanilla aroma, and cinnamaldehyde is the principal component of cinnamon flavor [103]. Ferulic acid is a potential precursor of off-flavors in stored citrus juice, and pasteurization increases both the release of free ferulic acid from bound forms and the formation of p-vinyl guaiacol [29].

B. Phenolic Acids as Antioxidants

Antioxidants play an important role in antioxidant defense mechanisms in biological systems, protecting lipids both in cells and in food products and having inhibitory effects on mutagenesis and carcinogenesis. Attention is now focused on natural antioxidants, since the use of synthetic antioxidants has been falling off because of their suspected action as cancer promotors [104]. Most natural antioxidants present a polyphenolic structure, and it is significant that most papers published since the early 1990s about the characterization of phenolic compounds, and especially phenolic acids, concern their antioxidant activity (see reviews 104–109; see also Ref. 110) and the different chapters of the present volume. Along with numerous other phenolic compounds, hydroxycinnamates and gallic acid derivatives (methyl and lauryl esters, propylgallate) act as free radical acceptors and show strong antioxidant properties [4,6,50,105,108].

Many fruits and vegetables (e.g., grape, citrus, olive, black pepper, spices, soya, cereals) and the derived foods and beverages are a good source of phenolic antioxidants and constitute an important part of our daily diet [4,5,25,26,107–109,111–115]. A good correlation between phenolic content and antioxidant activity is often observed, as reported for monomeric and dimeric hydroxycinnamates of rye bran [75] and various caffeoyl quinic esters in peach puree [116], tart cherries [117], and prunes and prune juice [69,118]. Regular consumption of phenolic antioxidants may provide protection against diseases, including cancer and cardio-and cerebrovascular diseases [110], and it increases the serum antioxidant capacity in humans, as shown after consumption of strawberries, spinach, or red wine [119].

In addition to flavonoids, HCA derivatives protect food products from oxidation: for example, the remarkable stability of virgin olive oil is directly related to its phenolic antioxidants [104,120,121]. They are also widely used as food antioxidant additives to protect lipid structures [4,6] and are good candidates for successful employment as topical protective agents against ultraviolet (UV) radiation–induced skin damage [122]. Agricultural and industrial residues and by-products of plant origin (e.g., potato peel waste, grape seeds, olive, apple or cranberry pomace, citrus peels and seeds) are attractive and cheap sources of natural antioxidants [123–125]. An extensive review of these last aspects, including the influence of processing conditions, has been published [108], and it clearly shows that antioxidant capacity is often associated with the presence of phenolic acids (Table 4).

The relationships between chemical structures of HCA conjugates and their antioxidant and free radical scavenging activity have been reported [5,50,106,108,110,126]. Although this may vary with temperature and the nature of the test used, antioxidant activity was always higher for free caffeic acid than

Table 4 Main Phenolic Acids in Crude Extracts from Agroindustrial Wastes

Residue	Identified compounds[a]
Durum wheat bran	*p*-HB, P, G, V, Sy, C, 5-CQ, *p*-C, F
Corn bran hemicellulose	*p*-C, F, diF
Oat hulls	*p*-HB, V, *p*-C, F
Potato peel extract	P, G, C, 5-CQ
Apple pomace	5-CQ
Lemon seeds	C, *p*-C, F, S
Sweet orange seeds and peel	C, *p*-C, F, S
Citrus peel molasses	*p*-C, F, S, FP
Grape marc	P, G, V
Grape pomace	G, G 3-glucoside, G 4-glucoside, *p*-CT, CT
Grape seeds	G, *p*-C
Cranberry pomace	*p*-HB, G, *p*-C, 5-CQ

[a] Phenolic acids: P, protocatechuic; *p*-HB, *p*-hydroxybenzoic; G, gallic; V, vanillic; Sy, syringic; *p*-C, *p*-coumaric; *p*-CT, coutaric; C, caffeic; 5-CQ, 5-caffeoylquinic (chlorogenic); CT, caftaric; F, ferulic; diF, diferulic; S, sinapic; FP, feruloylputrescine.
Source: Refs. 108,124,125.

for its glucose or quinic esters (e.g., chlorogenic acid), whereas it was lower for ferulic acid [50,75]. For the less active HCAs, p-coumaric and ferulic acids, esterification to tartaric acid may enhance ability to inhibit low-density lip-oprotein (LDL) oxidation [127]. The presence of hydroxyl groups in the ortho position increases antioxidant activity, as shown for caffeic and rosmarinic acids. Nevertheless, a limitation of the utilization of HCAs and their natural esters as lipid protectors is their low liposolublity, and propyl esters have been proposed to prevent oxidative deterioration of edible oils [128].

Although phenolics are generally considered good antioxidants, gallic acid and its derivatives have also been reported simultaneously to exert pro-oxidant effects on various biological molecules, and their consumption should be regarded with caution [108,129].

VI. CHANGES DURING FOOD PROCESSING

As reported, phenolic acids contribute to the organoleptic and nutritional qualities of fruits and vegetables, but they also play an important role in the acquisition of the sensory properties of derived products, fermented or not (e.g., juices, wines, beers, ciders, purees, jam, jelly, minimally processed fruits or vegetables, oils). In all cases, the phenolic profile is different from that of the original organ, as qualitative and quantitative modifications occur during processing. They mainly result from three set of reactions: (1) the oxidative degradation of phenolic acids, including enzymatic browning; (2) the release of free acids from conjugate forms; and (3) the formation of complex structures of phenolic acids and other chemicals, in particular proteins, tannins, and anthocyanins.

A. Phenolic Acids and Enzymatic Browning

Enzymatic browning is often observed during fruit and vegetable processing, and it results from an initial cellular decompartmentation. The formation of yellow and brown pigments is controlled by phenolic compound levels, the presence of oxygen, and polyphenoloxidase (PPO) activities. PPO is a membrane-bound enzyme that catalyzes the hydroxylation of monophenols to o-diphenols and their oxidation to quinones. Chlorogenic acid and its isomers in numerous fruits, and other caffeic esters in many others (e.g., caffeoylshikimic in date, caffeoyltartaric in grape), are easily degraded by PPO [2,27,65,99], and it seems that esterification of the carboxyl group of caffeic acid with quinic acid or tartaric acid leads to an increase in PPO activity in practically all cases. In fruit juice processing, unit operations such as crushing, prepressing, enzymatic treatments, and pressing provide opportunities for PPO activities, and HCAs and catechins are mainly affected [130]. The first PPO reaction products are quinone-colored substances, which then rapidly condense and do or do not combine with amino or sulfhydryl groups of proteins [65]. The relatively insoluble brown polymers formed are generally eliminated when they disturb the quality of the product, such as fruit juice.

The oxidation of grape caftaric acid during vinification or juice preparation has drawn the attention of many researchers. It is now established that oxidation of caftaric acid by PPO leads to the formation of 2-S-glutathionyl-caftaric acid [27]. This compound results from a four-component reaction involving caftaric acid, PPO, oxygen, and glutathione. It seems likely that the presence of glutathione or cysteine derivatives of sinapyl alcohol in pineapple juice results from similar oxidative reactions during juice processing [131]. As S-glutathionylcaftaric acid is not inherently colored and is not a substrate for PPO, its formation is therefore believed to limit most browning by trapping caftaric acid quinones in the form of a stable glutathione substituted product. Nevertheless, it may also be oxidized by fungus laccase during vinification and may induce coupled oxidations of other grape phenolics.

In comparison to that of caffeoyl conjugates, the contribution of p-coumaric, ferulic, and sinapic derivatives to browning is rather low, and they even have been reported to act as PPO inhibitors [2]. Many other phenolic compounds present in large quantities in fruits and vegetables (e.g., anthocyanins, flavonol glycosides) do not appear to be direct substrates of PPO, but chlorogenic acid or other caffeoyl conjugates may participate in their coupled oxidation, which increases the browning intensity [27,65,99]. Thus, apple procyanidins are not themselves substrates for PPO, but they are oxidized by a chlorogenic acid/chlorogenoquinone redox shuttle while the amount of chlorogenic acid remains almost constant. Enzymatic oxidation of chlorogenic acid may also be combined with nonenzymatic degradation of anthocyanins as shown in eggplant, sweet cherry, and d'Ente plum.

In addition to the enzymatic oxidation of phenolic acids, their auto-oxidation can also result in brown-colored pigments that are detrimental to food quality. It is dependent on the physicochemical environment and strongly increases with pH. For example, auto-oxidation of phenolic acids was responsible for color loss of carrot puree, and processing treatments that reduce residual oxygen may result in better color retention after processing and storage [132].

Enzymatic oxidation or auto-oxidation may also affect potato tuber either during cold storage or during processing and cooking. Reviews on the chemical, biochemical, and dietary roles of potato polyphenols have been published [11,133]. They focus on chlorogenic acid, as this compound constitutes up to 90% of the total phenolic content in potato tuber. No correlation was found between bruising-induced browning of potato tuber and chlorogenic acid, as tyrosine and PPO activity plays the major role ([11] for a review). On the contrary, chlorogenic acid seemed to be responsible for the bluishgray discoloration of boiled or steamed potatoes after exposure to air. This after-cooking darkening appears to be due to the oxidation of a colorless ferrous ion-chlorogenic acid complex to a dark ferric one [133].

Oxidation of chlorogenic acid during processing results in a dramatic loss in concentration: only traces were recovered in an oxidized apple juice, whereas 14, 23, or 32 mg/100 mL was present in conventionnal juice, a rapidly pasteurized juice, and nonoxidized juice, respectively [2]. As the antioxidant capacity of apple and apple products is due to phenolic constituents rather than to ascorbic acid [111,134], the chain-breaking activity decreased in apple puree when chlorogenic acid was oxidized and brown polymers accumulated [135]. Inhibition of browning during processing may then protect both the natural coloration of the product and its antioxidant properties. This may be achieved by physical, chemical, genetic, or molecular methods, and numerous treatments have been proposed to prevent browning [27,65,99,136–138]. The most widely used in industry are heat, ascorbic acid, and sulfite adjunctions, but use of sulfiting agents has been limited and now banned because of health risks. These aspects are not discussed further here, as they do not specifically concern phenolic acids.

In contrast to the previous examples, oxidative treatments may promote positive consequences in some other cases, as they decrease concentrations in phenolic acids (and other phenolics) and limit subsequent oxidations. Therefore, hyperoxidation of musts during wine processing has sometimes been proposed, and this process particularly affected HCA derivatives [27]. In addition, a peroxidasic treatment allowed the dimerization of ferulic acid in wheat bran tissues and consequently decreased arabinoxylan solubility and increased mechanical strength of the tissue [58].

B. Changes in the Phenolic Acid Profile

Although free phenolic acids are rarely present in plant organs, they have often been detected in derived products, e.g., fruit juices, wines, ciders, and oils [2,27,69,118,130,139,140]. It is assumed that these free acids are formed by partial degradation of the combined forms during extraction or processing. For example, removal of olive bitterness requires treatments with sodium hydroxyde, and, during this process, caffeic acid has appeared, directly derived from alkaline hydrolysis of verbascoside [141]. Hydrolysis of esterified forms of phenolic acids may also be catalyzed by esterases, as shown during the preparation of prunes and prune juice [69]. A decrease in bound phenolic acids (mainly ferulic acid) was also observed during malting of cereals, and it may due to the action of esterases that are induced during germination and act on various phenolic acid esters linked to wall polysaccharides [142,143]. In addition to hydrolytic enzymes issued either from the plant material or from the microrganisms naturally present on the plant cuticle, commercial pectolytic enzymes are often used in fruit juice processing to improve clarification, and they also can cause hydrolysis of HCA esters [130]. HCA conjugates may also be transformed by endogenous micro-organisms to cause off-flavor in food products. For example, yeasts isolated from unpasteurized apple juice are responsible for the formation of 4-vinyl guaicol as they contain both feruloyl esterase and ferulic acid decarboxylase activity [144].

Analysis of wine phenolic acids reveals two important differences in comparison with the original grapes: (1) lower content of HCA esters and (2) appearance of free p-hydroxybenzoic, gallic, p-coumaric, and caffeic acids [2,27]. The final phenolic composition of a wine mainly depends on three sets of parameters: (1) the phenolics initially present in grapes; (2) the extraction and winemaking techniques, including fermentations; and (3) the chemical reactions that occur during aging either in wood barrels or in bottles. In particular, when seed crushing and long extraction times were employed, high levels of HBA derivatives and S-glutathionylcaftaric acid were present in wines [145]. Whereas red wines contain high levels of HCA conjugates, anthocyanins, and condensed tannins, the phenolic profile of white wines essentially results from the presence of flavonols and phenolic acid derivatives [27]. Twelve new benzoic and cinnamic derivatives (caffeol ethyl tartrates, glucose esters of cinnamic, p-coumaric, ferulic, or vanillic acids; 4-O-glucosides of ferulic acid; ethyl protocatechuate, or gallate) were identified in Riesling wine, where they are responsible for antioxidant activity [146]. Nevertheless, the highest antioxidant capacity was obtained for red wines, because of their high content of catechins and related structures [147,148].

In addition to microbial or chemical processes of clarification, physical ones may also have a great influence on the phenolic profile of fruit juices, especially for HCA. A significantly higher oxidation level was observed on ultrafiltration membranes than on microfiltration membranes of apple juices [149]. Furthermore, chlorogenic acid content was one of the most relevant variables to differentiate juices clarified according to the two membrane filtration technologies. Important changes in the composition of the phenolic fraction of kiwi was observed during processing [150]. In contrast to juice subjected to a high-temperature short-period treatment, in pressed juice low levels of HCA and HBA were found (respectively, 1.11 and 0.17 mg/L chlorogenic acid). Reduction in the level of phenolic acid derivatives also occurred after clarification (0.71 mg/L in comparison to 1.11 before clarification).

Thermal treatments also cause changes in the phenolic acids of food products, as they accelerate chemical degradation and either promote or block enzymatic activities. For example, the roasting of coffee beans and the baking of potatoes or cereal flour induce a significant loss in chlorogenic content [4,11]. In the same way, the red-brown color of semidried pickled plums is obtained by thermal processing that induces the oxidation of the three isomers of chlorogenic acid [151]. In the case of quince, jellies contained lower concentrations of chlorogenic acid than jams, because of the more severe thermal treatment to which the fruit is subjected during the preparation of jelly [152,153]. An increase in ellagic acid, explained by a release from ellagitannins, was also observed during the thermal treatment of red raspberry to prepare jams [18], whereas freezing and long-term frozen storage of raspberries induced a significant decrease in the ellagic content [17].

Qualitative and quantitative changes in phenolic acids also occurred during long-term storage of fruit juices or juice concentrates and extensive degradation was observed at 25 °C [130]. Refrigerated storage of orange juice also induced modifications in the phenolic pattern, as HCA derivatives, in contrast to flavanones, remained in soluble form and were not present in the cloud that appears in juice [154].

Phenolic compounds have sometimes been used to detect adulterations of fruit juices and jams. Most of them are flavonols, but some HCA derivatives can be used since they are typical of some fruit species such as tartaric derivatives in grape. For example, grape juice can be detected by the presence of caffeoyl-, *p*-coumaroyl-, and feruloyl-tartaric acids, whereas the presence of quinic esters of HCA would imply adulteration with other fruits. A method has also been developed for detection and quantitation of pulpwash, a lower-quality juice product, in orange juice [9]: feruloyl and sinapoyl glucose, in addition to different other phenolic compounds, were present in much larger amounts in pulpwash.

VII. CONCLUSIONS

Considerable progress in the analysis of phenolic acid derivatives has been made since the 1980s, and the diversity observed in fleshy fruits and vegetables with regard to their distribution is found in terms of quality and quantity. The raison d'être of phenolic acids and, more generally, of the totality of phenolic compounds, is now better understood, and their role in the interface between the plant and its environment has now been well established. Furthermore, phenolic acids have great importance, on the one hand, as precursors for many other phenolic molecules often found in fruits and vegetables (e.g., anthocyanins, tannins, lignin) and, on the other hand, to the organoleptic and nutritional quality of plant products that play an important role in the daily diet. The most important biological activity of phenolic acids, along with other phenolic compounds, is probably that of their antioxidant property and, consequently, their inhibitory effects on mutagenesis, carcinogenesis, and human heart diseases. As shown, many papers published since the early 1990s about phenolic acids in fruits and vegetables concern their antioxidant activity. Nevertheless, frequently only global activity has been reported, and more data are still necessary to explain individual effects and synergistic interactions, e.g., by determining relative health-promoting properties of the different chlorogenic acid isomers.

Modification of the phenolic pattern of fruits and vegetables can be envisaged by means of plant selection both by conventional hybridization and by use of new genetic engineering techniques in the near future. In light of their potential role as antioxidants, as well as their inhibitory effect on plant pathogens, phenolic acids may be viewed as positive constituents of fruits and vegetables. Nevertheless, high phenolic acid concentrations may also be undesirable, as some phenolic acids are good substrates to polyphenoloxidases and participate in browning during storage or processing.

Genetic engineering, which is being pursued in plants for the manipulation of plant secondary metabolism [64], might lead to the production of fruits and vegetables in which the phenolic metabolism is over-or underexpressed. However, it must first be determined whether plants with higher or lower phenolic content are desired. The answer may depend on which of the following is preferred: resistance to pathogens, improvement of organoleptic qualities, or accumulation of one or several phenolics showing antioxidant properties and interactions affecting human health. Suppression of genes involved in the biosynthesis of polyphenoloxidase through antisense ribonucleic/acid (RNA), which have already been obtained for potato [136], might lead to fruits and vegetables with low browning capacity and a high level of monomeric phenolic compounds with high antioxidant properties.

REFERENCES

1. Hermann K. Occurence and content of hydroxycinnamic and hydroxybenzoic acid compounds in foods. Crit Rev Food Sci Nutr 1989; 28:315–347.
2. Macheix JJ, Fleuriet A, Billot J. Fruit phenolics. Boca Raton, FL: CRC Press, 1990.
3. Macheix JJ, Fleuriet A. Phenolic acids in fruits. In: Rice-Evans CA, Packer L, eds. Flavonoids in Health and Disease. New York: Marcel Dekker, 1998:35–59.
4. Clifford MN. Chlorogenic acids and other cinnamates—nature, occurrence and dietary burden. J Sci Food Agric 1999; 79:362–372.
5. Robards K, Prenzler PD, Tucker G, Swatsitang P, Glover W. Phenolic compounds and their role in oxidative processes in fruits. Food Chem 1999; 66:401–436.
6. Kroon PA, Williamson G. Hydroxycinnamates in plants and food: current and future perspectives. J Sci Food Agric 1999; 79:355–361.
7. Ibrahim R, Barron D. Phenylpropanoids. In: Harborne JB, ed. Methods in Plant Biochemistry: Vol. 1. Plant Phenolics. London: Academic Press, 1989:73–111.
8. Shobha SV, Ravindranath B. Supercritical carbon dioxide and solvent extraction of the phenolic lipids of cashew nut (*Anacardium occidentale*) shells. J Agric Food Chem 1991; 39:2214–2217.
9. Cancalon PF, Bryan CR. Use of capillary electrophoresis for monitoring citrus juice composition. J Chromatogr A 1993; 652:555–561.
10. Tomas-Barberan FA. Capillary electrophoresis: a new technique in the analysis of plant secondary metabolites. Phytochem Anal 1995; 6:177–192.

11. Friedman M. Chemistry, biochemistry, and dietary role of potato polyphenols: a review. J Agric Food Chem 1997; 45:1523–1540.
12. Gorinstein S, Zemser M, Weisz M, Halevy S, Deutsch J, Tilis K, Feintuch D, Guerra N, Fishman M, Bartnikowska E. Fluorometric analysis of phenolics in persimmons. Biosci Biotech Biochem 1994; 58:1087–1092.
13. Yoshiki Y, Okubo K, Onuma M, Igarashi K. Chemiluminescence of benzoic and cinnamic acids, and flavonoids in the presence of aldehyde and hydrogen peroxide or hydroxyl radical by Fenton reaction. Phytochemistry 1995; 39:225–229.
14. Rapisarda P, Carollo G, Fallico B, Tomaselli F, Maccarone E. Hydroxycinnamic acids as markers of italian blood orange juices. J Agric Food Chem 1998; 46: 464–470.
15. Sukrasno N, Yeoman MM. Phenylpropanoid metabolism during growth and development of *Capsicum frutescens* fruits. Phytochemistry 1993; 32:839–844.
16. Ayaz FA, Kadioglu A. Changes in phenolic acid contents of *Diospyros lotus* L. during fruit development. J Agric Food Chem 1997; 45:2539–2541.
17. Ancos de B, Gonzalez EM, Cano MP. Ellagic acid, vitamin C, and total phenolic contents and radical scavenging capacity affected by freezing and frozen storage in raspberry fruit. J Agric Food Chem 2000; 48:4565–4570.
18. Zafrilla P, Ferreres F, Tomas-Barberan FA. Effect of processing and storage on the antioxidant ellagic acid derivatives and flavonoids of red raspberry (*Rubus idaeus*) jams. J Agric Food Chem 2001; 49:3651–3655.
19. Gil MI, Holcroft DM, Kader AA. Changes in strawberry anthocyanins and other polyphenols in response to carbon dioxide treatments. J Agric Food Chem 1997; 45:1662–1667.
20. Sakushima A, Coskun M, Maoka T. Hydroxybenzoic acids from *Boreava orientalis*. Phytochemistry 1995; 40:257–261.
21. Ng A, Parker ML, Parr AJ, Saunders PK, Smith AC, Waldron KW. Physicochemical characteristics of onion (*Allium cepa* L.) tissues. J Agric Food Chem 2000; 48:5612–5617.
22. Mattila P, Könkö K, Eurola M, Pihlava JM, Astola J, Vahteristo L, Hietaniemi V, Kumpulainen J, Valtonen M, Piironen V. Contents of vitamins, mineral elements, and some phenolic compounds in cultivated mushrooms. J Agric Food Chem 2001; 49:2343–2348.
23. Bradfield AE, Flood AE, Hulme AC, Williams AH. Chlorogenic acid in fruit trees. Nature 1952; 170:168.
24. Ding CK, Chachin K, Ueda Y, Imahori Y, Wang CY. Metabolism of phenolic compounds during loquat fruit development. J Agric Food Chem 2001; 49:2883–2888.
25. Kähkönen MP, Hopia AI, Vuorela HJ, Rauha JP, Pihlaja K, Kujala TS, Heinonen M. Antioxidant activity of plant extracts containing phenolic compounds. J Agric Food Chem 1999; 47:3954–3962.
26. Kähkönen MP, Hopia AI, Heinonen M. Berry phenolics and their antioxidant activity. J Agric Food Chem 2001; 49:4076–4082.
27. Macheix JJ, Sapis JC, Fleuriet A. Phenolic compounds and polyphenoloxidase in relation to browning in grapes and wines. Crit Rev Food Sci Nutr 1991; 30:441–486.

28. Oleszek W, Amiot MJ, Aubert SY. Identification of some phenolics in pear fruit. J Agric Food Chem 1994; 42:1261–1265.

29. Naim M, Zehavi U, Nagy S, Rouseff RL. Hydroxycinnamic acids as off-flavor precursors in *Citrus* fruits and their products. In: Ho CT, Lee CY, Huang MT, eds. Series 506: Phenolic Compounds in Food and Their Effects on Health I. Analysis, Occurrence, and Chemistry. Washington, DC: ACS, 1992:180–191.

30. Mouly PP, Gaydou EM, Faure R, Estienne JM. Blood orange juice authentication using cinnamic acid derivatives: variety differentiations associated with flavanone glycoside content. J Agric Food Chem 1997; 45:373–377.

31. Sakushima A, Coskun M, Tanker M, Tanker N. A sinapic acid ester from *Boreava orientalis*. Phytochemistry 1994; 35:1481–1484.

32. Sakushima A, Coskun M, Maoka T. Sinapinyl but-3-enylglucosinolate from *Boreava orientalis*. Phytochemistry 1995; 40:483–485.

33. Yoshikawa K, Sugawara S, Arihara S. Phenylpropanoids and other secondary metabolites from fresh fruits of *Picrasma quassioides*. Phytochemistry 1995; 40: 253–256.

34. Sanbongi C, Osakabe N, Natsume M, Takizawa T, Gomi S, Osawa T. Antioxidative polyphenols isolated from *Theobroma cacao*. J Agric Food Chem 1998; 46:454–457.

35. Mazza G, Miniati E. Anthocyanins in fruits, vegetables and grains. Boca Raton, FL: CRC Press, 1993.

36. Price CL, Wrolstad RE. Anthocyanin pigments of royal Okanogan huckleberry juice. J Food Sci 1995; 60:369–374.

37. Bloor SJ. Deep blue anthocyanins from blue *Dianella* berries. Phytochemistry 2001; 58:923–927.

38. Özipek M, Calis I, Ertan M, Rüedi P. Rhammetin 3-*p*-coumaroylrhamninoside from *Rhamnus petiolaris*. Phytochemistry 1994; 37:249–253.

39. Glässgen WE, Metzger JW, Heuer S, Strack D. Betacyanins from fruits of *Basella rubra*. Phytochemistry 1993; 33:1525–1527.

40. Alasalvar C, Grigor JM, Zhang D, Quantick PC, Shahidi F. Comparison of volatiles, phenolics, sugars, antioxidant vitamins, and sensory quality of different colored carrot varieties. J Agric Food Chem 2001; 49:1410–1416.

41. Plumb GW, Proce KR, Rhodes MJC, Williamson G. Antioxidant properties of the major polyphenolic compounds in broccoli. Free Radic Res 1997; 27:429–435.

42. Bilia AR, Fumarola M, Gallori S, Mazzi G, Vincieri FF. Identification by HPLC-DAD and HPLC-MS analyses and quantification of constituents of fennel teas and decoctions. J Agric Food Chem 2000; 48:4734–4738.

43. Azuma K, Nakayama M, Koshioka M, Ippoushi K, Yamaguchi Y, Kohata K, Yamauchi Y, Ito H, Higashio H. Phenolic antioxidants from the leaves of *Corchorus olitorius* L. J Agric Food Chem 1999; 47:3963–3966.

44. Ben Hod G, Basnizki Y, Zohary D, Mayer AM. Cynarin and chlorogenic acid content in germinating seeds of globe artichoke (*Cyanara scolymus* L.). J Gen Breed 1992; 46:63–68.

45. Chuda Y, Suzuki M, Nagata T, Tsushida T. Contents and cooking loss of three quinic acid derivatives from Garland (*Chrysanthemum coronarium* L.). J Agric Food Chem 1998; 46:1437–1439.

46. Kweon MH, Hwang HJ, Sung HC. Identification and antioxidant activity of novel chlorogenic acid derivatives from bamboo (*Phyllostachys edulis*). J Agric Food Chem 2001; 49:4646–4655.

47. Bergman M, Varshavsky L, Gottlieb HE, Grossman S. The antioxidant activity of aqueous spinach extract: chemical identification of active fractions. Phytochemistry 2001; 58:143–152.

48. Strack D, Hartfeld F, Austenfeld FA, Grotjahn L, Wray V. Coumaroyl-, caffeoyl- and feruloyltartronates and their accumulation in mung bean. Phytochemistry 1985; 24:147–150.

49. Cuvelier ME, Richard H, Berset C. Antioxidative activity and phenolic composition of pilot-plant and commercial extracts of sage and rosemary. J Am Oil Chem Soc 1996; 73:645–652.

50. Chen JH, Ho C-T. Antioxidant activities of caffeic acid and its related hydroxycinnamic acid compounds. J Agric Food Chem 1997; 45:2374–2378.

51. Zheng W, Wang SY. Antioxidant activity and phenolic compounds in selected herbs. J Agric Food Chem 2001; 49:5165–5170.

52. Bokern M, Heuer S, Wray V, Witte L, Macek T, Vanek T, Strack D. Ferulic acid conjugates and betacyanins from cell cultures of *Beta vulgaris*. Phytochemistry 1991; 30:3261–3265.

53. Kujala TS, Loponen JM, Klika KD, Pihlaja K. Phenolics and betacyanins in red beetroot (*Beta vulgaris*) root: distribution and effect of cold storage on the content of total phenolics and three individual compounds. J Agric Food Chem 2000; 48:5338–5342.

54. Yu J, Vasanthan T, Temelli F. Analysis of phenolic acids in barley by high-performance liquid chromatography. J Agric Food Chem 2001; 49:4352– 4358.

55. Faulds CB, Williamson G. The role of hydroxycinnamates in plant cell wall. J Sci Food Agric 1999; 79:393–395.

56. Iiyama K, Lam TBT, Stone BA. Covalent cross-links in the cell wall. Plant Physiol 1994; 104:315–320.

57. Parr AJ, Ng A, Waldron KW. Ester-linked phenolic components of carrot cell walls. J Agric Food Chem 1997; 45:2468–2471.

58. Peyron S, Abecassis J, Autran JC, Rouau X. Enzymatic oxidative treatments of wheat bran layers: effects on ferulic acid composition and mechanical properties. J Agric Food Chem 2001; 49:4694–4699.

59. Garcia E, Filisetti TMCC, Udaeta JEM, Lajolo FM. Hard-to-cook beans (*Phaseolus vulgaris*): involvement of phenolic compounds and pectates. J Agric Food Chem 1998; 46:2110–2116.

60. Norton RA. Isolation and identification of steryl cinnamic acid derivatives from corn bran. Cereal Chem 1994; 71:111–117.

61. Dimberg LH, Theander O, Lingnert H. Avenanthramides: a group of phenolic antioxidants in oats. Cereal Chem 1993; 70:637–641.

62. Lapierre C, Pollet B, Negrel J. The phenolic domain of potato suberin: structural comparison with lignins. Phytochemistry 1996; 42:949–953.

63. Yan B, Stark RE. Biosynthesis, molecular structure, and domain architecture of potato suberin: a ^{13}C NMR study using isotopically labeled precursors. J Agric Food Chem 2000; 48:3298–3304.

64. Dixon RA, Paiva NL. Stress-induced phenylpropanoid metabolism. Plant Cell 1995; 7:1085–1097.

65. Nicolas J, Richard-Forget FC, Goupy PM, Amiot MJ, Aubert SY. Enzymatic browning reactions in apple and apple products. Crit Rev Food Sci Nutr 1994; 34:109–157.

66. Golding JB, McGlasson WB, Wyllie SG, Leach DN. Fate of apple peel phenolics during cool storage. J Agric Food Chem 2001; 49:2283–2289.

67. van der Sluis A, Dekker M, de Jager A, Jongen WMF. Activity and concentration of polyphenolic antioxidants in apple: effect of cultivar, harvest year, and storage conditions. J Agric Food Chem 2001; 49:3606–3613.

68. Podsedek A, Wilska-Jeszka J, Anders B, Markowski J. Compositional characterisation of some apple varieties. Eur Food Res Technol 2000; 210:268–272.

69. Donovan JL, Meyer AS, Waterhouse AL. Phenolic composition and antioxidant activity of prunes and prune juice (*Prunus domestica*). J Agric Food Chem 1998; 46:1247–1252.

70. Tomas-Barberan FA, Gil MI, Cremin P, Waterhouse AL, Hess-Pierce B, Kader AA. HPLC-DAD-ESIMS analysis of phenolic compounds in nectarines, peaches, and plums. J Agric Food Chem 2001; 49:4748–4760.

71. Chang S, Tan C, Frankel EN, Barrett DM. Low-density lipoprotein antioxidant activity of phenolic compounds and polyphenol oxidase activity in selected Clingstone peach cultivars. J Agric Food Chem 2000; 48:147–151.

72. Mayr U, Treutter D, Santos-Buelga C, Bauer H, Feucht W. Developmental changes in the phenol concentrations of "*Golden delicious*" apple fruits and leaves. Phytochemistry 1995; 38:1151–1155.

73. Bocco A, Cuvelier ME, Richard H, Berset C. Antioxidant activity and phenolic composition of citrus peel and seed extracts. J Agric Food Chem 1998; 46:2123–2129.

74. Talcott ST, Howard LR, Brenes CH. Antioxidant changes and sensory properties of carrot puree processed with and without periderm tissue. J Agric Food Chem 2000; 48:1315–1321.

75. Andreasen MF, Landbo AK, Chritensen LP, Hansen A, Meyer AS. Antioxidant effects of phenolic rye (*Secale cereale* L.) extracts, monomeric hydroxycinnamates, and ferulic acid dehydrodimers on human low-density lipoproteins. J Agric Food Chem 2001; 49:4090–4096.

76. McCallum JA, Walker JRL. Phenolic biosynthesis during grain development in wheat: changes in phenylalanine ammonia-lyase activity and soluble phenolic content. J Cereal Sci 1990; 11:35–49.

77. Régnier T, Macheix JJ. Changes in wall-bound phenolic acids, phenylalanine and tyrosine ammonia-lyases, and peroxidases in developing durum wheat grains (*Triticum turgidum* L. var. durum). J Agric Food Chem 1996; 44:1727–1730.

78. Wende G, Waldron KW, Smith AC, Brett CT. Developmental changes in cell-wall ferulate and dehydroferulates in sugar beet. Phytochemistry 1999; 52: 819–827.

79. Bewley JD, Black M. Seeds, Physiology of Development and Germination. 2nd ed. New York: Plenum Press, 1994.

80. Wu H, Haig T, Pratley J, Lemerle D, An M. Allelochemicals in wheat (*Triticum*

aestivum L.): cultivar difference in the exudation of phenolic acids. J Agric Food Chem 2001; 49:3742–3745.

81. Griffiths DW, Bain H. Photo-induced changes in the concentration of individual chlorogenic isomers in potato (*Solanum tuberosum*) tubers and their complexation with ferric ions. Potato Res 1997; 40:307–315.

82. Percival GC, Baird L. Influence of storage upon light-induced chlorogenic acid accumulation in potato tubers (*Solanum tuberosum* L.). J Agric Food Chem 2000; 48:2476–2482.

83. Bureau SM, Baumes RL, Razungles AJ. Effects of vine or bunch shading on the glycosylated flavor precursors in grapes of *Vitis vinifera* L. cv. Syrah. J Agric Food Chem 2000; 48:1290–1297.

84. Oufedjik H, Mahrouz M, Amiot MJ, Lacroix M. Effect of γ-irradiation on phenolic compounds and phenylalanine ammonia-lyase activity during storage in relation to peel injury from peel of *Citrus clementina* Hort. Ex. Tanaka. J Agric Food Chem 2000; 48:559–565.

85. Lewis CE, Walker JRL, Lancaster JE. Changes in anthocyanin, flavonoid and phenolic acid concentrations during development and storage of coloured potato (*Solanum tuberosum* L.) tubers. J Sci Food Agric 1999; 79:311–316.

86. Nicholson RL, Hammerschmidt R. Phenolic compounds and their role in disease resistance. Annu Rev Phytopathol 1992; 30:369–389.

87. Lattanzio V, De Cicco V, Di Venere D, Lima G, Salerno M. Antifungal activity of phenolics against fungi commonly encountered during storage. Ital J Food Sci 1994; 6:23–30.

88. Saltveit ME. Physical and physiological changes in minimally processed fruits and vegetables. In: Tomas-Barberan FA, Robin RD, eds. Phytochemistry of Fruits and Vegetables. Oxford: Oxford University Press, 1997:205–220.

89. Babic I, Amiot MJ, Nguyen-The C, Aubert S. Changes in phenolic content in fresh ready-to-use shredded carrots during storage. J Food Sci 1993; 58: 351–356.

90. Ferreres F, Gil MI, Castaner M, Tomas-Barberan FA. Phenolic metabolites in red pigmented lettuce (*Lactuca sativa*): changes with minimal processing and cold storage. J Agric Food Chem 1997; 45:4249–4254.

91. Cantos E, Espin JC, Tomas-Barberan FA. Effect of wounding on phenolic enzymes in six minimally processed lettuce cultivars upon storage. J Agric Food Chem 2001; 49:322–330.

92. Bostock RM, Wilcox SM, Wang G, Adaskaveg JE. Suppression of *Monilinia fructicola* cutinase production by peach fruit surface phenolic acids. Physiol Mol Plant Pathol 1999; 54:37–50.

93. Kuc J. Phytoalexins, stress metabolism, and disease resistance in plants. Annu Rev Phytopathol 1995; 33:275–297.

94. Lyon GD. The biochemical basis of resistance of potatoes to soft rot *Erwinia spp.*—a review. Plant Pathol 1989; 38:313–339.

95. Phukan SN, Baruah CK. Chlorogenic acid content of potato plant tissues in relation to infection by *Phytophtora infestans* (Mont.) De Bary. Adv Plant Sci 1991; 2:218–222.

96. Johansen OP, Andersen OM, Nerdal W, Aksnes DW. Cyanidin 3-[(6-*p*-

coumaroyl)-2-(xylosyl)-glucoside—and other anthocyanins from fruits of *Sambucus canadensis*. Phytochemistry 1991; 30:4137–4141.

97. Nakatani N, Kikuzaki H, Hikida J, Ohba M, Inami O, Tamura I. Acylated anthocyanins from fruits of *Sambucus canadensis*. Phytochemistry 1995; 38:755–757.

98. Brouillard R, Figueiredo P, Elhabiri M, Dangles O. Molecular interactions of phenolic compounds in relation to the colour of fruit and vegetables. In: Tomas-Barberan FA, Robins RJ, eds. Proceeding of the Phytochemical Society of Europe: Phytochemistry of Fruit and Vegetables. Oxford: Clarendon Press, 1997:29–49.

99. Amiot MJ, Fleuriet A, Cheynier V, Nicolas J. Phenolic compounds and oxidative mechanisms in fruit and vegetables. In: Tomàs-Barberàn FA, Robins RJ, eds. Phytochemistry of Fruit and Vegetables. Oxford: Clarendon Press, 1997:51–85.

100. Clifford MN. Astringency. In: Tomas-Barberan FA, Robins RJ, eds. Proceeding of the Phytochemical Society of Europe: Phytochemistry of fruit and vegetables. Oxford: Clarendon Press, 1997:87–107.

101. Shimazaki N, Mimaki Y, Sashida Y. Prunasin and acetylated phenylpropanoic acid sucrose esters, bitter principles from the fruits of *Prunus jamasakura* and *P. maximowiczii*. Phytochemistry 1991; 30:1475–1480.

102. Crouzet J, Sakho M, Chassagne D. Fruit aroma precursors with special reference to phenolics. In: Tomas-Barberan FA, Robins RJ, eds. Proceedings of the Phytochemical Society of Europe: Phytochemistry of Fruit and Vegetables. Oxford: Clarendon Press, 1997:109–123.

103. Friedman M, Kozukue N, Harden LA. Cinnamaldehyde content in foods determined by gas chromatography-mass spectrometry. J Agric Food Chem 2000; 48:5702–5709.

104. Shahidi F, Janitha PK, Wanasundara PD. Phenolic antioxidants. Crit Rev Food Sci Nutr 1992; 32:67–103.

105. Huang MT, Ferraro T. Phenolic compounds in food and cancer prevention. In: Huang MT, Ho CT, Lee CY, eds. ACS Symposium Series 507: Phenolic Compounds in Food and Their Effects on Health. II. Antioxidants and Cancer Prevention. Washington, DC: ACS, 1992:8–34.

106. Rice-Evans CA, Miller NJ, Paganga G. Structure-antioxidant activity relationships of flavonoids and phenolic acids. Free Rad Biol Med 1996; 20:933–956.

107. Vinson JA, Hao Y, Su X, Zubik L. Phenol antioxidant quantity and quality in foods: vegetables. J Agric Food Chem 1998; 46:3630–3634.

108. Moure A, Cruz JM, Franco D, Dominguez JM, Sineiro J, Dominguez H, Nunez MJ, Parajo JC. Natural antioxidants from residual sources. Food Chem 2001; 72:145–171.

109. Vinson JA, Su X, Zubik L, Bose P. Phenol antioxidant quantity and quality in foods: fruits. J Agric Food Chem 2001; 49:5315–5321.

110. Rice-Evans CA, Packer L. Flavonoids in Health and Disease. New York: Marcel Dekker, 1998.

111. Wang H, Cao G, Prior RL. Total antioxidant capacity of fruits. J Agric Food Chem 1996; 44:701–705.

112. Heinonen IM, Lehtonen PJ, Hopia AI. Antioxidant activity of berry and fruit wines and liquors. J Agric Food Chem 1998; 46:25–31.

113. Rapisarda P, Tomaino A, Lo Cascio R, Bonina F, De Pasquale A, Saija A.

Antioxidant effectiveness as influenced by phenolic content of fresh orange juices. J Agric Food Chem 1999; 47:4718–4723.

114. Kaur C, Kapoor HC. Antioxidants in fruits and vegetables—the millennium's health. Int J Food Sci Technol 2001; 36:703–725.

115. Richelle M, Tavazzi I, Offord E. Comparison of the antioxidant activity of commonly consumed polyphenolic beverages (coffee, cocoa, and tea) prepared per cup serving. J Agric Food Chem 2001; 49:3438–3442.

116. Talcott ST, Howard LR, Brenes CH. Contribution of periderm material and blanching time to the quality of pasteurized peach puree. J Agric Food Chem 2000; 48:4590–4596.

117. Wang H, Nair MG, Strasburg GM, Booren AM, Gray JI. Antioxidant polyphenols from tart cherries (*Prunus cerasus*). J Agric Food Chem 1999; 47:840–844.

118. Nakatani N, Kayano SI, Kikuzaki H, Sumino K, Katagiri K, Mitani T. Identification, quantitative determination, and antioxidative activities of chlorogenic acid isomers in prune (*Prunus domestica* L.). J Agric Food Chem 2000; 48:5512–5516.

119. Cao G, Russell RM, Lischner N, Prior RL. Serum antioxidant capacity is increased by consumption of strawberries, spinach, red wine or vitamin C in elderly women. J Nutr 1998;128: 2383–2390.

120. Gutierrez F, Jimenez B, Ruiz A, Albi MA. Effect of olive ripeness on the oxidative stability of virgin olive oil extracted from the varieties Picual and Hojiblanca and on the different components involved. J Agric Food Chem 1999; 47:121–127.

121. McDonald S, Prenzler PD, Antolovich M, Robards K. Phenolic content and antioxidant activity of olive extracts. Food Chem 2001; 73:73–84.

122. Saija A, Tomaino A, Lo Cascio R, Trombetta D, Proteggente A, De Pasquale A, Uccella N, Bonina F. Ferulic and caffeic acids as potential protective agents against photooxidative skin damage. J Sci Food Agric 1999; 79:476–480.

123. Rodriguez de Sotillo D, Hadley M, Holm ET. Phenolics in aqueous potato peel extract: extraction, identification and degradation. J Food Sci 1994; 59:649–651.

124. Zheng Z, Shetty K. Solid-state bioconversion of phenolics from cranberry pomace and role of *Lentinus edodes*. β-glucosidase. J Agric Food Chem 2000; 48:895–900.

125. Manthey JA, Grohmann K. Phenols in citrus peel byproducts: concentrations of hydroxycinnamates and polymethoxylated flavones in citrus peel molasses. J Agric Food Chem 2001; 49:3268–3273.

126. Natella F, Nardini M, Di Felice M, Scaccini C. Benzoic and cinnamic acid derivatives as antioxidants: stucture-activity relation. J Agric Food Chem 1999; 47:1453–1459.

127. Meyer AS, Donovan JL, Pearson DA, Waterhouse AL, Frankel EN. Fruit hydroxycinnamic acids inhibit human low-density lipoprotein oxidation in vitro. J Agric Food Chem 1998; 46:1783–1787.

128. Silva FAM, Borges F, Ferreira MA. Effects of phenolic propyl esters on the oxidative stability of refined sunflower oil. J Agric Food Chem 2001; 49:3936–3941.

129. Aruoma OI, Murcia A, Butler J, Halliwell B. Evaluation of the antioxidant and pro-oxidant actions of gallic acid and derivatives. J Agric Food Chem 1993; 41: 1880–1885.

130. Spanos GA, Wrolstad RE. Phenolics of apple, pear, and white grape juices and their changes with processing and storage—a review. J Agric Food Chem 1992; 40:1478–1487.

131. Wen L, Wrolstad RE, Hsu VL. Characterization of sinapyl derivatives in pineapple (*Ananas comosus* L. Merill) juice. J Agric Food Chem 1999; 47:850–853.

132. Talcott ST, Howard LR. Phenolic autoxidation is responsible for color degradation in processed carrot puree. J Agric Food Chem 1999; 47:2109–2115.

133. Putz B. The current state of knowledge on blue and black spot or injury. Kartoffelbau 1995; 46:284–286.

134. Miller NJ. Flavonoids and phenylpropanoids as contributors to the antioxidant activity of fruit juices. In: Rice-Evans CA, Packer L, eds. Flavonoids in Health and Disease. New York: Marcel Dekker, 1998:387–403.

135. Nicoli MC, Calligaris S, Manzocco L. Effect of enzymatic and chemical oxidation on the antioxidant capacity of catechin model systems and apple derivatives. J. Agric. Food Chem., 2000; 48:4576–4580.

136. Coetzer C, Corsini D, Love S, Pavek J, Tumer N. Control of enzymatic browning in potato (*Solanum tuberosum* L) by sense and antisense RNA from tomato polyphenol oxidase. J Agric Food Chem 2001; 49:652–657.

137. Lee CY, Whitaker JR. Enzymatic Browning and Its Prevention. ACS Symposium Series, 600. Washington, DC: ACS, 1995.

138. Son SM, Moon KD, Lee CY. Inhibitory effects of various antibrowning agents on apple slices. Food Chem 2001; 73:23–30.

139. Pirisi F, Angioni A, Cabras P, Garau VL, Sanjust di Teulada ML, Kaim dos Santos M, Bandino G. Phenolic compounds in virgin olive oils. I. Low-wavelength quantitative determination of complex phenols by high-performance liquid chromatography under isocratic elution. J Chromatogr A 1997; 768:207–213.

140. Brenes M, Garcia A, Garcia P, Rios JJ, Garrido A. Phenolic compounds in Spanish olive oils. J Agric Food Chem 1999; 47:3535–3540.

141. Brenes-Balbuena M, Garcia-Garcia P, Garrido-Fernandez A. Phenolic compounds related to the black color formed during the processing of ripe olives. J Agric Food Chem 1992; 40:1192–1196.

142. Maillard MN, Berset C. Evolution of antioxidant activity during kilning: role of insoluble bound phenolic acids of barley and malt. J Agric Food Chem 1995; 43:1789–1793.

143. Subba Rao MVSST, Muralikrishna G. Non-starch polysaccharides and bound phenolic acids from native and malted finger millet (Ragi, *Eleusine coracana*, Indaf-15). Food Chem 2001; 72:187–192.

144. Donaghy JA, Kelly PF, McKay A. Conversion of ferulic acid to 4-vinyl guaiacol by yeasts isolated from unpasteurised apple juice. J Sci Food Agric 1999; 79:453–456.

145. Meyer AS, Yi OS, Pearson DA, Waterhouse AL, Frankel EN. Inhibition of human low-density lipoprotein oxidation in relation to composition of phenolic antioxidants in grapes (*Vitis vinifera*). J Agric Food Chem 1997; 45:1638–1643.

146. Baderschneider B, Winterhalter P. Isolation and characterization of novel benzoates, cinnamates, flavonoids, and lignans from Riesling wine and screening for antioxidant activity. J Agric Food Chem 2001; 49:2788–2798.

147. Waterhouse AL, Walzem RL. Nutrition of grape phenolics. In: Rice-Evans CA, Packer L, eds. Flavonoids in Health and Disease. New York: Marcel Dekker, 1998:359–385.

148. Landrault N, Poucheret P, Ravel P, Gasc F, Cros G, Teissedre P-L. Antioxidant capacities and phenolics levels of french wines from different varieties and vintages. J Agric Food Chem 2001; 49:3341–3348.

149. Mangas JJ, Suarez B, Picinelli A, Moreno J, Blanco D. Differentiation by phenolic profile of apple juices prepared according to two membrane techniques. J Agric Food Chem 1997; 45:4777–4784.

150. Dawes HM, Keene JB. Phenolic composition of kiwifruit juice. J Agric Food Chem 1999; 47:2398–2403.

151. Fu HY, Huang TC, Ho CT. Changes in phenolic compounds during plum processing. In: Huang MT, Ho CT, Lee CY, eds. ACS Symposium Series 506: Phenolic Compounds in Food and Their Effects on Health. I. Analysis, Occurrence, and Chemistry. Washington, DC: ACS, 1992:287–295.

152. Silva BM, Andrade PB, Mendes GC, Valentao P, Seabra RM, Ferreira MA. Analysis of phenolic compounds in the evaluation of commercial quince jam authenticity. J Agric Food Chem 2000; 48:2853–2857.

153. Silva BM, Andrade PB, Valentao P, Mendes GC, Seabra RM, Ferreira MA. Phenolic profile in the evaluation of commercial quince jellies authenticity. Food Chem 2000; 71:281–285.

154. Gil-Izquierdo A, Gil MI, Ferreres F, Tomas-Barberan FA. In vitro availability of flavonoids and other phenolics in orange juice. J Agric Food Chem 2001; 49:1035–1041.

2
Flavonoids in Herbs

Piergiorgio Pietta and Claudio Gardana
Institute of Biomedical Technologies–National Council of Research
Segrate, Italy

Annamaria Pietta
Farmacia Dr. Carlo Bravi
Brescia, Italy

I. INTRODUCTION

Flavonoids are a large group of polyphenolic compounds that occur commonly in plants. This group contains more than 8000 known compounds, and this number is constantly growing because of the great structural diversity arising from the various hydroxylation, methoxylation, glycosylation, and acylation patterns.

Flavonoids are the pigments responsible for the shades of yellow, orange, and red in flowering plants. They are also important factors for plant growth, development, and defense. Many flavonoids are endowed with biological activities, such as anti-inflammatory, antiallergic, antischemic, antiplatelet, immunomodulatory, and antitumoral activities [1–3]. Flavonoids have also been shown to inhibit several enzymes, including lipoxygenases and cyclooxygenases, mono-oxygenases, xanthine oxidase, mitochondrial succinoxidase, reduced nicotinamide-adenine dinucleotide (NADH) oxidase, phospholipase A_2, topoisomerases, and protein kinases [4–6]. The biological activities of flavonoids are thoughy to be due mainly to their antioxidant properties [7–8], which are displayed by limiting the production of reactive oxygen species (ROS) and/or scavenging them.

Flavonoids are components of the diet of numerous herbivores and omnivores, including humans [9]. They are principally found in fruits, vegetables, and popular drinks, such as red wine, tea, beer, and their intake may reach 1 g/day [10]. In addition, flavonoids are present in various herbs* [11]. Approximately 50 species, from *Achillea millefolium* to *Viola tricolor*, have been used as herbal remedies for their flavonoid content; some are listed in Table 1. These preparations have been reported to be effective for the treatment of disorders of peripheral circulation and for the improvement of aquaresis. In addition, flavonoid-based herbal medicines are available in different countries as anti-inflammatory, antispasmodic, antiallergic, and antiviral remedies [12–14]. The pharmacological effects of these phytomedicines are ascribed either to their functions as radical scavengers, reductants, and metal chelators or to alternative nonantioxidant functions, including the interaction with different enzymes, the inhibition of calcium ion influx into the cells, and the regulation of cell signaling [15] and gene expression [16]. However, it should be remembered that the health benefit properties of most medicinal plants high in flavonoids cannot be assigned exclusively to these compounds, since other components present in the phyto-complex may either directly contribute to or display a "permissive" role that enhances the effects of flavonoids. When examining different examples including aquaretic, anti-inflammatory, sedative, and antispasmodic herbs, it is found that the observed pharmacological effect is due to flavonoidic and nonflavonoidic constituents [17].

As natural products, herbs can greatly differ in their composition as a result of genetic factors, climate, soil quality, and other external factors. Therefore, controlled cultivation and selection represent the first steps to ensuring the most consistent concentration of specific ingredients or groups of compounds. Second, the production of the herbal ingredients by extracting the herbs with solvents must be carefully monitored to select the components that are important to the action and the efficacy of the product. To achieve consistent pharma-ceutical quality, the analytical quality control is essential. This is not an easy task, as herbs and related extracts are complex mixtures of constituents with different physicochemical (i.e., analytical) characteristics. With flavonoid-containing herbs, however, phytochemical data are largely available: i.e., the chemical nature of flavonoids present in these herbs is known. Almost all the flavonoid classes are present in herbs with proven therapeutic activity, including flavonols, flavones, and their dihydroderivatives; isoflavones; catechins; flava-

---

* In botanical nomenclature, the word *herb* refers to non woody seed-producing plants that die down at the end of the growing season. Currently, in the culinary arts, herbs are defined as vegetable products used to add flavor or aroma to food and beverages. However, in botanical medicine the word *herb* refers to plants or plant parts used in freshed, dried, or extracted form for the treatment of disease states, often of a chronic nature, or for improvement or maintenance of health. So here, for the purpose of this chapter, herbs are considered as healing herbs (medicinal plants), as distinguished from vegetables with aromatic and savory properties, that is, culinary herbs.

nolignans; and anthocyanins. Some of the additional phytochemicals are closely related to flavonoids such as phenolic and hydroxycinnamic acids, whereas others have different chemical natures, including various terpenes (mostly present in volatile oils), coumarin derivatives, phytosterols, and other species-characteristic constituents.

The analysis of the flavonoid fraction in the raw herbs and in standardized (i.e., having known potency) extracts may be accomplished by using different approaches [18–19], including high-performance liquid chromatography (HPLC), capillary electrophoresis (CE), and mass spectrometry. HPLC coupled with "online" ultraviolet (UV) detection and/or mass spectrometry (MS) allows data on the chromatographic, UV, and MS behavior of the analytes to be obtained from a single run. This approach remains the method of choice (1) to obtain typical "fingerprints" for the herbal ingredient; (2) to assay single flavonoids; and (3) to detect evidence the presence of adulterants. CE has been proved a valuable alternative to HPLC, because of its high selective power, which allows detection of some flavonoids not separable by HPLC. Unfortunately, CE has not become as popular as HPLC, which remains the technique of choice for routine quality control of flavonoid-containing vegetables [20].

Typical examples of flavonoid herbs examined by HPLC or CE have already been described in a previous contribution of this volume series [11]. This chapter aims to describe three mass techniques, the electrospray ionization MS, (ESI-MS), atmospheric pressure chemical ionization MS (APCI-MS), and ion trap MS (ITMS) techniques, and their application to the analysis of flavonoids in some standardized herbal extracts with proven therapeutic efficacy. In addition, the flavonoid composition of some commonly consumed vegetables with aromatic or savory properties (culinary herbs) is described.

A. Mass Spectrometry of Flavonoids

Flavonoid herbs have been increasingly studied by mass spectrometry (MS) since the introduction of the thermospray (TSP), ESI, and APCI interfaces, which allow direct coupling of MS with HPLC. Being characterized by "soft" ionization, these techniques permit the analysis of flavonoids in their native form without derivatization [21]. TSP-MS was used, at first, to analyze flavonoids in different plant extracts, such as *Arnica montana*, Gentianaceae species, *Ginkgo biloba, Calendula officinalis*, and *Hypericum perforatum* [11]. Unfortunately, TSP-MS fails in the case of thermolabile compounds, such as the flavonol-glycosides. These compounds undergo fragmentation and yield mainly the aglycone fragment $[A+H]^+$, with the molecular ion $[M+H]^+$ present in very low quantity. This was one reason to switch to the ESI and APCI interfaces, which involve a low level of fragmentation. Both ESI and APCI produce mainly molecular ions, and they are particularly suitable for detecting intact molecular species present

Table 1 Selected Flavonoid Herbs

Herb	Purported active components[a]	Alleged activity	Uses	Reference
Artichoke leaf (*Cynara scolymus*)	Luteolin-glycosides (<1%); caffeoyl-quinic acids (<2%); sesquiterpene lactones (<4%)	Choleretic, hepatoprotective, spasmolytic	Dyspeptic problems	54
Bilberry fruit extract (*Vaccinium myrtillus*)	Anthocyanins (mainly malvidin, cyanidin, and delphinidin glycosides) (<24%)	Vasoprotective, antioxidant, astringent	Capillary weakness, venous insufficiency, diarrhea	55
Calendula flowers (*Calendula officinalis*)	Isoquercitrin, rutin, narcissin; triterpene saponins	Anti-inflammatory	Oral mucosa inflammation, wound healing	56
Chamomile flowers (*Matricaria chamomilla*)	Apigenin and luteolin glycosides (>8%); volatile oil (α-bisabolol, chamazulene)	Antispasmodic, anti-inflammatory, antibacterial	Gastrointestinal complaints, skin and mucosa irritations	57,58
Elder flower (*Sambucus nigra*)	Rutin, astragalin, hyperoside, isoquercitrin (<3%); triterpene acids (about 1%)	Anti-inflammatory, aquaretic	Feverish conditions, common cold	59
Elderberry extract	Anthocyanins (>15%)	Antioxidant, anti-inflammatory, antiallergic		60
Ginkgo leaf extract (*Ginkgo biloba*)	flavonol glycosides (22–27%); terpene lactones (5–7%)	Antioxidant, neuroprotective, PAF inhibitory	Cognitive deficits	61
Goldenrod (*Solidago virgaurea*)	Rutin, quercitrin, isoquercitrin, astragalin, hyperoside (1.5–2.4%); triterpene saponins (>8%)	Aquaretic, analgesic, anti-inflammatory	Irritation of urinary tract in cases of inflammation and renal gravel	62
Hawthorn leaf/flower/fruit (*Crataegus monogyna, oxyacantha*)	Procyanidins (>3%); flavonol and flavone derivatives (>1%)	Cardioactive, hypotensive, coronary vasodilator	Cardiotonic	63,64
Licorice root (*Glycyrrhiza glabra*)	Glycyrrhizin-glycosides (4–24%); flavanones, chalcones, and isoflavones (≈1%)	Anti-inflammatory, demulcent, expectorant	Dyspeptic problems	65,66

Herb	Content	Action	Indication	Ref.
Linden flowers (*Tilia cordata*)	Procyanidin dimers (about 2%); quercetin and kaempferol glycosides (about 1%)	Diaphoretic, sedative	Cold relief, nervous tension	67
Meadowsweet flower (*Filipendula ulmaria*)	Quercetin glycosides, mostly quercetin-4'-glucoside (0.5–1%); hydrolyzable tannins (10–15%); salicylates (about 0.15%)	Diaphoretic	Cold relief	68
Milk thistle fruit (*Sylibum marianum*)	Flavanolignanes, collectively called *sylimarin* (1.5–3%)	Hepatocellular protection	Toxic liver damage, supportive treatment in chronic liver diseases, dyspeptic complaints	69
Orange peel, bitter (*Citrus aurantium*)	Volatile oil (>2%); flavones, flavanones	Aromatic bitter, choleretic	Dyspeptic ailments, appetite stimulant	70
Passionflower aboveground parts (*Passiflora incarnata*)	Vitexin, isovitexin, and their c-glycosides, luteolin glycosides (>2.5%)	Sedative, anxiolytic	Nervous unrest, mild sleep disorders	71
St. John's wort (*Hypericum perforatum*)	Hyperoside (0.5–2%), rutin (0.3–1.6%), isoquercitrin (0.3%); phoreglucinols (>4%); naphthodianthrones (>0.75%)	Antidepressant	Mild to moderate depression	72
Willow bark (Salix spp.)	Salicin and its esters (>10%); naringenin glucosides (>1.5%)	Anti-inflammatory, antipyretic, analgesic	Mild neuralgic pains, osteoarthritis	73,74
Witch hazel leaf/bark (*Hamamelis virginiana*)	Hamamelitannin, gallotannins (>12); catechins, quercetin, and kaempferol glycosides	Astringent, anti-inflammatory, vasoconstrictive, antioxidant	Skin injuries, varicose veins, hemorrhoids	75,76

[a] The content refers to the dried herb.
PAF, platelet activating factor.

in herbal extracts. For further structural information, these interfaces may be coupled to an ITMS analyzer to promote mass fragmentation, and this arrangement provides data helpful in identifying the flavonoids of interest.

1. Electrospray Ionization Mass Spectrometry

ESI-MS produces ions as a result of the application of a potential to a flowing liquid, which causes the liquid to charge and spray. Electrospray forms very small droplets of solvent containing the analytes. Usually, the solvent is removed by heat and multicharged ions are produced. As previously stated, ESI has the advantage over TSP of producing low fragmentation of flavonoid derivatives.

To elucidate the difference between TSP and ESI, the following example is illustrative. TSP-MS of kaempferol-3-O-rutinoside (molecular weight [MW] 594 da) mainly produces kaempferol fragment (m/z 286) resulting from the loss of rutinose ([M-rutinose+H]$^+$), and the kaempferol-rhamnoside fragment (m/z 433, [M-glucose + H]$^+$). The molecular ion (m/z 595 = [M+H]$^+$) is present in very low quantity. Conversely, ESI-MS of the same glycoside predominantly yields the sodium and potassium adducts of the molecular ions: ([M+Na]$^+$), m/z 617; ([M+K]$^+$), m/z 633. Fragmentation ions are almost absent.

For this reason, ESI-MS is particularly suitable for direct analysis of samples without preliminary chromatographic separation. As a result, specific fingerprints of complex natural mixtures are easily and rapidly obtained. This information on the overall components is particularly valuable for herbal medicines, because they are in toto regarded as the active principle rather than single constituents.

Closely related to ESI is APCI in that the source operates at near-atmospheric pressure. APCI produces almost molecular ions with very little fragmentation, and it provides fingerprints of herbs [22].

2. Ion Trap Mass Spectrometry

MS spectra with fragmentation of molecules require collision-activated disso-ciation (CAD) and triple quadrupole analyzers. In these instruments, the analysis is performed as follows: the first quadrupole selects the interesting ion (parent ion), the second produces the fragments from the isolated ion, and the third quadrupole analyzes the fragmentation products (daughter ion spectrum). These steps (ion isolation, fragmentation, and analysis) can be repeated by addition of n quadrupole devices (multisector mass spectrometer) to allow multiple MS/MS experiments (MS^n) to be performed.

As an alternative, MS^n analysis can be carried out in the same physical space by means of ITMS. This approach involves using combinations of direct and rf-field pulse sequences on trapped ions in a helium reagent gas atmosphere.

Besides being simple, ITMS offers significant advantages in terms of sensitivity over a triple quadrupole and it may play an important role in flavonoid analysis.

To exemplify, ITMS is capable of isolating the ions m/z 271, 301, 353, 447, and 609 from the negative ESI-MS spectrum of naringenin, quercetin, chlorogenic acid, quercitrin, and rutin and fragmenting them to produce an "ion map," which shows both isolated ions (parent m/z axis) and their fragments (product m/z axis). In practice, rutin and quercitrin are identified from their molecular ions (m/z 609 and 447) and from the same fragment (the aglycone quercetin, m/z 300). Similarly, the identity of chlorogenic acid is given by the molecular ion (m/z 354) and the fragment m/z 191, which represents quinic acid.

The techniques described allow three different analytical approaches: infusion, direct injection, and injection after a separation step. The infusion is the simplest method of sample introduction (by means of a syringe) into the mass spectrometer. High sample volumes (50–150 µL) at flow rates (3–10 µL/min) are required, and these conditions facilitate the structural investigation of the analytes subjected to a continuous infusion into the spectrometer.

In the second approach, the sample solution is injected by means of a HPLC injector directly into the mass spectrometer, without using any chromatographic column. Direct injection involves low sample volumes (1–10 µL) and has the advantage over the infusion method that no cleaning is needed after each analysis. In addition, the analysis times are very short (1–2 min), thereby permitting rapid screening of many samples. Furthermore, the direct injection approach allows minimization of the ion suppression effects due to the matrix by adding different concentrations of flavonoid standards to the herbal sample, and it may be considered for semiquantitative and rapid screening of herbal extracts.

The third approach involves coupling of a separative system (usually HPLC) with the mass spectrometer. This procedure simultaneously provides chromatographic, ultraviolet, and mass spectrometric data, and this range of information may be very helpful when assessing the identity of principles present in herbs. Further, HPLC coupled to MS (LC-MS) permits discrimination of compounds with the same molecular masses and exclusion of interference from the herbal matrix. Therefore, this approach remains the method of choice for quantitative analyses.

3. Sample Preparation

The herb is usually extracted with methanol or aqueous methanol at room temperature or at 40–50 °C, depending on the stability of its components. The resulting crude extract may be purified to remove undesired constituents, such as lipids, chlorophyll, sugars, organic acids, and salts. In the case of commercial extracts, which are normally enriched in specific compounds, this step may be eliminated. Similarly, the purification may not be necessary in the case of LC-MS, since the analytes of interest are separated from the interfering compounds during the chromatographic elution. By contrast, purification of the sample is recommended in the infusion and direct injection approaches. Indeed, the presence in the herbal matrix of different molecular species at concentrations ranging from 1 to 10 mM can cause ion suppression: i.e., the MS analyzer fails to detect the ions. In some circumstances the matrix effect may be reduced by diluting the sample and/or lowering the flow rate. These expedients appear to be successfull when highly sensitive and salt-compatible MS instruments are used.

4. Alkali Adducts

In positive ESI-MS, some molecular species can form adducts with alkali cations (sodium and potassium). In particular, potassium adducts are typical of raw herbal samples, because vegetable matrices are rich in potassium salts. Alkali adduct formation may be diminished by desalting the samples through solid phase extraction (SPE). Diluting the sample solutions is a simple way to replace potassium ions with sodium ions. The latter are the most common in commercial extracts of herbs.

Not all flavonoids are capable of yielding alkali adducts. Thus, flavonol-3-O-glycosides generate sodium or potassium adducts. By contrast, these adducts are not obtained from flavonol-4'-O-glycosides and flavone glycosides.

So the spectrum of rutin (quercetin-3-O-rutinoside, MW 610 da) is characterized by the presence of the sodium adduct (m/z 633); the molecular ion ([MH]$^+$, m/z 611) is almost absent. Also, the abundance of the aglycone residue (m/z 302) is low, indicating that the removal of the glucose residue is hindered. The same behavior can be observed for other flavonol-3-O-glycosides, as described for *Ginkgo biloba* and St. John's wort extracts (see Secs. II.A and II.B).

Conversely, in the case of spiraeoside (quercetin-4'-O-glucoside) no adduct is formed. Its spectrum mainly presents the molecular ion ([MH]$^+$ = 465) and a relevant fragmentation occurs. Likewise, the flavone-glycoside (lacking the 3-OH group) rhoifolin (apigenin-7-O-neohesperidoside, MW 578) produces the m/z values 579 ([MH]$^+$) and 270, corresponding to molecular and aglycone residue ions, respectively. This finding may suggest that the presence of the hydroxyl group at position 3 may be important for adduct formation. However, this hypothesis is not appropriate, since flavonol aglycones, such as quercetin, kaempferol, and isorhamnetin, do not produce sodium adducts. What seems crucial is the 3-O-glycosylation, which may favor the formation of a crown-embedded stable cation.

Isoflavones form molecular ion adducts. Likely, the isoflavone ring at position 3 plays the same role as the sugar moiety in 3-O-glycosyl flavonols. In fact, the ESI-MS spectrum of the isoflavone biochanin A (MW 284 da) shows as main ion the sodium adduct ($[M+Na]^+$, 307 m/z).

Alkali adducts are also formed with other flavonoid classes (see Secs. II.D. and II.G).

II. MASS SPECTROMETRY FINGERPRINTS OF SELECTED FLAVONOID HERBS

A. *Ginkgo biloba*

The dried green leaves of the ginkgo tree (*Ginkgo biloba* L., Fam. Ginkgoaceae) are the crude herb from which ginkgo extracts are obtained. According to the monograph published by ESCOP [23], the only acceptable ginkgo extracts are those obtained with an average herb-to-extract ratio of 50:1. The extraction procedure is strictly standardized to eliminate unwanted constituents, such as fats, tannins, biflavonoids, ginkgol, and ginkgolic acids. The resulting extract consists of 22–27% flavonol-glycosides, 5–7% terpene lactones (2.8–3.4% ginkgolides A, B, C, and J and 2.6–3.2% bilobalide), small amounts of phenolic acids, and fewer than 5 parts per million (ppm) ginkgolic acids (allergenic).

More than 300 papers have been published on the pharmacological properties of this extract [23]. Experimentally documented activity includes increased tolerance to hypoxia in brain tissue, improved learning capacity and memory, inhibition of platelet activator factor (PAF) [24], improved cerebral and peripheral circulation, neuroprotective effects [25], and reduction of retinal edema. Although all the constituents of the ginkgo extract are considered to contribute to the therapeutic effects, the ginkgo flavonoids are assumed to play an important role that is due to their free radical scavenging capacity.

The fingerprint of *Ginkgo biloba* extracts obtained by direct infusion in ESI-MS is shown in Figure 1. The ions (m/z 617–779) refer to different flavonol-glycosides. In particular, the ions m/z 617, 633, and 647 are due to the sodium adducts of kaempferol-3-*O*-rutinoside, quercetin-3-*O*-rutinoside, and isorhamnetin-3-*O*-rutinoside, respectively. The ions m/z 763, 779, and 793 correspond to the sodium adducts of the 3-*O*-[rhamnosyl-(1→2)-rhamnosyl-(1→6)-glucoside] derivatives of kaempferol, quercetin, and isorhamnetin, respectively. Ions m/z 763 and 779 account also for the 3-*O*-[6‴-p-coumaroylglucosyl-(1→2)rhamno-2)rhamnoside] derivatives of kaempferol and quercetin. Finally, the ions m/z 431, 447, and 463 correspond to sodium adducts of ginkgolide A, ginkgolide B, ginkgolide J, and ginkgolide C, respectively. Interestingly this approach has allowed simultaneous detection of flavonoid and terpene compounds of *Ginkgo biloba*, thereby permitting rapid screening of many samples.

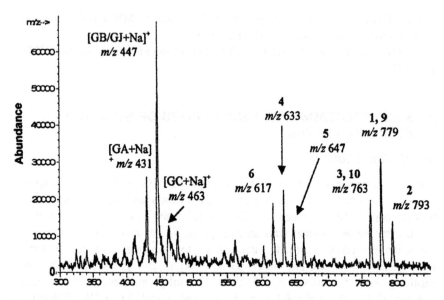

Figure 1 Typical positive ESI-MS mass spectrum of *Ginkgo biloba* extract. ESI-MS, electrospray ionization mass spectrometry; GA, ginkgolide A; GB, ginkgolide B; GC, ginkgolide C; GJ, ginkgolide J.

B.   St. John's Wort

St. John's wort consists of the dried aboveground parts of *Hypericum perforatum* L. (Fam. Hypericaceae) gathered during the flowering season. St. John's wort contains numerous compounds with documented biological activity. Components that have received most attention are the naphthodianthrones hypericin and pseudohypericin (up to 0.15%) and the phloroglucinols hyperforin and adhyperforin (up to 4%). Besides these specific components, St. John's wort contains significant amounts of some very common plant metabolites [26–27], such as flavonol derivatives, mainly hyperoside (0.5–2%), rutin (0.3–1.6%), quercitrin, and isoquercitrin (up to 0.3%); biflavones (amentoflavone); procyanidins; and a volatile oil (about 1%). Clinical proof of therapeutic efficacy for mild to moderate depression [28–29], restlessness, and irritability has been documented for St. John's wort ased on alcoholic extracts, with an herb-to-extract ratio in the range of 4:1 to 7:1 (i.e., 0.2–0.3% hypericins and 2–6% hyperforins, respectively).

ESI-MS in negative mode of *Hypericum perforatum* extracts provides typical fingerprints showing three different classes of compounds, i.e., flavonoids, hypericins, and hyperforins (Fig. 2). The abundant ions at m/z 535 and 549 are due to $[M-H]^-$ ions of hyperforin and adhyperforin, respectively. Deprotonated molecules ($[M-H]^-$) of hypericin (m/z 503), pseudohypericin (m/z 519), and

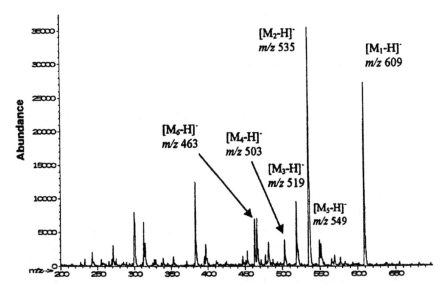

Figure 2 Typical negative ESI-MS mass spectrum of *Hypericum perforatum* extract. ESI-MS, electrospray ionization mass spectrometry; M_1, rutin; M_2, hyperforin; M_3, pseudohypericin; M_4, hypericin; M_5, adhyperforin; M_6, isoquercitrin/hyperoside.

rutin (m/z 609) are present in lower abundance. $[M-H]^-$ ions of other known flavonoids can be detected, such as those of isoquercitrin and hyperoside (m/z 463), quercitrin (m/z 447), and quercetin (m/z 301). The ions m/z 466, 397, 383, and 313 are produced by fragmentation of $[M-H]^-$ ions of hyperforin, presumably by collisions in the skimmer region. As expected, ESI-MS of St. John's wort extracts in positive mode evidences produces cation adducts of isoquercitrin ($[M+Na]^+ = 487$; $[M+K]^+ = 503$) and rutin ($[M+Na]^+ = 639$; $[M+K]^+ = 649$).

ESI-MS may be applied to confirm differences due to environmental conditions. For example, in the case of *H. perforatum* flowers collected in northern and southern Italy, the northern sample was higher in rutin (m/z 609) and pseudohypericin (m/z 519) and had low levels of hyperforin (m/z 535).

C. Bilberry

Bilberry consists of the dried ripe fruit of *Vaccinium myrtillus*, a dwarf shrub of the Ericaceae family. Dried bilberries contain 1–5% catechins, approximately 30% invert sugar, and small amounts of flavonol glycosides (e.g., astragalin, quercitrin, isoquercitrin, hyperoside), phenolic acids (e.g., caffeic and chlorogenic acids), and anthocyanins, particularly glycosides of malvidin, cyanidin, and delphinidin [30].

Bilberry has shown vasoprotective, antiedematous, antioxidant, anti-inflammatory, and astringent actions. In particular, a standardized extract enriched with anthocyanins and their aglycons (anthocyanidins) is endowed with constant biological activity useful in ophthalmology and treatment of vascular disorders including capillary weakness, venous insufficiency, and hemorrhoids [12].

The positive mass spectrum of this standardized bilberry extract is shown in Figure 3. The most abundant ions are due to the molecular ions of anthocyanidins. In particular, the ions with m/z 287, 303, 317, and 331 are related to cyanidin, delphinidin, petunidin, and malvidin, respectively. In addition, the anthocyanins (glycoside derivatives) are also present as molecular ions: m/z 419, 449, 463, 479, and 493 correspond to cyanidin-3-O-arabinoside, cyanidin-3-O-glucoside/petunidin-3-O-arabinoside, delphinidin-3-O-glucoside/malvidin-3-O-arabinoside, petunidin-3-O-glucoside, and malvidin-3-O-glucoside, respectively. By this approach, ESI-MS in the positive mode permits simultaneous detection of different anthocyanidins and anthocyanins in bilberry extract, and the results are in good agreement with those previously reported [30]. Interestingly, the fingerprint of *Vaccinium myrtillus* differs from those of *Catharanthus roseus*, *Vaccinium macrocarpon* (cranberry fruit), and *Sambucus nigra* (elder fruit),

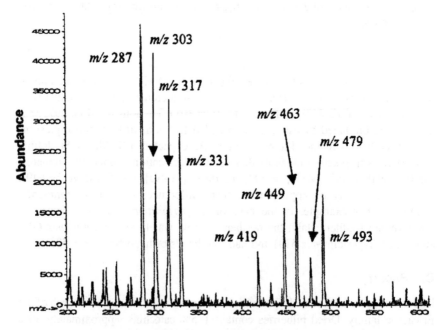

Figure 3 Typical positive ESI-MS mass spectrum of *Vaccinium myrtyllus* extract. ESI-MS, electrospray ionization mass spectrometry.

indicating that extracts containing the same class of compounds may be easily differentiated by ESI-MS.

D. *Camellia sinensis*

Tea (*Camellia sinensis*, Fam. Theaceae) can be considered one of the most important herbal antioxidants [31,32]. Indeed, fresh leaves of tea contain flavanols (catechins), flavonols, catechin tannins (procyanidins), and phenolic acids, which are all endowed with antioxidant properties. This composition is almost preserved in green tea, i.e., in the variety obtained by steaming and drying the leaves immediately after harvesting to prevent enzymatic modifications (fermentation). In green tea the group of flavanols usually accounts for 15–30% of dried leaves, and it includes different catechins, e.g., epigallocatechin, epicatechin, epigalloca-techin-3-*O*-gallate, gallocatechin-3-*O*-gallate, methyl-epigallocatechin-3-*O*-gallate, and epicatechin-3-*O*-gallate. Green tea as black (fermented) and oolong (partially fermented) teas also contains caffeine (2.5–5.5%). Because of the presence of this methylxanthine, the intake of large doses of tea is not advisable, and this property reduces the healthy effects, that are thought to be associated with high consumption of tea [33]. To overcome this limitation a caffeine-free green tea extract containing about 80% catechins has been developed.

Figure 4 shows the ESI-MS positive mass spectrum of this decaffeinated green tea extract. The most abundant ion, m/z 481, is due to the sodium adduct of epigallocatechin gallate (EGCg). The ions m/z 465, 329, and 291 are present in lower abundance and correspond to the sodium adducts ([M+Na]$^+$) of epica-techin gallate (ECg), epigallocatechin (EGC), and epicatechin/catechin (EC/C), respectively.

E. Soybean and Red Clover

The primary isoflavones in soybeans (*Glycine max*) are genistein and daidzein and their respective 7-*O*-β-glucosides. There are also small amounts of a third isoflavone, glycitein, and its glucoside, glycitin. In soybeans and in nonfermented soy foods, isoflavone glucosides are esterified with malonic or acetic acid [34].

The isoflavone pattern of red clover (*Trifolium pratense*) differs from that of soybean. In addition to genistein, daidzein, and their glucosides, red clover contains formononetin (a precursor of equol) and biochanin-A in free and glycosylated forms [35].

Isoflavones have a chemical structure similar to that of mammalian estro-gens and are referred as *phytoestrogens* [36]. Isoflavones are quite weak, possessing 1/1000 to 1/10,000 the estrogenic activity of 17-*O*-β-estradiol. How-

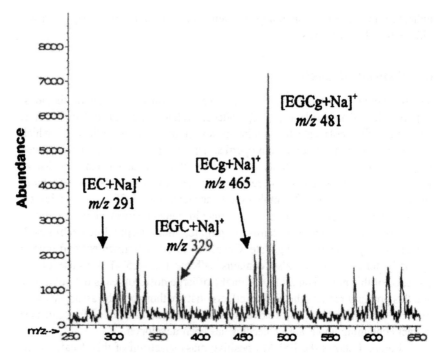

Figure 4 Typical positive ESI-MS mass spectrum of green tea extract. ESI-MS, electrospray ionization mass spectrometry.

ever, their circulating concentrations in subjects consuming a moderate amount of soy foods may be 1000-fold higher than peak levels of endogenous estrogens. Furthermore, isoflavones have a stronger binding affinity for estrogen receptor β (Erβ) than Erα [37], suggesting that the estrogenic effects of isoflavones may be tissue-selective. Indeed, some tissues predominantly contain one form of the receptor or the other, and this may partly explain the health benefits of isoflavones on bone and breast tissues [38]. Specifically, genistein and daidzein are thought to decrease osteoporosis, diminish the risk of estrogen-enhanced carcinogenesis, relieve menopausal symptoms, and decrease heart disease [39]. In addition, isoflavones possess antioxidant capacity and have effects on protein synthesis, enzyme activity, growth factor action, and angiogenesis. Unfortunately, most of these health benefits are based on epidemiological and laboratory evidence, and only a few small and contradictory trials are presently available. What seems necessary are some larger clinical studies, comparing placebo with low and high isoflavone intake [40,41]. As an example of an isoflavone-containing herb, Figure 5 shows the positive ESI mass spectrum of a

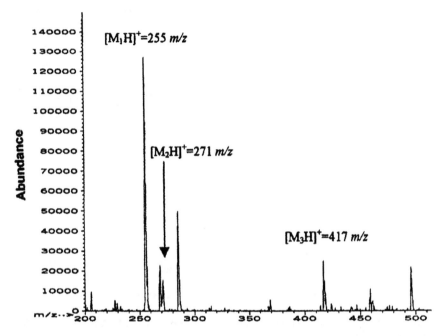

Figure 5 Typical positive ESI-MS mass spectrum of soybean extract. ESI-MS, electrospray ionization mass spectrometry.

standardized soybean extract. The main ion (*m/z* 255) is due to the aglycone daidzein, whereas the glycoside daidzin (*m/z* 417) and genistein (*m/z* 271) are present in lower abundance.

F. Propolis

Propolis is a resinous substance collected by honeybees from leaf buds and cracks in the bark of various plants, mainly from poplar (*Populus*) species and, to a lesser extent, beech, horse chestnut, birch, and conifer. Bees mix the original propolis with beeswax and β-glucosidase they secrete during the propolis collection. The resulting material is used by bees to seal holes in the hives, exclude drafts, protect against external invaders, and mummify their carcasses. Propolis has been used extensively in folk medicine for many years, and there is substantial evidence indicating that propolis has antiseptic, antifungal, antibacterial, antiviral, anti-inflammatory, and antioxidant properties [42].

Propolis cannot be used as raw material, it has to be purified by either extraction with alcoholic solvents or supercritical fluid extraction. Depending on the geographical source and extraction processes, propolis extracts may differ

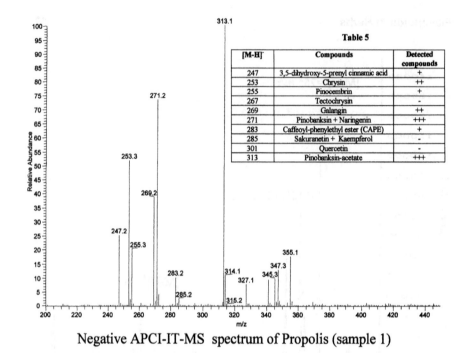

Table 5

[M-H]⁻	Compounds	Detected compounds
247	3,5-dihydroxy-5-prenyl cinnamic acid	+
253	Chrysin	++
255	Pinocembrin	+
267	Tectochrysin	-
269	Galangin	++
271	Pinobanksin + Naringenin	+++
283	Caffeoyl-phenylethyl ester (CAPE)	+
285	Sakuranetin + Kaempferol	-
301	Quercetin	-
313	Pinobanksin-acetate	+++

Negative APCI-IT-MS spectrum of Propolis (sample 1)

Figure 6 Negative APCI-IT-MS spectrum of propolis. APCI-IT-MS, atmospheric pressure ionization ion trap mass spectrometry.

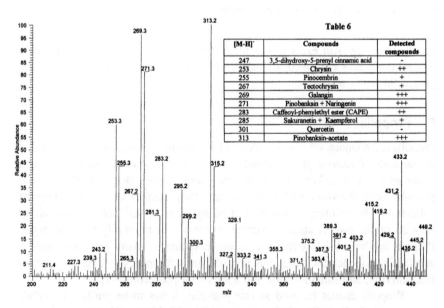

Table 6

[M-H]⁻	Compounds	Detected compounds
247	3,5-dihydroxy-5-prenyl cinnamic acid	-
253	Chrysin	++
255	Pinocembrin	+
267	Tectochrysin	+
269	Galangin	+++
271	Pinobanksin + Naringenin	+++
283	Caffeoyl-phenylethyl ester (CAPE)	++
285	Sakuranetin + Kaempferol	+
301	Quercetin	-
313	Pinobanksin-acetate	+++

Negative APCI-IT-MS spectrum of Propolis (sample 2)

Figure 7 Negative APCI-IT-MS spectrum of propolis. APCI-IT-MS, atmospheric pressure ionization ion trap mass spectrometry.

from each other. Unfortunately and unlike for other herbal extracts used to prepare phytomedicines, a generally accepted standardized extract of propolis has been not yet assessed. For this reason, three different samples of propolis are considered, and their APCI mass spectra recorded in negative mode are shown in Figures 6–8. In contrast to HPLC, which permits identification only a limited number of flavonoids (those with a reference standard) and may suggest the presence of other flavonoid like compounds (from the UV spectra), APCI-IT-MS of demonstrates a larger array of flavonoids, as summarized in Tables 5–7 of Figs. 6–8. The identity of the ions is confirmed through their MS-MS fragmentation, as exemplified for pinocembrin, pinobanksin, and its acetate (Fig. 9). A comparison of MS fingerprints indicates that sample 2 has a broader pattern of compounds than sample 3, which in turn is more rich than sample 1.

MS may also provide a rapid comparison of samples obtained by different extraction procedures. So, propolis obtained by supercritical fluid extraction produces a negative ESI mass spectrum mainly showing chrysin (m/z 253). Apigenin/galangin (m/z 269) and naringenin (m/z 271) are also present at significant levels, whereas pinocembrin (m/z 255) is less abundant. In contrast, the MS spectrum of propolis sample obtained by ethanol extraction evidences higher levels of this flavanone.

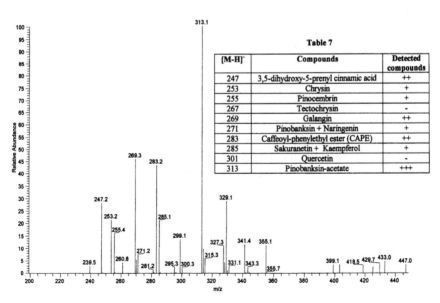

Table 7

[M-H]⁻	Compounds	Detected compounds
247	3,5-dihydroxy-5-prenyl cinnamic acid	++
253	Chrysin	+
255	Pinocembrin	+
267	Tectochrysin	-
269	Galangin	++
271	Pinobanksin + Naringenin	+
283	Caffeoyl-phenylethyl ester (CAPE)	++
285	Sakuranetin + Kaempferol	+
301	Quercetin	-
313	Pinobanksin-acetate	+++

Negative APCI-IT-MS spectrum of Propolis (sample 3)

Figure 8 Negative APCI-IT-MS spectrum of propolis. APCI-IT-MS, atmospheric pressure ionization ion trap mass spectrometry.

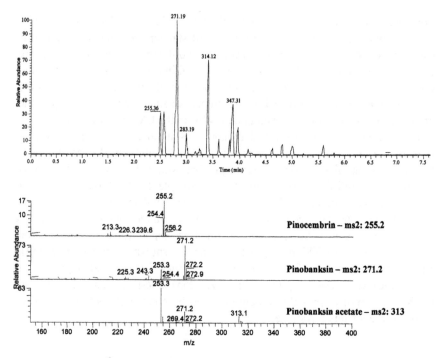

Figure 9   MS/MS² of a propolis extract.

G. *Vitis vinifera*

Polyphenolic oligomers of the flavanoid type result from the condensation of two or more flavan-3-ols and are known as *catechin tannins* or *oligomeric cyanidins* (OPCs). The most abundant sources of OPC are extracts of grape seed (*Vitis vinifera*), standardized to contain 80–85% procyanidins, and extracts of the bark of the maritime pine (*Pinus pinaster*), concentrated to contain about 85% procyanidins [43]. Both these extracts are characterized by the presence of procyanidins (oligomers of catechin epicatechin and their 3-*O*-gallates) [44]. Procyanidins are potent antioxidants [32] and active inhibitors of collagenase, elastase, hyaluronidase, and β-glucuronidase [45]. All these enzymes are involved in the degradation of the main structural components of the extravascular matrix, and their inhibition may aid maintenance of normal capillary permeability.

The negative mass spectrum of *Vitis vinifera* permits identification of the ions related to proanthocyanidins, i.e., the monomer, dimer, trimer, tetramer, and pentamer (Table 2), as previously reported for positive ESI-MS equipped with a triple quadrupole [45]. The ion map of the same sample allows the detection of the different parent ions and their fragment products. For example, the ion *m/z* 729 (dimer gallate) produces the fragments *m/z* 711 (loss of water),

Table 2 Procyanidin Ions Detected in a Standardized *Vitis vinifera* Seed Extract by Negative Electrospray Ionization–Ion Trap Mass Spectrometry

[M-H]$^-$ (*m/z*)	Compound
169	Gallic acid
289	Monomer (catechin-epicatechin)
441	Monomer-gallate
577	Dimer
729	Dimer-gallate
865	Trimer
881	Dimer-digallate
1017	Trimer-gallate
1153	Tetramer
1169	Trimer-digallate
1305	Tetramer-gallate
1441	Pentamer
1593	Pentamer-gallate

577 (loss of gallate), 441 (loss of a monomer), and 289 (loss of a monomer gallate). All the other oligomers yield analogous results.

H. Flavonoids in Herbs with Aromatic or Savory Properties

The presence of polyphenols (e.g., flavonoids and phenolic acid) has been reported in different species of the Labiatae family, including *Ocimum basilicum* (basil), *Origanum vulgare* (oregano), *Origanum majorana* (marjoram), *Mentha* var. (mint), *Rosmarinus officinalis* (rosemary), and *Thymus vulgaris* (thyme), and *Umbelliferae* family, such as *Anethum graveolens* (dill), *Coriandrum sativum* (coriander), *Cuminum sativum* (cumin), *Foeniculum vulgare* (fennel), and *Petroselinum crispum* (parsley) [46]. Many of these species are consumed as ingredients in foods and beverages for their flavoring and aromatic properties and are regarded as culinary herbs. In most of these culinary herbs, the flavor is provided by a variety of compounds (Table 3) present in their volatile oils [46,47]. In contrast, the flavonoid and nonvolatile phenolic (e.g., carnosol, rosmarinic acid, carnosic acid, ferulic and caffeic acids) fractions are not relevant for sensory effects, but they may contribute antioxidant effects. However, this contribution is quite disputed, because only modest amounts (just for flavoring) of these vegetables are considered safe. This limitation is suggested by the presence in these species of some phytochemicals with potential adverse effects [46], such as apiole and

Table 3 Selected Flavonoid Culinary Herbs

	Volatile oil main components [46]	Flavonoids[a]	Other phenolics [46]
Labiatae			
Mentha var. (mints)	≈ 0.4% (menthol, 30–40%; menthone, 20–30%; pulegone, menthylacetate, limonene, cineole)	Apigenin and luteolin derivatives Apigenin, 18–99 mg Luteolin, 11–42 mg [51]	Rosmarinic acid
Ocimum basilicum (basil)	≈ 0.08% (linalool, ≈ .45%; estragol, ≈ 50%; methylcinnamate, geraniol, cineole, safrole)	Apigenin, nevadensin, salvigenin [77]	Rosmarinic acid
Origanum vulgare (oregano)	≈ 0.6% (thymol, carvacrol, monoterpene hydrocarbons, borneol, linalool)	Apigenin and luteolin derivatives [51] Apigenin, 18–99 mg Luteolin, 11–42 mg	Rosmarinic acid
Origanum majorana (marjoram)	≈ 1% (terpinenes, *p*-cymene, carvacrol, estragole)	Apigenin and luteolin derivatives [46]	Labiatic acid, rosmarinic acid
Rosmarinus officinalis (rosemary)	≈ 0.5% (α-pinene, geraniol, cineol, borneol)	Luteolin derivatives [51] Luteolin, ≈ 4 mg	Rosmarinic acid, labiatic acid, carnosic acid, carnosol
Satureya hortensis (summer savory) and *Satureya montana* (winter savory)	≈ 1% (carvacrol, α-pinene, cineole, borneol)	Apigenin derivatives [46]	Labiatic acid
Thymus vulgaris (thyme)	≈ 1% (thymol, 30–70%; carvacrol, 3–15%; α-pinene, α-terpineol, linalool)	Apigenin and luteolin derivatives [51] Apigenin, 18–99 mg Luteolin, 11–42 mg	Caffeic acid, labiatic acid

Umbelliferae

Species (common name)	Essential oil	Flavonoids	Coumarins
Anethum graveolens (dill seeds)	2.5–4% (carvone, 35–60%; α-pinene limonene, phellandrene, eugenol, cineole)	Quercetin, kaempferol, isorhamnetin derivatives [51]; Quercetin, 48–110 mg; Kaempferol, 16–24 mg; Isorhamnetin, 15–72 mg	Coumarins
Carum carvi (caraway seeds)	4–7% (carvone, 50–60%; limonene, ≈ 40%)	Quercetin derivatives [46]	Not detected
Coriandrum sativum (coriander seeds)	0.4–1% (linalool, decyl aldehyde, cineole)	Quercetin derivatives [46]	Coumarins
Cuminum cyminum (cumin seeds)	2–5% (cuminaldehyde, ca. 50%; monoterpene and sesquiterpene hydrocarbons)	Apigenin, luteolin derivatives; hesperidin [46]	Not detected
Foeniculum vulgare (fennel seeds)	2–6% (anethole, 50–70%; fenchone, 10–20%; estragole, trace; limonene, α-pinene)	Quercetin kaempferol derivatives [46]	Coumarins
Petroselinum crispum (parsley)	0.05–0.3% (myristicin, ≈ 80%; apiole, terpenes)	Apigenin and luteolin derivatives [51]; Apigenin, 510–630 mg; Luteolin, 0–4 mg	Coumarins
Pimpinella anisum (anise seeds)	1–4% (anethole, 75–90%; estragole, anise ketone, monoterpene hydrocarbons)	Quercetin, luteolin, apigenin derivatives [46]	Coumarins

a The content of total flavonoids of fresh vegetables, measured by high-performance liquid chromatography–diode array detection (HPLC-DAD) after acid hydrolysis, is expressed as milligrams per 100 g fresh weight [51]. The flavonoid content of savory and coriander, cumin, fennel, and anise seeds is not known.

Table 4 m/z Values of Ions Produced by Liquid Chromatography Mass Spectrometry of Flavonoid Glycosides Present in Some Culinary Herbs

HERB	[M-H]⁻	Product ions of [M-H]⁻	Product ions of aglycone	Suggested name[a]
Chives (*Allium*	463	301	ND	Q-glu
schoenoprasum)	477	315	300,271,151,107	I-glu
	447	285	ND	K-glu
Dill (*Anethum*	609	301	ND	Q-rut
graveolens)	477	301,255, 179,151	179,151,121,107	Q-gluc
	461	285	—	K-gluc
	491	315,300	300,271,164, 151,136,107	I-gluc
Mint (*Mentha* var.)	593	285	175,151,133,107	L-rut
	579	271	ND	N-rut
	609	301	286,242,164, 151,125	H-rut
	577	269	151,149,117,107	A-rut
	607	299	284,256,177, 151,121	D-rut
	795	283	268,240,(151)	Ac-acetyl-glu-rut
	591	283	268,240,151,107	Ac-rut
Oregano (*Origanum*	799	285	199,175,151,133	L-gly
vulgare)	635	269	ND	A-acetyl-diglu
	635	593,299	284,137	D-acetyl-apiosylglu
Parsley (*Petroselium*	563	269	151,149,117,107	A-apiosylglu
crispum)	593	299	284,256,(107)	D-apiosylglu
	605	563,269	151,149,117,107	A-acetyl-apiosylglu
	635	593,299	284,256,(107)	D-acetyl-apiosylglu
Rosemary (*Rosmarinus* officinalis)	503	285	L-acetyl-glu	
Thyme (*Thymus*	463	287	151,135,107	E-gluc
vulgaris)	447	285	175,151,149, 133,107	L-glu
	461	285	175,151,149, 133,107	L-gluc
	489	285	175,151,(149), 133,107	L-acetyl-glu
	445	269	151,149,117,107	A-gluc
	609	285		L-diglu

[a] Q, quercetin; I, isorhamnetin; K, kaempferol; L, luteolin; N, naringenin; H, hesperidin; A, apigenin; Ac, acacetin; D, diosmetin; E, eriodictyol; Glu, glucoside; rut, rutinoside; gluc, glucuronide; gly, glucoside.

Source: Adapted from Ref. 53.

myristicin (parsley), coumarins (anise, coriander, fennel, parsley), estragol (anise, basil, fennel), safrole (basil), thuyone (sage), and thymol and carvacrol (thyme, oregano, marjoram). Namely, coumarins are phototoxic; apiole and myristicin are uterine stimulants and may cause damage to kidney; estragole and safrole are alkylbenzenes with weak hepatotoxic potential; thujone is a bicyclic diterpene (related to camphor) with neurotoxic potential; thymol and carvacrol may induce nausea [48–50,17]. Hence, only low consumption of these culinary herbs (that means also low amounts of all phytochemicals, including the antioxidant phenolics) is advisable. In terms of flavonoid composition, quercetin and kaempferol are the most abundant, followed by luteolin, apigenin, hesperetin, and methoxylated flavones [51]. The majority of these flavonoids are present as glycosides and in most studies are measured as aglycones after acid hydrolysis [52]. In 2000, structural information on flavonoid glycosides present in commonly eaten herbs was obtained by negative APCI-MS [53]. This approach provides the deprotonated molecular ions (Mr) of glycosylated flavonoids. In addition, source fragmentation of Mr ions produces the aglycones, whose identity is confirmed by comparison of the product-ion spectra obtained through low-energy collision-induced dissociation (CID) MS-MS with those of aglycone standards (Table 4).

III. CONCLUSIONS

It may be concluded that mass spectrometry allows the analysis of flavonoids in complex herbal matrices. Characterized by high specificity and sensitivity (up to 10 nM), ESI-MS and APCI-MS can produce fingerprints of herbal extracts without prepurification steps. In addition, MS-MS ion trap methodology provides reliable assignment of the compounds present in the extracts.

ACKNOWLEDGMENTS

Research stimulating this review was supplied by grants from Indena S.p.A—Italy and Specchiasol S.r.I—Italy.

REFERENCES

1. Prior RL, Cao G. Flavonoids: diet and health relationships. Nutr Clin Care 2000; 3:279–288.
2. Ielpo MTL, Basile A, Mirando R, Moscatello V, Nappo C, Sorbo S, Laghi E, Ricciardi MM, Ricciardi L, Vuotto ML. Immunopharmacological properties of flavonoids. Fitoterapia 2000; 71:S101–S109.

3. Craig WJ. Health-promoting properties of common herbs. Am J Clin Nutr 1999; 70:491S–499S.
4. Valerio LG, Kepa JK, Pickwell GV, Quattrochi LC. Induction of human NADPH:quinone oxidoreductase (NQO1) gene expression by the flavonol quercetin. Toxicol Lett 2001; 119:49–57.
5. Samman S, Wall PML, Farmakalidis E. Flavonoids and other phytochemicals in relation to coronary disease. In: Basu TK, Temle NJ, Garg ML, eds. Antioxidants in Human Health and Disease. Wallingford, England: CABI, 1999:175–183.
6. Dugas AJ, Castaneda J, Bonin GC, Price KL, Fisher NH, Winston GW. Evaluation of the total peroxyl radical scavenging capacity of flavonoids: structure-activity relationships. J Nat Prod 2000; 63:327–331.
7. Zandi P, Gordon MH. Antioxidant activity of extracts from old tea leaves. Food Chem 1999; 64:285–288.
8. Pietta PG. Flavonoids as antioxidants. J Nat Prod 2000; 63:1035–1043.
9. Karakaya S, Nehir SEL. Quercetin, luteolin, apigenin and kaempferol contents of some foods. Food Chem 1999; 66:289–292.
10. Petersen J, Dwyer J. Flavonoids: dietary occurrence and biochemical activity. Nutr Res 1998;18:1995–2018.
11. Pietta PG. Flavonoids in medicinal plants. In: Rice-Evans C, Packer L, eds. Flavonoids in Health and Disease. New York: Marcel Dekker, 1998:61–109.
12. Mills S, Bone K. Principles and Practice of Phytotherapy-Modern Herbal Medicine, Edinburgh: Churchill Livingston, 2000.
13. Hudson J, Towers GHN. Phytomedicines as antivirals. Drugs Future 1999; 24:295–320.
14. Pietta PG, Pietta A. Fitomedicine e Nutrienti. 2d ed. Verona: Ricchiuto GM, 1998.
15. Lin JK, Liang YC, Lin-Shiau SY. Cancer chemoprevention by tea polyphenols through mitotic signal transduction blockade. Biochem Pharmacol 1999; 58:911–915.
16. Ferguson LR. Role of plant polyphenolics in genomic stability. Mutat Res 2001; 47:89–111.
17. Pietta PG. Phytomedicines: creating safer choices. In: Watson RR, ed. Vegetables, Fruits and Herbs in Health Promotion. Boca Raton, FL: CRC Press, 2001:73–84.
18. Pietta PG, Mauri PL. Analysis of flavonoids in medicinal plants. Methods Enzymol 2001; 335:26–46.
19. Pietta PG, Gardana C. Capillary electrophoresis and herbal drug analysis. In press.
20. Merkem HM, Beecher GR. Measurement of food flavonoids by high performance liquid chromatography. J Agric Food Chem 2000; 48:577–599.
21. Mauri PL, Pietta PG. Electrospray characterization of selected medicinal plant extracts. J Pharm Biomed Anal 2000; 23:61–68.
22. Pietta PG, Gardana C, Cristoni S. Analytical methods for quality control of propolis. Phytother Res. In press.
23. DeFeudis FV. Ginkgo biloba extract (EGb 761): from chemistry to the clinic. Wiesbaden: Ullstein Medical, 1998.

24. Dutta-Roy AK. Inhibitory effect of ginkgo biloba extract on human platelet aggregation. Platelets 1999; 10:298–305.
25. Luo Y. Ginkgo biloba neuroprotection: Therapeutic implications in Alzehimer's disease. J Alzheimer Dis 2001; 3:401–407.
26. Mauri PL, Pietta PG. Analysis of Hypericum perforatum extracts by high performance liquid chromatography\electrospray mass spectrometry. Rapid Commun Mass Spectrom 2000; 14:95–99.
27. Pietta PG, Gardana C, Pietta AM. Comparative evaluation of St. John's wort from different Italian regions. II Farmaco 2001; 56:491–496.
28. Schrader E. Equivalence of St John's wort extract (ZE 117) and fluoxetine: a randomized, controlled study in mild to moderate depression. Int Clin Psychopharmacol 2000; 15:61–68.
29. Stevenson C, Ernst E. Safety of Hypericum in patients with depression: a comparison with conventional antidepressant. CNS drugs 1999; 11:125–132.
30. Morazzoni P, Bombardelli E. Vaccinium myrtillus. Fitoterapia 1996; 67:3–29.
31. Liu F, Ng TB. Antioxidative and free radical scavenging activities of selected medicinal herbs. Life Sci 2000; 66:725–735.
32. Pietta PG, Simonetti PL, Mauri PL. Antioxidant activity of selected medicinal plants. J Agric Food Chem 1998; 46(11):4487–4490.
33. Mukhtar H, Ahmad N. Tea polyphenols: prevention of cancer and optimizing health. Am J Clin Nutr 2000; 71(suppl):1698S–1702S.
34. Mellenthin O, Galena R. Analysis of polyphenols using capillary zone electrophoresis and HPLC: detection of soy, lupin and pea protein in meat products. J Agric Food Chem 1999; 47:594–602.
35. Newall CA, Anderson LA, Phillison JD. Herbal medicines: a guide for health-care professionals. London: Pharmaceutical Press, 1996.
36. Setchell, KD, Cassidy, A. Dietary isoflavones: biological effects and relevance to human health. J Nutr 1999; 129:4252–42639.
37. Kuiper GG, Lemmen JC, Carlsson B, Corton JC, Safe SH, van der Saag PT, van der Burg B, Gustafsson JA. Interaction of estrogenic chemicals and phytoestrogens with estrogen receptor beta. Endocrinology 1998; 138:4252–4263.
38. Hendrich S, Murphy PA. Isoflavones, source and metabolism. In: Wildman REC, ed. Handbook of Nutraceuticals and Functional Foods. Boca Raton, FL: CRC Press, 2001:55–75.
39. Bidlack WR, Wang W. Designing functional foods to enhance health. In: Bidlack WR, Omaye ST, Meskin MS, Topham DKW, eds. Phytochemicals as Bioactive Agents. Lancaster, PA: Technomic, 2000:241–270.
40. Mitchell JH. Phytoestrogens: involvement in breast and prostate cancer. In: Wildman REC, ed. Handbook of Nutraceuticals and Functional Foods. Boca Raton, FL: CRC Press, 2001:99–112.
41. Banes S. What is the safe level of phytoestrogen intake? Phytoestrogens in food and bone health. Final Meeting of the European Concerted Action on Phytoestrogens in Food and Bone Health, Versailles, Oct 4–6, 2001.
42. Pietta PG, Gardana C, Pietta A. Propolis: a review. Fitoterapia 2002; 73(S1):7–20.
43. Packer L, Rimbach G, Virgili F. Antioxidant activity and biological properties of a

procyanidin-rich extract from pine bark, pycnogenol. Free Radic Biol Med 1999; 27:704–724.

44. Lazarus SA, Adamson GE, Hammerstone JF, Schmitz HH. High performance liquid chromatography\mass spectrometry analysis of procyanidins in foods and beverages. J Agric Food Chem 1999; 47:3693–3701.

45. Gabetta B, Fuzzati N, Griffino A, Lolla E, Pace R, Ruffilli T, Peterlongo F. Characterization of procyanidins from grape seeds (Leucoselect). Fitoterapia 2000; 71(2):162–175.

46. Leung AJ, Foster S. Encyclopedia of Common Natural Ingredients, 2nd ed. New York: John Wiley, 1996.

47. Craig WJ. Health-promoting properties of common herbs. Am J Clin Nutr 1999; 70:491S–499S.

48. Towers GHN, Page JE, Hudson JB. Light-mediated biological activities of natural products from plants and fungi. Curr Org Chem 1997; 1:395–414.

49. Guffin M, Hobbs C, Upton R, Goldberg A. Botanical safety handbook. Boca Raton, FL: CRC Press, 1997.

50. Brinker F. Herb contraindications and drug interactions. Sandy, OR: Eclectic Medical, 1998.

51. Justesen U, Knuthsen P. Composition of flavonoids in fresh herbs and calculation of flavonoid intake by use of herbs in traditional Danish dishes. Food Chem 2001; 73:245–250.

52. Justesen U, Knuthsen P, Leth T. Quantitative analysis of flavonols, flavones and flavanones in fruits, vegetables and beverages by HPLC with photo-diode array and mass spectrometric detection. J Chromatogr A 1998; A799:101–110.

53. Justesen U. Negative atmospheric pressure chemical ionisation low-energy collision activation mass spectrometry for the characterisation of flavonoids in extracts of fresh herbs. J Chromatogr A 2000; 902:369–379.

54. Weggener T, Fintelmann V. Pharmacological properties and therapeutic profile of artichoke. Wien Med Wochenschr 1999; 149:241–247.

55. Bilberry fruit. In: Blumenthal M, Goldberg A, Brinckmann J, eds. Herbal Medicine-Expanded Commission E Monographs. Newton, MA: Integrative Medicine Communications, 2000:16–21.

56. Calendula flower. In: Blumenthal M, Goldberg A, Brinckmann J, eds. Herbal Medicine-Expanded Commission E Monographs. Newton, MA: Integrative Medicine Communications, 2000:44–46.

57. Chamomile flower. In: Blumenthal M, Goldberg A, Brinckmann J, eds. Herbal Medicine-Expanded Commission E Monographs. Newton, MA: Integrative Medicine Communications, 2000:57–61.

58. Licorice root. In: Blumenthal M, Goldberg A, Brinckmann J, eds. Herbal Medicine-Expanded Commission E Monographs. Newton, MA: Integrative Medicine Communications, 2000:233.

59. Elder flower. In: Blumenthal M, Goldberg A, Brinckmann J, eds. Herbal Medicine-Expanded Commission E Monographs. Newton, MA: Integrative Medicine Communications, 2000:103–105.

60. Youdim KA, Martin A, Joseph JA. Incorporation of the elderberry anthocyanins by

endothelial cells increases protection against oxidative stress. Free Radic Biol Med 2000; 29:51–60.

61. Ginkgo biloba leaf. In: Blumenthal M, Goldberg A, Brinckmann J, eds. Herbal Medicine-Expanded Commission E Monographs. Newton, MA: Integrative Medicine Communications, 2000:160–169.

62. Goldenrod. In: Blumenthal M, Goldberg A, Brinckmann J, eds. Herbal Medicine-Expanded Commission E Monographs. Newton, MA: Integrative Medicine Communications, 2000:178–171.

63. Hawtorn leaf with flower: Analytical, quality control and therapeutic monograph. American Herbal Pharmacopoeia February 1999:1–29.

64. Tauchert M, Gildor A, Lipinski J. High-dose crataegus extract WS 1442 in the treatment of NYHA stage II heart failure. Herz 1999; 24:465–474.

65. Shibata S. A drug over the millenia: pharmacognosy, chemistry, and pharmacology of licorice. J Pharm Sci Jpn 2000; 120:849–962.

66. Schultz V, Hansel R, Tyler VE. Rational Phytotherapy 200. Berlin: Springer Verlag, 2001.

67. Linden flower. In: Blumenthal M, Goldberg A, Brinckmann J, eds. Herbal Medicine-Expanded Commission E Monographs. Newton, MA: Integrative Medicine Communications, 2000:240–243.

68. Meadowsweet. In: Blumenthal M, Goldberg A, Brinckmann J, eds. Herbal Medicine-Expanded Commission E Monographs. Newton, MA: Integrative Medicine Communications, 2000:253–255.

69. Milk thistle fruit. In: Blumenthal M, Goldberg A, Brinckmann J, eds. Herbal Medicine-Expanded Commission E Monographs. Newton, MA: Integrative Medicine Communications, 2000:257–263.

70. Orange peel. In: Blumenthal M, Goldberg A, Brinckmann J, eds. Herbal Medicine-Expanded Commission E Monographs. Newton, MA: Integrative Medicine Communications, 2000:287–289.

71. Berdonces JL. Attention deficit and infantile hyperactivity. Rev Enfermeria 2001; 24:11–14.

72. Nathan PJ. Hypericum perforatum: a non-selective reuptake inhibitor? A review of the recent advances in its pharmacology. J Psychopharmacol 2001; 15:47–54.

73. Willow bark. Analytical, quality control and therapeutic monograph. American Herbal Pharmacopoeia December 1999:1–15.

74. Schmid B, Ludtke R, Selbmann HK, Kotter I, Tschirdewahn B, Schaffner W, Heide L. Efficacy and tolerability of a standardized willow bark extract in patients with osteoarthritis: randomized placebo-controlled, double blind clinical trial. Phytother Res 2001; 15:344–350.

75. MakKay D. Hemorroids and varicose veins: a review of treatments options. Altern Med Rev 2001; 6:126–140.

76. Grimme H, Augustin M. Phytotherapy in chronic dermatoses and wounds: what is the evidence? Forsch Komplementarmed 1999; 6:5–8.

77. Grayer RJ, Veitch NC, Kite KG, Price AM, Kobubun T. Distribution of 8-oxygenated leaf-surface flavones in the genus *Ocimum*. Phytochemistry 2001; 56:559–567.

3

The Relationship Between the Phenolic Composition and the Antioxidant Activity of Fruits and Vegetables

Anna R. Proteggente and Catherine A. Rice-Evans
King's College London
London, England

Sheila Wiseman and Frans H. M. M. van de Put
Unilever Health Institute
Vlaardingen, The Netherlands

I. INTRODUCTION

Over the years, several beneficial properties (anti-inflammatory, anticarcinogenic, antimutagenic) have been attributed to phenolic compounds [1–2], and a number of studies have suggested that consumption of fruits and vegetables can reduce the risk of cardiovascular diseases and cancer, potentially through the biological actions of the phenolic components [3–8] as well as antioxidant vitamins.

The antioxidant activity is probably the most extensively studied aspect of the bioactivity of phenolic compounds. Pure phenolic compounds are powerful free radical scavengers in vitro, and this property has been demonstrated both with synthetic free radicals [e.g., 2,2'-azinobis-(3-ethylbenzothiazoline-6-sulfonic acid) (ABTS) and (2,2-diphenyl-l-picrylhydrazyl) (DPPH)] and with physiologically relevant peroxyl radicals, hydroxyl radicals, and superoxide [9–11]. However, although increasing evidence for the role of oxidative damage in pathological disorders suggests that antioxidant ability might be an important factor in the effects of flavonoids in vivo [12], one cannot infer that the potential beneficial effects of phenolics exclusively depend on their antioxidant activity. In this chapter, the relationships among the phenolic content and composition of regularly consumed fruits and vegetables, their vitamin C concentrations, and antioxidant potentials in vitro are reviewed [13–20]. In addition, a comparison of their H-donating antioxidant activities is made through three assays: Trolox equivalent antioxidant capacity (TEAC), oxygen radical absorbance capacity (ORAC), and ferric reducing ability of plasma (FRAP). The information that is provided represents a useful compendium of the composition and the content of major phenolic compounds in some common fruit and vegetables. This information can enhance understanding of the potential for uptake of specific phenolic compounds and determine whether they may be present in vivo at concentrations able to exert protective effects. Thus, by identifying the major components in foods and by characterizing their biological activities and bioavailable forms, it will be possible to provide useful information to illuminate the potential role of these dietary components in protection against chronic degenerative diseases.

II. ANTIOXIDANT ACTIVITY IN VITRO OF FRUITS AND VEGETABLES

Over the years, a variety of assays to measure the total antioxidant capacity of pure substances, biological fluids, food extracts, and beverages have been developed. Among the most commonly used are the Trolox equivalent antioxidant capacity (TEAC), oxygen radical absorbance capacity (ORAC), and ferric reducing ability of plasma (FRAP) assays.

The TEAC assay, as in the modified version by Re and associates [21], is based on the scavenging of the ABTS, 2,2′-azinobis-(3-ethylbenzothiazoline-6-sulfonic acid) diammonium salt by hydrogen-donating compounds. The ORAC assay, according to the method of Cao and colleagues [22], is based on the oxidation of β-phycoerythrin, a fluorescent protein, by 2,2′-azobis(2-methyl-propionamidine dihydrochloride) (ABAP), a peroxyl radical generator, and the inhibition of the formation of the fluorescent protein by antioxidants. The FRAP, according to the method of Benzie and Strain [23], is based on the reduction of a ferric 2,4,6-tripyridyl-s-triazine complex to the ferrous form by antioxidants. Although the three assays differ in the radical used, the radical-generating system, and the kinetic of reaction, all are based on reactions occurring in a hydrophilic system and refer the radical-scavenging ability of test mixtures to a Trolox standard.

A comparison of antioxidant activities reported in literature, for a number of regularly consumed fruits and vegetables, and those obtained by using the TEAC, ORAC, and FRAP assays is presented in Tables 1, 2, and 3, respectively. A 2002 study utilized these assays to assess comparatively the antioxidant

Table 1 Antioxidant Capacity, Trolox Equivalent Antioxidant Capacity of Fruits and Vegetables[a]

Fruit/vegetable	Proteggente et al. [13]	Other studies
Strawberry	2591 ± 68	
Raspberry	1846 ± 10	1725 ± 103 [25]
Red plum	1825 ± 28	
Grapefruit	861 ± 53	
Orange	849 ± 25	
Red cabbage	1377 ± 49	
Broccoli	648 ± 25	
Onion	532 ± 29	580 ± 320 [17]
Green grape	594 ± 72	
Spinach	757 ± 54	
Green cabbage	492 ± 18	
Pea	440 ± 18	
Cauliflower	295 ± 16	
Leek	240 ± 11	
Lettuce	171 ± 12	
Pear	282 ± 19	
Apple	343 ± 13	640 ± 270 [17]
Peach	244 ± 9	
Banana	181 ± 39	
Tomato	255 ± 14	160 ± 60 [17] 270 ± 30 [24][b]

[a] Micromolar Trolox equivalents/100 g fresh weight (FW).
[b] Extrapolated from figure in original Ref. 24.

potentials of polyphenol-rich extracts of fruits and vegetables [13]. The hierarchy of the antioxidant activities (μmol Trolox equivalents/100 g [FW] uncooked portion size) for some common fruits and vegetables, as determined by these authors by using the TEAC assay, was as follows: strawberries ≫ raspberries = red plums ≫ red cabbages ⋙ grapefruit = oranges > spinach > broccoli > green grapes ≅ onions > green cabbages > peas > apples > cauliflowers ≅ pears > tomatoes ≅ peaches = leeks > bananas ≅ lettuce. The hierarchy of antioxidant potentials determined by the ORAC assay was somewhat different from the one obtained by using the TEAC assay (red plums ≅ strawberries > red cabbage > oranges ≅ raspberries ≫ spinach > grapefruit > broccoli > green cabbage > onions > green grapes > peaches > peas ≫ pears ≅ apples > cauliflower ≅ tomatoes ≅ leeks > bananas ≅ lettuce). However, as for the TEAC's hierarchy, the anthocyanin-rich fruits exerted the highest antioxidant activity, followed by the citrus fruits and the majority of the flavonol-rich fruit and vegetables, whereas the hydroxycinnamate-containing fruits and vegetables consistently

Table 2 Antioxidant Capacity, Oxygen Radical Absorbance Capacity, of Fruits and Vegetables[a]

Fruit/Vegetable	Proteggente et al. [13]	Other studies
Strawberry	2437 ± 95	1536 ± 238 [28]
		2060 ± 233 [26]
		2680 [27]
Raspberry	1849 ± 232	1820 ± 80 [20]
		2140 ± 224 [26]
Red plum	2564 ± 185	949 ± 67 [28]
Grapefruit	1447 ± 67	483 ± 18 [26]
Orange	1904 ± 259	750 ± 101 [26]
		1970 [27]
Red cabbage	2124 ± 68	
Broccoli	1335 ± 62	
Onion	988 ± 30	1270 [27]
Green grape	872 ± 48	446 ± 106 [28]
Spinach	1655 ± 115	1940 [27]
Green cabbage	1180 ± 68	490 [27]
Peas	704 ± 62	
Cauliflower	425 ± 44	790 [27]
Leek	413 ± 15	
Lettuce	319 ± 37	400 [27]
Pear	587 ± 50	134 ± 6 [28]
		460 [27]
Apple	560 ± 18	218 ± 35 [28]
		490 [27]
Peach	764 ± 49	
Banana	331 ± 59	221 ± 19 [28]
		460 [27]
Tomato	420 ± 39	189 ± 12 [28]
		450 [27]

[a] Micromolar Trolox equivalents/100 g fresh weight (FW).

elicited lower antioxidant activity. Also the hierarchy of the antioxidant activities determined by the FRAP assay mirrored those observed by using the TEAC and ORAC assays: strawberries ⋙ raspberries > red plums > red cabbage ≳ oranges ≅ spinach > broccoli = grapefruits ≅ green cabbage > green grapes > apples ≅ onions ≅ tomatoes ≅ peaches ≅ pears > cauliflower = peas > bananas = leeks > lettuce.

Previously reported TEAC values for onion and tomato extracts [17,24] and for raspberry extracts [25] were consistent with those obtained by Proteggente and coworkers [13]; the TEAC value for apple was different because of varietal differences. As for reports on antioxidant potentials that

Table 3 Antioxidant Capacity, Ferric Reducing Ability of Plasma, of Fruits and Vegetables[a]

Fruit/Vegetable	Proteggente et al. [13]
Strawberry	3352 ± 38
Raspberry	2325 ± 53
Red plum	2057 ± 25
Grapefruit	829 ± 6
Orange	1181 ± 6
Red cabbage	1870 ± 18
Broccoli	833 ± 16
Onion	369 ± 13
Green grape	519 ± 48
Spinach	1009 ± 35
Green cabbage	694 ± 14
Pea	251 ± 9
Cauliflower	259 ± 5
Leek	160 ± 1
Lettuce	124 ± 7
Pear	315 ± 24
Apple	394 ± 8
Peach	336 ± 4
Banana	164 ± 32
Tomato	344 ± 7

[a] Micromolar Fe^{2+} equivalents/100 g fresh weight (FW).

have been determined by using the ORAC assays, the values for strawberry and raspberry measured by Kalt and associates [26] are in agreement with those indicated by Proteggente and colleagues [13]. Similar ORAC values for a number of other fruits and vegetables, such as strawberry, orange, apple, pear, banana, onion, spinach, green cabbage, cauliflower, and lettuce, have been indicated [27]. However, previously reported ORAC values of plum, orange, grapefruit, tomato, white grape, apple, pear, and banana [28] are much lower. Thus, the literature data for the antioxidant activities of some fruits and vegetables, obtained by the same or similar in vitro assays, consistently indicate a high potential of certain foods such as berries, green leafy vegetables, and citrus fruits. However, the findings of different authors are not always comparable because of the enormous variability of the antioxidant content in fruits and vegetables, not only between varieties, but also in the same variety, depending on the cultivation site, the climate, the stage of maturity of the fruit or vegetable, and the sample preparation and extraction procedures. However, the data reviewed here present a compendium of the available information on the antioxidant potentials of some of the most commonly consumed fruits and vegetables to enhance understanding of the nutritional importance of some foods in terms of antioxidant defense.

III. PHENOLIC CONTENT OF FRUITS AND VEGETABLES

When evaluating the antioxidant potential of fruits and vegetables, it is of interest to consider their total phenolic content, as presented in Table 4. Strawberry extracts have been reported to possess the highest total phenolic content (400 mg/100 g FW), with respect not only to the other anthocyanin-rich fruits but also to all the other fruit and vegetable extracts (Table 4) [13]. However, total

Table 4 Total Phenolic Content of Fruits and Vegetables[a]

Fruit/Vegetable	Proteggente et al. [13]	Others
Strawberry	330 ± 4	161–265 [29]
		86 [26]
		95–152, 91–278 [20]
Raspberry	228 ± 6	265–303 [29]
		121 [26]
Red plum	320 ± 12	
Grapefruit	150 ± 4	
Orange	126 ± 6	
Red cabbage	158 ± 4	
Broccoli	128 ± 4	104[b] [18]
Onion	88 ± 1	70–116 [18]
Green grape	80 ± 4	
Spinach	72 ± 1	49 [18]
Green cabbage	58 ± 1	52 [18]
Pea	32 ± 1	
Cauliflower	30 ± 1	
Leek	22 ± 1	
Lettuce	14 ± 1	23 [18]
Pear	60 ± 3	
Apple	48 ± 1	110–600 [30]
Peach	38 ± 1	43–77 [31]
Banana	38 ± 4	
Tomato	30 ± 1	38 [18]

[a] Milligrams gallic acid equivalents/100 g fresh weight (FW), unless otherwise stated.
[b] Catechin equivalents.

phenolics in strawberry and raspberry extracts have also been indicated to range from 161 to 265 mg and from 265 to 303 mg gallic acid equivalents (GAE)/100 g FW, respectively, depending on the method of extraction [29]. Furthermore, Kalt and coworkers [26] have determined the total phenolic content of fresh strawberries and raspberries and found even lower levels of phenolics: 86 and 121 mg GAE/100 g FW, in strawberries and raspberries, respectively. These large differences, however, should not be surprising when considering that the phenolic content can vary enormously, depending not only on the variety of the fruit or vegetable but also on their stage of maturity. For example, Wang and Lin [20] have reported that the total phenolic content ranged from 95 to 152 mg GAE/100 g FW in different varieties of ripe strawberries and from 91 to 278 mg GAE/100 g FW, depending on ripeness.

For orange and grapefruit, which are rich in flavanones, Proteggente and associates [13] have found that the total phenolic content (126 mg and 150 mg/ 100 g FW, respectively) was similar and much lower than that measured for berry fruits. Other data on the phenolic content of these fruits were not available in the literature.

The total phenolic content of apples of different varieties has been reported to be in the range 110–600 mg epicatechin equivalents/100 g FW [30], which is much higher than the level of 48 mg GAE/100 g FW [13]. However, this large difference may depend on the use of different standards. The total phenolic content of the fruits such as pear, peach, and tomato was found to range between 30 and 60 mg GAE/100 g FW by Proteggente and colleagues [13], considerably lower than the levels measured by the same authors for the berry and citrus fruits and most of the green leafy vegetables. The measurements of total phenolic content of peach and tomato extracts by Proteggente and coworkers [13] are essentially in the same range of levels previously reported by Chang and associates [31] for peach and by Vinson and colleagues [18] for tomato; no other literature data were available for a comparison of the phenolic content of pears (Table 4). Among the vegetables, Proteggente and coworkers [13] observed that broccoli extracts had the highest total phenolic content (128 mg GAE/100 g FW) and lettuce extracts had the lowest phenolic content (14 mg GAE/100 g FW). Similarly, Vinson and associates [18] have found a total phenolic content, as catechin equivalents, of 104 and 23 mg/100 g for broccoli and lettuce, respectively, and their values for onion, spinach, and green cabbage are consistent with those reported by Proggente and coworkers [13] (Table 4).

Furthermore, Proteggente and colleagues [13] have found that cauliflower, peas, and banana had a low total phenolic content, 30, 32, and 38 mg GAE/100 g FW, respectively, whereas red cabbage extracts showed an appreciable phenolic content (158 mg GAE/100 g FW), higher than that of some of the flavanone- and flavanol-rich fruits and vegetables (Table 4).

A

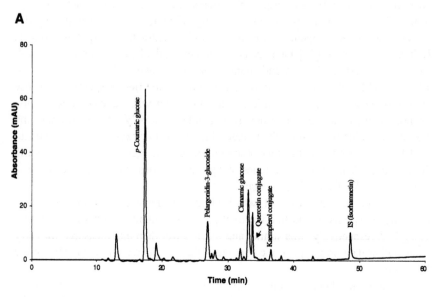

B

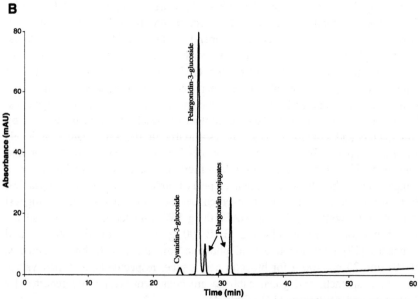

Figure 1 Representative HPLC-DAD (monitored over the range 200–750 nm) chromatogram of an anthocyanin-rich extract, strawberry. Peaks were derived at (A) 320 nm and (B) 520 nm. HPLC-DAD, high-performance liquid chromatography–diode array detection.

IV. PHENOLIC COMPOSITION OF FRUITS AND VEGETABLES

Some of the most common fruits and vegetables containing four specific families of phenolic compounds, anthocyanins, flavanones, flavonols, and hydroxycinnamates, are characterized for their individual major components and compared with reports from the leading literature of Macheix and coworkers [32] and other researchers.

Anthocyanins have been found to be the major phenolics in strawberry, raspberry, and red plum extracts (hydroxycinnamic acid derivatives and quercetin and kaempferol conjugates are also contained in these fruits) [13], consistently with previous reports by Macheix and associates [32]. More specifically, pelargonidin-3-glucoside was indicated as the major phenolic compound in strawberry extracts, together with appreciable amounts of cyanidin-3-glucoside (Fig. 1B), and cyanidin-3-glucoside and cyanidin-3-sophoroside were the major components in red plum and raspberry extracts, respectively; red plum extracts also contained minor amounts of cyanidin-3-rutinoside [13]. A summary of the total anthocyanin content reported in the literature for strawberries, raspberries, and red plums is presented in Table 5, which generally indicates good agreement in the findings of various authors [13,14,26,32,33].

Comparative total amounts of flavanones in orange and grapefruit reported in the literature are given in Table 6. Proteggente and colleagues [13] have reported hesperidin as the major phenolic component in orange extracts, accompanied by appreciable amounts of narirutin and neohesperedin (Fig. 2), and naringin as the most important flavanone and narirutin the second major identifiable component in grapefruit extracts. The data by Proteggente and

Table 5 Anthocyanin Content of Red Fruits[a]

Reference	Red plum	Strawberry	Raspberry
Proteggente et al. [13]	19.4	11.9	28.9
Clifford [33]	2–25	15–35	10–60
Kahkonen et al. [14][b]	—	20.4	22.9
Kalt et al. [26][c]	—	7.6	41.4
Macheix et al. [32]	1.9–53	28–70	23.6

[a] Milligrams/100 grams fresh weight.
[b] Data converted from milligrams/100 g dry weight reported in the original reference, using a 9.7% and 14.9% dry matter value for strawberry and raspberry, respectively.
[c] Data converted from micromoles malvidin-3-glucoside/g fresh weight reported in the original reference into milligrams malvidin-3-glucoside/100 g fresh weight.

Table 6 Flavanone Content of Citrus Fruits[a]

Reference	Orange	Grapefruit
Proteggente et al. [13]	130.9	226.3
Tomás-Barberán and Clifford [35][b]	140–319	96–543
Justesen et al. [44][c]	67.3	—
Bronner and Beecher [34][d]	144–180	262–268
Macheix et al. [32][e]	270–600	170–280

[a] Milligrams/100 g fresh weights.

[b] Data obtained from the sum of hesperidin and narirutin for fresh and hand-squeezed orange juices and naringin and narirutin for grapefruit juices reported in the original reference.

[c] Data obtained from the sum of hesperetin and naringenin (the aglycone forms of hesperidin and narirutin, respectively) for orange pulp reported in the original reference.

[d] Data obtained from the sum of hesperidin and narirutin for orange juice concentrates and naringin and narirutin for grapefruit juice concentrates reported in the original reference.

[e] Hesperidin only for oranges and naringin only for grapefruits (whole fruit) reported in the original reference.

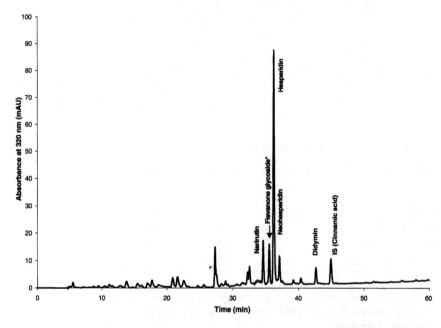

Figure 2 Representative HPLC-DAD (monitored over the range 200–750 nm) chromatogram of a flavanone-rich extract, orange. Peaks were derived at 320 nm. HPLC-DAD, high-performance liquid chromatography–diode array detection.

coworkers [13] for citrus fruits are reasonably consistent with other reports on the flavanone content of fresh juices and concentrates [34,35], when accounting for the diversity of the material analyzed (whole fruit, pulp only, fresh juice or concentrates, etc.) and the method of analysis.

The flavonol levels found in the literature for some fruits and vegetables are reported in Table 7. Proteggente and associates [13] indicated quercetin-3,4'-diglucoside and quercetin-4'-glucoside, together with traces of quercetin-3-glucoside, as the major phenolic compounds in onion (Fig. 3). The amount of flavonols in onion reported by these authors is significantly higher than that indicated by Hollman and Arts [36] but similar to that reported by Moon and colleagues [37]. It should, however, be noted that Proteggente and coworkers [13] and Moon and associates [37] quantified on the basis of quercetin glucoside, whereas Hollman and Arts [36] quantified on the basis of the aglycone quercetin after acid hydrolysis. Furthermore, Proteggente and colleagues [13] quantified all peaks relative to quercetin conjugates as quercetin-3-glucoside, and this method may have resulted in an overestimation of the relative amounts of the three major glucosides, particularly of the diglucoside. Therefore, the flavonol levels reported by these authors for broccoli, spinach, and green cabbage, quantified on the basis of quercetin-3-glucoside, are also likely to be overestimated. Of course there would also be differences due to the use of different varieties of onions.

Also, Proteggente and coworkers [13] have found quercetin conjugates, as identified from the spectral characteristics, to be the main components in lettuce, which also contained small amounts of chlorogenic acid and anthocyanin

Table 7 Flavonol Content of Fruits and Vegetables[a]

Reference	Green grape	Onion	Leek	Lettuce	Broccoli	Spinach	Green cabbage
Proteggente et al. [13]	1.3	65.9	91.2	16.8	7.1	52.5	15.0
Hollman and Arts [36][b]	—	34	—	1.4–7.9	3.7	—	—
Moon et al. [37][c]	—	50.5	—	—	—	—	—
Hertog et al. [64][d]	1.2	34.7 ± 6.3	3 ± 2.3	1.4 ± 1.4	10.2	<3	<3

[a] Milligrams/100 g fresh weight.
[b] Data obtained for quercetin, the aglycone, after acid hydrolysis, in the original reference.
[c] Data obtained from the sum of the individual amounts of quercetin-3,4'-diglucoside, quercetin-4'-glucoside, and quercetin-3-glucoside reported in the original reference.
[d] Data obtained for quercetin and kaempferol aglycones after acid hydrolysis in the original reference.

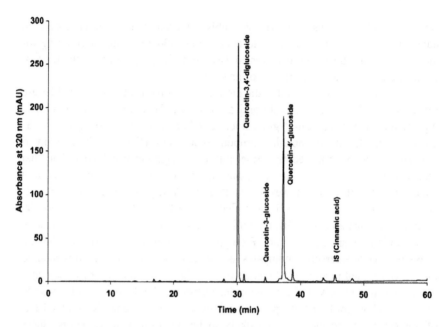

Figure 3 Representative HPLC-DAD (monitored over the range 200–750 nm) chromatogram of a flavonol-rich extract, onion. Peaks were derived at 320 nm. HPLC-DAD, high-performance liquid chromatography–diode array detection.

conjugates in the red varieties, and kaempferol conjugates in leek, consistently with findings by Fattorusso and associates [38]. However, the former authors, who were using high-performance liquid chromatography (HPLC) with diode array detection, could not precisely identify the individual components of these vegetables, because of the complexity of their flavonol glycoside mixtures, in which glycosides are often acylated with hydroxycinnamic acids. In fact, Fattorusso and colleagues [38], by using ultraviolet (UV) and nuclear magnetic resonance (NMR) spectroscopy, have identified, but not quantified, five flavonoid glucosides based on the kaempferol aglycone, some of which were acylated with a 3-methoxy-4-hydroxycinnamoyl moiety, in leek extracts. Others have also reported the presence of 3-O-β-D-sophoroside-7-O-β-D-glucosides of kaempferol and quercetin, with and without further acylation with ferulic, sinapic, and caffeic acids, in cabbage by using HPLC with mass spectrometric detection but with no indication of the amounts [39,40]. Similarly, acylated flavonol glycosides and gentiobiosides, derivatives of spinacetin and patuletin, have been isolated from spinach leaves [41].

A summary of the levels of hydroxycinnamate derivatives and other components among the hydroxycinnamate-containing fruits and vegetables is presented in Table 8. Chlorogenic acid was identified as one of the major phenolic components in pear, apple, tomato, and peach/nectarine by Proteggente

Table 8 Hydroxycinnamates and Miscellaneous Phenolic Content of Certain Fruits[a]

		Proteggente et al. [13]	Schieber et al. [65]	Tomás-Barberán and Clifford [35]	Stewart et al. [43]	Clifford [66]	Justesen et al. [44]	Hertog et al. [64]	Macheix et al. [32]
Pear	Hydroxycinnamates	5.1	1.8[b]			6–28[d]			6.4–51.8
	Flavonols	2.6	0.8[b]				4.5[f]	<0.4[f]	4.20–60[g]
	Others								
Apple	Hydroxycinnamates	4.5	7.2[c]			6.2–38.5[d]			6.2–134
	Flavonols	5.4	13.5[c]				2[f]	3.6 ± 1.9[f]	17.8–40.5[e]
	Others	3.1	15.0[c]			8.7–33[h]			
Tomato	Hydroxycinnamates	3.5							
	Flavonols	2.6			0.1–2.2		1.4[f]	0.8[f]	
	Others	3.1		6.4[h]					
Peach/ Nectarine	Hydroxycinnamates	4.5							
	Flavonols	0.8						<0.3[f]	
	Others	3.5							

[a] Milligrams/100 g fresh weight.
[b] Average of three different cultivars.
[c] In apple pomace, data converted from milligrams per kilogram dry weight reported in the original reference, based on a 14.3% dry matter value.
[d] In whole fruit.
[e] In peel.
[f] Quercetin only, after acid hydrolysis.
[g] In the flesh and in the peel, respectively.
[h] In the skin.

and coworkers [13]. These authors also found that pear extracts contained quercetin-3-glucoside and that other major components in apple extracts were rutin, flavonol conjugates, and phloridzin (Fig. 4). Literature data indicate a great variability in phenolic content of apples and pears (Table 8), which may be due to the lack of appropriate standards and the consequent unspecific quantification of the various chlorogenic acid isomers, such as 4'- and 3'-caffeoylquinic acids, and of p-coumaroylquinic acids, feruloylquinic acids, p-coumaroylmalic esters, and hydroxycinnamic acid glucose derivatives that have been reported to be present in these fruits [32,42].

As for tomatoes, Proteggente and associates [13] indicated that they also contained rutin and chalconaringenin, but no naringenin or naringenin glucosides, in agreement with other reports [35,43,44].

Thus, although the available information on the individual phenolic composition of fruits and vegetables is very fragmentary and still insufficient, from the data summarized in this section it is possible to form an idea of which are those components of fruit and vegetables that are most likely to have a dietary importance and whose potential biological actions would therefore be of interest.

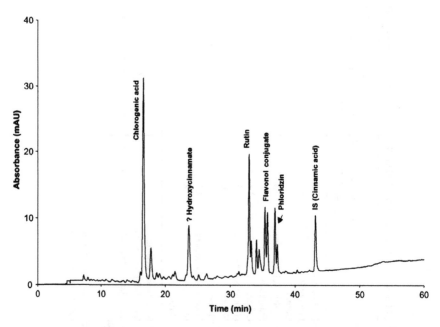

Figure 4 Representative HPLC-DAD (monitored over the range 200–750 nm) chromatogram of a hydroxycinnamate-rich extract, apple. Peaks were derived at 320 nm. HPLC-DAD, high-performance liquid chromatography–diode array detection.

V. VITAMIN C CONTENT OF FRUITS AND VEGETABLES

When considering the antioxidant potential of fresh fruits and vegetables, it is important also to evaluate the vitamin C content of these foods, because ascorbic acid is a very effective scavenger of free radicals both in vitro and in vivo. A comparison of the total vitamin C content of some common fruits and vegetables is shown in Table 9. As can be seen, the levels from the different reports [13,45] are generally consistent, except for most of the green vegetables, such as broccoli, spinach, lettuce, and cauliflower. This may be accounted for by losses of vitamin C during the freeze-drying used in the preparation of the material, although this effect would be expected to apply in all cases.

According to Table 9, strawberries contained the highest total vitamin C content (61 mg/100 g FW) of the fruits and vegetables analyzed, followed by the citrus fruits (52 and 46 mg/100 g FW for grapefruit and orange, respectively). Broccoli had a vitamin C content comparable to that of the citrus fruits, 45 mg/100 g FW, and red cabbage also showed an appreciable amount of vitamin C (37

Table 9 Total Vitamin C in Fruits and Vegetables[a]

Fruit/Vegetable	Proteggente et al. [13]	Holland et al. [45]
Strawberry	61	40–90
Raspberry	26	32
Red plum	5	4
Grapefruit	52	36
Orange	46	44–79
Red cabbage	37	40–60[b]
Broccoli	45	87
Onion	6	5
Green grape	2	3
Spinach	7	26
Green cabbage	28	40–60[a]
Pea	22	24
Cauliflower	15	43
Leek	16	17
Lettuce	<2	10–30
Pear	3	3
Apple	6	3
Peach	6	31–37
Banana	10	10
Tomato	18	10–30

[a] Milligrams/100 g fresh weight.
[b] Includes both red and green cabbage.

mg/100 g FW). For the other fruits and vegetables reported by Proteggente and colleagues [13], the vitamin C content ranged from <2 mg/100 g FW for lettuce to 28 mg/100 g FW for green cabbage, with pear, green grape, apple, peach, red plum, onion, spinach, and banana at the lower end and raspberry, pea, tomato, leek, and cauliflower at the higher end of the range.

VI. RELATIONSHIPS AMONG ANTIOXIDANT ACTIVITY, PHENOLIC CONTENT AND COMPOSITION, AND VITAMIN C CONTENT OF FRUITS AND VEGETABLES

A number of studies have shown that the antioxidant capacity of fruits and vegetables is strongly correlated to both vitamin C concentration and total phenolic content. For example, Gardner and coworkers [19] have assessed the antioxidant potential of various fruit juices, through their ability to reduce a synthetic free radical, potassium nitrosodisulfonate (by using electron spin resonance (ESR)), and Fe(III) (by using FRAP) and found a strong correlation with both phenolics and vitamin C ($r = 0.97$–0.99 and 0.90–0.93, respectively, depending on the procedure used to determine the antioxidant capacity). Also, Prior and colleagues [15] observed a good correlation between total phenolic concentrations ($r = 0.84$), as well as the anthocyanin content ($r = 0.77$), and antioxidant activity (ORAC) of different cultivars of *Vaccinium* species. Furthermore, Guo and colleagues [27] and Velogliu and associates [16] have observed a strong correlation between total phenolic content and antioxidant activity among a variety of fruits, vegetables, and grain products. Consistently, Proteggente and colleagues [13] have shown a good correlation between total phenolic and vitamin C content and all the three measurements of antioxidant activity they have used, particularly with the TEAC assay (Fig. 5). These authors not only indicated the existence of a correlation between the phenolic content and the antioxidant activity but also strongly suggested that the antioxidant capacity is closely related to the phenolic composition, i.e., to the prevalence of a particular class of phenolic components in a specific fruit or vegetable. In fact, the hierarchy of antioxidant activities of the fruits and vegetables studied showed that fruits that are rich in anthocyanins have a higher antioxidant

Figure 5 Correlation between total phenolics (milligrams gallic acid equivalents/ 100 g and measurements of antioxidant activity, TEAC, FRAP, and ORAC, of fruit and vegetables extracts. The correlation coefficients (r) and the correspondent significance values (*P*) are indicated. TEAC, Trolox equivalent antioxidant capacity; FRAP, ferric reducing ability of plasma; ORAC, oxygen radical absorbance capacity.

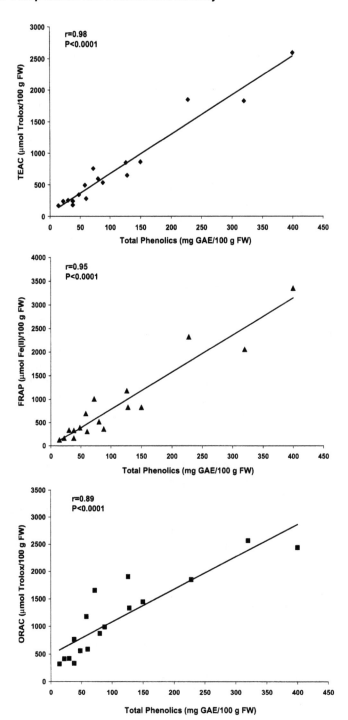

potential than those rich in flavanones, flavanols, and hydroxycinnamates. This pattern appears to be consistent with the hierarchy of antioxidant potential observed for the diverse classes of phenolic compounds in a number of in vitro systems. In fact, the anthocyanins generally have shown higher antioxidant activity than other phenolic compounds, such as flavonols and flavanones, when tested in in vitro systems [9,21,46,47].

However, when considering the antioxidant potential of fruits and vegetables, other aspects than the phenolic content must be taken into account, such as the contribution of vitamin C to the total antioxidant activity of hydrophilic extracts and the combined actions by the phenolic and the vitamin C components. For example, the data from Proteggente and colleagues [13] showed that the antioxidant activities of orange and grapefruit were much lower than those measured for berry fruits, as expected by the lower antioxidant efficiency of flavanones, in which citrus fruits are rich, compared to that of anthocyanins and flavonols when evaluated in in vitro systems [9,11,21]. However, the antioxidant potential of orange and grapefruit was appreciably higher than that observed for flavonol-rich fruits and vegetables, although pure flavonols, which possess all the structural features, such as the o-dihydroxy structure in the B ring, the 2,3 double bond in conjugation with a 4-oxo function in the C ring, and the 3- 5-OH groups with 4-oxo function in the A and C rings, that have been identified as determinants for maximal radical scavenging ability [9], have demonstrated higher antioxidant potential than flavanones in in vitro systems. The finding that the antioxidant activity observed for the citrus fruits is higher than expected on the basis of the phenolic quality and quantity is likely to be due to the high vitamin C content of citrus fruits with respect to flavonol-rich fruit and vegetables. The vitamin C content of flavonol-containing fruits and vegetables reported by Proteggente and coworkers [13] was in fact very low, and therefore it might be deduced that the phenolic components almost exclusively accounted for the antioxidant activities of these foods. Consistently, these authors reported that extracts from hydroxycinnamate-containing fruits and vegetables had the lowest phenolic and vitamin C content, reflected in the lowest antioxidant activity, as predicted by the low antioxidant potential of the hydroxycinnamic acids [9].

VII. DIETARY BURDEN OF PHENOLIC-RICH FRUITS AND VEGETABLES

In 1983 the UK total diet study estimated that the total average intake for selected flavonols, quercetin, kaempferol, apigenin, and luteolin, from six food groups, green and other vegetables, canned vegetables, fresh fruits, fruit products and beverages, was 30 mg/day, and quercetin contributed 64% of this total. Beverages contributed 82% of the total intake of these flavonoids; in particular, tea provided up to 92% of flavonoids from beverages. More recent estimates of flavonoid dietary intake range from 10 to 100 mg/day, depending on the population studied and the technique used [48]. Estimated average intakes of selected flavonols and flavones from the UK total diet study are similar to the 23 mg/day estimated average intakes of selected flavonols, flavones, and flavanones from a Dutch food survey [49].

More recently, Scalbert and Williamson [50] have reviewed the main dietary sources of phenolic compounds and have calculated that the total daily intake, for a given diet containing some common fruits (i.e., apple and cherry), vegetables (i.e., potato, tomato, lettuce, and onion), and beverages (i.e., orange juice, red wine, coffee, and black tea) as well as other foods (i.e., wheat bran and dark chocolate), was equivalent to ~ 1 g/day. These authors also indicated that, as a general trend, phenolic acids account for approximately one-third of the total phenolics, and flavonols, flavanols, flavanones, and anthocyanins for two-thirds. However, the proportion in which the different classes of phenolics are ingested varies greatly, according to the foodstuffs consumed. Therefore, comprehensive surveys of the content of the important phenolic classes in common foods are essential in estimating the dietary phenolic intake and in investigating the impact that dietary phenolics might have in terms of disease prevention and/or treatment.

VIII. IS THERE A POTENTIAL FOR ANTIOXIDANT ACTIVITY OF COMPONENTS OF FRUITS AND VEGETABLES IN VIVO?

The data presented here indicate that red fruits, strawberries in particular, have an enormous antioxidant potential in vitro, when compared with other types of fruits such as citrus or apples, for both high anthocyanin and high vitamin C content, and therefore could represent an excellent source of dietary antioxidants.

Recent studies are starting to build evidence that these foods help to improve the antioxidant potential in vivo. For example, Pedersen and associates [51] have shown that consumption of cranberry juice was able to increase the plasma antioxidant capacity of healthy female volunteers, by determining a 30% increase in vitamin C and a small but significant increase in total phenols plasma concentration.

Furthermore, a study by Manuel y Keenoy and colleagues [52] has shown that supplementation with diosmin (90%) and hesperidin (10%), flavonoids found in lemons and oranges, respectively, induced a significant decrease in hemoglobin A_{1c} in type 1 diabetic patients that was not related to the glycemic control. In these patients, the decrease in hemoglobin A_{1c} was accompanied by an increase in glutathione peroxidase activity and in the lag time of the copper-induced in vitro oxidability of non–high-density lipoproteins (non-HDLs), suggesting that the flavonoid-induced decrease in glycation was associated with an improvement of the antioxidant status of the patients, specifically of the antioxidant component represented by the thiol-containing proteins. However, a more recent study has shown that supplementation with 500 mg/day of rutin, quercetin-3-rutinoside, had no effect on plasma antioxidant status or markers of oxidative stress in healthy female volunteers, although the plasma levels of quercetin, kaempferol, and isorhamnetin were significantly elevated [53].

Other studies have evaluated the antioxidant potential in vivo of phenolic-rich beverages and foods. It has been found that the antioxidant capacity of plasma was significantly enhanced in humans, after the ingestion of apple juice, whose main antioxidant components were identified as ascorbic acid and chlorogenic acid [54]. Furthermore, it has been observed that plasma total antioxidant capacity increased significantly after ingestion of red wine, green tea, and black tea in healthy volunteers [55]. Moreover, an increase in plasma antioxidant capacity and a reduction in plasma concentration of thiobarturic acid reactive substances (TBARS) have been observed in healthy subjects after the consumption of a meal of procyanidin-rich chocolate [56]. Also, it has been shown that the total antioxidant capacity of serum can be increased significantly, after consumption of red wine, strawberries, and spinach, in elderly women [57].

Thus, there is abundant evidence to indicate that phenolic-rich foods and beverages can have a significant impact on the antioxidant capacity of plasma, ultimately suggesting that flavonoids may act as antioxidants in vivo without, of course, excluding other biological actions or the effects of other components.

IX. CONCLUSIONS

In this chapter, the phenolic composition and content of regularly consumed fruits and vegetables are extensively discussed through the analysis of the current relevant literature, in order to provide a comprehensive summary of the current compositional and quantitative data on some flavonoid-rich foods. Furthermore, the formal relation of the in vitro antioxidant potential of these fruits and vegetables to the quality of the phenolic and, to a lesser extent, vitamin C content is emphasized. The potential for antioxidant activity of flavonoid-rich fruits and vegetables in vivo is also discussed. The data described here allow identification of the potentially most effective fruits and vegetables in terms of phenolic content and antioxidant activity. However, much research is still needed: the elucidation of the metabolism and bioavailability of flavonoids in vivo, as well as of the amounts and the forms in which they are taken up into cells and tissues, is crucial in order to establish the mechanisms and the forms in which dietary phenolics may act in vivo [58]. Finally, it remains to be established to what extent phenolics act as antioxidants or enzymes and gene modulators, and how their actions may be relevant to disease prevention [59–63].

REFERENCES

1. Middleton E Jr, Kandaswami C, Theoharides TC. The effects of plant flavonoids on mammalian cells: implications for inflammation, heart disease, and cancer. Pharmacol Rev 2000; 52:673–751.
2. Galati G, Teng S, Moridani MY, Chan TS, O'Brien PG. Cancer chemoprevention and apoptosis mechanisms induced by dietary polyphenolics. Drug Metabol Drug Interact 2000; 17:311–349.
3. Joseph JA, Shukitt-Hale B, Desinova NA, Prior RL, Cao G, Martin A, Taglialatela G, Bickford PC. Long-term dietary strawberry, spinach or vitamin E supplementation retards the onset of age-related neuronal signal-transduction and cognitive behavioural deficits. J Neurosci 1998; 18:8047–8055.
4. Cao Y and Cao R. Angiogenesis inhibited by drinking tea. Nature 1999; 398:381.
5. Fuhrman B, Levy A, Aviram M. Consumption of red wine with meals reduces the susceptibility of human plasma and LDL to lipid peroxidation. Am J Clin Nutr 1996; 61:549–554.
6. Gaziano JM, Hennikens CH. Update on dietary antioxidants and cancer. Pathol Biol 1996; 44:42–45.
7. Hertog MG, Kromhout D, Aravinis C, Blackburn H, Buzina R, Fidanza F, Giampaoli S, Jansen A, Menotti A, Nedeljkovic S, Pekkarinen M, Simic BS, Toshima H, Feskens EJM, Hollman PCH, Katan MB. Flavonoid intake and long-term risk of coronary heart disease and cancer in the seven countries study. Arch Intern Med 1995; 155:381–386.
8. Gandini S, Merzenich H, Robertson C, Boyle P. Meta-analysis of studies on breast cancer risk and diet: the role of fruit and vegetable consumption and the intake of associated micronutrients. Eur J Cancer 2000; 36:636–646.
9. Rice-Evans CA, Miller NJ, Paganga G. Structure-antioxidant activity relationships of flavonoids and phenolic acids. Free Radic Biol Med 1996; 20:933–956.
10. Visioli F, Bellomo G, Galli C. Free radical-scavenging properties of olive oil polyphenols. Biochem Biophys Res Commun 1998; 247:60–64.
11. Jovanovic SV, Steenken S, Simic MC, Hara Y. Antioxidant properties of flavonoids: reduction potentials and electron transfer reactions of flavonoids radicals. In: Rice-Evans CA, Packer L, eds. Flavonoids in Health and Disease. New York: Marcel Dekker 1998:137–161.
12. Halliwell B, Gutteridge JMC, eds. Free Radicals in Biology and Medicine, 3rd ed. New York: Oxford University Press, 1999.
13. Proteggente AR, Pannala AS, Paganga G, van Buren L, Wagner E, Wiseman S, van de Put F, Dacombe C, Rice-Evans CA. The antioxidant activity of regularly consumed fruit and vegetables reflects their phenolic and vitamin C composition. Free Radic Res 2002; 36:217–233.

14. Kahkonen MP, Hopia AI, Vuorela HJ, Rauha J-P, Pihlaja K, Kujala TS, Heinonen M. Antioxidant activity of plants extracts containing phenolic compounds. J Agric Food Chem 1999; 47:3954–3962.

15. Prior RL, Cap G, Martin A, Sofic E, McEwen J, O'Brien C, Lischner N, Ehlenfeldt M, Kalt W, Krewer G, Mainland CM. Antioxidant capacity as influenced by total phenolic and anthocyanin content, maturity, and variety of Vaccinium species. J Agric Food Chem 1998; 46:2686–2693.

16. Velogliu YS, Mazza G, Gao L, Oomah BD. Antioxidant activity and total phenolics in selected fruits, vegetables and grain products. J Agric Food Chem 1998; 46:4113–4117.

17. Paganga G, Miller N, Rice-Evans C. The polyphenolic content of fruit and vegetables and their antioxidant activities: what does a serving constitute? Free Radic Res 1999; 30:153–162.

18. Vinson JA, Hao Y, Su X, Zubik L. Phenol antioxidant quantity and quality in foods: vegetables. J Agric Food Chem 1998; 46:3630–3634.

19. Gardner PT, White TAC, McPhail DB, Duthie GG. The relative contributions of vitamin C, carotenoids and phenolics to the antioxidant potential of fruit juices. Food Chem 2000; 68:471.

20. Wang SY, Lin H-S. Antioxidant activity in fruits and leaves of blackberry, raspberry, and strawberry varies with cultivar and developmental stage. J Agric Food Chem 2000; 48:140–146.

21. Re R, Pellegrini N, Proteggente AR, Pannala AS, Rice-Evans CA. Antioxidant activity applying an improved ABTS radical cation decolorization assay. Free Radic Biol Med 1999; 26:1231–1237.

22. Cao G, Verdon CP, Wu AHB, Wang H, Prior RL. Automated-assay of oxygen radical absorbance capacity with the COBAS FARA-II. Clin Chem 1995; 41:1738–1744.

23. Benzie IFF, Strain JJ. Ferric reducing/antioxidant power assay: direct measure of total antioxidant activity of biological fluids and modified version for simultaneous measurement of total antioxidant power and ascorbic acid concentration. Methods Enzymol 1999; 299:15–27.

24. Djuric Z, Powell LC. Antioxidant capacity of lycopene-containing foods. Int J Food Sci Nutr 2001; 52:143–149.

25. Deighton N, Brennan R, Finn C, Davies HV. Antioxidant properties of domesticated and wild Rubus species. J Sci Food Agric 2000; 80:1307–1313.

26. Kalt W, Forney CF, Martin A, Prior RL. Antioxidant capacity, vitamin C, phenolics and anthocyanins after fresh storage of small fruits. J Agric Food Chem 1999; 47:4638–4644.

27. Guo C, Cao G, Sofic E, Prior RL. High-performance liquid chromatography coupled with coulometric array detection of electroactive components in fruits and vegetables: relationship to Oxygen Radical Absorbance Capacity. J Agric Food Chem 1997; 45:1787–1796.

28. Wang H, Cao G, Prior RL. Total antioxidant capacity of fruits. J Agric Food Chem 1996; 44:701–705.

29. Heinonen IM, Meyer AS, Frankel EN. Antioxidant activity of berry phenolics on

human low-density lipoprotein and liposome oxidation. J Agric Food Chem 1998; 46:4107–4112.

30. Sanoner P, Guyot S, Marnet N, Molle D, Drilleau J-F. Polyphenol profiles of French cider apple varieties (Malus domestica sp.). J Agric Food Chem 1999; 47:4847–4853.

31. Chang S, Tan C, Frankel EN, Barrett DM. Low-density lipoprotein antioxidant activity of phenolic compounds and polyphenol oxidase activity in selected clingstone peach cultivars. J Agric Food Chem 2000; 48:147–151.

32. Macheix J-J, Fleuriet A, Billot J, eds. Fruit Phenolics. Boca Raton, FL: CRC Press, 1990.

33. Clifford MN. Anthocyanins—nature, occurrence and dietary burden. J Sci Food Agric 2000; 80:1063–1072.

34. Bronner WE, Beecher GR. Extraction and measurement of prominent flavonoids in orange and grapefruit concentrates. J Chromatogr A 1995; 705:247–256.

35. Tomás-Barberán FA, Clifford MN. Flavanones, chalcones and dihydrochalcones—nature, occurrence and dietary burden. J Sci Food Agric 2000; 80:1073–1080.

36. Hollman PCH, Arts ICW. Flavonols, flavones and flavanols—nature, occurrence and dietary burden. J Sci Food Agric 2000; 80:1081–1093.

37. Moon J-H, Nakata R, Oshima S, Inakuma T, Terao J. Accumulation of quercetin conjugates in blood plasma after short-term ingestion of onion by women. Am J Physiol Regul Integr Comp Physiol 2000; 279:R461–R467.

38. Fattorusso E, Lanzotti V, Taglialatela-Scafati O, Cicala C. The flavonoids of leek, Allium porrum. Phytochemistry 2001; 57:565–569.

39. Nielsen JK, Olsen CE, Petersen MK. Acylated flavonol glycosides from cabbage leaves. Phytochemistry 1993; 34:539–544.

40. Nielsen JK, Nørbæk R, Olsen CE. Kaempferol tetraglucosides from cabbage leaves. Phytochemistry 1998; 49:2171–2176.

41. Ferreres F, Castañer M, Tomás-Barberán FA. Acylated flavonol glycosides from spinach leaves (Spinacia Oleracea). Phytochemistry 1997; 45:1701–1705.

42. Oleszek W, Amiot MJ, Aubert SY. Identification of some phenolic in pear fruit. J Agric Food Chem 1994; 42:1261–1265.

43. Stewart AJ, Bozonnet S, Mullen W, Jenkins GI, Lean MEJ, Crozier A. occurrence of flavonols in tomatoes and tomato-based products. J Agric Food Chem 2000; 48:2663–2669.

44. Justesen U, Knuthsen P, Leth T. Quantitative analysis of flavonols, flavones and flavanones in fruits, vegetables and beverages by high-performance liquid chromatography with photo-diode array and mass spectrometric detection. J Chromatogr A 1998; 799:101–110.

45. Holland B, Welch AA, Unwin ID, Buss DH, Paul AA, Southgate DAT. McCance and Widdowson's the composition of Foods, 5th ed. London: Royal Society of Chemistry, 1991.

46. Cao G, Sofic E, Prior RL. Antioxidant and prooxidant behavior of flavonoids: structure-activity relationships. Free Radic Biol Med 1997; 22:749–760.

47. Ghiselli A, Nardini M, Baldi A, Scaccini C. Antioxidant activity of different

phenolic fractions separated from an Italian red wine. J Agric Food Chem 1998;
46:361–367.

48. Wearne S. Estimating dietary intakes of flavonoids. In: Rice-Evans C, ed. Wake Up
to Flavonoids. London: Royal Society of Medicine Press, 2000:35–43.

49. Hertog MG, Hollman PC, Katan MB, Kromhout D. Intake of potentially
anticarcinogenic flavonoids and their determinants in adults in The Netherlands.
Nutr Cancer 1993; 20:21–29.

50. Scalbert A, Williamson G. Dietary intake and bioavailability of polyphenols. J Nutr
2000; 130:2073S–2085S.

51. Pedersen CB, Kyle J, Jenkinson AM, Gardner PT, McPhail DB, Duthie GG. Effects
of blueberry and cranberry juice consumption on the plasma antioxidant capacity of
healthy female volunteers. Eur J Clin Nutr 2000; 54:405–408.

52. Manuel y Keenoy B, Vertommen J, De Leeuw I. The effect of flavonoid treatment
status in Type 1 diabetic patients. Diabetes Nutr Metab 1999; 12:256–263.

53. Boyle SP, Dobson VL, Duthie SJ, Hinselwood DC, Kyle JA, Collins AR.
Bioavailability and efficiency of rutin as an antioxidant: a human supplementation
study. Eur J Clin Nutr 2000; 54:774–782.

54. Bitsch R, Netzel M, Carlé E, Strass G, Kesenheimer B, Herbst M, Bitsch I.
Bioavailability of antioxidative compounds from Brettacher apple juice in humans.
Inn Food Sci Emerg Tech 2001; 1:245–249.

55. Serafini M, Laranjinha JAN, Almeida LM, Maiani G. Inhibition of human LDL
lipid peroxidation by phenol-rich beverages and their impact on plasma total
antioxidant capacity in humans. J Nutr Biochem 2000; 11:585–590.

56. Rein D, Lotito S, Holt RR, Keen CL, Schmitz HH, Fraga CG. Epicatechin in human
plasma: in vivo determination and effect of chocolate consumption on plasma
oxidation status. J Nutr 2000; 130:2109S–2114S.

57. Cao G, Russell RM, Lischner N, Prior RL. Serum antioxidant capacity is increased
by consumption of strawberries, spinach, red wine or vitamin C in elderly women.
J Nutr 1998; 128:2383–2390.

58. Rice-Evans C, Spencer JPE, Schroeter H, Rechner AR. Bioavailability of flavonoids
and potential bioactive forms in vivo. Drug Metabol Drug Interact 2000; 17:291–310.

59. Gerritsen ME, Carley WW, Ranges GE, Shen CP, Phan SA, Ligon GF, Perry CA.
Flavonoids inhibit cytokine-induced endothelial cell adhesion protein gene
expression. Am J Pathol 1995; 147:278–292.

60. Panes J, Gerritsen ME, Anderson DC, Miyasaka M, Granger DM. Apigenin inhibits
tumor necrosis factor-induced intercellular adhesion molecule-1 upregulation in
vivo. Microcirculation 1996; 3:279–286.

61. Soriani M, Rice-Evans CA, Tyrrell RM. Modulation of the UVA activation of haem
oxygenase, collagenase and cyclooxygenase gene expression by epigallocatechin in
human skin cells. FEBS Lett 1998; 439:253–257.

62. Schroeter H, Williams RJ, Matin R, Iversen L, Rice-Evans CA. Phenolic
antioxidants attenuate neuronal cell death following uptake of oxidized low density
lipoprotein. Free Radic Biol Med 2000; 29:1222–1233.

63. Spencer JPE, Schroeter H, Kuhnle G, Srai SKS, Tyrrell RM, Hahn U, Rice-Evans C.
Epicatechin and its in vivo metabolite, 3'-O-methyl epicatechin, protect human

fibroblasts from oxidative-stress-induced cell death involving caspace-3 activation. Biochem J 2001; 354:493–500.

64. Hertog MGL, Hollman PCH, Katan MB. Content of potentially anticarcinogenic flavonoids of 28 vegetables and 9 fruits commonly consumed in The Netherlands. J Agric Food Chem 1992; 40:2379–2383.

65. Schieber A, Keller P, Carle R. Determination of phenolic acids and flavonoids of apple and pear by high-performance liquid chromatography. J Chromatogr A 2001; 910:265–273.

66. Clifford MN. Chlorogenic acids and other cinnamates—nature, occurrence and dietary burden. J Sci Food Agric 1999; 79:362–372.

4

Applications of Flavonoid Analysis and Identification Techniques: Isoflavones (Phytoestrogens) and 3-Deoxyanthocyanins

Ewald E. Swinny
University of Adelaide
Adelaide, South Australia

Kenneth R. Markham
New Zealand Institute for Industrial Research and Development
Lower Hutt, New Zealand

I. INTRODUCTION

The flavonoids are among the most numerous and widespread natural products found in plants and have many diverse applications and properties. There have been several texts published on flavonoids, some comprising collections of specialist chapters [1–5], and a number of more practically oriented texts have appeared independently [6–11]. A previous edition of this book contains a detailed treatise on the analysis and identification of flavonoids in practice [11]. Rather than simply describe further techniques of flavonoid analysis, we have chosen in this chapter to present two significant examples of the application of techniques previously described. These applications demonstrate how the various available techniques are used in concert to produce pure flavonoids and to establish their structures. Selected for this purpose are two classes of flavonoids that we see as having contemporary emerging importance, the isoflavones and the 3-deoxyanthocyanins.

The isoflavones have recently enjoyed widespread attention as potential therapeutic agents, particularly in the area of women's health. They have acquired the phytoestrogen label because they have estrogenic activity and have been associated with prevention of breast and prostate cancer in addition to cardio-vascular disease. There are several reports in the literature that describe the analysis of phytoestrogens that occur in a variety of dietary legumes such as red clover, soybeans, and kudzu root [12–25]. There are a large number of dietary phytoestrogen supplements available on the phytopharmaceutical market, and assessments of some of these for their estrogenic activity and phytoestrogen content are also well described in the literature [12,16].

The 3-deoxyanthocyanins are a rare group of anthocyanins that are different from the normal anthocyanins in that they lack oxygenation at carbon-3 of the flavonoid heterocyclic ring. Generally anthocyanins are found in many flowers, fruits, vegetables, and red wines, etc., and they are responsible for the array of impressive colors associated with many of these. Anthocyanin-rich foods have attracted much attention in the health sector with regard to their reputed beneficial pharmacological and biological effects. The 3-deoxyanthocyanins are a unique subgroup of these colored pigments. Their chemical and biological properties have not been widely explored and the limited research on them may be attributed to their very restricted occurrence in nature.

The analysis and identification of flavonoids from natural sources, as exemplified by isoflavones and 3-deoxyanthocyanins, usually follows a series of sequential steps. Each step constitutes the subject of a section of this chapter:

Section II. Structure and distribution
Section III. Preparation of plant material for extraction
Section IV. Extraction procedures
Section V. Chromatographic analysis: high-performance liquid chroma-tography (HPLC)
Section VI. Flavonoid quantification: ultraviolet/visible- (UV/VIS) spec-troscopy and HPLC
Section VII. Isolation and purification procedures, e.g., 3-deoxyantho-cyanins
Section VIII. Structure analysis, e.g., 3-deoxyanthocyanins

II. STRUCTURE

The distribution of the isoflavones in the plant kingdom is largely restricted to the family Leguminoseae. There are over 350 known isoflavones, making them the largest group of compounds in the class of isoflavonoids. The four most common isoflavones associated with phytoestrogenic herbs or extracts are daidzein,

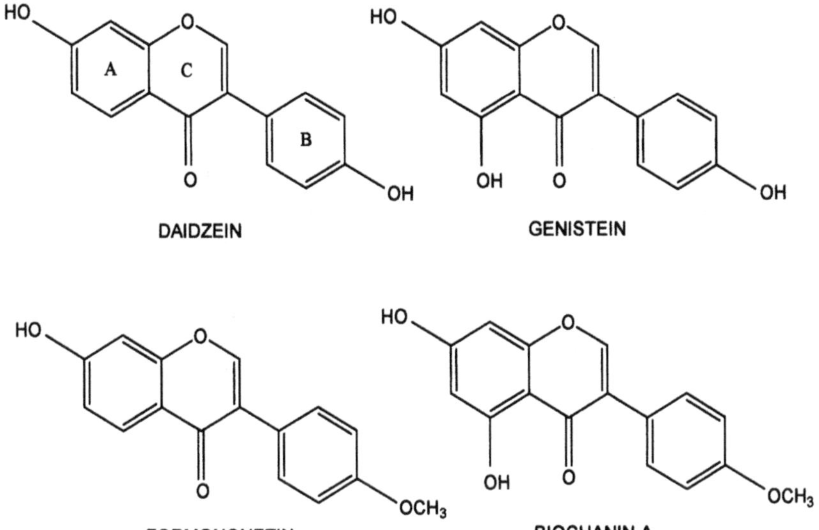

Figure 1 Examples of isoflavones.

genistein, formononetin, and biochanin A (see Fig. 1). These are commonly found to occur as the glucosides, glucoside malonate esters, or free aglycones.

The pool of known 3-deoxyanthocyanins is very small. The two better known 3-deoxyanthocyanidins (aglycones) are apigeninidin and luteolinidin. Their 5-O-glycoside derivatives occur in the brightly colored petals of *Sinningia cardinalis* [26,27]. An acetylated diglucoside was found in the red fronds of the New Zealand fern *Blechnum novae-zealandiae* [28], and a series of 5-di-, 5,7-di-, 7-di-, and 7-O-glycosides were reported to occur in the fern *Blechnum procerum* [29]. More highly oxygenated pigments include tricetinidin, found in black tea *Camellia sinensis* [30], and the methoxylated carajurin found in the anti-inflammatory leaves of *Arrabidea chica* [31]. Another one of these highly oxygenated pigments, columnin, was reported to occur in the gesneriad *Columnea banksii* [26]. However, the structure of this compound has never been proved. The structures of some of the aglycones are shown in Fig. 2.

III. PREPARATION OF PLANT MATERIAL FOR EXTRACTION

Some flavonoids are unstable and are degraded by enzyme action in undried freshly harvested plant material. The safest method for drying fresh material,

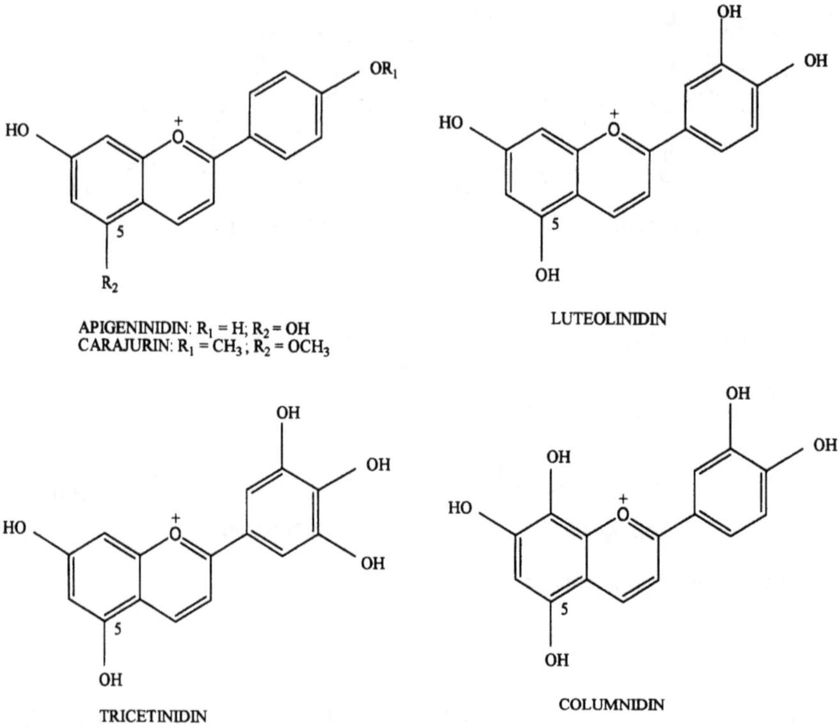

APIGENINIDIN: R_1 = H; R_2 = OH
CARAJURIN: R_1 = CH_3; R_2 = OCH_3

LUTEOLINIDIN

TRICETINIDIN

COLUMNIDIN

Figure 2   Examples of 3-deoxyanthocyanidins.

particularly flower petals and fern fronds containing 3-deoxyanthocyanins, is freeze-drying. The powder so obtained can be stored for future use in sealed containers in a freezer. If flavonoid quantification is the objective of the subsequent analysis, snap freezing in liquid nitrogen immediately after harvest is advisable. Drying of well spread out plant material in an oven at 100 °C is also an acceptable method for larger amounts, such as isoflavone-containing material in red clover. Subsequent storage of the dried material in a sealed container under refrigeration prevents further significant flavonoid losses. Air-drying plant material at room temperature is not recommended, as this process can give rise to enzymic degradation, e.g., the conversion of glyco-sides to aglycones. A satisfactory alternative, especially with anthocyanin-containing material, is to extract the freshly harvested plant material by chopping up the sample in a blender with the appropriate solvent. Enzyme action is not a problem here if alcohol is included in the extracting medium to denature plant enzymes.

IV. EXTRACTION PROCEDURES

In flavonoid analysis the extraction procedure selected is usually determined by the types of flavonoids to be extracted and the purpose of the extraction, whether for qualitative or quantitative use. The literature abounds with different methods that have been used for the extraction of isoflavones. These include the use of aqueous methanolic, ethanolic, and acetonitrile mixtures [13–25]. Some of these also contain a portion of hydrochloric acid and involve a refluxing process that results in subsequent hydrolysis and conversion of the glycosides to the aglycones. In general, the isoflavone aglycones together with their glycosides and malonate esters present in herbs such as red clover are conveniently extracted by soaking the finely ground and dried plant material in $MeOH:H_2O$ or $EtOH:H_2O$ (80:20 to 90:10) over a period of about 24 h at room temperature. The extraction may be enhanced by sonication in an ultrasonic bath. Quantitative yields of the constituent flavonoids are best obtained when two to three sequential extractions of the weighed and dried plant material are pooled. The less polar aglycones may be obtained qualitatively from the leaf surface simply by rinsing the intact plant material in an organic solvent such as diethyl ether or ethyl acetate.

   3-Deoxyanthocyanins are readily extracted with acidified aqueous or alcoholic solvents at room temperature from ground or mashed material such as petals, leaves, or fronds. Formic acid, acetic acid, and trifluoroacetic acid (TFA) are popular acid components as these are least likely to cause hydrolysis or deacylation on work-up. The 3-deoxyanthocyanins apigeninidin-5-O-glucoside and luteolinidin-5-O-glucoside are conveniently extracted into 0.5% TFA [27] from freeze-dried *Sinningia cardinalis* petals.

V. CHROMATOGRAPHIC ANALYSIS

Quantitative or qualitative analysis of flavonoid extracts generally involves some form of chromatography. In the case of isoflavones in red clover and other legumes, high-performance liquid chromatography (HPLC) is the popular method of choice. HPLC is also very useful for the analysis of 3-deoxyanthocyanins and for monitoring of the sequential progress involved in preparative isolation of these pigments. Quantitative analysis of the 3-deoxyanthocyanins is best carried out by using absorption spectroscopy (discussed later).

A. High-Performance Liquid Chromatography

The identification and quantification of health-promoting flavonoids in herbal extracts and plant-derived food supplements have gained increasing importance in the phytopharmaceutical and health-food industries. HPLC technology has developed rapidly over the years to the point where it is now undoubtedly the analytical method of choice for these flavonoids. It has the advantage that it is fast and reproducible, requires little sample, and can be used for both qualitative and quantitative analysis as well as for preparative work.

1. High-Performance Liquid Chromatography Instrumentation

The fundamental components of any modern-day HPLC system are a solvent delivery system, a sample injector, a column, a detector, and a computer with the appropriate data acquisition and processing software. There are numerous HPLC methods described in the literature for isoflavones [13–25] and for the common anthocyanins, each method invokes different combinations of solvent systems, columns, and detectors. HPLC has been interfaced with a variety of detection methods such as ultraviolet/visible (UV/vis) spectrocopy and liquid chromatography–mass spectrometry (LC-MS) [21,22]. In this chapter, however, discussion is restricted to the most commonly used pairing in flavonoid analysis, that of a reverse-phase (RP-18) column and a UV/visible detector.

Standard solvent delivery systems now offer a choice of high-pressure or low-pressure pumping units capable of delivering at least a binary solvent gradient. The predominant column used is the reversed-phase octadecyl (RP-18) column. The trend toward shorter analysis times and more economical solvent consumption has led to an array of alternative column dimensions such as narrow-bore, or shorter and thicker columns, or even variable particle size and type. Generally a particle size of 5 μm or smaller and column dimension of 125 or 750 mm suffice for most flavonoid analyses. Multichannel, fast scanning photodiode array (PDA) detectors coupled to computer-based data acquisition and processing software have now become the norm. PDA detection is extremely useful for flavonoid analysis because it is able to provide a UV/ visible absorption spectrum for each peak. For the detection of anthocyanins, it is essential that the detection extend into the visible region (to at least 600 nm). Spectral libraries are a useful software feature that allow automatic tentative identification of components.

2. Solvent System

The solvent mixture eluting through the column can remain the same throughout the chromatographic run (an isocratic system), or the mixture may change in a predetermined way (a gradient system). Flavonoid analysis is best achieved by using a gradient. Typical binary gradient solvent programs used with RP-18 columns start with a high proportion of a polar solvent and gradually increase the proportion of a less polar solvent [32,33]. In this way the analytes are initially concentrated at the top of the column, progressively desorbed, and eventually refocused into narrower bands that are detected as sharp peaks on elution. The polar solvent is normally water-based and the less polar solvent methanol- or acetonitrile-based. The aqueous solvent is commonly acidified to prevent ionization of phenolic compounds, which can yield broadened peaks for some compounds. This sort of solvent system is appropriate for all flavonoids, including anthocyanins, which require a strongly acidic environment to ensure complete conversion to the stable flavylium ion [34]. In most cases poor peak separation and resolution result when an ordinary linear increase in the proportion of the nonpolar solvent is used. Better chromatography can usually be achieved when a more complex gradient system involving linear, stepped, or curved increases is used. Such complex gradients vary from application to application and are best determined by trial and error.

Solvent degassing or sparging is an essential step in solvent preparation. There are two main reasons for degassing the mobile phase. Dissolved gas levels are lowered by degassing, thereby reducing the risk of air bubbles' entering and remaining in the detector cell. Air bubbles trapped in the detector cell produce an undesirable sawtooth detector output. Another reason for degassing is that oxygen absorbs strongly at low UV wavelengths, resulting in unsuitable detector response during the analysis. Newer HPLC systems have built in degassing units. Helium sparging is a popular and efficient method of degassing. Alternatively solvents may be degassed by prior ultrasonic agitation, although this method is not as effective as helium sparging.

Various solvent systems have been described in the literature for the analysis of phytoestrogens [13–25]. In general, good chromatography of clover isoflavones may be obtained by using an acidic water and acetonitrile gradient. In our experience a combination of aqueous formic or acetic acid with acetonitrile produces very satisfactory results. A typical chromatogram obtained in our laboratory showing the common red clover isoflavone glycosides and aglycones detected at 260 nm is depicted in Fig. 3.

The 3-deoxyanthocyanins are conveniently analyzed by using a low-pH gradient that is also very suitable for most flavonoid types, including normal 3-oxygenated anthocyanins (see Fig. 4) [27]. A gentle linear gradient gives good separation over the retention period in which most flavonol glycosides are encountered. The 3-deoxyanthocyanins are detected in the wavelength range 470–490 nm. Detection in the 320 to 350-nm range is suitable for most common flavones and flavonols, whereas normal anthocyanins are detected at 510–530 nm.

3. Peak Identification

A good starting point when identifying the flavonoid peaks in a chromatogram is to examine the associated online UV/visible absorption spectrum captured by the PDA detector in the range from ca. 210 to 600 nm. Most flavonoids exhibit absorption peaks in two regions, one in the low-wavelength region, 210–290 nm (band II), and one in the longer-wavelength region, 320–380 nm, or 470–540 nm

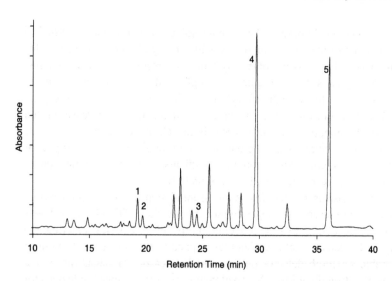

Figure 3 High-performance liquid chromatography (HPLC) chromatogram of red clover extract recorded at 260 nm. Column: Merck Supersphere Lichrocart 125-4 RP-19 endcapped (4 μm, 4 mm × 119 mm). Elution (0.8 mL/min) performed by using a solvent system comprising solvent A (5% HCOOH) and solvent B (CH₃CN), mixed by using a linear gradient starting with 90% A, decreasing to 60% A at 30 min, 20% A at 33 min, and 0% A at 39 min. Peak identities: 1, ononin; 2, daidzein; 3, genistein; 4, formononetin; 5, biochanin A.

for anthocyanins (band I). The exact position of λ_{max} of band I can give a good indication of the type of flavonoid. Thus, for example, the band I λ_{max} for isoflavones appears as a low-intensity shoulder in the range 310–330 nm and band II is usually prominent in the region 250–265 nm. The band I λ_{max} for the 3-deoxyanthocyanins, however, appears in the range 470–490 nm and band II in the range 270–280 nm. Acylation with hydroxycinnamic acids can be recognized by a third absorption band at about 310–330 nm. Variation within the ranges mentioned is due primarily to the effect of varying oxygenation. Additional oxygenation in either ring shifts the band I (or II) absorption to longer wavelength. The absorption spectra of a wide selection of flavonoids are now available in the literature [6,8–10,35]. Online spectra obtained from HPLC runs, however, may vary somewhat from those in the references, as they are measured in the HPLC solvent rather than the usual methanol or ethanol. The other point of caution is that sometimes a single peak on the chromatogram may actually represent more than one compound, thereby resulting in a distorted spectrum. The purity of a peak can be checked by measuring the spectrum at different points across the peak.

The retention times on RP-18 are dependent on the relative affinity of the compound for the stationary and mobile phases [32,36]. Those parts of the molecule that are capable of forming hydrogen bonds such as the C-4 carbonyl

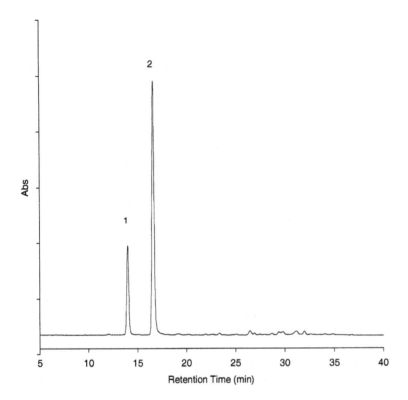

Figure 4 High-performance liquid chromatography (HPLC) chromatogram of a *Sinningia cardinalis* flower petal extract recorded at 470 nm. Column as for Fig. 3. Elution (0.8 mL/min) performed by using a solvent system comprising solvent A [1.5% H_3PO_4] and solvent B [HOAc-CH$_3$CN-H$_3$PO$_4$-H$_2$O (20:24:1.5:54.5)], mixed by using a linear gradient starting with 80% A, decreasing to 33% A at 30 min, 10% A at 33 min, and 0% A at 39.3 min. Peak identities: 1, luteolinidin-5-*O*-glucoside; 2, apigeninidin-5-*O*-glucoside.

group and free hydroxyl groups have affinity for the more polar mobile phase. Flavonoids that have more of these groups elute earlier than those with fewer. Alternatively, an increase in the number of hydrophobic groups (e.g., methylation or acylation) increases affinity for the stationary phase and so increases retention time. However, although genistein and biochanin A both have one more hydroxyl than daidzein and formononetin, respectively, this hydroxyl is strongly hydrogen-bonded to the C-4 carbonyl, resulting in an overall decrease in polarity and hence longer relative retention times. The glucoside of each of these isoflavones has a shorter retention time than the corresponding aglycone as a result of the increase in hydrophilic hydroxyls. Acylation, as seen with the malonates found in red clover, increases the retention time by effectively deleting a hydrophilic hydroxyl group. In the 3-deoxyanthocyanin group, luteolinidin 5-*O*-glucoside has one more free hydroxyl than apigeninidin 5-*O*-glucoside, causing it to be more polar and hence elute earlier.

Although it is possible to arrive at some reasonable conclusion regarding the identity of a peak on the basis of a comparison of the retention time and online spectrum with published data, it is always best to confirm the identity by cochromatography against an authentic standard. A simple way to do this is to inject a known concentration and volume of the standard and, separately, of the plant extract. An amount of each of these is then combined and injected, and a comparison is made between the detector response (peak height or peak area) obtained and the response of the plant extract and the standard solution. A corresponding increase in the resultant response usually confirms the identity.

4. Sample Preparation

The key priorities associated with the preparation of a sample for HPLC analysis are cleanup, matrix removal or modification, and optimal analyte concentration.

Cleanup of the sample is essential since impurities in the extract or sample matrix may interfere with analyte detection and measurement. It is also vital that the sample to be injected is free of solid matter to prevent clogging of the injector or the guard column and the eventual deterioration of the analytical column. Filtration or centrifugation of the extract is usually an effective first step in the cleanup process.

Sometimes the analyte concentration is too dilute for direct injection and measurement and it becomes necessary to increase the concentration per unit volume. Solid phase extraction (SPE) is a very effective and elegant sample preparation procedure for most phenolics that can be used simultaneously to achieve cleanup, matrix removal or modification, and optimal analyte concentration.

Typically the process involves retention and concentration of the analytes on a RP-18 solid phase cartridge, while the impurities are washed off. To speed up this process the cartridge or column may be attached to a vacuum manifold processor that is connected to a vacuum source. The vacuum manifold processor also allows a large number of samples to be processed simultaneously. The analytes are selectively eluted as a concentrated solution by using a minimal volume of the appropriate solvent.

A preparation of a red clover sample for HPLC injection would typically involve the following steps. The ethanolic extract (typically 1 mL) is diluted with 3–4 mL water and introduced to a water preconditioned disposable RP-18 500-mg cartridge. Unbound sugars and polar impurities are washed out with water (3–5 mL), followed by 3–5 mL of 10% MeOH. The retained isoflavones are eluted with about 3–5 mL 80–90% MeOH. This extract is dried down and reconstituted in the appropriate volume of mobile phase or other solvent and injected directly.

This procedure works equally well for the 3-deoxyanthocyanins except that it is desirable to maintain an acidic medium (the water component in the eluant is replaced by 0.1% HCOOH or TFA) throughout the process. SPE is particularly useful when extracts or selected analytes are subjected to chemical reactions such as acid or base hydrolysis. By-products formed and excess reagents such as HCl and NaOH used in hydrolysis of sugars are conveniently removed.

VI. FLAVONOID QUANTIFICATION

UV/visible absorption spectroscopy and HPLC with UV/visible detection are both very useful for the determination of the quantities of individual or total flavonoids in plant extracts and herbal preparations. Quantification by UV/visible spectroscopy is especially suitable for 3-deoxyanthocyanins and other anthocyanins because they have a visible maximum that is usually free of interference from other phenolic compounds.

A. Quantification Using Ultraviolet/Visible Spectroscopy

Quantification of flavonoids using UV/vis spectroscopy involves application of the Beer-Lambert law:

$$A = \epsilon c l$$

In this relationship A = absorbance (read from the spectrophotometer), ϵ = extinction coefficient, c = concentration in moles per liter, and l is the path length (centimeters) of the cell used. The ϵ value is taken from the literature or calculated by using a standard solution. Thus the concentrations of standard solutions of the 3-deoxyanthocyanidins apigeninidin and luteolinidin may be determined by using the published extinction coefficients of 18000 $M^{-1} \cdot cm^{-1}$ and 13,800 $M^{-1} \cdot cm^{-1}$, respectively [37]. A solution of an unknown or unknowns can also be quantified by using these ϵ values. In this case, however, the level of the unknown(s) is expressed in terms of equivalents of the standard relating to the ϵ value used. In order to maintain the linear relationship as required for the Beer-Lambert law to hold, solutions should be prepared such that the absorption (A) is between 0.05 and 1.00 absorption units. Anthocyanin concentrations must always be measured in acidic solution, usually 0.1 N HCl (aqueous or alcoholic), to ensure that the anthocyanin is totally in the flavylium form [6,38].

The concentration of isoflavonoid standard solutions used for phytoestrogen quantification can be accurately and conveniently determined by UV/vis spectroscopy using known ϵ values [13,14,23]. A known volume of such a solution is injected into an HPLC (see Sec. VI. B.), and the peak area relativities obtained from the standard and the sample are used to calculate the concentration of analyte. ϵ Values for flavonoids are available in the literature, and a good compilation is reported by Jurd [35].

B.   Quantification Using High-Performance Liquid Chromatography

HPLC has become a widely used and reliable technique for the accurate quantification of flavonoids, for example, for the determination of phytoestrogens in herbal supplements and plant extracts [13–25]. Typically a good chromatogram shows well separated peaks representing those components of the plant extract that absorb at a selected wavelength. For example, analysis of an isoflavone extract would ideally involve detection at the band II maximum of about 260 nm. However, not all the peaks appearing in the chromatogram would necessarily represent isoflavones. An online spectral facility is very useful at this stage. With this facility a complete absorption spectrum for each peak may be obtained, thereby allowing the grouping of peaks into their appropriate flavonoid classes.

When using a UV/vis detector the peak height, or peak area produced by an electronic integrator, is a response of the UV/vis detector signal, which in turn is a function of the molar absorptivity of the flavonoid. Thus both the height and the area of a well-resolved peak in a chromatogram are proportional to the concentration of the flavonoid detected. For quantification, either peak height or peak area can be used. Peak area quantification is very popular for most flavonoid analysis. However, it is not always the most accurate. When peaks are well resolved and consistently symmetrical, peak height quantification can be far more accurate. This is also the preferred option when trace amounts are to be determined. A number of operating conditions affect the peak height and area measurement differently. For example, a small change in the flow rate affects the peak area measurement more than it affects the peak height measurement. When the column is not thermostatted, the preferred option is peak area measurement. Whatever the choice, accurate quantitative analysis is reliant on the detector's operating in the linear range. For a more detailed discussion on this topic and other aspects of HPLC theory and practice the reader is referred to more specialized HPLC texts [39,40]. Discussion here is restricted to the use of peak area measurement in two common methods of flavonoid quantification, the external standard method and the internal standard method.

1. External Standard Method

The external standard method is the simplest quantification method and is usually used for straightforward analysis, especially those not requiring extensive sample preparation, and for analyses in which the chromatography is highly reproducible and the instrumentation is very dependable. The compound chosen as external standard (ES) should ideally be the same flavonoid as that being quantified. Otherwise it should be closely related to the flavonoids of interest and show similar solvent solubility and spectral and chromatographic properties. The ES should also be readily available in pure form and economically priced.

For isoflavone analysis the four common isoflavones biochanin A, formononetin, daidzein, and genistein are all very suitable as external standards. They are found in many herbal extracts exhibiting estrogenic properties. The levels of isoflavones may thus be conveniently expressed as equivalents of one of these standards. Other standards used in our laboratory for general flavonoid quantification include rutin, naringenin, catechin, apigenin-7-glucoside, quercetin, and kaempferol.

With the ES method a number of concentrations of the standard solution, also referred to as a *calibration solution* (CS), are prepared. The exact concentration of each CS can be established by UV/vis spectroscopy by using the extinction coefficient of the compound. A known volume of each CS is injected and the response (peak area) vs. concentration is plotted to produce a calibration plot. This plot should be linear and have a zero intercept. The sample to be analyzed is prepared in the same solvent as the CS and injected, chromatographed, and detected exactly as the CS. The concentration of the relevant components in the sample may then be determined graphically from the peak integrals by using the calibration plot, or numerically by using a response factor RF (or calibration factor). It is essential that the concentrations of the CS cover the concentration range expected for the unknown sample. If the unknown sample concentration falls outside the range of the CS injected, the sample should be diluted or concentrated, whichever is applicable.

The RF method is a convenient numerical alternative to reading a value from the calibration plot. The response factor for the CS is determined by

$$RF = \frac{CS_{area}}{CS_{concentration}}$$

For greater accuracy it is advisable to determine an average RF of a CS over a range of concentrations. If the calibration plot is linear and the intercept is zero, the RF is equivalent to the slope of the plot. After the RF is obtained, it can then be used to calculate the flavonoid level in the unknown sample or plant extract:

$$\text{Flavonoid concentration} = \frac{\text{Flavonoid}_{area}}{RF}$$

Conversion of the result to a more appropriate unit of concentration such as milligrams per gram (mg/g) can be made if required, by incorporating factors such as amount of dry plant material extracted, volume of extraction solvent, sample dilution, and injection volume. A practical example is the extraction of 98 mg of dry red clover leaves with 5 mL of 80% MeOH. Injection of 10 μL of the prepared sample produces a chromatogram shown in Figure 3. The biochanin A area in this chromatogram is 6,886,805 units. Injection of 10 μL of biochanin A (CS), which has a concentration of 0.04 mg/mL, produces a peak with area 4,282,266 units (peak 5). The response factor is

$$RF = \frac{4,282,266}{0.04}$$

The biochanin A concentration in the extract is therefore

$$\text{biochanin A}_{conc} = \frac{6,886,805}{\left(\frac{4,282,266}{0.04}\right)}$$

$$= 0.06 \text{ mg/mL}$$

Conversion to milligrams of biochanin A per gram of plant material would be

$$0.06(\text{mg/mL}) \times 5(\text{mL}) \times \frac{1000(\text{mg})}{98(\text{mg})}$$

$$= 3.06 \text{ mg/g}$$

The total isoflavone content may be calculated in a similar way by summing the total area of all the isoflavone peaks and expressing the value in this case as biochanin A equivalents. Alternatively, several standards may be used to determine the total isoflavone level. In commercial supplements the total phytoestrogen content is often expressed as milligrams per gram or micrograms per gram (mg/g or μg/g), and is a measure of the total free aglycone content [13,14,16,18,23,25]. This total content is actually a sum of the glucosides and aglycones, but the level is expressed as free aglycone units because isoflavones are absorbed by the gut as aglycones. The total aglycone content is arrived at by correcting for the molecular weight of the glucosides, which is almost twice that of the aglycones. Hence the total isoflavone concentrations indicated on labels of phytoestrogen supplements are never just an expression of the arithmetic sum of the individual conjugate forms.

2. Internal Standard Method

An internal standard (IS) is used primarily to monitor the reliability of extraction, sample preparation, chromatographic, and instrumentation procedure. The use of an IS is strongly recommended if the sample preparation method consists of several steps. In such a case a known amount of an internal standard would be added to the sample at an early stage on the assumption that the internal standard and the flavonoid analytes will behave similarly during sample work-up and throughout the chromatographic process. Alternatively the IS could be added to the prepared sample immediately before HPLC if the sole concern is instrument reliability, for example, injection volume reproducibility.

The choice of IS is usually not easy. The compound chosen should have properties similar to those of the flavonoids being quantified, and consideration should be given to factors such as solubility in the extraction solvent and the chromatographic and detection characteristics. Ideally this compound should elute in a gap in the pattern of peaks, somewhere in the middle of the chromatogram or in the region of the peaks being quantified. Some chromatograms of plant extracts are crowded with peaks, making finding a gap for an IS difficult. The chromatogram of red clover (Fig. 3) recorded at 260 nm has an abundance of glucoside and aglycone peaks, some minor and others major. Such chromatograms can be simplified by including an acid hydrolysis step in the sample preparation. This produces a chromatogram showing only the isoflavone aglycones. Extreme care, however, should be exercised when employing an acid hydrolysis step in a quantification procedure since hydrochloric acid can degrade flavonoids. Several examples of the use of acid hydrolysis for this purpose have been reported in the literature [13,24,25]. Enzymic hydrolysis is milder than acid hydrolysis, although many attempts have been described as not very successful. Cellulase from *Aspergillus niger* is reported to yield good results [41]. Several examples of internal standards used to quantify isoflavones accurately have appeared in the literature. These include compounds such as 4-hydroxybenzophenone [18], flavone [13], equilenin [16], and 2,4,4'-trihydroxybenzoin [14].

Once a suitable compound has been chosen as an IS, the procedure would be to add an equal amount of the IS to the sample and to the calibration standard CS (essentially this is the external standard), which is made up to different concentrations. Unlike in the external standard method, here peak area or peak height *ratios* are compared. The ratio of the peak area of the CS to the IS for each CS prepared is determined. This ratio is plotted vs. the concentration of the CS to produce a calibration plot. The flavonoid concentration in the analyte can then be determined directly from the calibration plot. As with the ES method the flavonoid concentration can also be determined numerically by using the response factor (providing that the plot is linear and has a zero intercept):

$$RF = \frac{\left(\frac{CS_{area}}{IS_{area}}\right)}{CS_{concentration}}$$

Therefore,

$$\text{Flavonoid concentration} = \frac{\left(\frac{Flavonoid_{area}}{IS_{area}}\right)}{RF}$$

VII. ISOLATION AND PURIFICATION PROCEDURES

Many plant extracts contain a complex mixture of flavonoids, some of which might contribute directly to particular biological properties of the extract, such as antioxidant, free radical scavenging, or estrogenic activity. The isolation of individual biologically active flavonoids from a complex mixture requires a systematic approach involving a carefully considered selection of solvents and chromatographic stationary supports.

In this section we discuss the isolation and purification of flavonoids, with a particular emphasis on 3-deoxyanthocyanins. Very few of these pigments have been isolated from natural sources, and presented here is a three-step procedure that we have found to be most useful: (1) initial cleanup of extract, (2) large-scale fractionation, (3) final purification.

A. Initial Cleanup

In most plant extractions a significant amount of carbohydrate ends up in the crude extract. A preliminary removal of these carbohydrates from the extract can be achieved by selection from the product range of non-ionic polystyrene (Amberlite XAD) [42] or polyacrylic resins (Diaion HP resins) [43]. The procedure with 3-deoxyanthocyanins involves passing the crude aqueous acidified (0.1% TFA) extract through a column containing one of these materials. The sugars are not adsorbed and are eluted from the column with additional amounts of acidified water. The retained less polar compounds, the 3-deoxyanthocyanins and other flavonoids, are then washed from the column with acidified aqueous methanol. If desired, at this stage some separation can be achieved by using a stepped gradient of water to methanol. The methanol is removed by rotary evaporation and the residual aqueous fraction is freeze-dried. If the crude extract contains some methanol, this should be removed before application to the resin. It is possible to employ a similar solvent regimen, with other adsorbents such as derivatized silica gel (e.g., RP-18, cyanopropyl) or polyamide [8] to achieve this initial cleanup. With flavonoids other than anthocyanins the water need not be acidified.

B. Large-Scale Fractionation

Polyamide and cellulose are both very economical and effective media for large-scale fractionation. The freeze-dried extract is dissolved in a minimal amount of 0.5% TFA and is applied to a polyamide column previously equilibrated in 0.5% TFA; elution is continued with the same solvent. A methanol component can be introduced to the eluant with a gradual incremental increase. The 3-deoxyanthocyanins are brightly colored in an acid medium, a feature that makes their separation in a glass column quite spectacular. For example, in the fractionation of a *Sinningia cardinalis* extract, the apigeninidin-5-*O*-glucoside and luteolinidin-5-*O*-glucoside are obvious as a yellow and reddish purple band, respectively. The most convenient way to collect the eluted fractions is to use an automatic fraction collector. Similar fractions can later be combined on the basis of HPLC analysis. At this stage some of the fractions may contain essentially pure compounds, which then require only minor subsequent cleanup.

When using cellulose for large-scale fractionation, the freeze-dried extract is dissolved in 5% acetic acid and then applied to a glass column containing cellulose preconditioned with 5% acetic acid. Depending on the rate and quality of separation, elution can be continued with 5% acetic acid or stepped up gradually to 15% acetic acid. Once the target fractions have been collected and suitably combined, an appropriate polymeric resin (e.g., Amberlite XAD or Diaion HP) can be used to concentrate the fractions in preparation for final purification.

C. Final Purification

The hydroxypropylated cross-linked dextran, Sephadex LH-20, is very effective for the final purification of 3-deoxyanthocyanins (or flavonoids in general). LH-20 separation is based on molecular size and H-bonding interactions. The fraction to be purified is dissolved in a minimal amount of MeOH:H_2O, 40:60, containing 0.1% TFA and applied carefully to a column containing LH-20 preconditioned with the same solvent. Elution is continued with this solvent. Although the 3-deoxyanthocyanins are strikingly visible to the naked eye, a glass column also allows the progress of the separation of colorless flavonoids to be monitored conveniently with a portable UV/visible lamp in a darkroom.

Medium-pressure (up to 6 bar) preparative chromatography employing ready-to-use glass columns (e.g., Lobar) with a variety of stationary phases is another excellent medium available for flavonoid purification. Using a peristaltic pump and a reverse-phase column, the separation emulates that of an analytical RP-18 column. By examining a chromatogram obtained with an analytical column it is possible to design a suitable solvent system and gradient and thereby obtain a similar chromatographic separation for a fraction needing purification. The other advantage of these columns is that they have a relatively large capacity and can tolerate larger quantities (up to a gram) than preparative HPLC, for instance. Solvent systems we have found to be most useful in our laboratories include 5–20% MeOH or CH_3CN acidified (0.1%) with TFA or formic or acetic acid. The fraction to be purified is dissolved in a minimal volume of solvent and is then introduced to the top of the preconditioned column via an in-line peristaltic pump. To obtain good chromatography the analytes need to be concentrated on the top of the column before changing of the gradient and solvent composition. The pump draws eluting solvent from the reservoir and passes it through the column; then appropriate fractions are collected. The glass casing of these columns makes monitoring the separation process with the naked eye possible for brightly colored pigments, or with a portable UV lamp for other flavonoids.

Preparative HPLC is often very useful for a difficult purification and situations in which only a small amount of a pure compound is required. A larger (10-mm diameter) column is needed and a higher flow rate than that used in analytical work is used. Since only small amounts can be collected from each run, the pure compounds are accumulated from several injections by collection of appropriate peaks as detected by UV/visible or other detection. This procedure can be automated through the use of automatic injection and time- or peak-based automated fraction collection.

One-dimensional paper chromatography is an inexpensive and convenient purification technique. Milligram quantities can be obtained by running several one-dimensional paper chromatograms. The appropriate bands are excised and eluted and the fractions combined [9]. Contaminant polysaccharide material may be removed subsequently by using a small RP-18 column.

VIII. STRUCTURE ANALYSIS

It is always a tremendously satisfying experience for the flavonoid scientist to toil through the intricacies of the fractionation of an extract and eventually obtain a pure flavonoid. Even more exhilarating are the eventual identification and characterizing of the compound, especially if it is novel. The 3-deoxyanthocyanins provide this sort of stimulation since they are very rare and not many structural variations are known. UV/vis spectroscopy, acid hydrolysis, nuclear magnetic resonance spectroscopy, and mass spectroscopy are valuable degradation and instrumental techniques that make structural identification of these pigments possible. The following section focuses on the latter three techniques.

A. Acid Hydrolysis

Acid hydrolysis of a flavonoid glycoside leads to the separation of the aglycone and the sugar entities, thereby enabling structural investigations to be carried out on each portion independently. Besides its use for pure flavonoids, it allows valuable information such as the aglycone ratio in a crude extract to be obtained. 3-Deoxyanthocyanins are cleaved under the conditions of acid hydrolysis, producing the stable 3-deoxyanthocyanidin aglycone and the liberated sugar in the process. The procedure involves refluxing the pure pigment or extract in 2N HCl : MeOH (1:1) on a boiling water bath for 30–40 min. The reaction may be monitored by HPLC, the aglycone has a longer retention time than the glycoside. The aglycone and sugars are conveniently separated on a small reverse-phase column such as that used for SPE (see Sec. V. A.4). Identification of the sugar liberated is commonly carried out by one-dimensional paper chromatography (1D-PC) by comparison with a mix of authentic sugars. The five most frequently encountered sugars associated with flavonoids, glucose, galactose, rhamnose, xylose, and arabinose, are well separated by 1D-PC by using n-BuOH:Pyr:H$_2$O (6:4:3), visualizing spots with an aniline phthalate spray reagent (ca. 3% in MeOH), or by thin-layer chromatography (TLC), using a dried silica plate impregnated with 0.3 M KH$_2$PO4, run in a BuOH:acetone:H$_2$O (4:5:1) mixture [9].

B. Nuclear Magnetic Resonance Spectroscopy

Nuclear magnetic resonance (NMR) spectroscopy is an invaluable instrumental technique for the structural determination of all flavonoids including 3-deoxyanthocyanins. As well as providing information on the chemical environment of each proton or carbon nucleus in the molecule, the technique can be employed to determine linkages among nearby nuclei, often enabling a complete structure to be assembled. The reader is referred to Refs. 44 and 45 for details of the principles of NMR and general interpretation of NMR spectra.

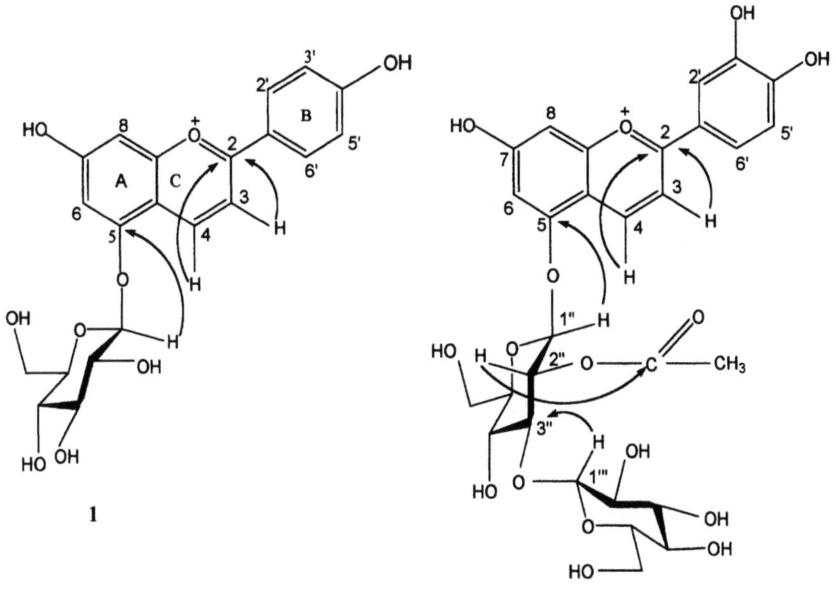

Figure 5 The structures of **1** and **2** showing the main HMBC correlations.

Dimethyl sulfoxide (DMSO-d_6) and methanol (CD$_3$OD) are both suitable solvents for 3-deoxyanthocyanins. Complete conversion to the flavylium ion requires a concentration of 1% TFA-d when methanol-d_4 is used and sometimes higher levels when DMSO-d_6 is used. A broad water peak in the proton NMR spectrum can obscure signals in the sugar region, so both the solvent and the sample should be as dry as possible. At the end of the NMR analysis the sample may be recovered by evaporation of the CD$_3$OD in vacuo. With DMSO the sample can be recovered by adding water and applying the mixture to a small RP-18 column.

The 3-deoxyanthocyanins **1** and **2**, shown in Fig. 5, are examples used later to show how NMR is used to establish the structure.

The first requirement in an NMR analysis is to obtain basic one-dimensional proton (^{1}H) and carbon (^{13}C) spectra, including a ^{13}C-DEPT (dispoportionless enhancement by polarization transfer) spectrum. Typical proton spectra for the pigments **1** and **2** are shown in Figs. 6 and 7, respectively [27,28]. As a result of the relatively low natural abundance of the ^{13}C isotope, ^{13}C experiments require considerably longer periods (hours) to acquire sufficient data to yield a presentable spectrum, whereas proton spectra can be obtained in a few minutes. A reasonable proton spectrum can be obtained with as little as 0.3 mg, whereas ^{13}C spectra generally require larger samples, typically more than 1 mg.

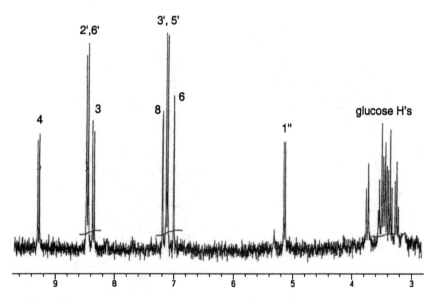

Figure 6 ^{1}H nuclear magnetic resonance (NMR) spectrum of **1**, apigeninidin-5-O-glucoside.

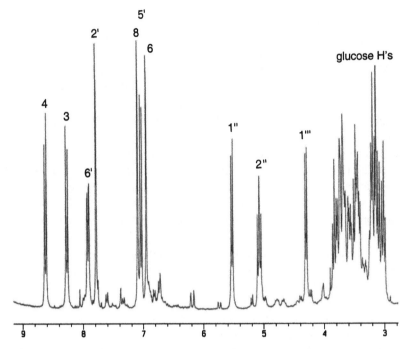

Figure 7 ^{1}H nuclear magnetic resonance (NMR) spectrum of 2, luteolinidin-5-*O*-β-D-[3-*O*-β-D-glucopyranosyl-2-*O*-acetylglucopyranoside].

The aglycone structures of **1** and **2** are determined from the patterns of signals in the aromatic region 6–9.5 ppm in the proton spectrum. The B-ring (H-2′,6′,3′,5′) and (H-2′,5′,6′) signal patterns define the apigenin and luteolin nature of **1** and **2**, respectively. A distinctive feature in these spectra is the presence of two downfield doublets that represent the protons H-3 and H-4. It is this feature that readily distinguishes 3-deoxyanthocyanins from normal C-3 oxygenated anthocyanins. The number of anomeric protons, two (1″ and 1‴) in the case of **2** (Fig. 7), indicates the number of sugars present. The number of carbon atoms and the number of hydrogens bonded to each carbon can be obtained from the ^{13}C spectrum and the DEPT spectrum. These carbon spectra are particularly useful in establishing the number of sugar carbons and eventually the type of sugar. The DEPT experiment allows identification of the carbons bearing two attached hydrogens. The methylene protons of the glucose unit in **1** and **2** are readily identified in this way, since their signals become inverted.

Following on from the one-dimensional proton and carbon spectra, the next step is to employ more sophisticated two-dimensional (2D) NMR techniques to help determine linkages within the molecule. In most of these experiments the instrument automatically combines the results of these experiments and the data are presented as a 2D contour plot. The more common of these techniques include homonuclear proton correlation spectroscopy (H,H-COSY), heteronuclear carbon correlation spectroscopy (H,C-COSY), heteronuclear multiple bond correlation (HMBC), and total correlation spectroscopy (TOCSY) [46].

The H,H-COSY or double quantum filtered correlation spectroscopy (DQF-COSY) is the simplest 2D experiment to run and is usually run first. This technique is useful for the determination of linkages between adjacent hydrogens. In this experiment the 1D proton spectrum is displayed along each axis with a contour display of the same spectrum along the diagonal axis. Off-diagonal peaks are seen where the corresponding protons are coupled, usually as a result of vicinal or geminal coupling. In both compounds **1** and **2** the H,H-COSY shows a characteristic correlation between H-3 and H-4 and the respective correlations between the protons in the A- and B-rings. Correlation between the relevant sugar protons is particularly helpful when allocating the signals in the sugar region of the proton spectrum. For instance, the downfield H-2″ signal in compound **2** is recognized from its correlation with the identifiable H-1″ signal.

A TOCSY experiment enables linkages to be made among all of the protons within a chain of coupled protons. This is especially useful in a diglycoside, because it is sometimes difficult to establish which protons belong to which of the sugar units. For example, in compound **2** the respective group of protons belonging to each of the sugars can be established by examining the respective correlations from the identifiable anomeric protons.

The C,H-COSY or heteronuclear single quantum coherence (HSQC) experiment shows linkages between carbon and hydrogen nuclei in the same general format as the H,H-COSY except that the ^{13}C spectrum is displayed on one axis and the proton spectrum on the other. The spectrum obtained shows correlations between protons and the specific carbons to which they are directly attached. For example, the C,H-COSY of **1** (Fig. 8) shows a very clear correlation between the protons H-3 and H-4 and the carbons C-3 and C-4, respectively. Similar correlations between the other protons and their respective carbons allow one to compile a fairly accurate structure of the compound. The point of linkage of the different moieties in the molecule can be determined by a long-range C,H-COSY experiment called an HMBC. This technique shows correlations between protons and carbons that are two, three, or four bonds away (depending on the instrument parameters used), as shown in the structures for **1** and **2**. This is particularly useful when determining the point of attachment of the sugar to the aglycone.

C. Mass Spectrometry

Mass spectrometry (MS) is used mainly in flavonoid analysis for the confirmation of molecular weight, and the technique is rarely used without prior recourse

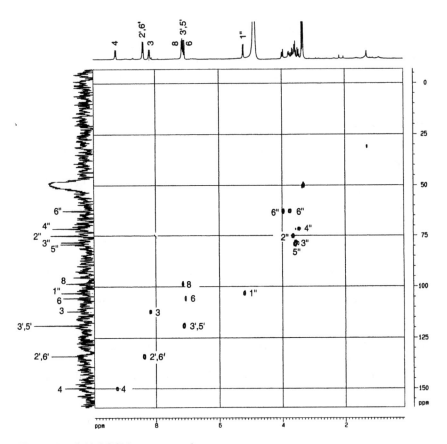

Figure 8 C,H-COSY spectrum of **1**.

to the other spectroscopic techniques described. There are various choices of mass spectrometric ionization methods available, including chemical ionization and electrospray mass spectrometry (ESMS). Compound **2** is a monoacetate of a luteolinidin diglycoside and typically a high-resolution ESMS of **2** gives a molecular ion $[M]^+$ at m/z 637.1747, which is consistent with that calculated, that is, 637.1763 for the formula $C_{29}H_{33}O_{16}^+$.

IX. CONCLUSIONS

The analysis and identification of flavonoids, in particular 3-deoxyanthocyanins and isoflavones, are conveniently executed by a series of sequential steps as discussed in this chapter. However, each class of flavonoids, and in fact each new compound discovered, usually requires a unique modification of the standard protocol. Of all the various techniques available to the flavonoid scientist, high performance liquid chromatography (HPLC) and nuclear magnetic resonance spectroscopy (NMR) have proven to be extremely useful and invaluable.

REFERENCES

1. Geissman TA, ed. The Chemistry of Flavonoid Compounds. Oxford: Pergamon Press, 1962.
2. Harborne JB, Mabry TJ, Mabry H, eds. The Flavonoids. London: Chapman and Hall, 1975.
3. Harborne JB, Mabry TJ, eds. The Flavonoids—Advances in Research. London: Chapman and Hall, 1982.
4. Harborne JB, ed. The Flavonoids—Advances in Research Since 1980. London: Chapman and Hall, 1988.
5. Harborne JB, ed. The Flavonoids—Advances in Research Since 1986. London: Chapman and Hall, 1994.
6. Harborne JB. Comparative Biochemistry of the Flavonoids. London: Academic Press, 1967.
7. Ribereau-Gayon P. Plant Phenolics. Edinburgh: Oliver and Boyd, 1972.
8. Mabry TJ, Markham KR, Thomas MB. The Systematic Identification of Flavonoids. New York: Springer-Verlag, 1970.
9. Markham KR. Techniques of Flavonoid Identification. London: Academic Press, 1982.
10. Markham KR. Flavones, flavonols and their glycosides. In: Dey PM, Harborne JB, eds. Methods in Plant Biochemistry. Vol. 1. Plant Phenolics. London: Academic Press, 1989:197–235.
11. Markham KR, Bloor SJ. Analysis and identification of flavonoids in practice. In: Rice-Evans CA, Packer L, eds. Flavonoids in Health and Disease. New York: Marcel Dekker, 1998:1–33.
12. Liu J, Burdette JE, Xu H, Gu C, van Breemen RB, Bhat KPL, Booth N, Constantinou AI, Pezzuto JM, Fong HHS, Farnsworth NR, Bolton JL. Evaluation of estrogenic activity of plant extracts for the potential treatment of menopausal symptoms. J Agric Food Chem 2001; 49:2472–2479.
13. Franke AA, Hankin JN, Yu MC, Maskarinec G, Low S, Custer LJ. Isoflavone levels in soy foods consumed by multiethnic populations in Singapore and Hawaii. J Agric Food Chem 1999; 47:977–986.
14. Murphy PA, Song T, Buseman G, Barua K, Beecher GR, Trainer D, Holden J. Isoflavones in retail and institutional soy foods. J Agric Food Chem 1999; 47:2697–2704.
15. Mazur W, Adlercreutz H. Naturally occurring oestrogens in food. Pure Appl Chem 1998; 9:1759–1776.
16. Setchell KDR, Brown NM, Desai P, Zimmer-Nechemias L, Wolfe BE, Brashear WT, Kirscher AS, Cassidy A, Heubi JE. Bioavailability of pure isoflavones in

healthy humans and analysis of commercial soy isoflavone supplements. J Nutr 2001; 131:1362S–1375S.

17. Hutabarat LS, Greenfield H, Mulholland M. Quantitative determination of isoflavones and coumestrol in soybean by liquid chromatography. J Chromatogr A 2000; 886:55–63.

18. Thomas BF, Zeisel SH, Busby MG, Hill JM, Mitchell RA, Scheffler NM, Brown SS, Bloeden LT, Dix KJ, Jeffcoat AR. Quantitative analysis of the principle soy isoflavones genistein, diadzein and glycetin, and their primary conjugated metabolites in human plasma and urine using reversed-phase high-performance liquid chromatography with ultraviolet detection. J Chromatogr B 2001; 760:191–205.

19. Kledjus B, Vitamvasova D, Kuban V. Reversed-phase high-performance liquid chromatographic determination of isoflavones in plant materials after isolation by solid-phase extraction. J Chromatogr A 1999; 839:261–263.

20. de Rijke E, Zafra-Gomez, Ariese F, Brinkman UA, Gooijer C. Determination of isoflavone glucoside malonates in *Trifolium pratense* L. (red clover) extracts: quantification and stability studies. J Chromatogr A. 2001; 932:55–64.

21. Lin L, He X, Lindenmaier M, Yang J, Cleary M, Qiu S, Cordell GA. LC-ESI-MS Study of the flavonoid glycoside malonates of red clover (*Trifolium pratense*). J Agric Food Chem 2000; 48:354–365.

22. He X, Lin L, Lian L. Analysis of flavonoids from red clover by liquid chromatography electrospray mass spectrometry. J Chromatogr A 1996; 755:127–132.

23. Murphy PA, Song T, Buseman G, Barua K. Isoflavones in soy-based infant formulas. J Agric Food Chem 1997; 45:4635–4638.

24. Petterson H, Kiessling KH. Liquid chromatographic determination of the plant estrogens coumestrol and isoflavones in animal feed. J Assoc Off Anal Chem 1984; 3:503–506.

25. Wang H, Murphy PA. Isoflavone content in commercial soybean foods. J Agric Food Chem 1994; 42:1666–1673.

26. Harborne JB. 3-Desoxyanthocyanins and their systematic distribution in ferns and gesnerads. Phytochemistry 1966; 5:589–600.

27. Swinny EE, Bloor SJ, Wong H. ^{1}H and ^{13}C NMR assignments for the 3-deoxyanthocyanins luteolinidin-5-O-glucoside and apigeninidin-5-O-glucoside. Magn Reson Chem 2000; 38:1031–1033.

28. Swinny EE. A novel acetylated 3-deoxyanthocyanidin laminaribioside from the fern Blechnum novae-zealandiae Z Naturforsch 2001; 56c:177–180.

29. Crowden RK, Jarman SJ. 3-Deoxyanthocyanins from Blechnum procerum. Phytochemistry 1974; 13:1947–1948.

30. Zorn B, Garcia-Piñeres AJ, Castro V, Murillo R, Mora G, Merfot I. 3-Desoxyanthocyanidins from Arrabidaea chica. Phytochemistry 2001; 56:831–835.

31. Ozawa T. Separation of the components in black tea infusion by chromatography on Toyopearl®. Agric Biol Chem 1982; 46:1079–1081.

32. Vande Casteele K, Van Sumere C, Geiger H. Separation of flavonoids by reversed-phase high-performance liquid chromatography. J Chromatogr A 1982; 240:81–94.

33. Van Sumere C, Fache P, Vande Casteele K, DeCooman L, Everaert E, De Loose R, Hutsebaut W. Improved extraction and reversed phase high-performance liquid chromatographic separation of flavonoids and the identification of Rosa cultivars. Phytochem Anal 1993; 4:279–292.

34. Strack D, Wray V. Anthocyanins. In: Harborne JB, ed. Methods in Plant Biochemistry. Vol. 1. Plant Phenolics. London: Academic Press, 1989:1–22.

35. Jurd L. Spectral properties of flavonoid compounds. In: Geissman TA, ed. The Chemistry of Flavonoid Compounds. Oxford: Pergamon Press, 1962:107–155.

36. Pietrogrande MC, Kahie YD. Effect of the mobile and stationary phases on rp-hplc retention and selectivity of flavonoid compounds. J Liquid Chromatogr 1994; 17:3655–3670.

37. Stafford HA. Regulatory mechanisms in anthocyanin biosynthesis in first internodes of sorghum vulgare: effect of presumed inhibitors of protein synthesis. Plant Physiol 1966; 41:953–961.

38. Francis JF, Analysis of anthocyanins. In: Markakis P, ed. Anthocyanins as Food Colors. New York: Academic Press, 1982:181–207.

39. Snyder LR, Kirkland JJ, Giajch JL. Practical HPLC Method Development. 2d ed. New York: John Wiley & Sons, 1997.

40. Kromidas S. Practical Problem Solving in HPLC. Germany: Wiley-VCH Verlag GmbH, 2000.

41. Liggins J, Bluck LJC, Coward WA, Bingham SA. Extraction and quantification of daidzein and genistein in food. Anal Biochem 1998; 264:1–7.

42. Tomas-Barberan FA, Blasquez MA, Garcia-Viquera C, Ferreres F, Tomas-Lorente F. A comparative study of different Amberlite XAD resins in flavonoid analysis. Phytochem Anal 1992; 3:178–181.

43. Lu TS, Saito N, Yokoi M, Shigihara A, Honda T. An acylated peonidin glycoside in the violet-blue flowers of Pharbilis nil. Phytochemistry 1991; 30:2387–2390.

44. Sanders JK, Hunter BK. Modern NMR Spectroscopy—a Guide for Chemists. 2d ed. Oxford: Oxford University Press, 1993.

45. Silverstein RM, Bassler GC, Morrill TC. Spectrometric Identification of Organic Compounds. 4th ed. New York: John Wiley & Sons, 1981.

46. Markham, KR, Geiger H. [1]H nuclear magnetic resonance spectroscopy of flavonoids and their glycosideshexadeuterodimethylsulfoxide. In: Harborne JB, ed. The Flavonoids—Advances in Research Since 1986. London: Chapman and Hall, 1994:441–497.

5

Synthesis, Identification, Quantification, and Chemical Reactivity of Methylated Flavan-3-ols

Cécile Cren-Olivé and Christian Rolando
Lille University of Science and Technology
Villeneuve d'Ascq, France

I. INTRODUCTION

Flavan-3-ols are a large class of flavanoïds ubiquitous in plants [1–5] and widely found in a number of foods [6,7]. They represent an integral part of the human diet and are considered to be key compounds in the relationship between health and diet. Indeed, they are known to combat aging pathologies in which oxidative stress is involved such as cancers [8–10], and cardiovascular [11–13] and neurodegenerative [14] diseases.

Because of the increasing significance of these potential beneficial roles, understanding the mechanism by which they behave as antioxidants is essential. However, polyphenols mainly circulate in blood as metabolites. For example, the most-studied flavan-3-ol, catechin, is present almost exclusively as methylated metabolites (3'-methylcatechin in the majority) as well as sulfate and glucuronide conjugates in plasma [15–17]. So it is not the native forms but the methylated forms that require further study.

Unfortunately, these metabolites are not commercially available and are difficult to extract from enzymatic synthesis in the quantities requested for biological or chemical studies. So whatever the study envisaged on the methylated flavan-3-ols (investigation of their chemical reactivity, development of an analytical tool to characterize each site involved in the metabolism), the first step is the synthesis of a whole family of methylated compounds.

II. SYNTHESES OF METHYLATED ANALOGUES OF FLAVAN-3-ols

Various strategies have been attempted for the chemical synthesis of methylated flavan-3-ols, which can be divided into two groups: total enantioselective synthesis [18–24] and hemisynthesis [25–31]. For total enantioselective synthesis, two strategies are predominant. The key step of the first, leading to either the catechin or the epigallocatechin gallate skeleton, consists of a stereospecific cyclization of the Sharpless asymmetrical dihydroxylation product [18,20,21] once the C6-C3-C6 skeleton is obtained either by base-catalyzed condensation of the appropriate oxygenated acetophenone and benzaldehyde [20,21] or by the coupling of cinnamyl alcohol derivative with a 3,5-dimethoxyphenol [18]. The second strategy requires four main steps for synthesizing the epigallocatechin gallate skeleton: (1) coupling of appropriate oxygenated acetophenone and benzaldehyde; (2) cyclization of the chalcone directly to 3-flaven; (3) hydroboration-oxidation, followed; (4) a two-step sequence involving oxidation with the Dess-Martin periodinane followed by selective reduction with lithium tri-sec-butyl-borohydride (L-Selectride) [19]. But neither the choice of the starting synthons nor the yields of the epigallocatechin gallate are satisfactory for accessing to catechin analogues and methylated metabolites. Since total enantioselective syntheses of methylated flavan-3-ols appear to be difficult, long, and expensive, this total synthesis strategy is applied more specifically when the initial natural compound is not available or is available only with difficulty in pure enantiomeric form, as in the case of epigallocatechin gallate.

When polyphenol precursors are available in pure enantiomeric form from the vegetal pool, strategies based on hemisyntheses seem much more appropriate. However, it is well known that partial methylation of catechin, for example, does not constitute a suitable method to synthesize methylated flavan-3-ols since it produces a complex untractable mixture of products in low yield [32–35]. So the hemisynthesis of methylated derivatives of flavan-3-ols is thus very quickly directed to strategies based on selective protection-deprotection of the A- and B-rings [25–30]. However, the choice of the reagent is rather limited, as catechin is known to undergo quite readily a base-catalyzed epimerization at C-2 to form ent-epicatechin through reversible opening of the C-ring via a B-ring quinone methide intermediate, which requires a free phenolic OH at the C4′ position (Scheme 1) [36,37].

Until recently, the use of benzyl carbonate [28] or cyclic borate [25,28] as protecting group led to the synthesis of essentially two dimethylated catechin

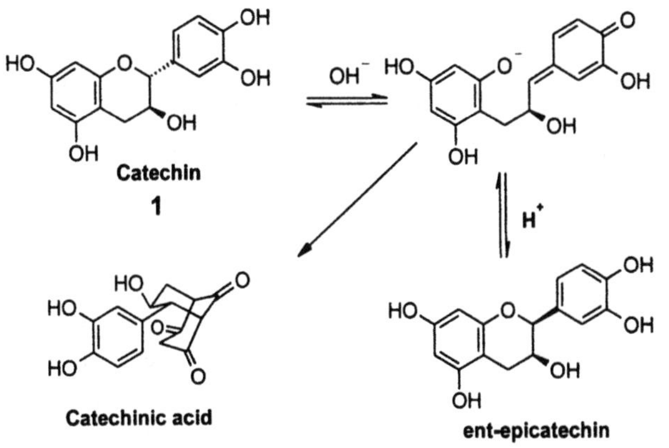

Scheme 1 Catechin epimerization and rearrangement in basic medium. (Adapted from Ref. 36.)

analogues, the 3',4'-dimethyl- and the 5,7-dimethylcatechin. But, these two protecting groups have important disadvantages in the case of flavan-3-ols. The first reagent, benzylchloroformate, is not, in fact, selective since it is not specific to the catechol moiety. So in the presence of polyphenol, its regioselectivity depends only on the difference of microscopic pKa between the different hydroxyl functions present in the molecule, which is, for catechin and epicatechin, too tiny ($pK_{3'-OH} = 9.02$; $pK_{4'-OH} = 9.12$; $pK_{5-OH} = 9.43$; $pK_{7-OH} = 9.58$ in water) [38] to obtain regiospecificity [28]. The second proposed protecting group proposed, cyclic borate, offers selective protection of the B-ring under mild basic conditions but it is rather delicate to use since the protected compound is not isolated [25,28].

Since the year 2000 a strategy for the selective protection of catechin based on the differentiation between catechol and other phenols has been developed to synthesize the four monomethylated isomers of (+)−catechin in position, respectively, 3', 4', 5, and 7, two dimethylated derivatives: the 5,7-dimethylcatechin and the 3', 4'-dimethylcatechin and two trimethylated isomers of (+)−catechin in position, respectively, 3', 5, 7, and 4', 5, 7 [29,30]. The key step is the differentiation of the catechol ring of catechin from the resorcinol-like ring by using reagents (dichlorodiphenylmethane or di-tert-butyldichlorosilane) leading to the formation of permanent or transient dioxolane cycle. These B-ring-protected compounds open access to two different pathways: the first leads to partial or total methylation of the A-ring, the second to the methylation of the same B-ring by means of specific protection of the A-ring and deprotection of the B-ring. So 5,7-dimethylcatechin and 5- and 7-methylcatechin were synthesized by dichlorodiphenyl methane protection of the catechol moiety, partial or total

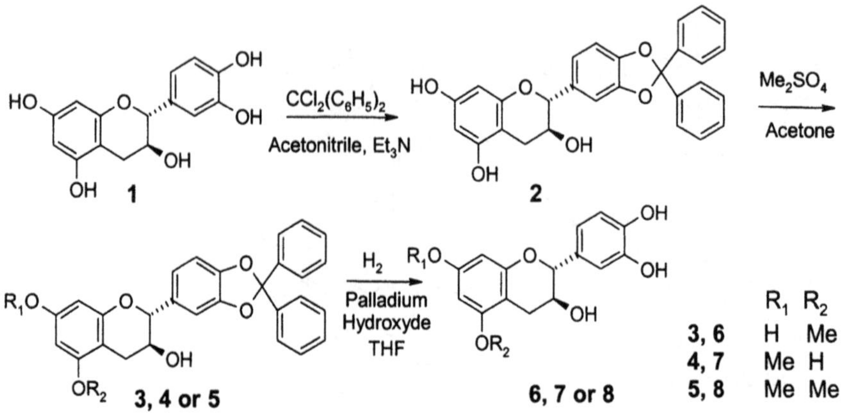

Scheme 2 Synthesis of catechin A-ring methylated analogues: 5,7-dimethyl-catechin, 5- and 7-methylcatechin. (Adapted from Ref. 30.)

methylation of the A-ring under standard conditions, followed by removal of the protection by hydrogenolysis (Scheme 2) [29,30].

Unlike A-ring methylated derivatives, the monomethylated and dimethylated B-ring compounds were synthesized in two different ways. Whereas the two hydroxysilyl monoethers (Scheme 3) obtained after reaction of catechin with di-*tert*-butyldichlorosilane led to the two B-ring monomethylated isomers, the 3′,4′-diphenylmethylenedioxycatechin (Scheme 4) initiated the synthesis of 3′,4′-dimethylcatechin [30]. The crucial step of these syntheses consists of the selective and successive protection of the B- and A-rings. Indeed, after selectively protecting the B-ring catechol moiety, protection of the A-ring, whose deprotection conditions differ from those of the B-ring, is required to allow selective deprotection of the B-ring; therefore, selective methylation of the B-ring is possible.

More precisely, the protection of catechin by di-*tert*-butyldichlorosilane is not stable and leads, after a rapid hydrolysis, to two hydroxysilyl monoesters of the parent catechol [30], providing material for the synthesis of the two B-ring monomethylated isomers on the B-ring (Scheme 3). The protection of the free phenolic functions of these two ethers is achieved by benzylation under standard conditions. After deprotection of the B-ring, i.e., desilylation induced by a fluoride ion source, the methylation of the position 3′ or 4′ is followed by the removal of the benzylic group by hydrogenolysis, giving the target compounds 3′- and 4′-methylcatechin [30].

The same synthetic approach ensures the synthesis of the two trimethylated isomers of (+)-catechin in position, respectively, 3′, 5, 7 and 4′, 5, 7 (Scheme 5).

In the case of 3′,4′-dimethylcatechin (Scheme 4), the most appropriate protection of the A-ring is acetylation, but the use of these protecting groups

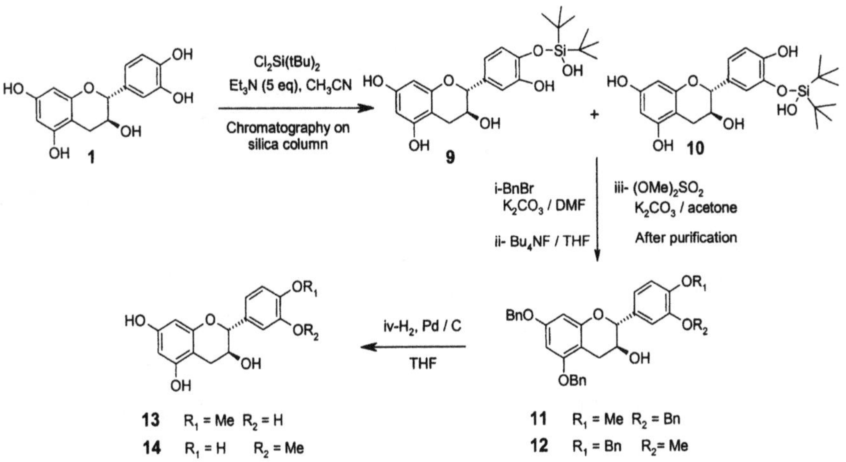

Scheme 3 Synthesis of catechin B-ring monomethylated analogues: 3'- and 4'-methylcatechin. (Adapted from Ref. 30.)

requires the development of a new deprotection method. Indeed, the usual deprotection of phenol acetate in slightly basic medium led to the formation of a complex mixture because of the high instability of catechin. The only way to achieve deprotection of 3,5,7-triacetyl-3',4'-dimethylcatechin consists of the action of a reagent that is both nucleophilic and reductive such as sodium sulfite. Unfortunately in these conditions the secondary alcohol is not deprotected;

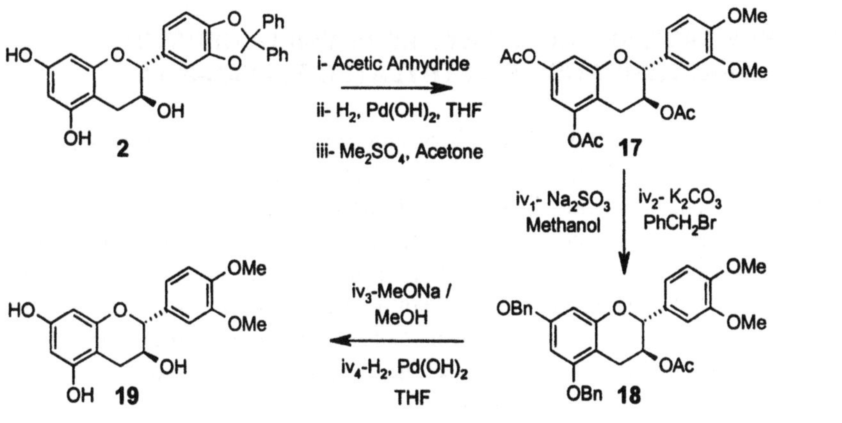

Scheme 4 Synthesis of catechin B-ring dimethylated analogues: 3',4'-dimethylcatechin. (Adapted from Ref. 30.)

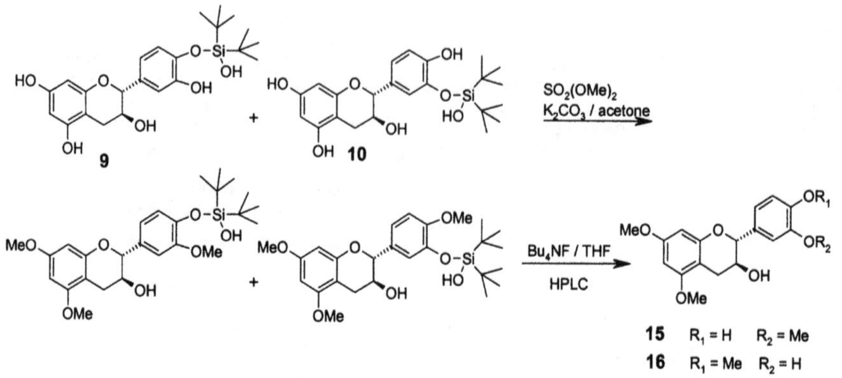

Scheme 5   Synthesis of two catechin trimethylated isomers. (Adapted from Ref. 30.)

therefore, a new cycle of protection/deprotection to obtain the 3′,4′-dimethyl-catechin is required (Scheme 4) [30].

   Synthetic routes that are now available provide access to all monomethylated flavan-3-ols, including the metabolites that are not available through natural sources, allowing the development of new analytical methodologies to characterize flavan-3-ols metabolites. Furthermore, the same strategy virtually gives access to any polymethylated flavan-3-ol analogues, especially the trimethylated series with only one free phenol, which are key model compounds for establishing reliable relation-structure activities in biological tests and for determining thermodynamic constants.

III.   FREE RADICAL CHEMISTRY AND PHYSICOCHEMISTRY CHARACTERISTICS OF METHYLATED FLAVAN-3-ols

The study of the free radical chemistry and physicochemistry characteristics of methylated flavan-3-ols is crucial to understand the mechanism by which flavan-3-ols behave as antioxidants. Indeed, on the one hand, physicochemical parameters such as redox potential, scavenging and decay constants, and pKa are of capital importance to the understanding of flavan-3-ols biological effects. On the other hand, it is the metabolites and not the native forms of flavan-3-ols that deserve further investigation since flavan-3-ols are present almost exclusively as methylated metabolites (3′-methylcatechin in the majority), as well as sulfate and glucuronide conjugates in plasma [15–17].

   However, until recently methylated flavan-3-ols were not readily available; so only a few studies [29,39] have examined their physicochemical parameters in contrast to the abundant literature dedicated to the chemical characteristics of flavan-3-ols radicals [40–49]. But there is much discussion and contradiction in this literature regarding the phenoxyl radicals structure [41–43,46–48], the reduction potentials [41,42,45,46,49], and therefore the structure-activity relationship to the antioxidant activity. Thus it is interesting to see how the study of the chemical reactivity of methylated flavan-3-ols enables us to understand the chemical reactivity of the native forms [29,39].

A. Free Radical Chemistry of Methylated Flavan-3-ols

The free radical reactivity of methylated flavan-3-ols has been investigated using a flash photolysis experiment for the photochemical generation of radicals and their characterization through the monitoring of their UV-visible spectra [29,31,39]. Phenoxyl radicals have been generated by different techniques: (1) by direct photoionization of the polyphenol derivatives in their basic form and (2) by H-atom abstraction from phenolic OH by *tert*-butoxyl radicals generated by the photoionization of *tert*-butyl peroxide in aprotic media (Fig. 1).

The study of the dimethylated compounds allows characterization of the intrinsic reactivity of each ring of flavan-3-ol. Whereas 3′,4′-dimethylcatechin presents an absorbency at 495 nm, 5,7-dimethylcatechin shows an absorption band at 380 nm (Table 1) [29]. During photo-oxidation experiments, another band at 550 nm is visible for the radical from 3′,4′-dimethylcatechin, an absorption band that has been ascribed to a fast further deprotonation of the neutral resorcinol-like radical to the corresponding radical anion in basic medium [29,39]. So the phenoxyl radicals of both rings have been described, which will enable us to understand the more complex behavior exhibited by catechin and monomethylated compounds (Table 1).

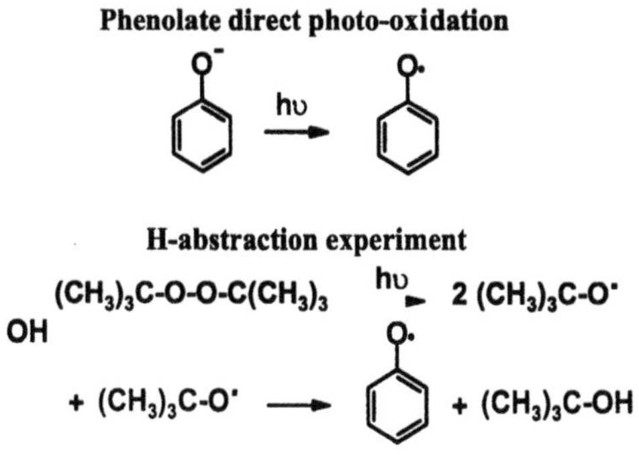

Figure 1   Techniques used to generate phenoxyl radicals.

Table 1 Spectral Characteristics of Phenoxyl Radicals Generated by Photo-Oxidation and H-Abstraction[a]

Compounds	Models of ring	Photo-oxidation on phenolate	H-abstraction on phenol	
		λmax (nm)	λmax (nm)	ϵ (mol^{-1}L^{-1}cm^{-1})
3′,4′-Dimethylcatechin	A	495/550	495	2700
5,7-Dimethylcatechin	B	380	380	6600
Catechin		380	380/495	—
3′-Methylcatechin		380/495	380/495	—
4′-Methylcatechin		495	495	2300

[a] Delay after pulse: 200 ns for photo-oxidation experiments, 5 μs for H-abstraction.
Source: Refs. 29 and 39.

Indeed, these results obtained on dimethylated compounds led to demonstration that H-abstraction of catechin by *tert*-butoxyl radicals is selective on the A-ring, since 70% of the reactivity occurs on the A-ring (absorption at 495 nm), whereas the 308-nm laser-induced photo-oxidation of catechin phenolate appears selective on the B-ring with an absorbency at 380 nm [39]. Moreover, the selectivity of direct irradiation experiments has been correlated to the deprotonation sequence since only phenolates absorb the laser light and are much more easily oxidized than neutral phenols. So although the protonation sequence between the two rings has not been clearly established and is still under discussion, these results indicate that the B-ring is slightly more acidic than the A-ring [29], as confirmed by a study of catechin deprotonation followed by NMR and affording the microscopic pKa [38] (Table 2). Indeed, the successive deprotonations of the different phenolic functions of flavan-3-ols

Table 2 Microscopic pKa of Each Catechin Phenolic Function

Phenolic function	Microscopic pKa ($\pm$0.05)
3′	9.02
4′	9.12
5	9.43
7	9.58

Source: Ref. 38.

induce great changes in the chemical environment of various carbons of the skeleton, as reflected in the ^{13}C-NMR chemical shifts [50–54]. More precisely, the deprotonation of a phenolic function induces deshielding for the *ipso* and *ortho* carbon atom and shielding for the *para* carbon atom. Such selective behavior allows, with the unambiguous assignment of ^{13}C-NMR signals of flavan-3-ols, the determination of the precise deprotonation site of catechin and epicatechin [38]. The quantification of each existing species allows determination of the intrinsic pKa. NMR studies are the only method yielding these microscopic pKa in the case of polyphenolic compounds.

It is now particularly interesting and possible to study the influence of the B-ring monomethylation on the reactivity of flavan-3-ol, comparing the results obtained for catechin and for two monomethylated compounds: 3′- and 4′-methylcatechin, first of which is the major metabolite currently identified as circulating in plasma [39]. Although the B-ring monomethylation only enhances the selectivity of the H-abstraction on the A-ring, it implies drastic changes in the behavior on photo-oxidation, indicating that the physicochemical parameter that is greatly affected by methylation is the relative acidity of each phenolic position. More precisely, the methylation of the 3′ position, the most acidic position in catechin (Table 2), reduces the pKa difference between the two rings since during photo-oxidation experiments, both radicals issuing from each ring appear equally (Table 1). Conversely, the methylation of the 4′ position drastically changes the protonation sequence of flavan-3-ols: only the radical absorbing at 495 nm exists (Table 1), indicating that in this case, the A-ring is more acidic.

Study of the reactivity of methylated flavan-3-ols leads to a better understanding of the free radical chemical processes of the whole flavan-3-ol family since it shows that the mechanisms of electron and H-atom transfer are radically different and are specific to one moiety of the flavan-3-ols: electron transfer involves the B-ring, whereas H-atom transfer involves the A-ring.

B. Redox Properties of Flavan-3-ols

The redox potential of interest to understand the biological effects of flavan-3-ols is the one related to phenoxyl radical–phenate couple, as this potential is roughly 1 V lower than the potential of the phenoxyl radical–phenol couple, which furthermore may transiently involve the oxidation of the aromatic atoms. Standard potential can be measured by electrochemistry [49] or pulse radiolysis [40–44]. However, determining the redox potential of polyphenolic compounds is a real challenge since for these methods the measurement must be faster than the subsequent reactions induced by the oxidation of the phenol group in order to obtain the thermodynamic value. By using ultramicroelectrodes (electrodes with a micrometer diameter), it has been shown that a very high scan rate, up to 1 million

V/s, can be reached; this implies that intermediates that have a lifetime in the microsecond range can be characterized by direct electrochemical methods [55]. So the use of fast cyclic voltametry allows determination of the E° value of coniferyl alcohol [56]; however, the E° of a simple catechol such as caffeic acid [57] cannot be determined even using ultramicroelectrodes.

Therefore, the only way to obtain precise values is to use fast cyclic voltametry on model compounds, for example, the two trimethylated flavan-3-ols, which offer only one free function hydroxyl. The redox potential of each phenolic function using fast cyclic voltametry has been investigated [39] (Fig. 2) and is summarized in Table 3.

As expected, the catechol B-ring is more oxidizable than the resorcinol A-ring, and these results are in good agreement with the acid-base properties discussed previously: on each ring, the more basic site is the most oxidable phenolic function.

C. Conclusion

The study of the physicochemistry and free radical chemistry characteristics of methylated flavan-3-ols allows [29,39] identification of the two flavan-3-ol

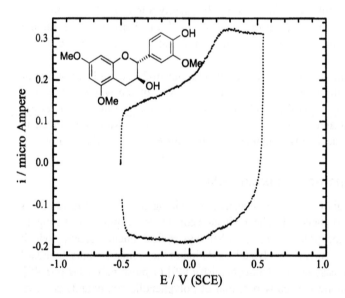

Figure 2 High scan rate cyclic voltametry of 3′,5,7-trimethylcatechin (C° = 0.98 mM) in acetonitrile-0,1M NBu$_4$BF$_4$ on a 10-μm-diameter glassy carbon ultramicroelectrode (scan rate v = 11500 Vs^{-1}). SCE, saturated calomel electrode reference.

Table 3 Redox Potentials of Different
Catechin Analogues

Catechin analogues	Free phenolic position	E° V (SCE)
3	7	-
4	5	0.285
14	4'	0.110
13	3'	0.135

Source: Ref. 39.

radical families, characterization of the intrinsic reactivity of each ring as well as determination of two thermodynamic values: redox potential and dissociation constant.

All these parameters permit the proposal of a new insight into the antioxidative properties of flavan-3-ols. Whereas in the past these properties were explained only by H-abstraction process, these results indicate that these activities also involve an electron transfer since the B-ring of flavan-3-ols has been shown to be the active moiety of the molecule.

IV. CHARACTERIZATION OF THE METHYLATION SITE OF FLAVAN-3-OLS BY MASS SPECTROMETRY

The characterization of the metabolism site via the methylation site of flavan-3-ols is essential for the identification of the metabolites circulating in the plasma and thus for the understanding of the bioavailability of flavan-3-ols. Indeed, in spite of extensive and detailed studies on their metabolism, little is known about the precise structures of the metabolites because of the lack of suitable methodology.

Indeed, until recently [35], the identification and quantification of metabolites in biological samples were often performed indirectly after initial hydrolysis conjugates with enzyme regardless of the technique used for the analysis (Table 4): high-performance liquid chromatography combined with UV [58,59], chemiluminescence [60], fluorescence [61], electrochemical [62], mass spectrometry [63] detection, capillary liquid chromatography coupled with electrospray mass spectrometry [64], or gas chromatography coupled with mass spectrometry [16,34,65]. The development of a new liquid chromatography electrospray ionization mass spectrometry (LC ESI-MS/MS) method that uses positive ion mode allows unambiguous characterization and differentiation of

Table 4 Characteristics of Various Analytical Methods for the Analysis of
Flavan-3-ols and Their Metabolites in Biological Medium[a]

Method		Limit of detection for catechin	Administered dose	Medium	Hydrolyze	Ref.
GC/MS		0.580 ng/mL 0.196 ng/mL*	0.46 mg/kg	Plasma	×	16, 17
HPLC	UV	500 ng/mL	0.21 mg/kg	Plasma	×	59
	EC	1–10 ng/mL	0.53 mg/kg	Plasma, urine, saliva	×	15, 62
	CL	0.996 ng/mL	1.61 mg/kg	Plasma	×	60
	FL	20 ng/mL	15 mg/kg	Plasma	×	61
	MS	Nd	100 μg EGCG per 1mL of plasma	Plasma	×	64

[a]Limit of detection of 3′-methylcatechin. GC/MS, gas chromatography/mass Spectrometry; UV,
ultraviolet; EC, electrochemical; HPLC, high-performance liquid chromatography; CL, chemiluminescence; FL, fluorescence; Nd, not determined; EGCG, epigallocatechin gallate.

each site involved in catechin metabolism [35] and thus allows analysis of crude
biological extracts [66].

A. Catechin Fragmentation Under Electrospray Ionization Mass Spectrometry Conditions

The fragmentation of catechin under ESI-MS/MS conditions can be rationalized
by three pathways depicted in Scheme 6 [35]. The first, **I**, leads to the unique
A-ring product ion and can be unambiguously attributed to the retro Diels
Alder fragmentation, well known and characteristic of the fragmentation of
flavan-3-ols under EI-MS conditions [67] and fast atom bombardment mass
spectrometry (FAB MS) using glycerol [67] or meta-nitrobenzyl-alcohol [67,68]
matrix. But whereas under EI and FAB MS conditions, both A- and B-ring
fragments appear in the spectrum, the MS/MS spectrum obtained using ESI-
MS/MS shows only one species, the o-hydroxylbenzylcation $^{1,3}A^+$ ion at m/z
139. The two other fragmentation pathways **II** and **III**, which first yield two
different product ions, lead finally to the same fragment ions after loss of water,
CO, and C_2H_4 (Scheme 6). The major product ions obtained in these two
pathways involve only the B-ring, the $^{1,2}B^+$ ion at m/z 123 and the $^{1,4}B^+$ ion at
m/z 165, which can further fragment by loss of water, giving rise to ($^{1,4}B^+$-
H_2O) ion at m/z 147.

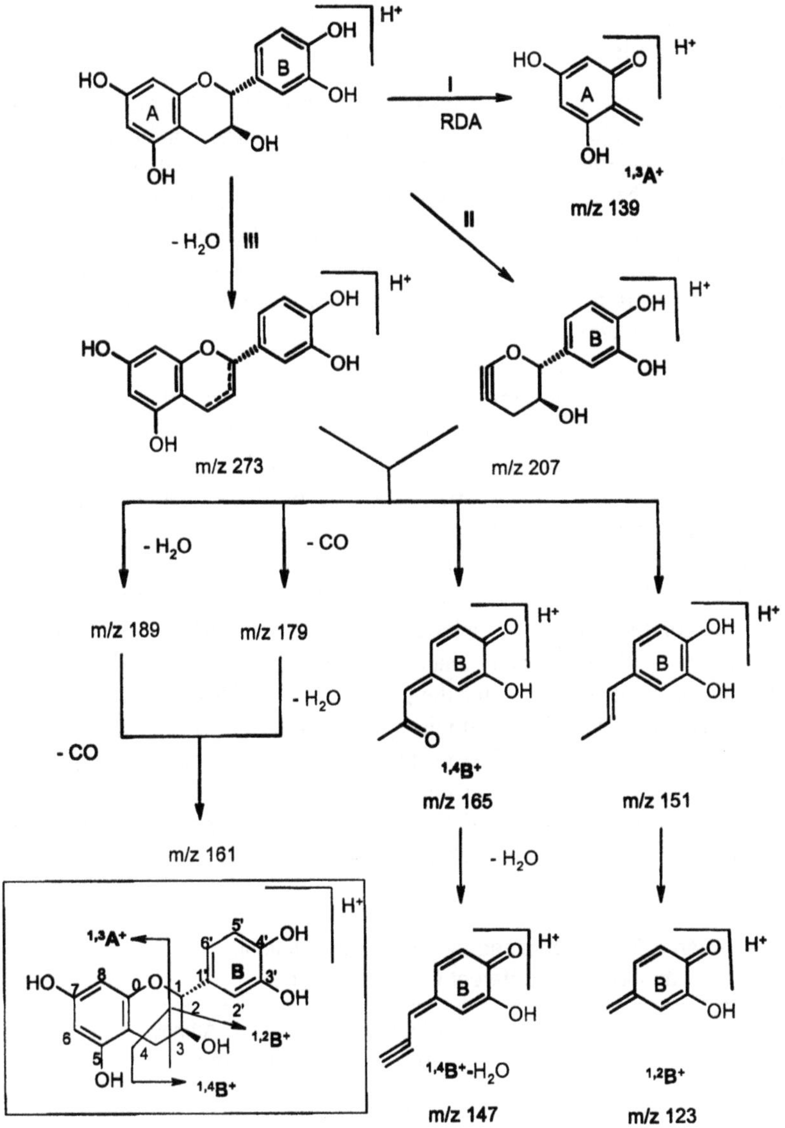

Scheme 6 Major fragmentations of protonated catechin. The insert gives the ion nomenclature for protonated flavan-3-ols. (From Ref. 35.)

B. Characterization of the Methylation Cycle of Flavan-3-ols

The most useful fragmentations in terms of characterization of the methylation or metabolism sites of catechin are those resulting in structurally informative $^{i,j}A^+$ and $^{i,j}B^+$ ions. Indeed, the study of the different mass shifts of the three diagnostic ions $^{1,3}A^+$, $^{1,2}B^+$, $^{1,4}B^+$ observed between catechin and its methylated analogues (Fig. 3) allows unambiguous identification of the methylated cycle [35]. For example, the 28-mass-unit shift of $^{1,2}B^+$, $^{1,4}B^+$ in the collision-activated dissociation (CAD) spectrum C of Figure 3 clearly indicates that this is the MS/MS spectrum of 3′,4′-dimethylcatechin.

More interesting is that the substitution, in this case methylation, changes the gas phase basicity of the substituted ring and creates a privileged fragmentation pathway [35]. For example, in Figure 3, whereas in the CAD spectra of catechin A, and of 5,7-dimethylcatechin B, the base peak is the A-ring product ion at respectively, $m/z = 139$ and 167 in the case of 3′,4′-dimethylcatechin B, there is an inversion and the B-ring product ion at $m/z = 151$ becomes the base peak. More precisely, the calculations of the proton affinity of the different models of A- and B-ring obtained using MNDO methods (Table 5) allow rationalization of these different behaviors. The resorcinol-like ring (A-ring) generally presents higher basicity than its isomeric catechol-like B-ring, as suggested by the results obtained for the proton affinity of 5,7-dihydroxybenzopyran (849 kJ mol^{-1}) and 1,2-dihydroxybenzene (768 kJ mol^{-1}). So the protonation of the more favorable ring leads to the formation of A-ring base peak at $m/z = 139$ for catechin. On the other hand, methylation, as expected, increases the gas phase basicity of the substituted ring: the proton affinities obtained for methylated models such as methoxybenzene, 1,2-dimethoxyben-

Table 5 Proton Affinities Obtained by MNDO Calculations Conducted at PM3 Level[a] for Different Catechin Models

Compounds	Position of protonation	Proton Affinity (kJ · mol^{-1})
5,7-Dihydroxybenzopyran	C6	849
1,2-Dihydroxybenzene	C4	768
1,2-Dihydroxy, 4-methylbenzene	C5	795
methoxybenzene		807
1,2-Dimethoxybenzene		804
1,2-Dimethoxy, 4-methylbenzene		813

[a] Calculated from proton affinity (PA) = $\Delta H°_f (H^+) + \Delta H°_f (R) - \Delta H°_f (R-H^+)$, where $\Delta H°_f (H^+) = 1528$ kJ · mol^{-1}.
Source: Ref. 35.

zene (807 kJ mol^{-1}) [69], 804, and 813 kJ mol^{-1}, are higher than those of phenol (786 kJ mol^{-1}) [69], 1,2-dihydroxybenzene (804 kJ mol^{-1}), and 1,2-dihydoxy, 4-methylbenzene (813 kJ mol^{-1}). So in the case of catechin methylated analogues, the protonation of the most basic ring leads for 3',4'-dimethylcatechin to the formation of a B-ring base peak and for 5,7-dimethyl-catechin to an A-ring base peak. So substitution induces modification in the spectrum of catechin-substituted analogues, which can be rationalized by the examination of the proton affinity of the substituted ring.

C. Characterization of the Methylation Site of Flavan-3-ols

The modifications observed in the relative intensity of the major product ions are specific to the substitution (nature and position). In the case of the isomeric B-ring monomethylcatechin, there is an inversion of base peak between the two isomers as already detected for the dimethylated ones: the $^{1,3}A^+$ ion at m/z 139 is the base peak for the 3'-methylcatechin (Fig. 3, spectrum D), whereas the $^{1,2}B^+$ ion at m/z 137 is for 4'-methylcatechin (Fig. 3, spectrum E). For the other couple of isomers (on the A-ring), differences appear on the major product ions $^{1,3}A^+$ and $^{1,4}B^+$. So these differences can be used to determine precisely the site of methylation on each cycle (3' versus 4' and 5 versus 7) [35].

In order to quantify these differences and propose criteria to localize precisely the substituent, branching ratios have been calculated: Σ relative intensity of $^{1,3}A^+/\Sigma$ relative intensity of $^{1,3}A^+$, $^{1,2}B^+$, and $^{1,4}B^+$ for position 3' and 4'; and Σ relative intensity of $^{1,4}B^+/\Sigma$ relative intensity of $^{1,3}A^+$ and $^{1,4}B^+$ for position 5 and 7, which gives 59 (3'-methyl) versus 36 (4'-methylated) for the B-ring pair of isomers and 34 (5-methylation) versus 28 (7-methylation) for the A-ring pair of isomers.

So isomeric methyl catechin with the same [M+H$^+$] ion can be charac-terized and differentiated on the basis of their CAD spectra alone: structurally informative fragmentation allows one to infer the substitution pattern in the A- or B-ring, whereas the study of the relative intensities of the major product ions through the determination of the branching ratio indicates the precise site of substitution on each ring since the higher branching ratio of each isomer pair is correlated with the substituent position (3' in the case of B-ring isomers and 5 in the case of A-ring ones) [35].

D. Application to Crude Biological Samples

Moreover, the sensitivity of this methodology appears to be excellent down to 30 pg/mL, which allows investigations of real biological samples. The LC-ESI-MS/MS methodology has indeed been applied with success to the analysis of crude samples obtained by rat liver homogenate [66]. The reconstructed mass chroma-

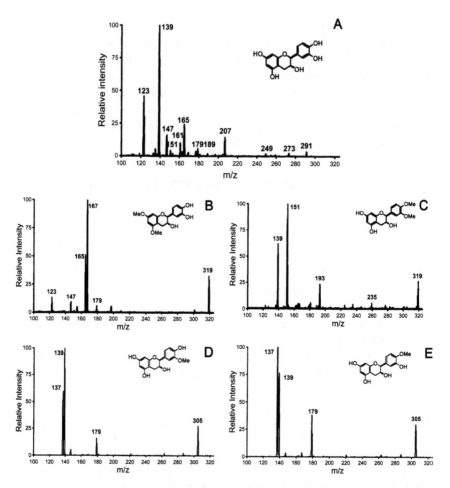

Figure 3 Collision-activated dissociation (CAD) spectra obtained for (A) catechin, (B) 5,7-dimethylcatechin, (C) 3′,4′-dimethylcatechin, (D) 3′-methylcatechin, (E) 4′-methylcatechin. Voltage cone, 30 V, collision energy, 10 eV; argon pressure, $2.8.10^{-3}$ mB. (From Ref. 35.)

togram at m/z 305 obtained on a C_{18} reversed-phase column exhibit two major peaks (Fig. 4A). The MS/MS spectrum of the major peak (Fig. 4B), which exhibits a high 139/137 ratio, can be confidently attributed to 3′-methylcatechin, whereas the more noisy MS/MS mass spectrum of the minor peak (Fig. 4C) can be attributed to 4′-methylcatechin.

Moreover, it has been shown that this method can be generalized to other series of flavan-3-ols such as epicatechin where no standard is available and it is

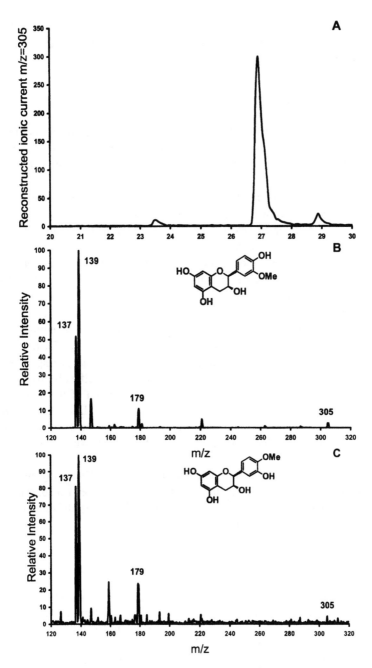

Figure 4 Mass chromatogram and CAD spectra obtained for the two major peaks of a mixture obtained from rat liver homogenate of catechin and analyzed by LC-ESI-MS/MS. Voltage cone, 30 V; collision energy, 10 eV; argon pressure, $2.8.10^{-3}$ mB.

expected that this kind of technology with a mild ionization technique (ESI) could be applied to other types of metabolites that are more polar and labile, such as sulfate or glucuronide metabolites.

V. CONCLUSIONS

The discovery of the numerous beneficial effects of flavan-3-ol metabolites on human health stimulates researchers to investigate more thoroughly the methylated flavan-3-ols chemistry: their synthesis, identification, quantification, and chemical reactivity.

Recently, a new strategy based on successive and selective protections of the various phenol functions present on flavan-3-ols has allowed synthesis of a whole family of methylated catechin analogues. These materials appear particularly useful for the development of a new analytical tool allowing the identification of all flavan-3-ol metabolites, for the study of their chemical reactivity, and thus for understanding the mechanism by which these methylated flavan-3-ols behave as antioxidants.

Indeed, with these selectively methylated catechin analogues, it has been possible to determine thermodynamic constants (redox potentials, pKa) and to study the intrinsic reactivity of each catechin moiety. Each moiety, the A- or B-ring, presents its own reactivity: whereas the A-ring is more specifically involved in H-atom transfer, the B-ring is specifically involved in electron transfer. These results lead to a new interpretation of the antioxidant properties of these molecules: since B-ring of flavan-3-ols has been shown to be the active moiety of the molecule [12], the antioxidant activity involves mainly an electron transfer and not only an H-atom transfer as is so often proposed.

Moreover, the access to a whole family of selectively methylated analogues opens new areas of research in the elucidation of the key role of flavan-3-ol metabolites in human health by offering the possibility of establishing reliable relation-structure activities in biological tests.

REFERENCES

1. Porter LJ. Flavans and proanthocyanidins. In: Harborne JB, ed. The Flavonoids: Advances in Research Since 1986. London, Chapman and Hall, 1994:23–56.
2. Bravo L. Polyphenols: chemistry, dietary sources, metabolism and nutritional significance. Nutr Rev 1998; 56:317–333.
3. Kühnau J. The flavonoids: a class of semi-essential food components: their role in human nutrition. World Rev Nutr Diet 1976; 24:117–191.
4. Ferreira D, Bekker R. Oligomeric proanthocyanidins: naturally occurring O-heterocycles. Nat Prod Rep 1996; 13:411–433.

5. Peterson J, Dwyer J. Flavonoids: dietary occurrence and biochemical activity. Nutr Res 1998; 18:1995–2018.

6. Arts IC, van de Putte B, Hollman PCH. Catechin contents of foods commonly consumed in the Netherlands. 1. Fruits, vegetables, staple foods and processed foods. J Agric Food Chem 2000; 48:1746–1751.

7. Arts IC, van de Putte B, Hollman PCH. Catechin contents of foods commonly consumed in the Netherlands. 2. Tea, wine, fruit juices, and chocolate milk. J Agric Food Chem 2000; 48:1752–1757.

8. Cao Y, Cao R. Angiogenesis inhibited by drinking tea. Nature 1999; 398:381.

9. Jankun J, Selman SH, Swiercz R, Skrypczak-Jankun E. Why drinking green tea could prevent cancer. Nature 1997; 387:561.

10. Vergote D, Cren-Olivé C, Chopin V, Toillon RA, Rolando C, Hondermarck H, Le Bourhis X. (-)-Epigallocatechin (EGC) of green tea induces apoptosis of human breast cancer cells but not their normal counterparts. Breast Cancer Treat 2002; 76:195–201.

11. Leake DS. Effects of flavonoids on the oxidation of low-density lipoproteins. In: Rice-Evans C, Packer L, eds. Flavonoids in health and disease. New York: Marcel Dekker, 1997:253–276.

12. Cren-Olivé C, Tessier E, Duriez P, Rolando C. Structure-activity relationship for the inhibition of LDL oxidation by catechin methylated metabolites and analogues. Free Rad Biol Med. In press.

13. Mangiapane H, Thomson J, Salter A, Brown S, Bell GD, White DA. The inhibition of the oxidation of low density lipoprotein by (+)-catechin, a naturally occurring flavonoid. Biochem Pharmacol 1992; 43:445–450.

14. Smith MA, Perry G, Richey PL, Sayre LM, Anderson VE, Beal MF, Kowall N. Oxidative damage in Alzheimer's. Nature 1996; 382:120–121.

15. Lee MJ, Wang ZY, Li H, Chen L, Sun Y, Gobbo S, Balentine DA, Yang CS. Analysis of plasma and urinary tea polyphenols in human subjects. Cancer Epidemiol, Biomarkers Prev 1995; 4:393–399.

16. Donovan JL, Bell JR, Kasim-Karadas S, German JB, Walzem RL, Hansen RJ, Waterhouse AL. Catechin is present as metabolites in human plasma after consumption of red wine. J Nutr 1999; 129:1662–1668.

17. Bell JR, Donovan JL, Wong R, Waterhouse AL, German JB, Walzem RL, Kasim-Karadas S. (+)-Catechin in human plasma after ingestion of a single serving of reconstituted red wine. Am J Clin Nutr 2000; 71:103–108.

18. Li L, Chan TH. Enantioselective synthesis of epigallocatechin-3-gallate (EGCG), the active polyphenol component from green tea. Org Lett 2001; 3:739–741.

19. Zaveri N. Synthesis of a 3,4,5-trimethoxybenzoyl ester analogue of epigallocatechin-3-gallate (EGCG): a potential route to the natural product green tea catechin, EGCG. Org Lett 2001; 3:843–846.

20. Van Rensburg H, van Heerden PS, Bezuidenhoudt BCB, Ferreira D. Enantioselective synthesis of the four catechin diastereomer derivatives. Tetrahedron Lett 1997; 38:3089–3092.

21. Van Rensburg H, van Heerden PS, Ferreira D. Enantioselective synthesis of flavonoids. Part 3. trans and cis-Flavan-3-ols methyl ether acetates. J Chem Soc Perkin Trans 1 1997; 3415–3421.

22. Nay B, Monti JP, Nuhrich A, Deffieux G, Merillon JM, Vercauteren J. Methods in synthesis of flavonoids. Part 2. High yield access to both enantiomers of catechin. Tetrahedron Lett 2000; 41:39049–39051.

23. Nay B, Arnaudinaud V, Peyrat JF, Nuhrich A, Deffieux G, Merillon JM, Vercauteren J. Total synthesis of isotopically labelled flavonoids, 2: ^{13}C-labelled (±)-catechin from potassium [^{13}C]cyanide. Eur J Org Chem 2000; 1279–1283.

24. Birch AJ, W. Clark-Lewis J, Robertson AV. Relative and absolute configurations of catechins and epicatechins. J Chem Soc 1957; 3586–3588.

25. Hathway DE, Seakins JWT. Autoxidation of polyphenols. Part III. Autoxidation in neutral aqueous solution of flavans related to catechin. J Chem Soc 1957; 1562–1566.

26. Yoshida K, Ikeda Y, Tsukamoto G. Japanese Patent 1982; 120584.

27. Akimoto K, Sugimoto I. Degradation of (+) cyanidanol-3 by sodium sulphite in aqueous solution. II. Reactivity of several (+) cyanidanol-3 dervatives with sodium sulphite. Chem Pharm Bull 1984; 32:3148–3154.

28. Van Dyk MS, Steynberg JP, Steynberg PJ, Ferreira D. Selective O-methylation of polyhydroxyflavan-3-ols via benzylcarbonates. Tetrahedron Lett 1990; 31:2643–2646.

29. Cren-Olivé C, Lebrun S, Hapiot P, Pinson J, Rolando C. Selective protection of catechin gives access to the intrinsic reactivity of the two phenol rings during H-abstraction and photo-oxidation. Tetrahedron Lett 2000; 41:5847–5851.

30. Cren-Olivé C, Lebrun S, Rolando C. An efficient synthesis of all mono and tri-O-methylated analogues of (+)-catechin including the major metabolites through sequential protection of catechol ring. J Chem Soc Perkin Trans 1 2002; 6:821–830.

31. Cren-Olivé C. Synthèse, Physico-Chimie et Analyse de Flavan-3-ols. Ph.D. dissertation, Université des Sciences et Technologies de Lille, Lille, 2001.

32. Sweeny JG, Iacobucci GA. Regiospecificity of (+)-catechin methylation. J Org Chem 1979; 44:2298–2299.

33. Steynberg JP, Burger JFW, Young DA, Desmond A, Brandt EV, Steenkamp JA, Ferreira D. Oligomeric flavanoids. Part 3. Structure and synthesis of phlobatannins related to (-)-Fisetinidol-(4α,6)-and (4α,8)-(+)-catechin profisetinidins. J Chem Soc Perkin Trans 1 1988; 12:3323–3329.

34. Donavan JL, Luthria DL, Stremple P, Waterhouse AL J. Analysis of (+)-catechin, (-)-epicatechin and their 3'- and 4'-O-methylated analogs: a comparison of sensitive methods. Chromatogr B 1999; 726:277–283.

35. Cren-Olivé C, Déprez S, Lebrun S, Coddeville B, Rolando C. Characterization of methylation site of flavan-3-ols by LC ESI MS/MS. Rapid Commun Mass Spectrom 2000; 14:2312–2319.

36. Kiatagrajai P, Wellons JD, Golob L, White JD J. Kinetics of epimerization of (+) catechin and its rearrangement to catechinic acid. Org Chem 1982; 47:2910–2912.

37. Kennedy JA, Munro MHG, Powell HKJ, Porter LJ, Foo Y. The protonation reactions of catechin, epicatechin and related compounds. Aust J Chem 1984; 38:885–892.

38. Cren-Olivé C, Wierulesky JM, Maes E, Rolando C. Catechin and epicatechin deprotonation followed by ^{13}C NMR. Tetrahedron Lett 2002; 43:4545–4549.

39. Cren-Olivé C, Hapiot P, Pinson J, Rolando C. Free radical chemistry of flavan-3-

ols: determination thermodynamics parameters and of kinetic reactivity from short (ns) to long (ms) time scale. J Am Chem Soc 2002; 124:14027–14038.

40. Steenken S, Neta P. One-redox potentials of phenols: hydroxy-and aminophenols and related compounds of biological interest. J Phys Chem 1982; 86:3661–3667.

41. Jovanovic SV, Steenken S, Tosic M, Marjanovic B, Simic MG. Flavonoids as antioxidants. J Am Chem Soc 1994; 116:4846–4851.

42. Jovanovic SV, Hara Y, Steenken S, Simic MG. Antioxidant potential of gallocatechins: a pulse radiolysis and laser photolysis study. J Am Chem Soc 1995; 117:9881–9888.

43. Jovanovic SV, Steenken S, Hara Y, Simic MG. Reduction potentials of flavonoid and model phenoxyl radicals: which ring in flavonoids is responsible for antioxydant activity? J Chem Soc Perkin Trans 2 1996; 11:2497–2504.

44. Jovanovic SV, Steenken S, Simic MG, Hara Y. Antioxidant properties of flavonoids: reduction potentials and electron transfer reaction reactions of flavonoid radicals. In: Rice-Evans C, Packer L, eds. Flavonoids in Health and Disease. New York: Marcel Dekker, 1997:137–161.

45. Bors W, Heller W, Michel C. The chemistry of flavonoids. In: Rice-Evans C, Packer L, eds. Flavonoids in Health and Disease. New York: Marcel Dekker, 1997:111–136.

46. Bors W, Michel C. Antioxidant capacity of flavanols and gallate esters: pulse radiolysis studies. Free Radic Biol Med 1999; 27:1413–1426.

47. Bors W, Michel C, Schikora S. Interaction of flavonoids with ascorbate and determination of their univalent redox potentials: a pulse radiolysis study. Free Radic Biol Med 1995; 19:45–52.

48. Bors W, Michel C, Stettmaier K. Electron paramagnetic resonance studies of radical species of proanthocyanidins and gallate esters. Arch Biochem Biophys 2000; 374:347–355.

49. Hodnick WF, Milosavljevic EB, Nelson JH, Pardini RS. Electrochemistry of flavonoids: relationships between redox potentials, inhibition of mitochondrial respiration, and production of oxygen radicals by flavonoids. Biochem Pharmacol 1988; 37:2607–2611.

50. Agrawal PK, Schneider HJ. Deprotonation induced 13 C NMR shifts in phenol and flavonoids. Tetrahedron Lett 1983; 24:177–180.

51. Jarosszewski JW, Matzen L, Frolund B, Krogsgaard P. Neuroactive polyamine wasp toxins: nuclear magnetic resonance spectroscopic analysis of the protolytic properties of philanthotoxin-343. J Med Chem 1996; 39:515–521.

52. Berger S. The pH dependence of phenolphthalein: a ^{13}C NMR study. Tetrahedron 1981; 37:1607–1611.

53. Haran R, Nepveu-Juras F, Laurent J.P. Les amines biogènes (phénol-et catéchol-amines): attribution des spectres de résonance ^{13}C et étude de la déprotonation. Org Magn Reson 1979; 12:153–158.

54. Maciel GE, James RV. Solvent effects on the ^{13}C chemical shift of the substituted carbon atom of phenol. J Am Chem Soc 1964; 86:3893–3894.

55. Andrieux CP, Hapiot P, Saveant JM. Fast kinetics by means of direct and indirect electrochemical techniques. Chem Rev 1990; 90:723–738.

56. Hapiot P, Pinson J, Neta P, Francesch C, Mhamdi F, Rolando C, Schneider S. Mechanism of oxidative coupling of coniferyl alcohol. Phytochemistry 1994; 36:1013–1020.

57. Hapiot P, Neudeck A, Pinson J, Fulcrand H, Neta P, Rolando C. Oxidation of caffeic acid and related hydroxycinnamic acids. J Electroanal Chem 1996; 405:169–176.

58. Goto T, Yoshida Y, Kiso M, Nagashima H. Simultaneous analysis of individual catechins and caffeine in green tea. J Chromatogr A 1996; 749:295–299.

59. Maiani G, Serafini M, Salucci M, Azzini E, Ferro-Luzzi A. Application of a new high-performance liquid chromatographic method for measuring selected polyphenols in human plasma. J Chromatogr B 1997; 692:311–317.

60. Nakagawa K, Miyazawa T. Chemiluminescence-high-performance liquid chromatographic determination of tea catechin, (-)-epigallocatechin-3-gallate, at picomole levels in rat and human plasma. Anal Biochem 1997; 248:41–49.

61. Ho Y, Lee YL, Hsu KY. Determination of (+)-catechin in plasma by high-performance liquid chromatography using fluorescence detection. J Chromatogr B 1995; 665:383–389.

62. Lee MJ, Prabhu S, Sheng S, Meng X, Li C, Yang CS. An improved method for the determination of green and black tea polyphenols in biomatrices by high-performance liquid chromatography with coulometric array detection. Anal Biochem 2000; 279:164–169.

63. Yong YL, Kwokei JNg, Shenjiang Y. Characterization of flavonoids by liquid chromatography-tandem mass spectrometry. J Chromatogr A 1993; 629: 389–393.

64. Dalluge JJ, Nelson BC, Thomas JB, Welch MJ, Sander LC. Capillary liquid chromatography/electrospray mass spectrometry for the separation and detection of catechins in green tea and human plasma. Rapid Commun Mass Spectrom 1997; 11:1753–1756.

65. Luthria DL, Jones AD, Donavan JL, Waterhouse AL. GC/MS determination of catechin and epicatechin levels in human plasma. J High Resol Chromatogr 1997; 20:621–623.

66. Cren-Olivé C, Lenoir M, Croq-Lemarrec F, Coddeville B, Salzet M, Rolando C. Characterization of metabolism site of flavan-3-ols by LC ESI MS/MS. 49th American Society for Mass Spectrometry Conference, Chicago, May 27–31, 2001.

67. Miketova P, Schram KH, Whitney JL, Kerns EH, Valvic S, Timmermann BN, Volk KJ. Mass spectrometry of selected components of biological interest in green tea extracts. J Nat Prod 1998; 61:461–467.

68. Stobneicki M, Popenda M. Flavan-3-ols from seeds of Lupinus angustifolius. Phytochemistry 1994; 37:1707–1711.

69. P.J. Linstrom and W.G. Mallard, eds., NIST Chemistry WebBook, NIST Standard Reference Database Number 69, July 2001, National Institute of Standards and Technology, Gaithersburg MD, 20899 (http://webbook.nist.gov).

6

Investigation of Flavonoids and Their In Vivo Metabolite Forms Using Tandem Mass Spectrometry

Gunter G. C. Kuhnle
King's College London
London, England

I. INTRODUCTION

The purpose of this chapter is to review different methods to investigate flavonoids and their in vivo metabolites using mass spectrometry (MS). In particular, the focus is the elucidation of the structure of in vivo flavonoid metabolite forms. Table 1 shows the structure and *m/z* values for the compounds described in this chapter.

The detection and characterization of flavonoid conjugates and metabolites are crucial for the investigation of their bioactivity. During their passage through the gastrointestinal tract into the blood circulation and potential target organs, several modifications occur through metabolism in the small intestine, the liver, and degrading enzymes of the colonic microflora. Further modifications may occur after cellular uptake. The major in vivo flavonoid conjugates and metabolites are *O*-glucuronidated, *O*-sulfated, and, in the case of catechol structures, *O*-methylated derivatives [1]. In addition, the colonic microflora generate breakdown products by ring cleavage, producing secondary metabolites such as phenolic acids with varying saturated chain lengths and other derivations. Mass spectrometric techniques provide a sensitive and powerful tool for both the detection and the structural elucidation of such metabolites.

Most methods developed for the mass spectrometric investigation of flavonoids have focused on biological material from plants, as a result of the importance of flavonoids as natural products and potential medicinal prepara-

Table 1   Structures and Names for Selected Flavones and Flavanols[a]

(i)   (ii)

	Trivial name	$[M+H^+]^+$	R_1	R_2	R_3	R_4
1	Apigenin	271	OH	H	H	H
2	Luteolin	287	OH	OH	H	H
3	Acacetin	285	OCH_3	H	H	H
4	Chrysoeriol	301	OH	OCH_3	H	H
5	Kaempferol	287	OH	H	H	OH
6	Quercetin	303	OH	OH	H	OH
7	Myricetin	319	OH	OH	OH	OH
8	Isorhamnetin	317	OH	OCH_3	H	OH
9	Chrysin	253	H	H	H	H
10	Galengin	269	H	H	H	OH
11	Kaempferid	299	OCH_3	H	H	OH
12	Eriodyctol	287	OH	OH	H	H
13	Naringenin	271	H	OH	H	H
14	Isosakuramnetin	285	H	OCH_3	H	H
15	(Epi)catechin	291	OH	OH	H	OH

[a] All structural information refers to structure i, except for number 15, (epi)catechin, which refers to structure ii.

tions. However, these compounds differ from the expected metabolites in in vivo mammalian systems in several ways: first, flavonoids occur in plants normally as glycosides with one or more sugar residues (with the exception of flavan-3-ols), and second, the glycosylation is not limited to O-linkages; indeed, C-glycosides are common.

Mass spectrometric investigations of flavonoids—especially regarding fragmentation reactions and structural elucidation—have been undertaken using electron impact (EI) and chemical ionization (CI) techniques [2–6]. Even though flavonoids and their glycosides are polar and nonvolatile, it has been possible to yield intense signals for the molecular ion of the aglycones. In contrast, it has been difficult to obtain data for glycosides, even by using derivatized compounds [7]. The introduction of soft ionization techniques, which allow the generation of intact ions of larger molecules without fragmentation, such as electrospray ionization (ESI) [8,9] and atmospheric pressure chemical ionization (APCI), in recent years increased the application of mass spectrometry to the analysis of these compounds. Using these techniques, it is possible to obtain intense signals for the quasi-molecular ions, $[M+H^+]^+$ and $[M-H^+]^-$, even for the glycosides. Furthermore, both techniques allow the hyphenation of mass spectrometric detectors with chromatographic separation devices such as high-performance liquid chromatography (HPLC/MS).

Structural information can be obtained by tandem MS experiments. The technique mainly used is low-energy collision-induced decomposition (CID MS/MS). With instruments equipped with a quadrupole ion trap (QIT) mass analyzer, several consecutive tandem MS experiments on sequential product ions can be performed (MS^n), permitting structural information from one single analysis. Characteristic fragmentation patterns can also be used for the identification of certain compounds in LC/MS.

II. MASS SPECTROMETRY OF FLAVONOIDS

There are several possibilities to investigate flavonoids and their in vivo metabolites by using the mass spectrometric methods described. The main two methods are direct infusion using a syringe and flow injection either with or without chromatographic separation. Using direct infusion, the sample solution is infused directly into the ion source of the mass spectrometer, normally by using a syringe and a syringe pump. This method allows a long and thorough investigation of the sample, including the acquisition of data for several consecutive fragmentation steps (MS^n experiments), and is therefore mainly used for structural characterization. However, this method normally requires a purified sample as a separation is not possible. Without purification, matrix effects and contaminants can lead to ion suppression effects and thereby decrease the sensitivity of the instrument. Furthermore, a large sample amount is necessary for longer investigations as the sample is normally infused with a flow rate of about 3 to 10 μL/min.

Using flow injection reduces the sample amount required and allows the separation of the sample components before the mass spectrometric analysis. However, flow injection allows only a short investigation time for each signal, which may be too short for a thorough analysis using tandem MS experiments.

A. Experimental

With respect to mass spectrometric analysis, the most interesting physicochemical properties of flavonoids are the proton affinity and pK_a values of the hydroxyl groups, as these groups are the main sites of protonation or deprotonation and therefore ion formation. Unfortunately, few data have been published so far. However, Cécile Cren-Olivé and Christian Rolando provide some data on flavan-3-ols in their chapter (see Chap. 5).

The large number of hydroxyl groups associated with flavonoids would suggest a basic environment and negative ion scan. However, many specific flavonoid structures are unstable under basic conditions and are thus likely to decompose at pH above neutrality. The decomposition of these compounds mainly involves the formation of quinonic structures from the catechol group at the B-ring. For this reason, a neutral or—better—acidic environment is normally necessary.

1. Ion Polarity

Flavonoids can be detected in both positive and negative ion modes, even under acidic conditions. Whereas the positive ion mode often generates higher yields, the noise level is lower in the negative ion mode, thus improving the quality of the signals. Furthermore, the fragmentation pathway can be influenced by the ion polarity [10], and it has been reported that phenolic compounds show less fragmentation in negative mode than in positive ion mode [11]. Thus, using positive ions can be advantageous for structure elucidation, whereas the negative ion mode is advantageous for the detection of compounds. Investigations show that the optimal ionization polarity depends very much on the compound used. For this reason, preliminary investigations regarding the polarity used are important.

2. Atmospheric Pressure Chemical Ionization and Electrospray Ionization

The most commonly applied ionization methods for LC-MS are electrospray ionization (ESI) and atmospheric pressure chemical ionization (APCI). In ESI, the ions are considered to be preformed in solution [12] and subsequently extracted in the spray; in APCI the molecules become ionized inside the source by using a corona discharge. In ESI, the sample is sprayed into a mist composed of small charged droplets by using a high voltage and an assisting nebulizing gas. In APCI sources, the sample is evaporated by using high temperatures of up to 600°C and ionized by using a corona discharge. Though flavonoids and their metabolites are unstable at high temperatures, the high temperatures in the APCI source do not seem to cause any damage to these compounds. Main advantages of the APCI source over the ESI source are the increased range for flow rates and the potential to obtain ions from aqueous solutions even at flow rates well above 1 mL/min.

However, there are no studies comparing the best conditions for the investigations of flavonoids using APCI or ESI. Investigations using pesticides

[13] showed that the response of the compounds depends largely on their chemical properties: i.e., APCI in the positive mode was more sensitive for basic nonionic compounds and positive ESI was more sensitive for positively charged ones. For acidic compounds, negative ESI proved to be the best ionization method. In general, it proved that compounds working well with positive APCI also worked well with positive ESI, but not necessarily vice versa. However, these data cannot be used unconditionally for the investigation of flavonoids and their metabolites, and further investigation is required.

3. Adduct Formation

Depending on the solvent used, the formation of adducts, for example, $[M+Na^+]^+$ or $[M+formiate^+]^+$ is frequently observed in the positive ion mode. For a larger abundance of the ion under investigation, but also for an accurate quantification, it is often necessary to remove those adducts. This can be done either by using in source fragmentation or by increasing either the transfer capillary temperature or the intensity of the drying gas.

B. Fragmentation

1. Fragment Nomenclature

To obtain structural information, compounds can be fragmented by several techniques. The technique mainly employed with soft ionization techniques is low-energy collision-induced dissociation (CID). To describe the fragments generated, the nomenclature proposed by Ma and colleagues [14] is generally used. Compared with the system introduced by Marby and Markham [7], this nomenclature is conceptually more similar to that commonly used for the description of carbohydrate fragmentation in CID spectra of glycoconjugates [15]. In this nomenclature, $^{i,j}A^+$ and $^{i,j}B^+$ labels are used to designate primary product ions containing intact A- and B-rings. The superscripts ^{i} and ^{j} indicate the bonds of the C-ring that have been cleaved (Fig. 1). Further fragmentations of the A- or B-ring products, in particular the loss of small molecules like H_2O or CO_2, are indicated by the combined use of the respective fragment and the lost molecules (e.g., $[^{0,2}A^+-H_2O]^+$).

2. Fragmentation Mechanisms and Major Fragments

With respect to the elucidation of unknown metabolites, the most useful fragments are those involving the cleavage of two bonds in the C-ring, in particular 1/3, 0/2, and 0/4 [14], leading to fragments shown in Fig. 2. Protonation on either the ether oxygen atom in the C-ring or the C-3 atom leads to two different quasi-molecular ions, which can both undergo a retro-

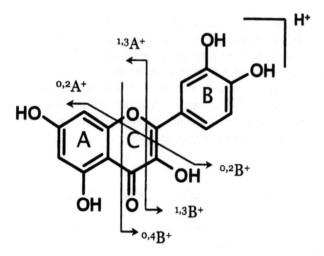

Figure 1 Nomenclature and selected product ions—derived from *retro-Diels-Alder* reaction - of protonated compounds under low-energy CID as proposed in Ref. 14. In ^{ij}X$^+$, the superscript number (^{ij}) denotes the C-ring bond cleaved, the letter **X** describes the ring remaining in the fragment.

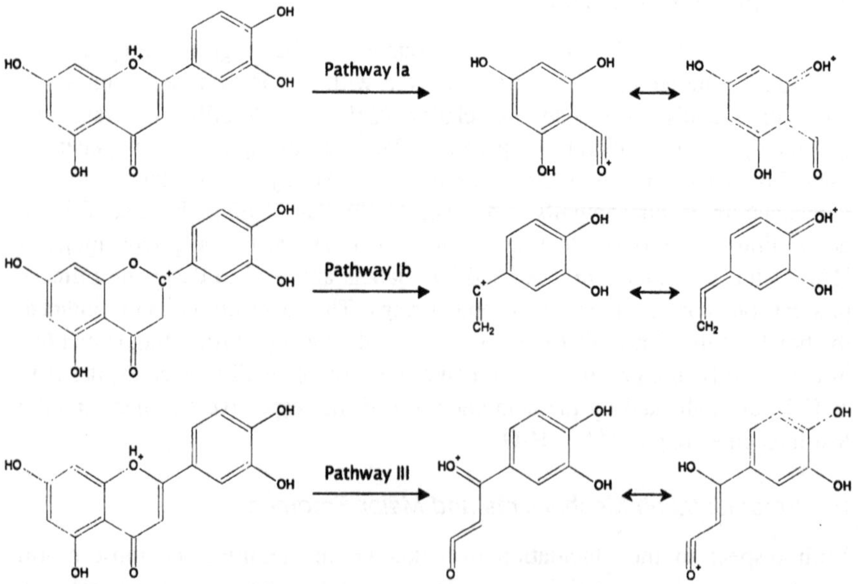

Figure 2 Mechanisms proposed for the fragmentation of flavonoid [M+H$^+$]$^+$ ions to generate ^{1,3}A$^+$ and ^{1,3}B$^+$ ions [14] via a *retro-Diels-Alder* reaction. An alternative is pathway III leading to a ^{2,4}B$^+$ fragment, also via a *retro-Diels-Alder* reaction.

Diels-Alder (RDA) reaction, leading to different fragmentation products ($^{1,3}A^+$ and $^{1,3}B^+$ ions, pathway Ia and Ib in Fig. 2). An alternative to this is pathway III, resulting in $^{0,4}B^+$ ions, which can undergo a further fragmentation by the loss of water, leading to $[^{0,4}B^+-H_2O]^+$. This pathway seems to be specific for CID as it has not been observed under EI conditions [7]. $^{0,2}A^+$ and $^{0,2}B^+$ ions can be formed after a proposed pathway II. Protonation at the C-3 and C-2 position may lead to a cleavage of bonds 0 and 2 in the C-ring and to the formation of these ions (Fig. 3) [14]. Other fragments include the loss of small molecules such as H_2O, CO, or CO_2. Table 2 and Table 3 show major fragments resulting from retro-Diels-Alder reactions and their intensities for selected compounds in positive and negative ion modes, which can be used for structure assignment and identification of these compounds and their metabolites [14,16].

Useful fragments for interpreting structures are those including only one possible site for metabolic modification of the A- or B-ring, showing the actual site of modification. However, few fragments are described with these properties, for example, an $[M-H^+-CO_2-CO]^-$ fragment of luteolin (2) after loss of the 5-hydroxyl group [16]. But most fragments described do not reveal the actual site of modification within the respective ring.

3. Fragmentation of In Vivo Metabolites

A main group of in vivo metabolites of flavonoids are conjugates with glucuronic acid, which demonstrate behavior in tandem MS experiments similar to that of other O-glycosides. O-Glycosides normally show the aglycone as an

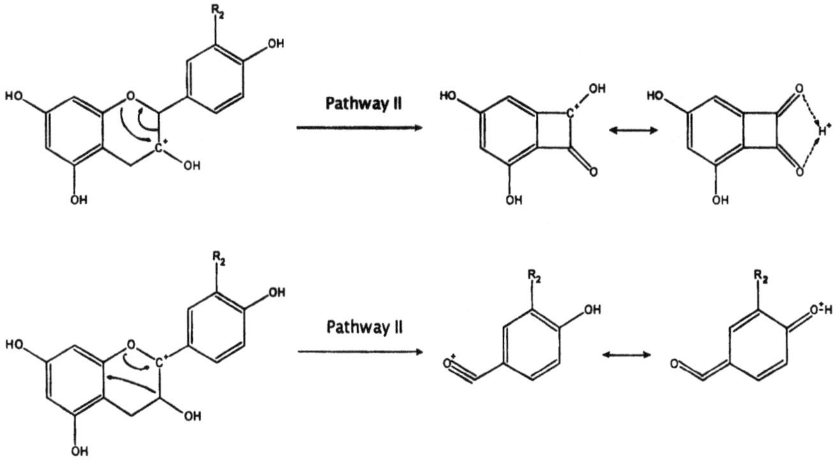

Figure 3 Mechanism proposed for the fragmentation of flavonoid $[M+H^+]^+$ ions leading to $^{0,2}A^+$ and $^{0,2}B^+$ ions, as proposed in Ref. 14.

Table 2 Principal Product Ions Including an Intact A- or C-Ring with Their Relative Intensities (Base Peak Equals 100) for Selected Flavonoids Generated from $[M+H^+]^+$ Ions by Low-Energy Collision-Induced Dissociation[a]

Compound	1	2	3	4	5	6	7	8
Base peak (100%)	$^{1,3}A^+$	$^{1,3}A^+$	$[M+H]^+$ $-CH_3]^+$	$[M+H]^+$ $-CH_3]^+$	$^{1,3}A^+$	$[M+H]^+$ $-CO]^+$	$[M+H^+ -H_2O$ $-CO]^+$	$[M+H]^+$ $-CH_3]^+$
$^{1,3}A^+$	153 (100)	153 (100)	153 (9.7)	153 (3.0)	153 (100)	153 (39)	153 (63)	153 (4.4)
$^{0,4}B^+$	163 (7)	179 (10)	177 (0.6)	-				
$[^{0,4}B+H_2O]^+$	145 (12)	161 (10)	159 (2.4)	175 (0.7)				
$^{0,2}B^+$	121 (10)	137 (9)	135 (1.1)	151 (0.9)	121 (8.1)	137 (4.8)	153 (39)	151 (5.3)
$^{1,3}B^+$	119 (24)	135 (21)	133 (4.3)	149 (1.0)				
$^{0,2}A^+$					165 (90)	165 (73)	165 (8.1)	165 (6.8)
$[^{0,2}A-CO]^+$					137 (20)	137 (48)	137 (12)	137 (1.4)
$[^{1,3}B-2H]^+$					133 (24)	149 (21)	165 (81)	163 (2.2)
$[^{1,4}A+2H]^+$					127 (14)	127 (11)	127 (11)	127 (1.4)
$[^{1,3}A-C_2H_2O]^+$					111 (22)	111 (16)	111 (20)	111 (1.4)

[a] See Table 1 for details.
Source: Ref. 14.

Table 3 Principal Product Ions Including Intact A- or C-Ring with Intensities (Base Peak Equals 100) for Selected Flavonoids Generated from $[M-H^+]^-$ Ions by Low-Energy Collision-Induced Dissociation[a]

Compound	1	2	5	6	9	10	11	12	13	14
Base Peak	$[M-H^+ -CO_2]^-$	$[M-H^+]^-$	$[M-H^+]^-$	$^{1,2}A^-$	$[M-H^+]^-$	$[M-H^+]^-$	$^{1,3}A^-$	$^{1,3}A^-$	$^{1,3}A^-$	$[M-H^+ -C_2H_2O]^-$
$^{1,2}A^-$				179 (100)						
$^{1,3}A^-$	151 (10)	151 (4)	151 (1)				151 (100)	151 (100)		151 (12)
$[^{1,4}B-2H]^-$	149 (36)									
$[^{1,3}A-CO_2]^-$	107 (1)							107 (1)	107 (3)	107 (1)
$^{1,3}B^-$	117 (1)	133 (1)					132 (1)	135 (3)		
$[^{1,2}A-CO]^-$				151 (67)						
$^{1,2}B^-$				121 (1)						
$[^{1,2}A-CO-CO_2]^-$				107 (1)			107 (4)			
$^{1,4}A^-$								125 (1)		125 (1)

[a] See Table 1 for details.
Source: Ref. 16.

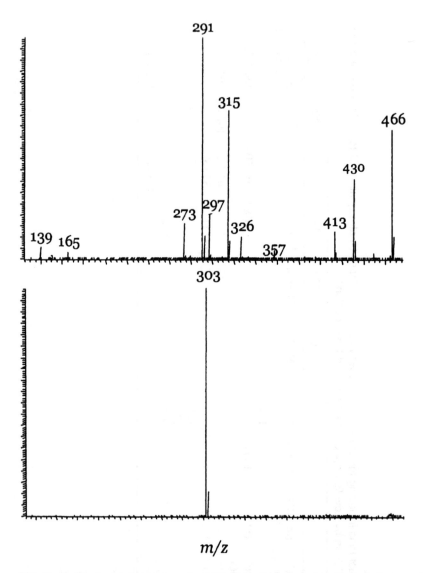

Figure 4 Product ion spectra of epicatechin-*O*-glucuronide (top) and quercetin-*O*-glucuronide. Whereas the spectrum for quercetin-*O*-glucuronide only shows the neutral loss of glucuronic acid (*m/z* 303), the spectrum for epicatechin-*O*-glucuronide shows—apart from the loss of water and other small molecules—several characteristic fragments like [^{1,3}A$^+$+glucuronic acid]$^+$ (*m/z* 315).

intense fragment ion. However, O-glycosides with more than one sugar residue can undergo a rearrangement with an internal sugar residue loss [17]. Mono-glycosides often show a fragmentation pattern similar to that of their aglycone, indicating that it is also possible to elucidate structures from the glycosides. The 5- and 7-O-glucuronidated epicatechin exhibits a fragmentation pattern very similar to that of the aglycone, showing similar fragments including the metabolic modification (e.g., [^{1,3}A$^+$+glucuronic acid]$^+$). However, other compounds produce a completely different product ion spectrum; for example, the product ion spectrum of the quercetin-O-glucuronide exhibits the neutral loss of only glucuronic acid, but no other fragments, even though they are present in the product ion spectrum of the aglycone (Fig. 4).

O-Methylated compounds often show the neutral loss of the methyl group, however, not all flavonoids behave in this manner. But—in contrast to the O-glucuronides—most O-methylated compounds generate several product ions, which allow identification and structure elucidation.

However, the influence of a modification of the aglycone, for example, by O-methylation or O-glucuronidation, must not be underestimated, as this has an influence on the fragmentation mechanisms and thereby the abundance of certain fragment ions.

III. STRUCTURE ELUCIDATION

Structural elucidation of metabolites focuses mainly on the kind and the site of modification. The main conjugations expected for in vivo metabolites are O-methylation, O-glucuronidation, and O-sulfation. They—and possible combinations—can be distinguished easily by the mass they add to the original molecule. In contrast to this, the detection of the site of modification is more difficult as most product ions still have an intact A- or C-ring, which is the main site of metabolism.

To elucidate the structure of flavanoid metabolites and investigate metabolic processes, tandem MS experiments (fragmentations) can be performed. O-Glucuronidated compounds normally exhibit a neutral loss of glucuronic acid, showing an intense signal for the aglycone. In contrast to this, O-methylated compounds often show only a very small signal for the original (unmethylated) compound. Table 4 gives a compilation of product ions and their intensities for some possible flavonoid metabolites. Metabolites in general often exhibit a fragmentation pattern similar to that of the original compounds, generating [^{i,j}**A**$^+$+**glucuronic acid**]$^+$ or [^{i,j}**B**$^+$+**CH₃**]$^+$ ions, for example, epicatechin-5O-β-D-glucuronide shows an intense signal at m/z 315, corresponding to the glucuronidated RDA product of the A-ring ([^{1,3}A$^+$+glucuronic acid]$^+$). Comparing these ions with the corresponding product ions of standard compounds can reveal a lot

Table 4 Molecular Ion and Major Fragments (Intensity > 10%) and Their
Intensities (Percentage of the Intensity of the Most Abundant Signal) of
Flavonoid Metabolites, Acquired by Direct Infusion by Using Electrospray Mass
Spectrometry and Low-Energy Collision-Induced Dissociation

Compound	$[M+H^+]^+$	Major fragments
Kaempferol-O-β-D-glucuronide	9	$[M-\text{Glucuronic acid}+H^+]^+$ (100)
Naringenin-O-β-D-glucuronide	8	$[M-\text{Glucuronic acid}+H^+]^+$ (100)
Hesperetin-O-β-D-glucuronide	30	$[M-\text{Glucuronic acid}+H^+]^+$ (100)
		$[M-H_2O+H^+]^+$ (30)
O-Methylcatechin	5	$[M-CH_3+H^+]^+$ (<1) $[M-H_2O+H^+]^+$ (12)
		$[^{1,2}B^++CH_3]^+$ (100) $^{1,3}A^+$ (28)
		$[^{1,5}B^++CH_3]^+$ (30)
Epicatechin-O-β-D-glucuronide	3	$[M-\text{Glucuronic acid}+H^+]^+$ (100)
		$[^{0,5}A^++\text{Glucuronic acid}]^+$ (35)
3′-O-Methylquercetin [14]		$[M-CH_3+H^+]^+$ (<1)

Source: Data from the authors unless indicated otherwise.

of information concerning the structure of the metabolites under investigation.
However, it remains difficult to discriminate between a modification within the A-
or the C-ring, respectively (for example, at the 5- or 7-hydroxy group), as these
rings appear to be very stable under the CID conditions applied; for example, the
binding moiety of glucuronic acid on quercetin could not be resolved by using a
tandem MS experiment [18]. Thus, the main differences between spectra are the
intensities for product ions of the respective compounds, as the modification
influences the stability of the fragment ions and thus their abundance. By having
standards available, it is often possible to compare the intensities of the respective
fragments to obtain further information concerning the actual site of modification.
For example, the comparison of the mass spectra for 3′-O-methylated and 4′-O-
methylated epicatechin show differences in the intensity for several product ions,
especially for the RDA product $^{1,3}A^-$ (see Fig. 5). The probable reason for the
higher intensity of this product is that the methylation prevents the 4′-hydroxyl
group from being deprotonized; thus the molecule becomes ionized on a
different site, leading to the altered fragmentation pattern (see Chap. 5).

   Complexes of flavonoids—for example, with transition metals—show a
fragmentation pattern different from that of the fragmentation pattern of $[M+H^+]^+$
or $[M-H^+]^-$ ions. For this reason using those complexes may be a solution for
this problem as they provide additional information regarding the structure of the
compound under investigation. Sodium complexes $[M+Na^+]^+$ generate a more
complex fragmentation pattern, but consistent pathways for the determination of
glycosylation sites have not yet been found. Ternary complexes with cobalt(II)

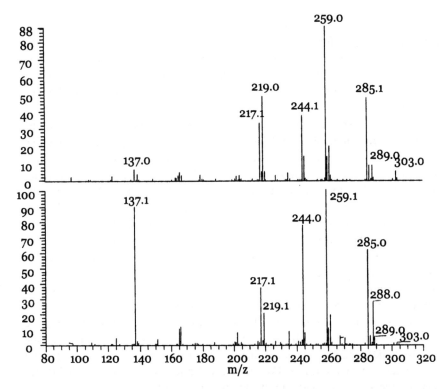

Figure 5 Product ion spectra of 3'-O-methylated (top) and 4'-O-methylated epicatechin, acquired in negative ion mode. The base peak (m/z 259) corresponds with the loss of CO_2. The main difference between both spectra is the intensity of the $^{1,2}A^-$ ion (m/z 137) which is much more abundant for the 4'-methylated compound.

and 2,2'-bipyridine of flavonoids have, however, proved to be very useful for the differentiation between glycoside attachment positions (3 vs. 5 position), as they generate a different fragmentation patterns [19]. This could also be useful to differentiate between different conjugates. Furthermore, complexation of flavonoids with transition metals increases the signal intensity by about one order of magnitude [20].

IV. LIQUID CHROMATOGRAPHY–MASS SPECTROMETRY

A. Acquisition Modes

Using LC-MS as an analytical tool for the identification of metabolites can be regarded as a process of structural elucidation of potential metabolites by interpretation of product ion mass spectra of selected precursor ions, guided by concepts of biotransformation pathways [21]. Using an MS detector allows the application of the information acquired by structure elucidation to the detection of these compounds in biological samples. The capabilities of this technique allow a selective and sensitive detection of the compounds (and their metabolites) under investigation.

The basic experiment that can be performed is a full ion scan during the complete chromatographic run to look for the original compound and possible metabolites by their calculated mass. As there is a large number of possible metabolites, searching manually for possible products can be very time-consuming. Therefore, metabolic profiling software such as Metabolite ID, which searches for all possible metabolites, can be very helpful. In addition, these programs often include sophisticated background subtraction algorithms, which reduce the noise resulting from the biological matrix (Fig. 6).

In spite of the abilities of a full scan to detect all metabolites in a sample, it does have major disadvantages: first, the detected mass is often insufficient for the definite identification of a compound; even together with a UV spectrum and the HPLC retention time, there is still some uncertainty regarding multiple mod-ifications or isomerization. Second, the signal-to-noise ratio is much higher in the full scan mode—even with noise reduction algorithms—thus decreasing the sensitivity. Depending on the mass analyzer used, different recording modes such as selected ion monitoring (SIM) or selected reaction monitoring (SRM) can increase the sensitivity and improve the results remarkably.

In the SIM mode, the mass spectrometer monitors only the signal for a single mass-to-charge ratio (m/z). This recording mode is more selective and therefore provides a higher signal-to-noise ratio. In instruments that apply a mass filter as analyzer—such as quadrupole instruments—this also directly increases the sensitivity. While these analyzers scan through the desired mass range for each full scan, monitoring each m/z only for a short time, they can spend much more time if they focus on a single ion, thus increasing the sensitivity. However, this advantage applies only for the kind of instruments in which only the selected ion(s) can reach the detector and trigger a signal. Other types of analyzers, such as ion traps or time-flight-mass spectrometers (TOF), can provide only a "simulated SIM." In normal ion traps, there is almost no advantage in using SIM, as the sensitivity of the instrument is still determined by the overall amount of ions trapped, regardless their m/z. As in the first step, all ions are trapped; a contaminant ion with a high intensity can still suppress the signal of the ion(s) under investigation, even though it does not appear in the spectrum. There are, however, techniques that lead to an improved sensitivity for SIM with quadrupole ion traps (QITs), for example, selected ion storage (SIS), in which selected ions are stored inside the trap, or the application of an RF voltage to assure that only selected ions are stored in the trap. With these

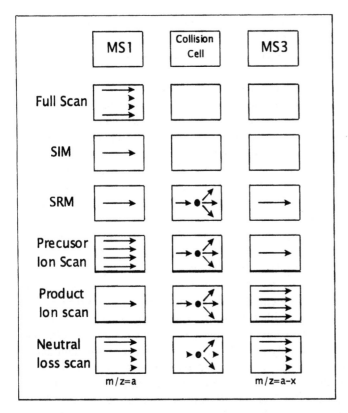

Figure 6 Acquisition modes in tandem MS with two mass analyzers. Multiple arrows indicate a full scan, a single arrow a single ion scan (mass filter fixed on a single *m/z*). In the neutral loss scan, a full scan with a constant mass difference is performed with both analyzers, selecting only ions which show a neutral loss. (Graphic adapted from Ref. 23.)

techniques, the advantage of SIM in quadrupole instruments can also be achieved with QIT.

SRM is an extended SIM mode, which uses the tandem MS capabilities of the instrument. As with SIM, ions of a certain *m/z* are selected and isolated, but these ions are fragmented, normally by collision-induced dissociation (CID). On the product ions formed, a selected ion monitoring is performed. Selecting specific fragments, this technique provides a very high specificity and thus a high signal-to-noise ratio, thereby increasing the sensitivity. In triple quadrupole instruments, selecting the parent ion with the first quadrupole and performing either a SIM or a full scan with the third quadrupole performs this. Performing a full scan on the product ions (product ion scan) provides fragment ion spectra for the compounds under investigation with the appropriate mass that allows structure elucidation.

A special form of this technique is the scan for neutral losses, which is a speciality of triple quadrupole instruments and only available with these instruments. Neutral losses occur frequently during CID of metabolites. For example, epicatechin-5-*O*-glucuronide loses glucuronic acid as a "neutral" during CID, with epicatechin as detectable ion. Therefore, neutral loss scans are an important and valuable tool for the investigation of metabolites. In this scan mode, both quadrupoles perform a full scan with a fixed mass-to-charge ratio difference: i.e., the second mass filtering quadrupole is always set to a mass-to-charge ratio lower than the settings of the first quadrupole, corresponding to the anticipated neutral loss. Thus, only ions that lose an appropriate "neutral," for example, glucuronic acid, are detected.

Another specialty of triple quadrupole instruments is the parent ion scan, which is the reverse form of a product ion scan. Whereas in a product ion scan the first quadrupole is fixed to a certain mass-to-charge ratio and the postfragmentation chamber quadrupole is either fixed or in scanning mode, the opposite arrangement is used for the parent ion scan: i.e., all ions undergo a collision-induced fragmentation, but only ions generating a fragment of a certain mass-to-charge ratio generate a signal. Applying a parent ion scan for a characteristic fragment, it is possible to detect yet unknown metabolites. For example, a precursor ion scan for the $^{1,3}A^+$-ion, the RDA product of epicatechin (*m/z* 139), can reveal all epicatechin metabolites with an unmodified A-ring [22]. This kind of acquisition can also be performed with a single quadrupole instrument by using the in-source fragmentation and monitoring a selected fragment by using SIM. But, in contrast to triple quadrupole instruments, it is not possible to determine the mass of the precursor ion with a single quadrupole instrument.

In contrast to the full scan, these acquisition modes require more information on the compounds under investigation. Especially for SRM mode acquisitions, information on the product ion(s) of the respective compounds is [16] necessary to perform an analysis. These techniques provide higher sensitivity, but information can be lost as unexpected metabolites remain undetected. Therefore, these methods are suitable only for the detection and quantification of known (or at least predicted) metabolites, even though a precursor ion scan or neutral loss scan can detect a large range of possible metabolites. However, as some metabolic reactions lead to an altered fragmentation pattern, there is still the chance that some products are missed.

Whereas a full scan reveals only the mass of a compound in the sample under investigation, a product ion scan is normally limited to a certain number of precursor ions, as a large number of simultaneously conducted product ion scans decreases the sensitivity. Sophisticated acquisition software allows "dependent" tandem MS experiments by selecting certain ions from a full scan (either the most intense ion from a list or from all ions in a full scan), thus providing product ion spectra for the most abundant ions in a sample. This is an advantage for the characterization of metabolites, as a large number of product ion spectra can be acquired without any effect on the sensitivity. Nevertheless, there are also disadvantages, especially with biological samples. Having a high background— either by a biological matrix or sample contamination—may confuse the ion selection algorithm and thereby prevent the fragmentation of the appropriate ions. For this reason, it is necessary either to remove all contaminations and background ions from the sample or to use algorithms to prevent unwanted ions from triggering the acquisition algorithm.

V. CONCLUSIONS

Mass spectrometry provides a very powerful and versatile tool for the investigation of flavonoid in vivo metabolites. This technique allows both structural characterization with tandem MS experiments and the detection of metabolites by using LC/MS.

Characteristic fragments, mainly deriving from a retro-Diels-Alder reaction within the C-ring, allow the elucidation of the structure of the metabolites. Additional information, such as proton affinity or pK_a values, allows one to distinguish between different isomers, such as 3'- and 4'-O-methylated compounds. The capabilities of quadrupole or quatrupole ion trap (QIT) instruments allow the detection of even small amounts of the metabolites in biological samples.

ACKNOWLEDGMENT

I am grateful to Anna Przyborowska for her collaboration.

REFERENCES

1. Kuhnle G, Spencer JPE, Schroeter H, Shenoy B, Debnam E, Srai S, Rice-Evans C, Hahn U. Epicatechin and catechin are O-methylated and glucuronidated in the small intestine. Biochem Biophys Res Commun 2000; 277(5):507–512.
2. Porter QN, Baldas J. Mass Spectrometry of Heterocyclic Compounds. New York: Wiley Interscience, 1971.

3.   Kingston DGI, Fales HM. Methane chemical ionization mass spectrometry of flavonoids. Tetrahedron 1973; 29:4083–4086.

4.   Kingston DGI. Mass spectrometry of organic compounds. VI. Electron-impact spectra of flavonoid compounds. Tetrahedron 1971; 27:2691–2700.

5.   Pelter A, Stainton P, Barber M. The mass spectra of oxygen heterocylces. II. The mass spectra of some flavonoids. J Heterocylc Chem 1965; 2:262–271.

6.   Grayer RJ. Flavonoids. Harborne JB, ed. In: Methods in Plant Biochemistry, London: Academic Press, 1989:283–323.

7.   Marby TJ, Markham KR. Mass spectrometry of flavonoids, In Harborne JB, Mabry TJ, Mabry H, eds. The Flavonoids. New York: Academic Press, 1975:p 79–126.

8.   Yamashita M, Fenn JB. Electrospray ion source: another variation of the free-jet theme. J Phys Chem 1984; 88:4451–4459.

9.   Aleksandrow ML, Gall LN, Krasnov NV, Nikolaev VI, Shkurow VA. Mass spectrometric analysis of thermally unstable compounds of low volatility by extraction of ions from solution at atmospheric pressure. J Anal Chem USSR 1985; 40:1227–1236.

10. Cuyckens F, Rozenberg R, de Hoffman E, Clacys M. Structure characterization of flavonoid O-diglycosides by positive and negative nano-electrospray ionization ion trap mass spectrometry. J Mass Spectrom 2001; 36(11):1203–1210.

11. McDonald S, Prenzler PD, Antolovich M, Robards K. Phenolic content and antioxidant activity in olive extracts. Food Chem 2001; 73:73–84.

12. Kebarle P, Tang L. From ions in solution to ions in the gas phase. Anal Chem 1993; 65(22):972A–986A.

13. Thurman EM, Ferrer I, Barcelo D. Choosing between atmospheric pressure chemical ionization and electrospray interfaces for the HPLC/MS analysis of pesticides. Anal Chem 2001; 73(22):5441–5449.

14. Ma Y-L, Li QM, Heuvel HVD, Claeys M. Characterization of flavone and flavonol aglycones by collision-induced dissociation tandem mass spectrometry. Rapid Commun Mass Spectrom 1997; 11:1357–1364.

15. Domon B, Costello C. Glycoconjugate. 1988; 5:397.

16. Fabre N, Rustan I, Hoffmann ED, Quetin-Leclercq J. Determination of flavone, flavonol, and flavanone aglycones by negative ion liquid chromatography electrospray ion trap mass spectrometry. J Am Soc Mass Spectrom 2001; 12(6):707–715.

17. Ma Y-L, Vedernikowa I, Heuvel HVD, Clayes M. Internal glucose residue loss in protonated O-diglycosyl flavonoids upon low-energy collision-induced dissociation. J Am Soc Mass Spectrom 2000; 11(2):136–144.

18. Wittig J, Hederich M, Graefe EU, Veit M. Identification of quercetin glucuronides in human plasma by high-performance liquid chromatography-tandem mass spectrometry. J Chromatogr B, 2001; 753:237–243.

19. Satterfield M, Brodbelt JS. Structural characterization of flavonoid glycosides by collisionally activated dissociation of metal complexes. J Am Soc Mass Spectrom 2001; 12(5):537–549.

20. Satterfield M, Brodbelt JS. Enhanced detection of flavonoids by metal complexation and electrospray ionization mass spectrometry. Anal Chem 2000; 72(24):5898–5906.

21. Bu H-Z, Poglod M, Micetich RG, Khan JK. Novel sample preparation method facilitating identification of urinary drug metabolites by liquid chromatography-tandem mass spectrometry. J Chromatogr B 2000; 738:259–265.

22. Zeeb DJ, Nelson BC, Albert K, Dalluge JJ. Separation and identification of twelve catechins in tea using liquid chromatography/atmospheric pressure chemical ionization-mass spectrometry. Anal Chem 2000; 72(20):5020–5026.

23. Griffiths WJ, Jonsson AP, Liu S, Rai DK, Wang Y. Electrospray and tandem mass spectrometry in biochemistry. Biochem J 2001; 355:545–561.

7

Effects of Flavonoids on the Oxidation of Low-Density Lipoprotein and Atherosclerosis

Michael Aviram and Bianca Fuhrman
Rambam Medical Center
Haifa, Israel

I. INTRODUCTION

Atherosclerosis is the leading cause of morbidity and mortality among people with a Western life-style. The early atherosclerotic lesion is characterized by the accumulation of arterial foam cells derived mainly from cholesterol-loaded macrophages [1,2]. Most of the accumulated cholesterol in foam cells originates from plasma low-density lipoprotein (LDL), which is internalized into the cells via the LDL receptor. Native LDL, however, does not induce cellular cholesterol accumulation, because the LDL receptor activity is downregulated by the cellular cholesterol content [3,4]. LDL has to undergo oxidative modification in order to be taken up by macrophages at an enhanced rate via the macrophage scavenger receptor pathway, which, unlike the LDL receptor, are not subjected to downregulation by cellular cholesterol [5–7]. The oxidative modification hypothesis of atherosclerosis proposes that LDL oxidation play a pivotal role in early atherogenesis [8–16]. This hypothesis is supported by evidence that LDL oxidation occurs in vivo [13,17] and contributes to the clinical manifestation of atherosclerosis. The uptake of oxidized LDL (Ox-LDL) via scavenger receptors promotes cholesterol accumulation and foam cell formation [5,7,16,18]. In addition, Ox-LDL atherogenicity is related to recruitment of monocytes into the intima [19], to stimulation of monocyte adhesion to the endothelium [20], and to cytotoxicity to arterial cells [21,22].

II. MECHANISMS OF LDL OXIDATION

A. Oxidation of LDL in a Cell-Free System

Oxidation of LDL involves free radical attack on the lipoprotein components, including cholesterol, phospholipids, fatty acids, and apolipoprotein B-100. LDL oxidation results, first, in the consumption of its antioxidants (mainly vitamin E and carotenoids) then in substantial loss of polyunsaturated fatty acids and of cholesterol, which is converted to oxysterols. A predominant oxysterol formed at early stages of oxidation is 7-hydroperoxycholesterol, and at later stages, 7-ketocholesterol is formed [23]. Both of these oxysterols are formed as a result of an oxygenation at the 7-position. The polyunsaturated groups of the esterified cholesteryl esters and of phospholipids are also major targets for oxidation. The primary products formed are hydroperoxides, which can undergo subsequent reduction to hydroxides and aldehydes. Nonenzymatic peroxidation of arachidonic acid results in the formation of isoprostanes and epoxyisoprostanes [24]. In the presence of transition metal ions, acyl hydroperoxides also undergo carbon-carbon bond cleavage to form reactive short-chain aldehydes.

During oxidation of LDL, apolipoprotein B also undergoes direct and indirect modifications. Direct attack of oxidants can oxidize amino acid side chains and fragment the polypeptide backbone. Reactive lipid peroxidation products, such as short-chain aldehydes, can form stable adducts with amino acid residues in the apolipoprotein B-100 [25], which can then lead to intermolecular cross-linking and to aggregation of lipoprotein particles [26]. Enrichment of LDL with lipid hydroperoxides appears to be an important first step in LDL oxidation. After depletion of LDL antioxidants, transition metal ions catalyze propagation reactions, breakdown of lipid hydroperoxides, and formation of reactive products, such as malondialdehyde and hydroxynonenal, which are responsible for apolipoprotein B-100 modification. All these reactions result in changes in the LDL structure, and the oxidatively modified LDL, which can no longer bind to the LDL receptor, interacts with the macrophage scavenger receptors, leading to the accumulation of cholesterol and oxidized lipids and to foam cell formation.

B. Macrophage-Mediated Oxidation of LDL

The process of LDL oxidation is unlikely to occur in plasma because plasma contains high concentrations of antioxidants and of metal ion chelators. It is more likely to occur within the artery wall, an environment, that is depleted of antioxidants and therefore where the LDL is exposed to oxidative stress. The identity of the cells responsible for the oxidation of LDL along atherogenesis in the arterial wall is uncertain. Monocyte-derived macrophages are likely candidates to induce the oxidation of LDL during early atherogenesis, because they are prominent in arterial lesions and because they generate reactive oxygen and nitrogen [27,28]. Macrophage-mediated oxidation of LDL is considerably affected by the oxidative state in the cells, which depends on the balance between cellular oxygenases and macrophage-associated antioxidants [29]. Macrophage binding of LDL to the LDL receptor initiates the activation of cellular oxygenases [30,31]. LDL oxidation by arterial wall cells was suggested to involve the activation of macrophage 15-lipoxygenase and of nicotinamide-adenine dinucleotide phosphate (NADPH) oxidase [31,32]. When NADPH oxidase is activated, the cytosolic components of the NADPH oxidase complex, P-47 and P-67, translocate to the plasma membrane, where they form, together with the membrane-bound cytochrome b558, the active NADPH oxidase complex. Both phospholipase A_2, and phospholipase D can induce macrophage NADPH oxidase–dependent oxidation of LDL [33]. On the other hand, macrophage antioxidants also contribute to the extent of cell-mediated oxidation of LDL. Cellular reduced glutathione (GSH) is a most potent antioxidant [34,35], and an inverse relationship was shown between the extent of macrophage-mediated oxidation of LDL and the cellular reduced glutathione content [35]. Macrophage-mediated oxidation of LDL can also result from an initial cellular lipid peroxidation. When cultured macrophages were exposed to ferrous ions, cellular lipid peroxidation took place [36,37]. These "oxidized macrophages" could easily oxidize the LDL lipids, even in the absence of additional transition metal ions. LDL oxidation by oxidized macrophages can also result from the transfer of peroxidized lipids from the cell membranes to the LDL particle.

C. Paraoxonase and LDL Oxidation

Human serum paraoxonase (PON 1) is an esterase that is physically associated with high-density lipoprotein (HDL) and is also distributed in tissues such as liver, kidney, and intestine [38,39]. Activities of PON 1, which are routinely measured, include hydrolysis of organophosphates, such as paraoxon (the active metabolite of the insecticide parathion); hydrolysis of arylesters, such as phenyl acetate; and lactonase activities. Human serum paraoxonase activity has been shown to be inversely related to the risk of cardiovascular disease [40,41], as shown in atherosclerotic, hypercholesterolemic, and diabetic patients [42–44]. In 1998 HDL-associated PON 1 was shown to protect LDL, as well as the HDL particle itself, against oxidation induced by either copper ions or free radical generators [45,46], and this effect could be related to the hydrolysis of the specific lipoproteins' oxidized lipids such as cholesteryl linoleate hydroperoxides and oxidized phospholipids. Protection of HDL from oxidation by PON 1 was shown to preserve the antiatherogenic effect of HDL in reverse cholesterol transport, as shown by its beneficial effect on HDL-mediated macrophage cholesterol efflux [45]. These effects of PON 1 may be relevant to its beneficial properties against cardiovascular disease [40,41]. Antioxidants were shown to preserve PON 1 activity as they decrease the formation of lipid peroxides that can inactivate PON 1 [47].

III. FLAVONOIDS INHIBIT LDL OXIDATION

Clinical studies investigated the antioxidative effects of antioxidant supplementation of humans on ex vivo LDL oxidation [48–52]. We have shown that dietary supplementation of β-carotene of healthy subjects resulted in a moderate inhibitory effect on the susceptibility of LDL to oxidative modification [53–55] in some, but not in all studied subjects. The combination of carotenoids with vitamin E, in contrast, demonstrated a synergistic inhibitory effect on LDL oxidation in all studied cases [56]. We showed that supplementation of vitamin E of atherosclerotic apolipoprotein E–deficient mice (25 μg/mouse/day for 3 months) inhibited LDL oxidation by 40% and the atherosclerotic lesion area by 35% [57]. In humans, unlike in animal models, both vitamin E and carotenoids did not significantly reduce atherosclerosis in primary prevention trials [58]. This result may be related to insufficient absorption, insufficient potency, and inappropriate tissue distribution.

Flavonoids are more potent antioxidants than carotenoids and vitamin E, and mortality rate can be attributed to differences in flavonoid intake of coronary heart disease across populations. Dietary consumption of flavonoids was shown to be inversely related to morbidity and mortality of coronary heart disease [59]. Flavonoids constitute the largest and most studied group of plant phenols. Over 4000 different flavonoids have been identified to date. They are usually found in plants as glycosides, and large compositional differences exist among different types of plants, even among different parts of the same plant. Flavonoids are grouped into anthocyanins and anthoxantins (Fig. 1). Anthocyanins are glycosides of anthocyanidin, and they are the most important group of water-soluble plant pigments, responsible for the red, blue, and purple colors of flowers and fruits. Anthoxantins are colorless or colored white to yellow; they include flavonols, flavanols, flavones, flavans, isoflavones, and isoflavans.

Flavonoids are powerful antioxidants, and their activity is related to their chemical structures [60–62]. Plant flavonoids are multifunctional and can act as reducing agents, as hydrogen atom–donating antioxidants, and as singlet oxygen quenchers. Some flavonoids also act as antioxidants via their metal ion chelation properties [63], thereby reducing the metal's capacity to generate

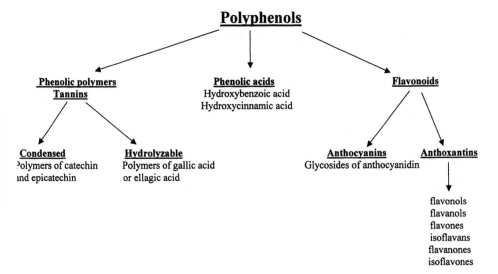

Figure 1 General classification of flavonoids.

free radicals. Flavonoids can act as potent inhibitors of LDL oxidation, via several mechanisms:

1. Scavenging of free radicals
2. Protection of the LDL-associated antioxidants α-tocopherol (vitamin E) and carotenoids from oxidation
3. Regeneration of vitamin E from oxidized α-tocopherol
4. Chelation of transition metal ions
5. Protection of cells against oxidative damage, with result, inhibition of cell-mediated oxidation of LDL, achieved via the potency of flavonoids to inhibit xanthine oxidase [64], NADPH oxidase [65], or lipoxygenase [66–68]
6. Preservation of serum paraoxonase (PON 1) activity, and as a result, hydrolysis of LDL-associated lipid peroxides

The protection of LDL against copper ion or free radical–induced oxidation by flavonoids depends on their structural properties in terms of their response to copper ion, their partitioning between the aqueous and the lipophilic compartments within the LDL particle, and their hydrogen donating antioxidant properties [63].

The flavanol catechin prevented plasma lipid peroxidation that was induced by azo compounds such as the water-soluble 2,2′-azobis, 2-amidinopropane hydrochloride (AAPH) or the lipid-soluble 2,2′-azobis, 2,4-dimethylvaleronitrile

(AMVN). This antioxidant effect of catechin depends on its plasma concentration, on the incubation time, and on the physical localization of the generated radicals. As expected from its hydrophilic structure, however, catechin showed a higher antioxidant capacity when the free radical reactions were initiated in the aqueous, rather than the lipid phase [69]. Catechin was also shown to inhibit LDL oxidation induced by copper ions, by cultured macrophages, or by vascular endothelial cells [70]. Quercetin, rutin, luteolin, and kaempferol also inhibited copper ion–induced LDL oxidation [63]; quercetin, rutin, and luteolin were more effective inhibitors of copper ion–induced LDL oxidation than kaempferol, as they also chelate copper ions. Morin, fisetin, quercetin, and gossypetin inhibited copper ion–induced LDL oxidation and macrophage-mediated LDL oxidation, with an IC_{50} of 1–2 μM [63]. Other flavonoids that were shown to inhibit LDL oxidation include the hydroxy-cinnamic acid–derived phenolic acid caffeic, ferulic, and p-coumaric acids [71] and the isoflavan glabridin [72,73]. Among the different groups of flavonoids, flavonols, flavanols, and the isoflavans are the most potent protectors of LDL against copper ion–induced oxidation. However, although possessing a similar OH group arrangement, the flavonol quercetin was a more potent antioxidant than the flavanol catechin, because of the 2-3 double bond and the 4-oxo structure present in the quercetin ring C. Similarly, studies on structural aspects of the inhibitory effect of glabridin on LDL oxidation revealed that the antioxidant effect of glabridin on LDL oxidation resides mainly in the 2′-hydroxyl group of the isoflavan B-ring [73]. The hydrophobic moiety of the isoflavan was also essential in order to obtain the inhibitory effect of glabridin on LDL oxidation, and the position of the hydroxyl groups at the B-ring significantly affected the ability of glabridin to inhibit LDL oxidation [73]. Flavonoids are also quite suitable for protecting cell membranes from free radical–induced oxidation, since they are both lipophilic and hydrophilic, thus resulting in reduced cell-mediated oxidation of LDL. Being partly inside and partly outside the cell's plasma membrane, flavonoids can scavenge free radicals that are generated within the cells as well as free radicals that attack the cell from the outside. Indeed, catechins from tea were shown to protect erythrocyte membranes and rat liver microsomes from lipid peroxidation [74]. Pretreatment of cells with flavanols or flavonols also protected the cells against damage induced by reactive oxygen species [75]. We have also demon-strated that enrichment of macrophages with the isoflavan glabridin protected the cells from lipid peroxidation under oxidative stress [65].

IV. PROTECTION OF LDL AGAINST OXIDATION BY FLAVONOID-RICH NUTRIENTS

The average daily human intake of flavonoids varies from as low as 25 mg to as high as 1 g [76–80]. After oral ingestion, some of the ingested flavonoids are absorbed from the gastrointestinal tract, and some of the absorbed flavonoids are metabolized by the gastrointestinal microflora. Differences in the bioavailability of different flavonoids exist and may be related to chemical structure differences [76,77]. The bioavailability and metabolic modifications of flavonoids determine the antioxidative capacity of these potent antioxidants in vivo. Major dietary sources of flavonoids and their chemical structures, along with their effects on LDL oxidation and on atherosclerosis, are discussed in the sections that follow (Fig. 2).

A. Wine

Wine has been part of the human culture for over 6000 years, serving dietary and socioreligious functions. Epidemiological studies of numerous populations reveal a significant lower cardiovascular mortality rate in individuals who have the habit of daily moderate wine consumption [81,82]. The "French paradox," i.e., the low incidence of cardiovascular events in spite of diet high in saturated fat, was attributed to the regular drinking of red wine in southern France [83]. The beneficial effect of red wine consumption against the development of atherosclerosis was attributed in part to its alcohol, but mostly to the antioxidant activity of its polyphenols. Red wine contains a range of polyphenols derived from the skin of the grape, with important biological activities [84,85]. Red wine contains the flavonols quercetin and myricetin (10–20 mg/L), the flavanols catechin and epi(gallo)catechin (up to 270 mg/L), gallic acid (95 mg/L), condensed tannins [catechin and epicatechin polymers (2500 mg/L)], and polymeric anthocyanidins. Phenolic compounds in red wine are derived from the grape's skin, as well as from the seeds, stems, and pulp, all of which are an important source of flavanols that are transferred to the wine during preparation together with the grape juice at the first stage of wine fermentation. On the contrary, white wines are usually made from the free running juice, without the grape mash, and have no contact with the grape's skin. This is thought to be the main reason for the relatively low polyphenol content and the low antioxidant activity of white wine, in comparison to those of red wine [86–91]. In previous studies, red wine, which contains a much higher concentration of polyphenols than white wine, was shown to be more effective in inhibiting LDL oxidation [92–95]. In 2001 we produced white wine with red wine–like antioxidant characteristics, by increasing the white wine polyphenol content [96]. This was achieved by imposing grape–skin contact for a short period in the presence of added alcohol, in order to augment the extraction of grape skin polyphenols into the wine. We have analyzed the antioxidant capacity of white wine samples obtained from whole squeezed grapes that were stored for increasing periods before grape skin removal or from whole squeezed grapes to which increasing concentrations of alcohol were added. White wine obtained from the whole squeezed grapes, which were incubated for 18 hours with

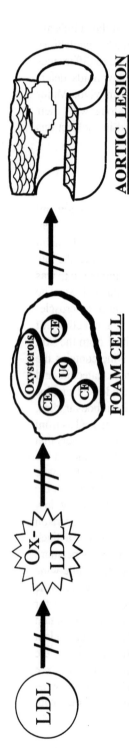

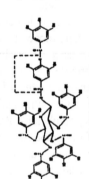

LDL → **Ox-LDL** → **FOAM CELL** → **AORTIC LESION**

Oxysterols, CE, UC, CE, CE

Tea
Catechin derivatives (Flavanol)

Red Wine
Catechin (Flavanol)
Quercetin (Flavonol)
Cyanidin (Anthocyanidin)

Pomegranate
(Hydrolyzable Tannins)
Gallotannin
Ellagitannin

Licorice
Glabridin (Isoflavan)

Olive Oil
Hydroxytyrosol Oleuropein (Phenolics)

18% alcohol, contained a 60% higher concentration of polyphenols than untreated white wine and exhibited significant antioxidant capacity against LDL oxidation, almost similar to that of red wine. The antioxidant capacity of the white wine was directly proportional to its polyphenol content (Fig. 3). Ingestion of red wine was previously shown to be associated with increased serum antioxidant activity [97]. Administration of 400 mL of red wine to healthy human volunteers for a period of 2 weeks reduced the propensity of the volunteers' LDL to lipid peroxidation in response to copper ions. On the contrary, the resistance to oxidation of LDL derived from subjects who consumed the same volume of white wine showed no significant change, in comparison to baseline LDL oxidation rates [85,92,98]. The administration of red wine to healthy human volunteers for a period of 2 weeks resulted in a substantial prolongation of the lag phase required for the initiation of LDL oxidation by as much as 130 minutes, whereas consumption of a similar volume of white wine had no significant effect on LDL oxidation. In parallel, the propensity of the volunteer LDL, which was obtained after red wine consumption, to copper ion–induced lipid peroxidation was reduced in comparison to that of LDL obtained at baseline, as measured by a 72% decrement in the content of the lipoprotein-associated lipid peroxides, whereas after white wine consumption no significant effect was observed [92]. The antioxidant effect of dietary red wine against LDL oxidation could be related to the elevation in polyphenol concentrations in the plasma and in the LDL particle. Thus, some phenolic substances that exist in red wine are absorbed, bind to plasma LDL, and protect the lipoprotein from oxidation [92]. The effect of red wine consumption on the susceptibility of LDL to oxidation was also studied in the postprandial state [99]. Five volunteers consumed 300 mL of California red wine, containing 1500 mg/L of total phenolic compounds. LDL isolated from plasma samples taken 1 and 2 h after red wine ingestion was significantly more resistant to copper ion–induced oxidation than LDL obtained before wine consumption, as shown by 50% and 66% inhibition in aldehyde formation, respectively. These results were further confirmed in a study [100] that demonstrated that red wine consumption increased concentration plasma- and LDL-associated polyphenols and protected LDL against copper ion–induced oxidation, as shown by increased lag time and decreased LDL content of lipid peroxides and thiobarbituric acid reactive substances (TBARSs) (by 34% and 22%, respectively).

Figure 2 Antiatherogenic effects of dietary polyphenolic flavonoids. Dietary consumption of nutrients rich in flavonoids inhibits LDL oxidation, foam cell formation, and development of aortic atherosclerotic lesions. Major flavonoid-rich nutrients are shown, along with the chemical structure of their flavonoids. CE, cholesteryl ester; Ox-LDL, oxidized low-density lipoprotein; UC, unesterified cholesterol.

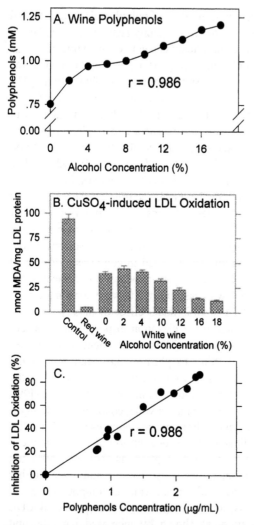

Figure 3 Alcohol augments grape skin polyphenol extraction into white wine and increases its antioxidant capacity. Whole squeezed Muscat grapes were incubated for 18 h with increasing concentrations of alcohol up to 18%, after which the juice was separated from the grape skin and allowed to ferment into wine. (A) Polyphenol concentration in wine samples was determined. (B) Wine samples at a final concentration of 2 μL/mL were added to LDL (100 mg of protein/L) and incubated with 5 μmol/L $CuSO_4$ for 2 h at 37 °C. LDL oxidation was measured by LDL TBARS assay. LDL, low-density lipoprotein; TBARS, thiobarbituric acid–reactive substance. (C) Linear regression analysis of the total polyphenol concentration of wine and the wine-induced inhibition of LDL oxidation.

The effect of the nonalcoholic components of red wine was also studied [101,102]. By using wine and alcohol-free red wine extract, it was shown that although the alcohol component of the wine may be important for a favorable lipid pattern, such potential health benefits may be independent of the proposed antioxidant effects of red wine [92,100,103,104]. In a 2001 study it was shown that polyphenols in dealcoholized red wine can reduce in vivo lipid peroxidation, as measured by F_2-isoprostanes, in smoking subjects, whereas no reduction in lipid peroxidation was observed after red or white wine consumption [102]. In 2001 human intervention study [102], it was shown that alcohol-free red wine extract can inhibit LDL oxidation ex vivo. A short-term ingestion of purple grape juice reduced LDL susceptibility to oxidation in patients with coronary artery disease [105,106].

Variations in the concentration and composition of flavonoids among red wines [107] may be responsible for the range of antioxidant potential exhibited by different red wines. We compared the composition of two red wines, which both increased the resistance of LDL ex vivo to oxidative modification in human supplementation studies [92,100]. Both studies used red wine from Cabernet Sauvignon cultivars, one grown in Israel [92] and the other in France [100]. After similar 400-mL daily consumption of red wine for 2 weeks, the inhibitory effect on LDL oxidation was found to be much higher in the Israeli study [92]. Comparison of the polyphenol composition of both wines revealed that even though the total polyphenol content of the wines was similar [1650 (Israeli) and 1800 (French) mg/L], the wines differed substantially in their flavonol and monomeric anthocyanin content. There is a wide variation in the flavonol content of different red wines throughout the world [108], and a major determinant factor for this phenomenon is related to the amount of sunlight to which the grapes are exposed during cultivation [109]. The flavonol synthesis in the skin of the grape is increased in response to sunlight so as to act as a yellow filter against the harmful effect of ultraviolet (UV) light. Thus, the climatic conditions under which grapes are grown may explain the increased content of flavonols in the Israeli red wines in comparison to that in the French red wine, studied by Nigdikar and associates [100]. Another comparison of two studies [103,110], both of which used alcohol-free red wine extracts, showed an increase in the resistance of LDL to oxidation in only one study [103]. A comparison of the wine composition showed that the concentrations of catechins and anthocyanins were double in the wine that showed an ex vivo inhibitory effect on LDL oxidation [103]. The direct effect of red wine consumption on the development of atherosclerotic lesions was further studied in E° mice that were supplemented with 0.5 mL of red wine/day per mouse for a period of 6 weeks [111,112]. LDL isolated after red wine consumption was less susceptible (by 30–80%) to oxidation induced either by copper ions, by the free radical generator AAPH, or by J-774 A.1 macrophages in culture, in comparison to

LDL isolated from the placebo (alcoholized water)-treated E^o mice [111,112]. The atherosclerotic lesion areas in E^o mice that consumed red wine were significantly reduced (by 40%), in comparison to the lesion areas in the placebo-treated E^o mice (Fig. 4). In contrast to wine supplementation of young E^o mice, in which development of atherosclerosis was substantially reduced, administration of red wine to old, already atherosclerotic E^o mice, for up to 26 weeks, did not significantly reduce the mature atherosclerosis development [113]. In another study, administration of red wine, dealcoholized red wine, or grape juice was shown to inhibit atherosclerosis development in hamsters. However, grape juice was calculated to be much more effective than red wine or dealcoholized red wine, at the same polyphenol dosage, in inhibiting atherosclerosis and improving plasma lipid pattern and antioxidant activity [114]. Phenolic substances in red wine were shown to inhibit LDL oxidation in vitro [95]. In previous studies, red wine–derived phenolic acids [115, 116], resveratrol [117], flavonols (quercetin, myricetin) [68,118,119], catechins [66,120], and the grape extract itself [121,122] have been shown to possess antioxidant properties. The finding that ethanol and wine stripped of phenols did not affect LDL oxidation further confirmed that the active antioxidant components in red wine are phenolic compounds [123]. Red wine fractionation revealed major antioxidative potency to monomeric catechins, procyanidins, monomeric anthocyanidins, and phenolic acids [123]. The flavonol quercetin and the flavonol catechin were both tested for antioxidative and antiatherogenic effects in the atherosclerotic E^o mice [111]. E^o mice at the age of 4 weeks were supplemented for up to 6 weeks in their drinking water with placebo (1.1% alcohol) or with catechin or quercetin (50 µg/day/mouse). The atherosclerotic lesion area was smaller by 39% or by 46%, respectively, in the treated mice than in E^o mice that were treated with placebo (Fig. 4A–E).

These results were associated with reduced susceptibility to oxidation (that was induced by different modes such as copper ions, free radical generators, or macrophages) of LDL isolated after quercetin, and to a lesser extent after catechin consumption, in comparison to LDL isolated from the placebo group. LDL isolated from E^o mice that consumed catechin or quercetin for 2 weeks was also found to be less oxidized in its basal, not induced, state, in comparison to LDL isolated from E^o mice that received placebo, as evidenced by 39% or 48% reduced content of LDL-associated lipid peroxides, respectively (Fig. 4F) [111]. The inhibitory effect of moderate red wine ingestion against LDL oxidation may also be explained in part by its effects on HDL-associated paraoxonase [124]. Daily moderate alcohol consumption was shown to increase serum paraoxonase activity in middle-aged men [125]. Catechin, quercetin, and red wine consumption also increased serum paroxonase activity by 14%, 13%, and 75%, respectively, in E^o mice (Fig. 4G) [111]. As paraoxonase protects against lipid peroxidation (probably by hydrolyzing oxidized lipids) the beneficial effect of

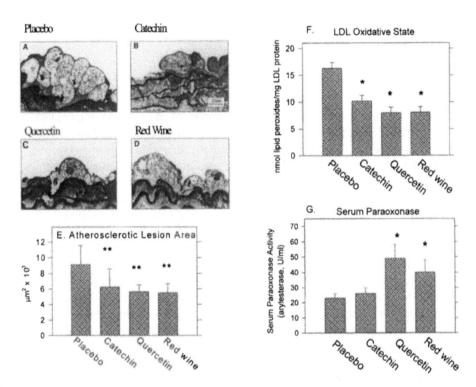

Figure 4 Consumption of red wine or its polyphenols catechin and quercetin by E° mice decreases the development of atherosclerotic lesions, inhibits oxidation of LDL, and increases serum paraoxonase activity. (A–E) The aortic arch from E° mice that consumed placebo, catechin, quercetin, or red wine was analyzed. Results are expressed as the mean of the lesion area in square micrometers ± SD. *$p < 0.05$ (vs. placebo). (F) LDL was isolated from E° mice that consumed placebo, red wine (0.5 mL/day per mouse), catechin or quercetin (50 µg/day per mouse) for a period of 6 weeks. LDL (100 µg of protein/mL) oxidative state (basal, not induced oxidation) was determined as lipid peroxide levels. Results are expressed as mean ± S.D. ($n = 3$), *$p < 0.01$ (vs. placebo). (G) Paraoxanase activity was measured as arylesterase activity in serum derived from E° mice that consumed placebo, catechin, quercetin, or red wine for 6 weeks. Results are expressed as mean ± S.D. of three separate determinations. *$p < 0.01$ vs. placebo. LDL, low-density lipoprotein.

red wine polyphenols on paraoxonase activity can be considered an additional antiatherogenic property of red wine.

B. Licorice

Glycyrrhiza glabra, the licorice plant, has a history of consumption of more than 3000 years. The licorice root has long been used as a flavoring and sweetening agent. Licorice root has also been used medicinally for a wide range of therapeutic functions, such as antibacterial, antiviral, anti-inflammatory, antiallergic, and antihepatotoxic functions. Minor components of licorice, mostly flavonoids from the isoflavan and chalcon subclasses, were shown to possess antioxidative properties. The antioxidative capability of licorice crude extract against LDL oxidation was investigated in vitro and ex vivo [126]. LDL oxidation induced by copper ions or by AAPH was inhibited by 90% with as little as 0.3 μg of licorice root extract/mL. Licorice ethanolic extract inhibited LDL oxidation by a mechanism that involves scavenging of free radicals. The protective effect of licorice root extract on the resistance of LDL to ex vivo oxidation was studied in normolipidemic humans, as well as in hypercholesterolemic patients and in atherosclerotic apolipoprotein E–deficient (E°) mice. LDL, which was isolated from the plasma of 10 healthy volunteers after consumption of 100 mg of licorice root ethanolic extract per day for a period of 2 weeks, was more resistant to copper ion–induced oxidation, as well as to AAPH-induced oxidation, by 44% and by 36%, respectively, in comparison to LDL isolated before licorice supplementation. Supplementation of licorice root extract (0.1 g/day) to hypercholesterolemic patients for a period of 1 month was followed by an additional 1 month of placebo consumption [127]. Licorice consumption resulted in a moderate reduction in the patients' plasma susceptibility to oxidation (by 19%) and in an increased resistance of the patients' plasma LDL to oxidation (by 55%). After an additional 1 month of placebo consumption, a reversal of the parameters studied to baseline levels was noted. Licorice extract supplementation resulted also in a 10% reduction in the patients' systolic blood pressure, which was sustained for an additional 1 month during the placebo consumption. Thus, dietary consumption of licorice root extract by hypercholesterolemic patients may provide a moderate hypocholesterolemic nutrient and a potent antioxidant agent, which confers a health benefit against cardiovascular disease. These effects were further supported by the antioxidative effects of licorice extract in the atherosclerotic apolipoprotein E–deficient mice. Dietary supplementation of licorice (200 μg/day/ mouse) to E° mice for a period of 6 weeks resulted in a 80% reduction in the susceptibility of their LDL to copper ion–induced oxidation in comparison to LDL isolated from placebo-treated mice [126].

Licorice root contains flavonoids with biological activities, several of which were isolated and purified. Licochalcone B and D, isolated from the roots of *Glycyrrhiza inflata*, were shown to inhibit superoxide anion production in the xanthine/xanthine oxidase system [128] and to possess free radical scavenging activity toward the 1,1-diphenyl-2-picrylhydrazyl (DPPH) radical. These phenolic compounds were also shown to be effective in protecting biological systems against various oxidative processes. They inhibited mitochondrial lipid peroxidation induced by Fe (III)–adenosine diphosphate/NADH (ADP/NADH), scavenged superoxide anions in microsomes, and protected red blood cells against oxidative hemolysis [128]. Other antioxidant constituents that were isolated from licorice were identified as the isoflavans glabridin, hispaglabridin A, hispaglabridin B, and 4-*O*-methyl glabridin and two chalcones, isoprenylchalcone and isolipuritegenin [72]. Among these compounds, glabridin constituted the major flavonoid in the licorice root extract (500 mg/kg of ethanolic root extract). On LDL incubation with glabridin, the latter was shown to bind to the LDL particles and subsequently to protect them from oxidation [73,129]. Glabridin inhibited AAPH-induced LDL oxidation in a dose-dependent manner, as shown by the inhibition of cholesteryl linoleate hydroperoxide (CL-OOH) formation, as well as by the inhibition of aldehyde and lipid peroxide formation. Addition of glabridin (30 µM) to LDL that was incubated with AAPH or with copper ions also inhibited the formation of oxysterols (7-hydroxycholesterol, 7-ketocholesterol, and 5, 5-epoxycholesterol) by 65%, 70%, and 45%, respectively. Glabridin inhibited the consumption of β-carotene and that of lycopene by 41% and 50%, respectively, after 1 h of LDL oxidation in the presence of AAPH but did not protect the major LDL-associated antioxidant, vitamin E, from oxidation [73] (Fig. 5). Finally, glabridin preserved the arylesterase activity of human serum paraoxonase (PON 1), including its ability to hydrolyze Ox-LDL cholesteryl linoleate hydroperoxides [47]. Administration of glabridin to E° mice in their drinking water was followed by analysis of its antioxidative effect against LDL oxidation ex vivo [126]. Analysis of LDL derived from E° mice after consumption of glabridin revealed that glabridin was absorbed and bound to the LDL particle. Although no glabridin could be detected in LDL of control mice, LDL of mice that consumed glabridin (20 µg/day/mouse) contained about 2 nmol of glabridin/mg LDL protein. LDL derived from E° mice after consumption of 20 µg glabridin/day/mouse for a period of 6 weeks was significantly more resistant (by 22%) than LDL derived from placebo-treated mice to copper ion–induced oxidation. Most importantly, inhibition of oxidative modification of LDL in E° mice after glabridin consumption was associated with a substantial reduction in the development of atherosclerotic lesions.

Glabridin consumption was shown to exert its antioxidative effects also at the cellular level. Enrichment of mouse peritoneal macrophages with glabridin either in vitro or in vivo (after its consumption by E° mice) resulted in 80% inhibition in macrophage-mediated oxidation of LDL, in comparison

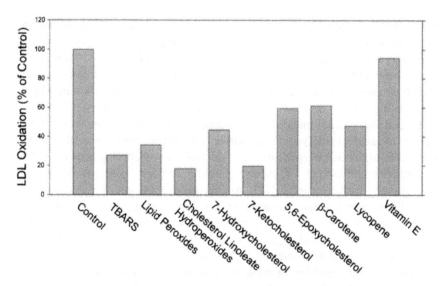

Figure 5 The antioxidative effect of glabridin on LDL endogenous constituents during AAPH-induced LDL oxidation. LDL (100 mg of protein/L) was incubated for 3 h at 37 °C with 5 mM AAPH in the presence of 30 μg glabridin. The extent of LDL oxidation was measured as formation of thiobarbituric acid reactive substances (TBARSs), lipid peroxides, cholesteryl linoleate hydroperoxide (CLOOH), 7-hydroxycholesterol, 7-ketocholesterol, 5,6-epoxycholesterol, or consumption (oxidation) of β-carotene, lycopene, and vitamin E. Results are expressed as percentage of inhibition relative to LDL incubated with AAPH in absence of glabridin. LDL, low-density lipoprotein; AAPH, 2,2′-azobis, 2-amidinopropane hydrochloride.

to that of control cells [65]. This effect was secondary to the inhibition of the macrophage NADPH oxidase, as reflected by the decrement in superoxide anion release. This latter effect was related to an inhibition in the translocation of the cytosolic component of NADPH oxidase P-47 to the plasma membrane. The effects of glabridin described were associated with the inhibition (by 70%) of macrophage protein kinase C activity, which is required for P-47phosphorylation and activation. Thus, glabridin-induced inhibition of P-47 phosphorylation may be the primary event in its inhibitory effect on NADPH oxidase–induced macrophage-mediated oxidation of LDL. All the inhibitory effects of glabridin on the events related to cell-mediated oxidation of LDL required the hydroxyl groups on the isoflavan B-ring. Since glabridin inhibited oxidative processes both in macrophages and in LDL, these mechanisms may be responsible for the attenuation of atherosclerosis in E° mice that consumed glabridin.

C. Pomegranate

The pomegranate tree, which is said to have flourished in the Garden of Eden, has been extensively used as a folk medicine in many cultures [130]. Edible parts of pomegranate fruits (about 50% of total fruit weight) comprise 80% juice and 20% seeds. Fresh juice contains 85% moisture, 10% total sugars, and a total of 1.5% for pectin, ascorbic acid, polyphenols, and flavonoids. Pomegranate seeds are a rich source of crude fibers, pectin, and sugars, and the pomegranate peel has been shown to contain phenols from the condensed and hydrolysable tannin class [131–133]. The dried pomegranate seeds contain the steroidal estrogen estrone [134,135], the isoflavonic phytoestrogens genistein and daidzein, and the phytoestrogen coumestrol [136]. Content of soluble polyphenols in pomegranate juice varies within the limits of 0.2% to 1.0%, depending on variety, and includes mainly anthocyanins (such as cyanidin-3-glycoside, cyanidin-3, 3-diglycoside, and delphindin-3-glucosid) and anthoxantins (such as catechins, ellagic tannins, and gallic and ellagic acids [131–133]). Pomegranate fermented juice and cold-pressed pomegranate seeds possess antioxidant activity and can reduce prostaglandin and leukotriene formation by inhibition of cyclooxygenases and lipoxygenases, respectively [137]. Pomegranate juice was shown to possess antioxidant activity that was three times higher than the antioxidant activity of red wine or of green tea [133]. The antioxidant activity was higher in commercial juices extracted from whole pomegranates than in juices obtained from arils only, suggesting that industrial processing extracts some of the hydrolyzable tannins present in the fruit rind. The effect of pomegranate juice on LDL oxidation was studied in vitro and ex vivo in healthy male volunteers and in the atherosclerotic apolipoprotein E–deficient (E^0) mice [138]. The in vitro studies demonstrated a significant dose-dependent antioxidant capability of pomegranate juice against LDL oxidation, as well as against oxidation of HDL. The mechanisms for the antioxidative effects of pomegranate juice against lipoprotein oxidation could be related to its capacity to scavenge free radicals. The water-soluble fractions of pomegranate's inner and outer peels, but not that of the seeds, were even stronger antioxidants against LDL oxidation than the juice [138].

LDL derived from human healthy volunteers after consumption of pomegranate juice ('Wonderful' cultivation, 50 mL/day of pomegranate juice, equivalent to 1.5 mmol total polyphenols/day), for a period of 2 weeks, was found to be more resistant to copper ion–induced oxidation than LDL obtained before pomegranate juice supplementation [138]. This effect was demonstrated by a 43% prolongation of the LDL oxidation lag time after 2 weeks of juice consumption, in comparison to that of LDL obtained before juice consumption, and this latter effect was accompanied by a significant 10% increment in plasma total antioxidant status.

Pomegranate juice consumption by healthy humans also resulted in increased activity of their serum paraoxonase. We have extended our studies on the antioxidative effects of pomegranate juice consumption to hypertensive patients [139,140]. The effect of pomegranate juice consumption for a period of 2 weeks by 10 hypertensive patients on their blood pressure was small but significant. In 7 of 10 hypertensive patients studied, serum angiotensin-converting enzyme (ACE) activity was significantly decreased by 36%. Pomegranate juice was shown to exhibit directly a dose-dependent inhibitory effect on serum ACE activity. Because ACE inhibitors are metabolized by cytochrome P-450 enzymes, serum ACE activity can be significantly affected by modulating P-450 enzyme activity [141]. Therefore, we next analyzed the effect of pomegranate juice on cytochrome P-450 enzymes [139]. Pomegranate juice decreased the activities of cytochrome P-450 3A4, 2D6, 2E1, and 2B6 by 40%, 30%, 20%, and 60%, respectively. In hypertensive patients treated with ACE inhibitors, the inhibitory effect of pomegranate juice consumption on the P-450 enzymes can possibly decrease P-450-mediated drug (the ACE inhibitor) breakdown, and hence, serum ACE activity may be further decreased in these patients. We have indeed observed in three hypertensive patients who were treated with the ACE inhibitor fosinopril (20 mg/day for 1 month) that their serum ACE activity decreased after 2 weeks of pomegranate juice consumption (50 mL of juice containing 1.5 mmol total polyphenols/day) by 26%. Taken together, the results on the inhibitory effect of pomegranate juice on serum ACE activity, on one hand, and on cytochrome P-450 enzymes, on the other hand, suggest that pomegranate juice may also affect ACE activity indirectly, secondary to its inhibitory effect on the cytochrome P-450 enzymes.

Consumption of pomegranate juice by E° mice has also been shown to have considerable antioxidative and antiatherogenic effects [138]. Pomegranate juice consumption substantially reduced the propensity of E° mice–derived LDL to copper ion–induced oxidation and to macrophage-mediated oxidation of LDL by reducing the oxidative capacity of the cells. The mechanism responsible for this effect was associated with inhibition of the NADPH oxidase cytosolic factor p-47 translocation to the macrophage plasma membrane and hence, inhibition of NADPH oxidase activation, reduction (by 49%) in superoxide anion release from the macrophages, and elevation in cellular glutathione content by 25% [138]. This effect could also be related to reduced levels of macrophage-associated lipid peroxides after pomegranate juice consumption, in comparison to levels of macrophages isolated from control mice that consumed placebo (Fig. 6). Most importantly, pomegranate juice supplementation to E° mice reduced the size of their atherosclerotic lesion and the number of foam cells in the lesion [138], in comparison to those of control placebo-treated E° mice that were supplemented with water. Furthermore, pomegranate juice supplementation to E° mice with already advanced atherosclerosis was still able to reduce the mice's atherosclerotic

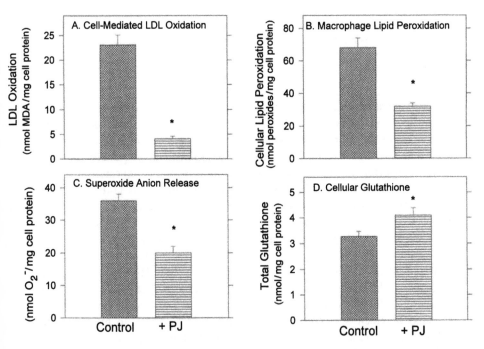

Figure 6 Pomegranate juice consumption by E° mice reduces macrophage-mediated LDL oxidation mechanisms. Mouse peritoneal macrophages (MPMs) were isolated from the peritoneal fluid of control E° mice or from mice that consumed 12.5 μL of pomegranate juice (PJ)/mouse/day, for a period of 2 months. (A) *Cell-mediated LDL oxidation.* The MPMs were incubated for 6 h at 37 °C with LDL (100 μg protein/mL) under oxidative stress (in the presence of 2 μM of $CuSO_4$). LDL oxidation was measured directly in the medium by the TBARS assay. Results are expressed as mean ± S.D. ($n = 3$). *$p < 0.01$ (vs. placebo). (B) MPM lipid peroxidation was determined as cellular lipid peroxides. (C) Superoxide anion release: The amount of superoxide anion release from the MPM to the medium in response to 50 ng/mL of PMA was determined. (D) Total glutathione was determined in MPM sonicate supernatant with the 5,5-dithiobis-2-nitrobenzoic acid-glutathione reductase (NADPH) recycling assay. (E) Results are expressed as mean ± S.D. ($n = 3$). *$p < 0.01$ (vs. control). LDL, low-density lipoprotein; TBARS, thiobarbituric acid reactive substance; PMA,

lesion size by 17%, in comparison to the size of atherosclerotic lesion in age-matched placebo-treated mice [142].

D. Tea

Tea drinking has been associated in epidemiological studies with a decreased risk for cardiovascular disease [59,143–145]. The terms *green tea* and *black tea* refer to products manufactured from the leaf of the tea plant, *Camellia sinensis*. Green tea is manufactured from fresh leaf and is rich in flavonoids, especially flavonols from the catechin group, of which epigallocatechin gallate, epicatechin gallate, and epicatechin account for 30–40% of the green tea solids [146]. Black tea manufacture includes an enzymatic step, in which most catechins are converted to complex condensation products, such as the aflavins or the arubigens. Green and black tea also contain small amount of flavonols, such as quercetin.

Absorption studies of tea polyphenols and their effects on LDL oxidation, and atherosclerosis have shown conflicting results [147,148]. The effect of green or black tea consumption on the resistance of LDL to oxidation was studied in 45 human volunteers who for a period of 4 weeks consumed 900 mL (6 cups) of green tea or black tea per day in comparison to mineral water [149]. Consumption of tea (green or black) had no effect on the ex vivo resistance of LDL to oxidation. Similar negative results have also been demonstrated in other studies [150–153]. Conversely, in another study [154], ingestion of tea (300 mL after an overnight fast) produced a significant increase of plasma antioxidant capacity, which peaked at 30–50 min after consumption. Similarly [155], ingestion of 400 mL of freshly prepared green tea resulted in a rapid absorption of the tea polyphenols, and it was associated with an increase in plasma total antioxidant state, peaking at 20–40 min post ingestion. Consumption of 750 mL of black tea/day for 4 weeks by 14 healthy volunteers revealed [156] that the lag time for LDL oxidation was significantly prolonged (from 54 to 62 min). In other studies [157,158], ingestion of black and green tea, in comparison to ingestion of alcohol-free red or white wine or water, resulted in a significant increase in plasma total antioxidant capacity at 30 min after consumption, and red wine and green tea were the most efficient in protecting LDL from oxidation [157], whereas black tea had a mild acute effect [158].

The effect of green or black tea on LDL oxidation and atherosclerotic lesion formation was also studied in animal models, including hypercholesterolemic rabbits [159] and hamsters [160], and apolipoprotein E–deficient mice [161]. These studies indicated that green tea consumption reduced the atherosclerotic plaque formation in hypercholesterolemic rabbits [159], whereas black tea showed no significant effect, although both green and black tea induced a 13% and 15% prolongation in the lag phase of LDL oxidation, respectively

[159]. Similar results were demonstrated in the hamster model [160], showing that green tea was significantly more effective than black tea in improving risk factors for heart disease, including hypolipidemic and antioxidant effects. Furthermore, supplementation of green tea extract (0.8 g/L) to apolipoprotein E–deficient mice significantly attenuated (by 23%) development of atherosclerotic lesions, without changing the plasma lipid level, probably through the potent antioxidative activity of the tea [161].

The potent antioxidative effects of tea are attributed to its polyphenols. Green tea extract and catechin-rich fractions from green tea were shown to inhibit the oxidation of LDL by endothelial cells [162]. Ingestion of 300 mg of green tea polyphenol extract twice daily for 1 week by 22 male volunteers resulted in increased resistance of LDL to oxidation [163]. Catechins or theaflavins (25–400 μmol/L) [164], as well as epicatechin, epigallocatechin, epicatechin gallate, epigallocatechin gallate, and gallic acid [165,166], which were added to LDL, dose-dependently inhibited its oxidation. Among the catechins, epigallocatechin gallate exerted the most marked effect in prolonging LDL oxidation lag time [163]. Furthermore, addition of 1.5 μM of epicatechin and epigallocatechin to a mixture of LDL and copper ion in the initiation phase inhibited LDL oxidation, whereas higher concentrations were needed (10 μM of epicatechin and 2 μM of epigallocatechin) for the inhibition of the LDL oxidation propagation phase [164]. The mechanisms responsible for the inhibition of LDL oxidation by tea include inhibition in the ability of macrophages to modify LDL oxidatively by decreasing macrophage production of superoxide and chelation of iron ions [167] as well as regeneration of vitamin E in human LDL [168].

E. Olive Oil

The Mediterranean diet, rich in fresh fruits and vegetables, was shown to be inversely related to the incidence of cardiovascular disease, as shown in the Seven Countries Studies [59,169–171]. Olive oil, the dietary fat of choice in the Mediterranean area, in comparison to other vegetable oils has a peculiar fatty acid composition. The monounsaturated oleic acid (C-18:1, $n = 9$) is the most abundant fatty acid in olive oil (56–84%), whereas the polyunsaturated linoleic acid (C-18:2, $n = 6$) ranges only from 3% to 21%. In addition, olive oil contains a variety of minor components, including polyphenols (up to 800 mg/kg), which provide the typical taste and aroma of extra virgin olive oil and confer on this oil its stability to oxidation [172].

The beneficial effects of the Mediterranean diet may stem from the high content of the monounsaturated oleic acid, as well as from the polyphenols, which are beneficial in reducing LDL oxidation. LDL isolated from Greek subjects consuming a diet naturally rich in olive oil was significantly less susceptible (by 12%) to oxidation as measured by conjugated diene formation, in comparison to LDL isolated from U.S. subjects consuming a typical American diet [173]. Furthermore, the proinflammatory potential of mildly oxidized LDL derived from Greek subjects measured as LDL promotion of monocyte chemotaxis and adhesion to endothelial cells, was decreased by 42%, in comparison to LDL derived from U.S. subjects. There was an inverse correlation between the LDL oleic acid content and the stimulation of monocyte chemotaxis and adhesion [173]. Consumption of a liquid diet supplemented with oleic acid for 8 weeks by U.S. subjects resulted in an oleic acid–enriched LDL, which subsequently promoted a 52% reduction in monocyte chemotaxis and a 77% reduction in monocyte adhesion, in comparison to linoleate-enriched LDL [173]. This study suggested that LDL enriched with oleic acid is less easily converted into the proinflammatory minimally modified LDL. These results are consistent with our study [174], which demonstrated that dietary supplementation of olive oil to healthy human subjects (50 g/day) for a period of 2 weeks increased the resistance of their LDL to oxidation [174]. This was shown by a significant reduction in LDL peroxides, thiobarbituric acid reactive substances (TBARSs), and conjugated diene content by 73%, 28%, and 32%, respectively. Furthermore, LDL obtained after olive oil supplementation demonstrated a 61% decrease in cellular uptake by macrophages [174]. These beneficial effects of olive oil consumption may be related, in addition to the oleic acid, to the olive oil polyphenols. To determine whether the minor polar components of virgin olive oil could have favorable effects on LDL composition and susceptibility to oxidation, 10 normolipidemic subjects received different oils in a crossover study (two diet periods of 3 weeks each). Subjects received either virgin olive oil or oleic acid–rich sunflower oil. LDL oxidation, which was measured as conjugated diene production, decreased only after intake of the virgin olive oil diet, suggesting a mild antioxidative effect against LDL oxidation for some minor components of virgin olive oil [175]. This was further evidenced by dietary administration of extra virgin olive oil versus refined olive oil to New Zealand white rabbits (NZW) [176]. This experiment resulted in increased resistance of the rabbits' LDL to oxidation only in the group of animals that consumed extra virgin olive oil [176], suggesting that the beneficial effect of extra virgin olive oil over that of the refined olive oil is due to the presence of phenolics, which are lost during the refining process [176]. Furthermore, administration of extra virgin olive oil, which contains 800 mg/ kg of phenolic compounds, in comparison to refined olive oil, which contains only 60 mg/kg of phenolic compounds, to patients with peripheral vascular disease resulted in a superior antioxidative effect of the extra virgin olive oil against LDL oxidation [177]. Consumption of olive oil together with a dietary supplement of fish oil by patients with peripheral vascular disease also resulted in protection of LDL from oxidation [178,179]. However, contradictory results were also presented. Consumption of 18 mg/day of phenols from extra virgin olive oil for 3 weeks did not affect LDL oxidation [180], suggesting that dietary studies in humans may vary according to the studied populations and the experimental conditions.

Studies performed in vitro also demonstrated the unique antioxidant properties of virgin olive oil phenolics against LDL oxidation [181–183]. These effects could be related to the ability of olive oil phenolics, such as tyrosol, to bind LDL in vitro and thus to protect LDL and other phenolic compounds previously bound to LDL from oxidation [182,183].

Some major phenols were extracted and purified from olive oil [172, 184], and their biological properties have been investigated. Oleuropein and hydroxy-tyrosol are two major representative phenols in olive oil. Direct antioxidant activities of oleuropein and hydroxytyrosol against LDL oxidation were demonstrated [185,186]. LDL incubation with oleuropein or with hydroxytyrosol resulted in the inhibition of copper ion–induced LDL oxidation. Oleuropein and hydroxytyrosol protection of LDL against oxidation can be related to the potency of these compounds in scavenging free radicals. Both hydroxytyrosol and oleuropein were shown to express scavenging capacity for superoxide radical in a cell-free system (based on the generation of urate and superoxide by xanthine/xanthine oxidase) and in a cellular system (based on the production of superoxide by phorbol myristate acetate (PMA)-challenged human neutrophils) [187]. Oleuropein was shown also to increase the resistance of LDL to oxidation after dietary supplementation to rabbits [188], and tyrosol could prevent oxidized LDL-induced cytotoxicity in Caco-2 cells when present during incubation [189]. Two other phenolic compounds of olive oil, 3,4-dihydroxy-phenylethanol-elenolic acid and protocatecuic acid, were shown to be able to inhibit LDL lipid peroxidation and to reduce the extent of its cytotoxic activity [190]. Biological active compounds with antioxidative properties against LDL oxidation were also extracted from the olive mill waste waters [191].

Olive oil, as well as olive oil phenolic constituents, were shown to affect biological processes involved in atherogenesis, as well as LDL oxidation. Polyphenols in olive oil possess beneficial effects on cellular processes, including inhibition of lipoxygenase [192], production of leukotriene B4 [193], and stimulation of nitric oxide (NO) release from macrophages [194]. The polyphenol-rich olive oil waste water was shown to depress cell production of superoxide anion [195], thus resulting in reduced cellular oxidative stress, which can subsequently reduce the capacity of the cells to modify LDL oxidatively. In addition, olive oil–enriched diet supplementation to mice resulted in a lower level of macrophage scavenger receptor (SRA-I, SRA-II, and CD36) gene expression, which can account for reduced macrophage uptake of oxidized LDL, reduced foam cell formation, and reduced atherosclerotic lesion development [196].

V. CONCLUSIONS

The beneficial health effects attributed to the consumption of fruits and vegetables are related at least in part to their antioxidant activity. Of special interest is the inverse relationship between intake of dietary nutrients rich in flavonoids and cardiovascular diseases. This effect is attributed to the flavonoids' capability to inhibit LDL oxidation, macrophage foam cell formation, and atherosclerosis. Fig. 7 summarizes some of our studies of total polyphenol concentration in several fruit juices or wines and the capacity of the juices, when compared on a similar total polyphenol concentration basis, to inhibit LDL oxidation. Pomegranate juice, red wine, and cranberry juice contained the

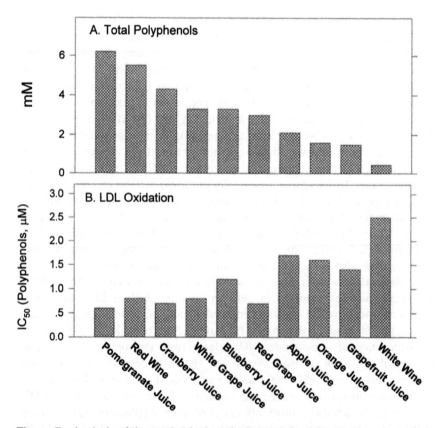

Figure 7 Analysis of the total polyphenol concentration of several common fruit juices, white wine, and red wine (A), and juice-induced inhibition of LDL oxidation (B) expressed as IC_{50}, which is the concentration needed to inhibit LDL oxidation by 50%. LDL, low-density lipoprotein.

highest polyphenol concentration (Fig. 7A), and they also showed the most potent antioxidant activity against LDL oxidation (Fig. 7B). As shown in Fig. 7B, the various juices exhibit different antioxidant capacities, which may be related to qualitative differences in the types of flavonoids present in the juices, which possess different antioxidant capabilities. Thus both flavonoid quantity and quality determine the antioxidant potency of the juices (Fig. 7).

Our current view on the major pathways by which flavonoids protect LDL against oxidative modifications, and thereby reduce macrophage foam cell formation and development of advanced atherosclerosis, is summarized in Fig. 8. Flavonoids can protect LDL against cell-mediated oxidation via two pathways: direct interaction of the flavonoids with the lipoprotein and flavonoid accumulation in arterial macrophages. Flavonoids were shown to accumulate in macrophages and to inhibit the activation of cellular NADPH oxidase via the

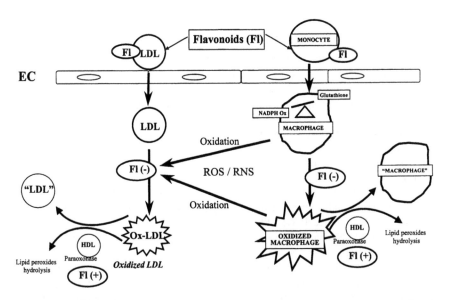

Figure 8 Major pathways by which flavonoids inhibit macrophage-mediated oxidation of LDL. Flavonoids (Fl) affect LDL directly by their interaction with the lipoprotein and inhibition (Fl −) of LDL oxidation. Flavonoids can also protect LDL indirectly, by their accumulation in arterial cells and protection of arterial macrophages against oxidative stress. This latter effect is associated with inhibition (Fl −) of the formation of oxidized macrophages and reduction in the capacity of macrophages to oxidize LDL. In addition, flavonoids preserve/increase (Fl +) paraoxonase activity, thereby increasing hydrolysis of lipid peroxides either in the LDL or in oxidized macrophages, resulting in further protection of LDL from oxidation. LDL, low-density lipoprotein.

inhibition of protein kinase C (PKC) activity. This effect results in reduced capacity of the cells to modify LDL oxidatively. Enrichment of macrophages with flavonoids inhibits macrophage lipid peroxidation, formation of lipid peroxide–rich macrophages, and cell-mediated LDL oxidation.

Furthermore, flavonoids increase serum paraoxonase activity, resulting in hydrolysis of lipid peroxides either in oxidized LDL or in lipid peroxide–rich macrophages and hence further prevent the formation of oxidized LDL. All these effects of flavonoids were demonstrated in vitro, as well as in vivo in humans and in the atherosclerotic apolipoprotein E–deficient mice, after dietary supplementation of nutrients rich in diverse flavonoids or of purified flavonoids. Dietary supplementation of pomegranate juice rich in flavonoids to atherosclerotic mice indeed resulted in a significant inhibition in the development of atherosclerotic lesions, along with the protection of LDL against oxidation [197–200].

REFERENCES

1. Schaffner T, Taylor K, Bartucci EJ, Fischer-Dzoga K, Beenson JH, Glagov S, Wissler R. Arterial foam cells with distinctive immunomorphologic and histochemical features of macrophages. Am J Pathol 1980; 100:57–80.
2. Gerrity RG. The role of monocytes in atherogenesis. Am J Pathol 1981; 103:181–190.
3. Goldstein JL, Brown MS. Regulation of the mevalonate pathway. Nature 1990; 343:425–430.
4. Brown MS, Goldstein JL. A receptor-mediated pathway for cholesterol homeostasis. Science 1986; 232:34–47.
5. Steinberg D, Parthasarathy S, Carew TE, Khoo JC, Witztum JL. Beyond cholesterol: modifications of low-density lipoprotein that increase its atherogenicity. N Engl J Med 1989; 320:915–924.
6. Aviram M. Modified forms of low density lipoprotein and atherosclerosis. Atherosclerosis 1993; 98:1–9.
7. Aviram M. Beyond cholesterol: modifications of lipoproteins and increased atherogenicity. In: Neri Serneri GG, Gensini GF, Abbate R, Prisco D, eds. Atherosclerosis Inflammation and Thrombosis. Florence, Italy: Scientific Press, 1993:15–36.
8. Jialal I, Devaraj S. The role of oxidized low density lipoprotein in atherogenesis. J Nutr 1996; 126:1053S–1057S.
9. Steinberg D. Low density lipoprotein oxidation and its pathobiological significance. J Biol Chem 1997; 272:20963–20966.
10. Berliner JA, Heinecke JW. The role of oxidized lipoproteins in atherosclerosis. Free Radic Biol Med 1996; 20:707–727.
11. Aviram M. Oxidative modification of low density lipoprotein and atherosclerosis. Isr J Med Sci 1995; 31:241–249.

12. Witztum JL, Steinberg D. Role of oxidized low density lipoprotein in atherogenesis. J Clin Invest 1991; 88:1785–1792.

13. Aviram M. Interaction of oxidized low density lipoprotein with macrophages in atherosclerosis and the antiatherogenicity of antioxidants. Eur J Clin Chem Clin Biochem 1996; 34:599–608.

14. Kaplan M, Aviram M. Oxidized low density during lipoprotein: atherogenic and proinflammatory characteristics during macrophage foam cell formation: an inhibitory role for nutritional antioxidants and serum paraoxonase. Clin Chem Lab Med 1999; 37:777–787.

15. Parthasarathy S, Santanam N, Auge N. Oxidized low-density lipoprotein, a two-faced janus in coronary artery disease? Biochem Pharmacol 1998; 56:279–284.

16. Parthasarathy S, Rankin SM. The role of oxidized LDL in atherogenesis. Prog Lipid Res 1992; 31:127–143.

17. Herttuala SY. Is oxidized low density lipoprotein present in vivo? Curr Opin Lipidol 1998; 9:337–344.

18. Aviram M. The contribution of the macrophage receptor for oxidized LDL to its cellular uptake. Biochem Biophys Res Commun 1991; 179:359–365.

19. Kim JA, Territo MC, Wayner E, Carlos TM, Parhami F, Smith CW, Haberland ME, Fogelman AM, Berliner JA. Partial characterization of leukocyte binding molecules on endothelial cells induced by minimally oxidized LDL. Arterioscler Thromb Vasc Biol 1994; 14:427–433.

20. Khan NBV, Parthasarathy S, Alexander RW. Modified LDL and its constituents augment cytokine-activated vascular cell adhesion molecule-1 gene expression in human vascular endothelial cells. J Clin Invest 1995; 95:1262–1270.

21. Rangaswamy S, Penn MS, Saidel GM, Chisolm GM. Exogenous oxidized low density lipoprotein injures and alters the barrier function of endothelium in rats in vivo. Circ Res 1997; 80:37–44.

22. Penn MS, Chisolm GM. Oxidized lipoproteins, altered cell function and atherosclerosis. Atherosclerosis 1994; 108:S21–S29.

23. Brown AJ, Leong SL, Dean RT, Jessup W. 7-hydroxycholesterol and its products in oxidized low density lipoprotein and human atherosclerotic plaque. J Lipid Res 1997; 38:1730–1745.

24. Lynch SM, Morrow JD, Roberts II LJ, Frei B. Formation of non-cyclooxygenase-derived prostanoids (F2-isoprostanes) in plasma and low density lipoprotein exposed to oxidative stress in vitro. J Clin Invest 1994; 93:998–1004.

25. Slatter DA, Paul RG, Murray M, Bailey AJ. Reactions of lipid derived malondialdehyde with collagen. J Biol Chem 1999; 274:19661–19669.

26. Jessup W, Mander EL, Dean RT. The intracellular storage and turnover of apolipoprotein B of oxidized LDL in macrophages. Biochim Biophys Acta 1992; 1126:167–177.

27. Chisolm GM, Hazen ST, Fox PL, Catchard MK. The oxidation of lipoproteins by monocyte-macrophages. J Biol Chem 1999; 274:25959–25962.

28. Parthasarathy S, Printz DJ, Boyd D, Joy L, Steinberg D. Macrophage oxidation of low density lipoprotein generates a modified form recognized by the scavenger receptor. Arteriosclerosis 1986; 6:505–510.

29. Aviram M, Fuhrman B. LDL oxidation by arterial wall macrophages depends on the antioxidative status in the lipoprotein and in the cells: role of prooxidants vs. antioxidants. Mol Cell Biochem 1998; 188:149–159.

30. Aviram M, Rosenblat M. Macrophage mediated oxidation of extracellular low density lipoprotein requires an initial binding of the lipoprotein to its receptor. J Lipid Res 1994; 35:385–398.

31. Aviram M, Rosenblat M, Etzioni A, Levy R. Activation of NADPH oxidase is required for macrophage-mediated oxidation of low density lipoprotein. Metabolism 1996; 45:1069–1079.

32. Herttuala YS, Rosenfeld ME, Parthasarathy S, Glass CK, Sigal E, Witztum JL, Steinberg D. Colocalization of 15-lipoxygenase mRNA and protein with epitopes of oxidized low density lipoprotein in macrophage-rich areas of atherosclerotic lesions. Proc Natl Acad Sci USA 1990; 87:6959–6963.

33. Aviram M, Kent UM, Hollenberg PF. Microsomal cytochrome P450 catalyze the oxidation of low density lipoprotein. Atherosclerosis 1999; 143:253–260.

34. Meister A, Anderson ME. Glutathione. Annu Rev Biochem 1983; 52:711–760.

35. Rosenblat M, Aviram M. Macrophage glutathione content and glutathione peroxidase activity are inversely related to cell-mediated oxidation of LDL. Free Radic Biol Med 1997; 24:305–313.

36. Fuhrman B, Oiknine J, Aviram M. Iron induces lipid peroxidation in cultured macrophages, increases their ability to oxidatively modify LDL and affect their secretory properties. Atherosclerosis 1994; 111:65–78.

37. Fuhrman B, Oiknine J, Keidar S, Kaplan M, Aviram M. 1997 Increased uptake of low density lipoprotein (LDL) by oxidized macrophages is the result of enhanced LDL receptor activity and of progressive LDL oxidation. Free Radic Biol Med 1994; 23:34–46.

38. Mackness MI, Mackness B, Durrington PN, Connelly PW, Hegele RA. Para-oxonases biochemistry, genetics and relationship to plasma lipoproteins. Curr Opin Lipidol 1996; 7:69–76.

39. La Du BN, Adkins S, Kuo CL, Lipsig D. Studies on human serum paraoxonase/arylesterase. Chem Biol Interact 1993; 87:25–34.

40. Aviram M. Does paraoxonase play a role in susceptibility to cardiovascular disease? Mol Med 1999; 5:381–386.

41. La Du BN, Aviram M, Billecke S, Navab M, Primo-Parmo S, Sorenson RC, Standiford TJ. On the physiological role(s) of the paraoxonases. Chem Biol Interact 1999; 119/120: 379–388.

42. Mackness MI, Harty D, Bhatnagar D, Winocour PH, Arrol S, Ishola M, Durrington PN. Serum paraoxonase activity in familial hypercholesterolaemia and insulin-dependent diabetes mellitus. Atherosclerosis 1991; 86:193–197.

43. Abbott CA, Mackness MI, Kumar S, Boulton AJ, Durrington PN. Serum para-oxonase activity, concentration, and phenotype distribution in diabetes mellitus and its relationship to serum lipids and lipoproteins. Arterioscler Thromb Vasc Biol 1995; 15:1812–1818.

44. Garin MC, James RW, Dussoix P, Blanche H, Passa P, Froguel P, Ruiz J. Paraoxonase polymorphism Met-Leu54 is associated with modified serum

concentrations of the enzyme: a possible link between the paraoxonase gene and increased risk of cardiovascular disease in diabetes. J Clin Invest 1997; 99:62–66.

45. Aviram M, Rosenblat M, Bisgaier CL, Newton RS, Primo-Parmo SL, La Du BN. Paraoxonase inhibits high density lipoprotein (HDL) oxidation and preserves its functions: a possible peroxidative role for paraoxonase. J Clin Invest 1998; 101:1581–1590.

46. Aviram M, Billecke S, Sorenson R, Bisgaier C, Newton R, Rosenblat M, Erogul J, Hsu C, Dunlp C, La Du BN. Paraoxonase active site required for protection against LDL oxidation involves its free sulfhydryl group and is different from that required for its arylesterase/paraoxonase activities: selective action of human paraoxonase allozymes Q and R. Arterioscler Thromb Vasc Biol 1998; 18:1617–624.

47. Aviram M, Rosenblat M, Billecke S, Erogul J, Sorenson R, Bisgaier CL, Newton RS, La Du B. Human serum paraoxonase (PON 1) is inactivated by oxidized low density lipoprotein and preserved by antioxidants. Free Radic Biol Med 1999; 26:892–904.

48. Keaney JF, Frei B. Antioxidant protection of low-density lipoprotein and its role in the prevention of atherosclerotic vascular disease. In: Frei B, ed. Natural Antioxidants in Human Health and Disease. San Diego: Academic Press, 1994: 303–352.

49. Aviram M. Antioxidants in restenosis and atherosclerosis. Curr Interven Cardiol Rep 1999; 1:66–78.

50. Vaya J, Aviram M. Nutritional antioxidants: mechanisms of action, analyses of activities and medical application. Curr Med Chem 2001; 1:99–117.

51. Aviram M. Review of human studies on oxidative damage and antioxidant protection related to cardiovascular diseases. Free Radic Res 2000; 33:S85–S97.

52. Aviram M. Macrophage foam cell formation during early atherogenesis is determined by the balance between pro-oxidants and antioxidants in arterial cells and blood lipoproteins. Antiox Redox Signal 1999; 1:585–594.

53. Levy Y, Ben-Amotz A, Aviram M. Effect of dietary supplementation of β-Carotene to humans on its binding to plasma LDL and on the lipoprotein susceptibility to undergo oxidative modification: comparison of the synthetic all trans isomer with the natural algae β-carotene. J Nutr Environ Med 1995; 5:13–22.

54. Levy Y, Kaplan M, Ben Amotz A, Aviram M. The effect of dietary supplementation of β-carotene on human monocyte-macrophage-mediated oxidation of low density lipoprotein. Isr J Med Sci 1996; 32:473–478.

55. Fuhrman B, Ben-Yaish L, Attias J, Hayek T, Aviram M. Tomato's lycopene and β-carotene inhibit low density lipoprotein oxidation and this effect depends on the lipoprotein vitamin E content. Nutr Metab Cardiovasc Dis 1997; 7:433–443.

56. Fuhrman B, Volkova N, Rosenblat M, Aviram M. Lycopene synergistically inhibits LDL oxidation in combination with vitamin E, glabridin, rosmarinic acid, carnosic acid, or garlic. Antiox Redox Signal 2000; 2:491–506.

57. Maor I, Hayek T, Coleman R, Aviram M. Plasma LDL oxidation leads to its aggregation in atherosclerotic apolipoprotein E-deficient mice. Arterioscler Thromb Vasc Biol 1997; 17:2995–3005.

58. Fuhrman B, Aviram M. Anti-atherogenicity of nutritional antioxidants. Drugs 2001; 4:82–92.

59. Hertog MG, Kromhout D, Aravanis C, Blackburn H, Buzina R, Fidanza F, Giampaoli S, Jansen A, Menotti A, Nadeljkovic S. Flavonoid intake and long-term risk of coronary heart disease and cancer in the seven countries study. Arch Intern Med 1995; 155:381–386.

60. Rice-Evans CA, Miller NJ, Bolwell PG, Bramley PM, Pridham JB. The relative antioxidant activities of plant-derived polyphenolic flavonoids. Free Radic Res 1995; 22:375–383.

61. Rice-Evans CA, Miller NJ, Paganga G. Structure-antioxidant activity relationships of flavonoids and phenolic acids. Free Radic Biol Med 1996; 20: 933–956.

62. Van Acker SABE, Van-den Berg DJ, Tromp MNJL, Griffioen DH, van Bennekom WP, Van der Vijgh WJF, Bast A. Structural aspects of antioxidants activity of flavonoids. Free Radic Biol Med 1996; 20:331–342.

63. Brown JA, Khodr H, Hider RC, Rice-Evans C. Structural dependence of flavonoid interactions with Cu^{2+} ions: implications for their antioxidant properties. Biochem J 1998; 330:1173–1178.

64. Chang WS, Chang YH, Lu FJ, Chiang HC. Inhibitory effects of phenolics on xanthine oxidase. Anticancer Res 1994; 14:501–506.

65. Rosenblat M, Belinky P, Vaya J, Levy R, Hayek T, Coleman R, Merchav S, Aviram M. Macrophage enrichment with the isoflavan glabridin inhibits NADPH oxidase-induced cell mediated oxidation of low density lipoprotein. J Biol Chem 1999; 274:13790–13799.

66. Hsiech R, German B, Kinsella J. Relative inhibitory potencies of flavonoids on 12-lipoxygenase of fish gill. Lipids 1988; 23:322–326.

67. Baumann J, Bruchhausen F, Wurm G. Flavonoids and related compounds as inhibitors of arachidonic acid peroxidation. Prostaglandins 1980; 20:627–639.

68. Luiz da Silva E, Tsushida T, Terao J. Inhibition of mammalian 15-lipoxygenase-dependent lipid peroxidation in low density lipoprotein by quercetin and quercetin monoglucosides. Arch Biochem Biophys 1998; 349:313–320.

69. Lotito SB, Fraga CG. (+) -Catechin prevents human plasma oxidation. Free Radic Biol Med 1998; 24:435–441.

70. Mangiapane H, Thomson J, Salter A, Brown S, Bell GD, White DA. The inhibition of the oxidation of low density lipoprotein by (+)-catechin, a naturally occurring flavonoid. Biochem Pharmacol 1992; 43:445–450.

71. De Whalley CV, Rankin SM, Hoult RS, Jessup W, Leake DS. Flavonoids inhibit the oxidative modification of low density lipoproteins by macrophages. Biochem Pharmacol 1990; 39:1743–1750.

72. Vaya J, Belinky PA, Aviram M. Antioxidant constituents from licorice roots: isolation, structure elucidation and antioxidative capacity toward LDL oxidation. Free Radic Biol Med 1997; 23:302–313.

73. Belinky PA, Aviram M, Fuhrman B, Rosenblat M, Vaya J. The antioxidative effects of the isoflavan glabridin on endogenous constituents of LDL during is oxidation. Atherosclerosis 1998; 137:49–61.

74. Namiki M, Osawa T. Antioxidants/mutagens in foods. Basic Life Sci 1986; 39;131–142.

75. Shimoi K, Masuda S, Furugori M, Esaki S, Kinae N. Radioprotective effect of antioxidative flavonoids in gamma-ray irradiated mice. Carcinogenesis 1986; 15:2669–2672.

76. Bravo L. Polyphenols: chemistry, dietary sources, metabolism, and nutritional significance. Nutr Rev 1998; 56:317–333.

77. Leibovitz BE, Mueller JA. Bioflavonoids and polyphenols: medical application. J Optim Nutr 1993; 2:17–35.

78. de Vries JH, Hollman PC, Meyboom S, Buysman MN, Zock PL, van Staveren WA, Katan MB. Plasma concentrations and urinary excretion of the antioxidant flavonols quercetin and kaempferol as biomarkers for dietary intake. Am J Clin Nutr 1998; 68:60–65.

79. Hollman PCH. Bioavailability of flavonoids. Eur J Clin Nutr 1997; 51(suppl 1):S66–S69.

80. Hertog MGL, Hollman POH, Katan MB, Kromhout D. Intake of potentially anticarcinogenic flavonoids and their determinants in adults in The Netherlands. Nutr Cancer 1993; 20:21–29.

81. German JB, Walzem RL. The health benefits of wine. Annu Rev Nutr 2000; 20:561–593.

82. Goldberg IJ, Mosca L, Piano MR, Fisher EA. Wine and your heart. Circulation 2001; 103:472–475.

83. Renaud S, de Lorgeril M. Wine alcohol, platelets and the French paradox for coronary heart disease. Lancet 1992; 339:1523–1526.

84. Soleas GJ, Diamandis EP, Goldberg DM. Wine as a biological fluid: history, production, and role in disease prevention. J Clin Lab Anal 1997; 11:287–313.

85. Hertog MGL, Hollman PCH, van de Putte B. Content of potentially anticarcinogenic flavonoids of tea infusions, wines, and fruit juices. J Agric Food Chem 1993; 41:1242–1246.

86. Serafini M, Maiani G, Ferro-Luzzi A. Alcohol-free red wine enhances plasma antioxidant capacity in humans. J Nutr 1998; 128:1003–1007.

87. Fuhrman B, Aviram M. White wine reduces the susceptibility of low density lipoprotein to oxidation. Am J Clin Nutr 1996; 63(3):403–404.

88. Lamuela-Raventos RM, de la Torre-Boronat MC. Beneficial effects of white wines. Drugs Exp Clin Res 1999; 25:121–124.

89. Paganga G, Miller N, Rice-Evans CA. The polyphenolic content of fruit and vegetables and their antioxidant activities: what does a serving constitute? Free Radic Res 1999; 30:153–162.

90. Rifici VA, Stephan EM, Schneider SH, Khachadurian AK. Red wine inhibits the cell-mediated oxidation of LDL and HDL. J Am Coll Nutr 1999; 18:137–143.

91. Vinson JA, Hontz BA. Phenol antioxidant index: comparative antioxidant effectiveness of red and white wines. J Agric Food Chem 1995; 43:401–403.

92. Fuhrman B, Lavy A, Aviram M. Consumption of red wine with meals reduces the susceptibility of human plasma and LDL to undergo lipid peroxidation. Am J Clin Nutr 1995; 61:549–554.

93. Frankel EN, Waterhouse AL, Teissedre PL. Principal phenolic phytochemicals in selected Californian wines and their antioxidant activity in inhibiting oxidation of human low-density lipoproteins. J Agric Food Chem 1995; 43: 890–893.

94. Lairon D, Amiot MJ. Flavonoids in food and natural antioxidants in wine. Curr Opin Lipidol 1999; 10:23–28.

95. Frankel EN, Kanner J, German JB, Parks E, Kinsella JE. Inhibition of oxidation human low-density lipoprotein by phenolic substances in red wine. Lancet 1993; 341:454–457.

96. Fuhrman B, Volkova N, Aviram M. White wine with red wine-like properties: increased extraction of grape skin's-polyphenols improves the antioxidant capacity of the derived white wine. J Agric Food Chem 2001; 49:3164–3168.

97. Whitehead TP, Robinson D, Allaway S, Syms J, Hale A. Effect of red wine ingestion on the antioxidant capacity of serum. Clin Chem 1995; 41:32–35.

98. Aviram M, Fuhrman B. Wine flavonoids protect against LDL oxidation and atherosclerosis. Annals NY Acad Sci 2002; 957:146–161.

99. Miyagi Y, Miwa K, Inoue H. Inhibition of human low density lipoprotein oxidation by flavonoids in red wine and grape juice. Am J Cardiol 1997; 80:1627–1631.

100. Nigdikar SV, Williams N, Griffin BA, Howard AH. Consumption of red wine polyphenols reduces the susceptibility of low density lipoproteins to oxidation in vivo. Am J Clin Nutr 1998; 68:258–265.

101. de Rijke YB, Demacker PN, Assen NA, Sloots LM, Katan MB, Stalenhoef AF. Red wine consumption does not affect oxidizability of low-density lipoprotein volunteers. Am J Clin Nutr 1996; 63:329–334.

102. Abu-Amsha CR, Burke V, Mori TA, Beilin LJ, Puddey IB, Croft KD. Red wine polyphenols, in the absence of alcohol, reduce lipid peroxidative stress in smoking subjects. Free Radic Biol Med 2001; 30:636–642.

103. Chopra M, Fitzsimons PEE, Strain JJ, Thurnham DI, Howard AN. Nonalcoholic red wine extract and quercetin inhibit LDL oxidation without affecting plasma antioxidant vitamin and carotenoid concentrations. Clin Chem 2000; 46(8):1162–1170.

104. Fremon L, Belguendouz L, Delpal S. Antioxidant activity of resveratrol and alcohol-free wine polyphenols related to LDL oxidation and polyunsaturated fatty acids. Life Sci 1999; 64:2511–2521.

105. Stein JH, Keevil JG, Wiebe DA, Aeschlimann S, Folts JD. Purple grape juice improves endothelial function and reduces the susceptibility of LDL cholesterol to oxidation in patients with coronary artery disease. Circulation 1999; 100:1050–1055.

106. van Golde PH, Sloots LM, Vermeulen WP, Wielders JP, Hart HC, Bouma BN, van de Wiel A. The role of alcohol in the anti low density lipoprotein oxidation activity of red wine. Atherosclerosis 1999; 147:365–370.

107. Ritchey JG, Waterhouse AL. A standard red wine: monomeric phenolic analysis of commercial Cabernet sauvignon wines. Am J Enol Viticul 1999; 50(1): 91–100.

108. McDonald MS, Hughes M, Burns J, Lean MEJ, Matthews D, Crozier A. Survey of free and conjugated myricetin and quercetin content of red wines of different geographical origin. J Agric Food Chem 1998; 46:368–375.

109. Price SF, Breen PJ, Valladao M, Watson BT. Clusters sun exposure and quercetin in Pinot noir grapes and wines. Am J Enol Viticul 1995; 46:187–194.

110. Carbonneau MA, Leger CL, Monnier L, Bonnet C, Michel F, Fouret G. Supplementation with wine phenolic compounds increases the antioxidant capacity of plasma and vitamin E of low-density lipoprotein without changing the lipoprotein Cu^{2+} oxidizability: possible explanation by phenolic location. Eur J Clin Nutr 1997; 51:682–690.

111. Hayek T, Fuhrman B, Vaya J, Rosenblat M, Belinky P, Coleman R, Elis A, Aviram M. Reduced progression of atherosclerosis in the apolipoprotein E deficient mice following consumption of red wine, or its polyphenols quercetin or catechin, is associated with reduced susceptibility of LDL to oxidation and aggregation. Arterioscler Thromb Vasc Biol 1997; 17:2744–2752.

112. Aviram M, Hayek T, Fuhrman B. Red wine consumption inhibits LDL oxidation and aggregation in humans and in atherosclerotic mice. Biofactors 1997; 6:415–419.

113. Bentzon JF, Skovenborg E, Hansen C, Moler J, Saint-Cricq de Gaulejac N, Proch J, Falk E. Red wine does not reduce mature atherosclerosis in apolipoprotein E-deficient mice. Circulation 2001; 103:1681.

114. Vinson JA, Teufel K, Wu N. Red wine, dealcoholized red wine, and especially grape juice, inhibit atherosclerosis in a hamster model. Atherosclerosis 2001; 156:67–72.

115. Nardini M, D'Aquino M, Tomassi G, Gentili V, Di Felice M, Scaccini C. Inhibition of human low density lipoprotein oxidation by caffeic acid and other hydroxycinnamic acid derivatives. Free Radic Biol Med 1995; 19:541–552.

116. Abu-Amsha R, Croft KD, Puddey IB, Proudfoot JM, Beilin LJ. Phenolic content of various beverages determines the extent of inhibition of human serum and low density lipoprotein oxidation in vitro: identification and mechanism of some cinnamic derivatives from red wine. Clin Sci 1996; 91:449–458.

117. Frankel EN, Waterhouse AL, Kinsella JE. Inhibition of human LDL oxidation by resveratrol. Lancet 1993; 341:1103–1104.

118. Manach C, Morand C, Texier O, Favier ML, Agullo G, Demigne C, Regerat F, Remesy C. Quercetin metabolites in plasma of rats fed diets containing rutin or quercetin. J Nutr 1995; 125:1911–1922.

119. Vinson JA, Dabbagh YA, Serry MM, Janj J. Plant flavonoids, especially tea flavonols, are powerful antioxidants using an in vitro model for heart disease. J Agric Food Chem 1995; 45:2800–2802.

120. Salah N, Miller NJ, Paganga G, Tijburg L, Bolwell GP, Rice Evans C. Polyphenolic flavanols as scavengers of aqueous phase radicals and as chain-breaking antioxidants. Arch Biochem Biophys 1995; 322:339–346.

121. Lanninghamfoster L, Chen C, Chance DS, Loo G. Grape extract inhibits lipid peroxidation of human low density lipoprotein. Biol Pharm Bull 1995; 18:1347–1351.

122. Rao AV, Shen H, Agarwal A, Yatcilla MT, Agarwal S. Bioabsorption and in vivo antioxidant properties of grape extract Biovin: a human intervention study. J Med Food 2000; 3:15–22.
123. Kerry NL, Abbey M. Red wine and fractionated phenolic compounds prepared from red wine inhibit low density lipoprotein oxidation in vitro. Atherosclerosis 1997; 135:93–102.
124. Fuhrman B, Aviram M. Paraoxonase activity is preserved by wine flavonoids: possible role in protection of LDL from lipid peroxidation. Annals NY Acad Sci 2002; 957:321–324.
125. van der Gaag MS, van Tol A, Scheek LM, James RW, Urgert R, Schaafsma G, Hendriks HFJ. Daily moderate alcohol consumption increases serum paraoxonase activity: a diet controlled, randomized intervention study in middle-aged men. Atheroclerosis 1999; 147:405–410.
126. Fuhrman B, Buch S, Vaya J, Belinky PA, Coleman R, Hayek T, Aviram M. Licorice extract and its major polyphenol glabridin protect low-density lipoprotein against lipid peroxidation: in vitro and ex vivo studies in humans and in athero-sclerotic apolipoprotein E-deficient mice. Am J Clin Nutr 1997; 66:267–275.
127. Fuhrman B, Volkova N, Kaplan M, Presser D, Attias J, Hayek T, Aviram M. Antiatherosclerotic effects of licorice extract supplementation to hypercholester-olemic patients: increased resistance of their LDL to atherogenic modifications, reduced plasma lipids levels, and decreased systolic blood pressure. Nutrition 2002; 18:268–273.
128. Haraguchi H, Ishikawa H, Mizutani K, Tamura Y, Kinoshita T. Antioxidative and superoxide scavenging activities of retrochalcones in Glycyrrhiza inflata. Bioorg Med Chem 1998; 6:339–347.
129. Belinky PA, Aviram M, Mahmood S, Vaya J. Structural aspects of the inhibitory effect of glabridin on LDL oxidation. Free Radic Biol Med 1998; 24:1419–1429.
130. Langley P. Why a pomegranate? Br Med J 2000; 321:1153–1154.
131. Ben Nasr C, Ayed N, Metche M. Quantitative determination of the polyphenolic content of pomegranate peel. Z Lebensm Unters Forsch 1996; 203:374–378.
132. El-Toumy SAA, Marzouk MSA. A new ellagic acid glycoside from *Punica granatum* L. Polyphenols Communications. Germany: Freising-Weihenstephan, 2000:127–128.
133. Gil MI, Tomas-Barberan FA, Hess-Pierce B, Holcroft DM, Kader AA. Antioxidant activity of pomegranate juice and its relationship with phenolic composition and processing. J Agric Food Chem 2000; 48:4581–4589.
134. Heftaman E, Bennett ST. Identification of estrone in pomegranate seeds. Phytochemistry 1966; 5:1337–1339.
135. Moneam NMA, El Sharasky AS, Badreldin MM. Oestrogen content of pomegranate seeds. J Chromatogr 1988; 438:438–442.
136. Sharaf A, Nigm SAR. The oestrogenic acitvity of pomegranate seed oil. J Endocrinol 1964; 29:91–92.
137. Shubert YS, Lansky EP, Neeman I. Antioxidant and eicosanoid enzyme inhibition properties of pomegranate seed oil and fermented juice flavonoids. J Ethno-pharmacol 1999; 66:11–17.

138. Aviram M, Dorenfeld L, Rosenblat M, Volkova N, Kaplan M, Hayek T, Presser D, Fuhrman B. Pomegranate juice consumption reduces oxidative stress, atherogenic modifications to LDL, and platelet aggregation: studies in humans and in the atherosclerotic apolipoprotein E deficient mice. Am J Clin Nutr 2000; 71:1062–1076.

139. Aviram M, Dornfeld L. Pomegranate juice consumption inhibits serum angiotensin converting enzyme activity and reduces systolic blood pressure. Atherosclerosis 2001; 158:195–198.

140. Aviram M, Dornfeld L, Kaplan M, Coleman R, Gaitini D, Nitecki S, Hoffman A, Rosenblat M, Volkova N, Presser D, Attias J, Hayek T, Fuhrman B. Pomegranate juice flavonoids inhibit LDL oxidation and cardiovascular diseases: studies in atherosclerotic mice and in humans. Proc Int Conf Mech Action Nutraceuticals (ICMAN). 2002; 28:49–62.

141. He K, Lyer KR, Hayes RN. Inactivation of cytochrome P-450 3A4 by bergamottin, a component of grapefruit juice. Chem Res Toxicol 1998; 11:252–259.

142. Kaplan M, Hayek T, Raz A, Coleman R, Dornfeld L, Vaya J, Aviram M. Pomegranate juice supplementation to atherosclerotic mice reduces macrophages lipid peroxidation, cellular cholesterol accumulation and development of atherosclerosis. J Nutr 2001; 131:2082–2089.

143. Thelle DS. Coffee, tea and coronary heart disease. Curr Opin Lipidol 1995; 6:25–27.

144. Yang CS. Tea and health. Nutrition 1999; 15:946–949.

145. Stensvold I, Tverdal A, Solvoll K, Foss PO. Tea consumption, relationship to cholesterol, blood pressure and coronary and total mortality. Prev Med 1992; 21:546–553.

146. Graham HN. Green tea composition, consumption and polyphenol chemistry. Prev Med 1992; 21:334–334.

147. Weisburger JH. Tea and health: the underlying mechanisms. Proc Soc Exp Biol Med 1999; 220:271–275.

148. Miyazawa T. Absorption, metabolism and antioxidative effects of tea catechins in humans. Biofactors 2000; 13:55–59.

149. Van het Hof KH, de Boer HS, Wiseman SA, Lien N, Westrate JA, Tijburg LB. Consumption of green or black tea does not increase resistance of low density lipoprotein to oxidation in humans. Am J Clin Nutr 1997; 66:1125–1132.

150. van het Hof KH, Wiseman SA, Yang CS, Tijburg LB. Plasma and lipoprotein levels of tea catechins following repeated tea consumption. Proc Soc Exp Biol Med 1999; 220:203–209.

151. McAnlis GT, McEneny J, Pearce J, Young IS. Black tea consumption does not protect low-density lipoprotein from oxidative modification. Eur J Clin Nutr 1998; 52:202–206.

152. Princen HM, van Duyvennvoorde W, Buytenhek R, Blonk C, Tijburg LB, Langius JA, Meinders AE, Pijl H. No effect of consumption of green and black tea on plasma lipid and antioxidant levels and on LDL oxidation in smokers. Arterioscler Thromb Vasc Biol 1998; 18:833–841.

153. O'Reilly JD, Mallet AI, McAnlis GT, Young IS, Halliwell B, Sanders TA,

Wiseman H. Consumption of flavonoids in onions and black tea: lack of effect on F2-isoprostanes and autoantibodies to oxidized LDL in healthy humans. Am J Clin Nutr 2001; 73:1040–1044.

154.   Serafini M, Ghiselli A, Ferro-Luzzi A. In vivo antioxidant effect of green and black tea in man. Eur J Clin Nutr 1996; 50:28–32.

155.   Benzie IFF, Szeto YT, Tomlinson B, Strain JJ. Drinking green tea leads to a rapid increase in plasma antioxidant potential. In: Kumpulainen JT, Salonen JT, eds. Natural Antioxidants and Anticarcinogens in Nutrition, Health and Disease. London: The Royal Society of Chemistry, 1999:280–282.

156.   Ishikawa T, Suzukawa M, Ito T, Yioshida H, Ayaori M, Nishiwaki M, Yonemura A, Hara Y, Nakamura H. Effect of tea flavonoid supplementation on the susceptibility of low density lipoprotein to oxidative modification. Am J Clin Nutr 1997; 66:261–266.

157.   Serafini M, Laranjinha JA, Almeida LM, Maiani G. Inhibition of human LDL lipid peroxidation by phenol-rich beverages and their impact on plasma total antioxidant capacity in humans. J Nutr Biochem 2000; 11:585–590.

158.   Hodgson JM, Puddey IB, Croft KD, Burke V, Mori TA, Caccetta RA, Beilin LJ. Acute effects of ingstion of black and green tea on lipoprotein oxidation. Am J Clin Nutr 2000; 71:1103–1107.

159.   Tijburg LB, Wiseman SA, Meijer GW, Weststrate JA. Effects of green tea, black tea and dietary lipophilic antioxidants on LDL oxidizability and atherosclerosis in hypercholesterolaemic rabbits. Atherosclerosis 1997; 135:37–47.

160.   Vinson JA, Dabbagh YA. Effect of green and black tea supplementation on lipids, lipid oxidation and fibrinogen in the hamster: mechanisms for the epidemiological benefits of tea drinking. FEBS Lett 1998; 14:44–46.

161.   Miura Y, Chiba T, Tomita I, Koizumi H, Miura S, Umegaki K, Hara Y, Ikeda M, Tomita T. Tea catechins prevent the development of atherosclerosis in apoprotein E-deficient mice. J Nutr 2001; 131:27–32.

162.   Yang TTC, Koo MWL. Inhibitory effect of Chinese green tea on endothelia cell-induced LDL oxidation. Atherosclerosis 2000; 148:67–73.

163.   Miura Y, Chiba T, Miura S, Tomita I, Umegaki K, Ikeda M, Tomita T. Green tea polyphenols (flavan 3-ols) prevent oxidative modification of low density lipoproteins: an ex vivo study in humans. J Nutr Biochem 2000; 11: 216–222.

164.   Yamanaka N, Oda O, Nagao S. Green tea catechins such as (−)-epicatechin and (−)-epigallocatechin accelerate Cu2+-induced low density lipoprotein oxidation in propagation phase. FEBS Lett 1997; 20:230–234.

165.   Liu ZQ, Ma LP, Zhou B, Yang L, Liu ZL. Antioxidative effects of green tea polyphenols on free radical initiated and photosensitized peroxidation of human low density lipoprotein. Chem Phys Lipids 2000; 106:53–63.

166.   Osada K, Takahashi M, Hoshina S, Nakamura M, Nakamura S, Sugano M. Tea catechins inhibit cholesterol oxidation accompanying oxidation of low density lipoprotein in vitro. Comp Biochem Physiol C Pharmacol Toxicol Endocrinol 2001; 128:153–164.

167.   Yoshida H, Ishikawa T, Hosoai H, Suzukawa M, Ayaori M, Hisada T, Sawada S, Yonemura A, Higashi K, Ito T, Nakajima K, Yamashita T, Tomiyasu K,

Nishiwaki M, Ohsuzu F, Nakamura H. Inhibitory effect of tea flavonoids on the ability of cells to oxidize low density lipoprotein. Biochem Pharmacol 1999; 58:1695–1703.

168. Zhu QY, Huang Y, Tsang D, Chen ZY. Regeneration of alpha-tocopherol in human low-density lipoprotein by green tea catechin. J Agric Food Chem 1999; 47:2020–2025.

169. Keys A. Mediterranean diet and public health: personal reflections. Am. J Clin Nutr 1995; 61:1321S–1323S.

170. Fidanza F, Puddu V, Imbimbo AB, Menotti A, Keys A. Coronary heart disease in seven countries. VII. Five-year experience in rural Italy. Circulation 1970; 41:163–175.

171. Lopez-Miranda J, Gomez P, Castro P, Marin C, Paz E, Bravo MD, Blanco J, Jimenez-Pereperez J, Fuentes F, Perez-Jimenez F. Mediterranean diet improves low density lipoprotein susceptibility to oxidative modifications. Med Clin (Barc) 2000; 115:361–365.

172. Visioli F, Galli C. Natural antioxidants and prevention of coronary heart disease: a potential role of olive oil and its minor constituents. Nutr Metab Cardiovasc Dis 1995; 5:306–314.

173. Tsimikas S, Philis-Tsimikas A, Alexopoulos S, Sigari F, Lee C, Reaven PD. 1999 LDL isolated from Greek subjects on a typical diet or from American subjects on an oleate-supplemented diet induces less monocyte chemotaxis and adhesion when exposed to oxidative stress. Arterioscler Thromb Vasc Biol 1995; 19:122–130.

174. Aviram M, Eias K. Dietary olive oil reduces low-density lipoprotein uptake by macrophages and decreases the susceptibility of the lipoprotein to undergo lipid peroxidation. Ann Nutr Metab 1993; 37:75–84.

175. Nicolaiew N, Lemort N, Adorni L, Berra B, Montorfano G, Rapelli S, Cortesi N, Jacotot B. Comparison between extra virgin olive oil and oleic acid rich sunflower oil: effects on postprandial lipemia and LDL susceptibility to oxidation. Ann Nutr Metab 19981; 42:251–260.

176. Wiseman SA, Mathot JN, de Fouw NJ, Tijburg LB. Dietary non-tocopherol antioxidants present in extra virgin olive oil increase the resistance of low density lipoprotein to oxidation in rabbits. Atherosclerosis 1996; 120:15–23.

177. Ramirez-Tortosa MC, Urbano G, Lopez-Jurado M, Nestares T, Gomez MC, Mir A, Ros E, Mataix J, Gil A. Extra-virgin olive oil increases the resistance of LDL to oxidation more than refined olive oil in free-living men with peripheral vascular disease. J Nutr 1999; 129:2177–2183.

178. Ramirez-Tortosa MC, Suarez A, Gomez MC, Mir A, Ros E, Mataix J, Gil A. Effect of extra-virgin olive oil and fish-oil supplementation on plasma lipids and susceptibility of low-density lipoprotein to oxidative alteration in free-living Spanish male patients with peripheral vascular disease. Clin Nutr 1999; 18:167–174.

179. Ramirez-Tortosa C, Lopez-Pedrosa JM, Suarez A, Ros E, Mataix J, Gil A. Olive oil and fish oil-enriched diets modify plasma lipids and susceptibility of LDL to oxidative modification in free-living male patients with peripheral vascular disease: the Spanish Nutrition Study. Br J Nutr 1999; 82:31–39.

180. Vissers MN, Zock PL, Wiseman SA, Meyboom S, Katan MB. Effect of phenol-rich extra virgin olive oil on markers of oxidation in healthy volunteers. Eur J Clin Nutr 2001; 55:334–341.

181. Caruso D, Berra B, Giavarini F, Cortesi N, Fedeli E, Galli G. Effect of virgin olive oil phenolic compounds on in vitro oxidation of human low density lipoproteins. Nutr Metab Cardiovasc Dis 1999; 9:102–107.

182. Covas MI, Fito M, Lamuela-Raventos RM, Sebastia N, de la Torre-Boronat C, Marrugat J. Virgin olive oil phenolic compounds: binding to human low density lipoprotein (LDL) and effect on LDL oxidation. Int J Clin Pharmacol Res 2000; 20:49–54.

183. Fito M, Fito M, Lamuela-Raventos RM, Vila J, Torrents L, de la Torre C, Marrugat J. Protective effect of olive oil and its phenolic compounds against low density lipoprotein oxidation. Lipids 2000; 35:633–638.

184. Visioli F, Galli C. The effect of minor constituents of olive oil on cardiovascular disease: new findings. Nutr Rev 1998; 56:142–147.

185. Visioli F, Bellomo G, Montedoro G, Galli C. Low density lipoprotein oxidation is inhibited in vitro by olive oil constituents. Atherosclerosis 1995; 117:25–32.

186. Visioli F, Galli C. Oleuropein protects low density lipoprotein from oxidation. Life Sci 1994; 55:1965–1971.

187. Visioli F, Bellomo G, Galli C. Free radical-scavenging properties of olive oil polyphenols. Biochem Biophys Res Commun 1998; 247:60–64.

188. Coni E, Di Benedetto R, Di Pasquale M, Masella R, Modesti D, Mattei R, Carlini EA. Protective effect of oleuropein, an olive oil biophenol, on low density lipoprotein oxidizability in rabbits. Lipids 2000; 35:45–54.

189. Giovannini C, Straface E, Modesti D, Coni E, Cantafora A, De Vincenzi M, Malorni W, Masella R. Tyrosol, the major olive oil biophenol, protects against oxidized-LDL-induced injury in Caco-2 cells. J Nutr 1999; 129:1269–1277.

190. Masella R, Cantafora A, Modesti D, Cardilli A, Gennaro L, Bocca A, Coni E. Antioxidant activity of 3,4-DHPEA-EA and protocatechuic acid: a comparative assessment with other olive oil biophenols. Redox Rep 1999; 4:113–121.

191. Visioli F, Romani A, Mulinacci N, Zarini S, Conte D, Vincieri FF, Galli C. Antioxidant and other biological activities of olive mill waste waters. J Agric Food Chem 1999; 47:3397–3401.

192. de la Puerta R, Ruiz Gutierrez V, Hoult JR. Inhibition of leukocyte 5-lipoxygenase by phenolics from virgin olive oil. Biochem Pharmacol 1999; 57:445–449.

193. Petroni A, Blasevich M, Papini N, Salami M, Sala A, Galli C. Inhibition of leukocyte leukotriene B4 production by an olive oil-derived phenol identified by mass-spectrometry. Thromb Res 1997; 87:315–322.

194. Visioli F, Bellosta S, Galli C. Oleuropein, the bitter principle of olives, enhances nitric oxide production by mouse macrophages. Life Sci 1998; 62:541–546.

195. Leger CL, Kadiri-Hassani N, Descomps B. Decreased superoxide anion production in cultured human promonocyte cells (THP-1) due to polyphenol mixtures from olive oil processing wastewaters. J Agric Food Chem 2000; 48:5061–5067.

196. Miles EA, Wallace FA, Calder PC. An olive oil-rich diet reduces scavenger receptor mRNA in murine macrophages. Br J Nutr 2001; 85:185–191.

197. Aviram M. Macrophage-mediated oxidation of LDL, antioxidants, and atherosclerosis. J Isr Heart Soc 1998; 8:6–8.
198. Aviram M, Fuhrman B. Polyphenolic flavonoids inhibit macrophage-mediated oxidation of LDL and attenuate atherogenesis. Atherosclerosis 1998; 137(suppl): S45–S50.
199. Fuhrman B, Aviram M. Flavonoids protect LDL from oxidation and attenuate atherosclerosis. Curr Opin Lipidol 2001; 12:41–48.
200. Fuhrman B, Aviram M. Polyphenols and flavonoids protect LDL against atherogenic modifications. In: Handbook of Antioxidants: Biochemical, Nutritional and Clinical Aspects, 2d ed. New York: Marcel Dekker, 2001:303–336.

8
Phytochemicals and Brain Aging: A Multiplicity of Effects

Kuresh A. Youdim
King's College London
London, England

James A. Joseph
Jean Mayer USDA Human Nutrition Center on Aging at Tufts University
Boston, Massachusetts, U.S.A.

> If we could give every individual the right amount of nourishment and exercise, not too little and not too much, we would have found the safest way to health.
>
> Hippocrates c. 460–377 B.C.E.

I. INTRODUCTION

Throughout the millennia, humankind has pursued the concept of achieving eternal youth. However, if an expanded existence is to be of benefit, it should be coupled with a high quality of life. Unfortunately, progress toward the goal of healthy longevity has often lagged well behind the aspirations of philosophers. Nonetheless, scientists have long been interested in the aging process, which has received renewed attention as a result of the ever-increasing number of aged persons and the significant burden of age-related disease on national expenditure.

Changes in the optimal performance of biological systems invariably impact health. However, alterations in the efficient functioning of the central nervous system (CNS) have perhaps the most devastating repercussions. Among those deficits having the greatest impact are those associated with dementia, a general term describing symptoms exhibited by people with various kinds of cognitive impairment. Common disorders in which dementia is observed include

Alzheimer's disease (AD); cerebrovascular disease, such as successive strokes or lesions; Parkinson's disease (PD); Creutzfeldt-Jakob disease; alcoholism; and certain traumas. Of greater concern perhaps are the changes in the optimal performance of the CNS that occur simply as a function of aging, possibly exacerbating common correlative motor and cognitive behavioral changes exhibited in these conditions. Age-related deficits in memory occur primarily in secondary memory systems and are reflected in the retrieval of newly acquired information. The impairments in retrieval are attributed to deficits in such encoding processes as motivation, attention, processing depth, and organizational skills and have been characterized in animals [1–3] and humans [4,5]. In contrast, motor performance deficits are thought to result from alterations in the striatal dopamine system [6] or in the cerebellum [7], whereas age-related memory decrements can result from alterations in either the hippocampus (which mediates allocentric spatial navigation or place learning) or the striatum (which mediates egocentric spatial orientation and response-cue learning). The mechanisms involved in the losses of sensitivity in the various receptor systems and the related loss in cognitive and motor behavior are the subjects of continued research. The critical issue is that additional research has shown that this sensitivity may increase further during aging [8].

Several factors contribute to losses of neuronal function and behavior observed in brain aging (Fig. 1). Examples include age-related changes in sensitivity to oxidative [9–11] and inflammatory stresses [12,13], endogenous antioxidant system [14] receptor sensitivity [15–18], membrane alterations [19], and calcium homeostasis [20]. With the involvement of these various pathophysiological processes, employing agents that nullify their actions could potentially provide effective neuroprotective therapies. Thus, there is a growing interest in the establishment of therapeutic strategies to combat oxidative stress–induced

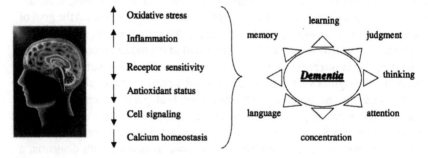

Figure 1 Common changes observed in the aging brain that are associated with dementia.

damage to the CNS, and attention is turning to the potential neuroprotective effects of dietary antioxidants, especially flavonoids. Only recently have studies been performed that focus on the potential for flavonoids per se to mediate neuroprotection. It is not clear whether the neuroprotective effects of flavonoids involve their reducing properties or some other mechanism independent of their antioxidant activities. Their precise mechanisms of action in vivo depend on the extent to which they are conjugated and metabolized during absorption (see Chap. 12) and the ability of bioavailable compounds to localize within the brain.

II. DE NOVO LOCALIZATION

The inherent difficulties in the characterization, distribution, and localization of flavonoids within the body pose a major hurdle when attempting to investigate mechanisms of flavonoid bioactivity in vivo. Compounds reported to localize within brain structures include epigallocatechin gallate [21], a major polyphenolic in tea and to a lesser degree in wine, and the citrus flavonoids hesperetin

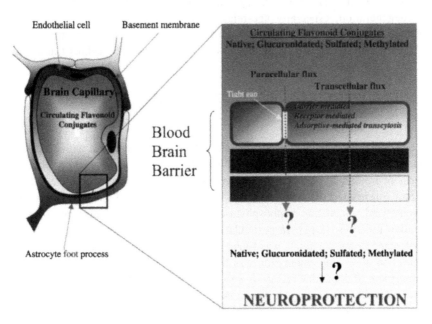

Figure 2 Potential interactions of bioavailable flavonoid conjugates with the blood brain barrier (BBB).

[22] and naringenin together with its glucuronide conjugate [23]. Schroder-van der Elst and associates [24,25] have also identified the synthetic flavonoid Emd 49209, which is able to localize both in the adult rat brain and in the developing fetal rat brain. The paucity of studies is due in part to a limited knowledge of polyphenolic bioavailability and characterization of the circulating forms that potentially interact with the CNS. Furthermore, knowledge about the physio-logical interaction between bioavailable flavonoids and/or their conjugates with the blood-brain barrier (BBB) is also limited (Fig. 2). This interaction is of fundamental importance, considering that the functional role of the BBB is to control the composition of extracellular fluid in the CNS, sealing off entry of all but the smallest molecules. Hence this barrier ultimately determines the fate of dietary components such as flavonoids within the CNS.

III. NUTRITIONAL COMPONENTS AND THEIR ROLE IN PREVENTING BRAIN AGING

Although evidence suggesting flavonoids are able to localize in the brain is scarce, growing awareness gained from epidemiological and dietary intervention studies of humans and animals suggests that flavonoid consumption may be important to neuronal "health." The contributory role of flavonoids to the modulation of neurodegeneration, especially age-related cognitive and motoric decline, in protection against oxidative stress, cerebral ischemia/reperfusion injuries, and other brain abnormalities is being extensively investigated (see Tables 1 and 2).

IV. NEUROPROTECTIVE EFFECTS OF FRUIT AND VEGETABLES

We and others have shown in rodents that dietary supplementation with extracts prepared from strawberry, spinach, and blueberry imparted significant protection against neurological parameters sensitive to oxidative stress. These included receptor sensitivity [16], cerebellar Purkinje cell activity and calcium buffering capacity [26], guanosine triphosphatase (GTPase) coupling/uncoupling, and cognitive functions [2,11]. In particular, blueberry extract supplementation to middle-aged or aged animals was found to have a profound effect. One striking observation after supplementation was the significant increase in oxotremorine enhancement of dopamine release from isolated striatal slices. This is especially important since maintaining the functional integrity of the striatal dopaminergic system has a major impact on certain behavioral parameters [27–29]. What remains to be examined is whether this modulation in dopamine release is attributable to an increase in neuronal sensitivity and/or neurogenesis of striatal neurons. It is also unclear whether these fruit and vegetable extracts might have promoted similar mechanisms toward age-related changes in cerebellar β-adrenergic function [30,31]. It has been postulated that the cerebellar noradrenergic system, which shows age-related changes in β-adrenergic function, may underlie certain age-related deficits in motor learning [7]. Norepinephrine potentiates GABA-induced inhibition of cerebellar Purkinje neurons via the β-adrenergic receptor. In aged rats, β-adrenergic potentiation occurs in only 30% of the recorded cerebellar Purkinje cells as compared with 70–80% of those in younger animals. However, despite these findings, current of studies investigating fruit and vegetable extracts have not investigated whether flavonoid components were able to localize within brain structures. Hence one can only hypothesize that the neuroprotective actions, such as modulation of receptor systems, are mediated from within the CNS or may reside in some kind of peripheral effect.

Although the site(s) where these extract mediate their effects are presently unknown, the findings suggest that in vitro antioxidant activities of these extracts were not predictive in assessing their potency against neurological deficits. In our early studies [32,33], extracts were supplemented in the diet at 1.36 mmol Trolox equivalents/kg diet, such that animals consumed an equal concentration of antioxidants per day. However, extracts differed in the ability to afford protection in parameters sensitive to oxidative stress. This illustrates that a simple measure of in vitro antioxidant activity alone may not be sufficient to argue potential health benefits, and that functional assessments need to be combined. More recently, in 2000, it was shown that two different cultivars of blueberry, when supplemented in the diet on an equal-weight basis (20 g/kg-diet), also exhibited different degrees of protection against memory and learning declines in aging rats [34]. Collectively these findings suggest that differences in the bioavailability of the diverse array of flavonoid components found in these extracts and their biological potency on entering the circulation play an important role. Moreover, it is not clear from these studies whether the neuroprotective effects of flavonoids against oxidative stress involve their reducing properties or some other mechanism independent of their antioxidant activities. In this regard, when two indices of antioxidant activity reactive oxygen species (ROS) production and glutathione levels in the striatum and cerebellum were examined [32], supplementations resulted in only modest effects that could not totally account for the efficacy of these extracts, especially with regard to their effects on motor and cognitive function.

Recent studies (Joseph, unpublished) appear to draw attention to a possible interaction with signaling molecules, although whether this is a direct or indirect property has yet to be elucidated. For example, blueberry supplementation was found to induce alterations in age- and calcium-sensitive signaling molecules associated with memory, especially the conversion of short-to long-term memory. These include calcium-dependent protein kinase C (PKC), for which studies have shown that its activity is important in formation of memory, particularly spatial

Table 1 The Neuroprotective Actions of Flavonoid-Rich Extracts Found in Dietary Sources

Flavonoid source	Concentration	Duration	Route	Species	Stress type	Observations	References
Fruits and vegetables							
Strawberry and spinach[a]	1.36 m mol Trolox equivalents/kg diet	8 mo	In diet	Rat	Normal aging	Improved striatal dopamine receptor sensitivity, cerebellar Purkinje cell activity, calcium buffering capacity, GTPase coupling/uncoupling, and cognitive functions	[33]
Strawberry, spinach, and blueberry[a]	1.36 m mol Trolox equivalents/kg diet	8 wk	In diet	Rat	Normal aging	Improved striatal dopamine receptor sensitivity, cerebellar Purkinje cell activity, calcium buffering capacity, GTPase coupling/uncoupling, and behavioral and cognitive functions	[129]
					Normal aging	Improved motor learning and cerebellar β-adrenergic receptor function and cerebellar glutathione concentrations	[31]
					Hyperoxia	Prevention of declines in and cerebellar β-adrenergic receptor function	[30]
Blueberry[a]	20 g/kg diet	8 wk	In diet	Rat	Normal aging	Improved striatal dopamine receptor sensitivity, striatal and cortical vitamin C concentrations, and behavioral and cognitive functions	[34]
Grape polyphenols	5 mg/dL	2 mo	In diet	Rat	5% Ethanol in drinking water	Prevention of decrease in synaptosomal Na, K-ATPase activity and dopamine uptake	[130]

Compound	Dose	Duration	Route	Species	Model	Effects	Ref.
Grape seed proanthocyanidin extract	25–100 mg/kg Body weight	1 wk	Gavage	Mouse	Intraperitoneal injection of TPA* (0.1 μg)	Dose-dependent reduction of reactive oxygen species formation, lipid peroxidation, and DNA fragmentation	[131]
Tea							
Green tea	0.5%	3 wk	Drinking water	Rat	Ischemia/ reperfusion	Reduced infarction volume, free radical production, lipid peroxidation, DNA damage, and apoptotic cell numbers in striatum	[132]
Green tea	0.5% And 2%	3 wk	Drinking water	Gerbil	Ischemia/ reperfusion	Reduced infarction volume, free radical production, lipid peroxidation, DNA damage, and apoptotic cell numbers in striatum	[133]
Green tea		Single dose	Intranigral infusion	Rat	Intranigral infusion of ferrous citrate (4.2nM)	Intranigral infusion decreased lipid peroxidation in substantia nigra. Co-infusion reduced elevation in lipid peroxidation in substantia nigra and associated decrease in striatal dopamine content. oral administration had no effect	[53]
Green tea	Chronic 0.5 and 1 mg/kg Body weight	2 wk 2×/day for 1 day	Orally Intraperitoneal	Mouse	MPTP (24 mg/kg body weight IP) for 4 days[c]	Protection against dopaminergic neuronal loss, maintenance of dopamine and tyrosine hydroxylase concentrations	[49]

(continued on next page)

Table 1 (*continued*)

Flavonoid source	Concentration	Duration	Route	Species	Stress type	Observations	References
Rooibos tea (*Aspalathus linearis*)	Chronic	21 mo	Drinking water	Rat	Normal aging	Reduced TBA reactive substance formation in frontal cortex, occipital cortex, hippocampus, and cerebellum	[134]
β-Catechin[b]	1 mL/kg body weight	1 mo	Drinking water	Rat	Cortical injection with 01 M ferric chloride	Increased mitochondrial superoxide dismutase activity of striatum and midbrain, decreased TBA reactive substances in cortex and cerebellum of aged rats and 8-OHdG formation in cortex	[135]
Kombucha[c]	chronic	3–yr longitudinal	Drinking water	Mouse	Normal aging	Inhibited weight gain, increased environmental awareness and responsiveness and life span	[136]
Black tea	400 mL (3 times/day)	1 day	Oral	Human	—	Caused transient improvements in performance, prevented steady decline throughout day in alertness and cognitive capacity, increased alertness and information processing capacity	[137]

[a] Aqueous extracts.
[b] 12-*O*-tetradecanoylphorbol-13-acetate.
[c] Green tea supplementation preceded each MPTP injection.
[d] Contains green tea extract as a main component, with ascorbic acid, sunflower seed extract, dunaliella carotene, and natural vitamin E.
[e] German form of the Japanese name for a lightly fermented tea beverage.
GTPase, guanosine triphosphatase; MPTP, *N*-methyl-4-phenyl-1,2,3,6-tetrahydropyridine; TPA, ATPase, adenosine triphosphatase; DNA, deoxyribonucleic acid; TBA, 8-OHdG.

Table 2 Neuroprotection and Neuroprotective Actions Afforded by Dietary and Phytochemical Mixtures to the Brain

	Properties	References
Herbs		
S-113[a] and DX-9386[b]	• Improves cognitive performance	[138–145]
Ding lang (*Policies fruticosum* L.)	• Reduce age-related deficits in behavioral performance • Modulates endogenous SOD levels	[146–150]
Qizhu tang[c]	• Prevents cerebral oxidative injury in rats	[151]
TJ-960 (baicalein[d])	• Reduces FeCl₃-induced epilepsy, mediates increases in peroxidation by-products • Inhibits cerebral ischemia–induced hippocampal neuronal death • Modulates neurotransmitter levels, NOS activity	[152–157]
Guilingji (baicalein[d])	• Protects against FeCl₃-induced epilepsy	[158]
Shou Xing Bu Zhi	• Reduces age-related increases in brain lipofuscin	[159,160]
Oren-gedoku-to (TJ-15) and Toki-shakuyaku-san (TJ-23)	• Ameliorates cerebral ischemia–induced and age-related learning and memory deficits • Prevents reduction of acetylcholine content in brain cerebral cortex, hippocampus, and striatum; modulates neurotransmitter levels	[161–169]
Aged garlic extract[e] (*Allium sativum*)	• Improves cognitive impairments in SAM • Protects against morphological changes in SAM • Reduces age-related increases in peroxidation by-products • Exhibits immunomodulatory properties	[170–176]
Spices		
Curcumin and tusmeric	• Protects against retinal induced peroxidation • Enhances glutathione levels	[177,178]
Saffron and crocin[f]	• Protects against ethanol-induced impairments in learning, inhibition of hippocampal LTP, and synaptic plasticity	[179,180]
Red bell pepper (*Capsicum annuum* L.)	• Improves memory and acquisition performance in SAM	[181]

[a] Consists of *Biota orientalis, Panax ginseng*, and *Schizandra chinesis*.
[b] Consists of *Panax ginseng, Polygala tenuifola, Acorus gamineus*, and *Paoria cocas*.
[c] Consists of *Rhizoma atractylodis, Poria, Radix notoginseng*, and *Radax astragali*.
[d] Major component found in extract believed to promote properties.
[e] Contains S-allycisteine, S-allymercaptocysteine, allicin, and diallosulfides.
[f] Crocetin di-gentobiose ester.
SOD, superoxide dismutase; NOS, nitric oxide synthase; SAM, senescence accelerated mice.
Source: Adapted from Ref. 182.

memory [35] and that treatment with PKC inhibitors impairs memory formation [36]. It appears that training induces the calcium-induced translocation [37] of PKC from the cytosol to the membrane subcellular fraction [38]. However, in aging, there appear to be alterations in this translocation, which are correlated with decrements in spatial memory. In this regard, Colombo and coauthors [37] showed that young rats with the best performance in spatial memory also had the highest PKC-γ in the membrane fraction of the hippocampus and PKC-β_2 in the soluble cytosolic fraction.

Also important are the mitogen-activated protein (MAP) kinases (MAPKs), critical in long-term memory formation. More specifically there is a great amount of data indicating that MAPKs are involved not only in hippocampal memory formation but also in memory modulation in other brain structures [39]. Moreover, recent studies have indicated that the activation of these molecules is sensitive to oxidative stress (see Chap. 9) and that they may serve as biochemical signal integrators and molecular detectors for modulating coordinated responses to extracellular signals in neurons, playing a critical role in synaptic plasticity. Particularly important in this regard are the extracellular signal regulated kinases (ERKs) 1/2 and the Jun kinases (JNKs). Studies have demonstrated the role of ERK signaling cascades in diverse types of learning and memory such as conditioned taste aversion [40], novel taste learning [41], spatial learning [42], and inhibitory avoidance [43]. Findings have shown that ERK activities were reduced in cortical brain slices of senescent rats (24 months) without declines in the corresponding proteins [44]. With respect to this, preliminary data from our lab (Joseph et al., unpublished) indicated that mice transgenic for amyloid precursor protein and presenilin-1 mutations, maintained on blueberry supplemented diet from the time of weaning up until 12 months of age, showed Y-maze performance equivalent to that of nontransgenic controls; that result correlated significantly with decreased sphingomyelin turnover and increased ERK, PKC, and GTPase activities in both the striatum and the hippocampus. These findings suggest that blueberry extract may benefit other parameters of cell signaling and synaptic plasticity that are involved in learning and memory. Although there is do direct evidence to suggest that blueberry flavonoids directly mediate these effects from within the CNS, recent in vitro studies appear to support the notion that MAPK signaling can be influenced by flavonoids, such as those found in tea and wine (see Chap. 9).

V. NEUROPROTECTIVE EFFECTS OF FLAVONOID-RICH BEVERAGES

The neuroprotective actions of polyphenolic-rich beverages, such as teas and red wine, have also been investigated. Studies that examined the neuroprotective properties of tea are shown in Table 1. These experiments were performed using either simple aging models or the cerebral ischemia/reperfusion model, which reproduces a number of pathophysiological features observed in age- and disease-related brain dysfunction [45]. From these findings (Table 1) one could argue that protection afforded by tea polyphenolics against the various types of deficits induced by ischemic damage, N-methyl-4-phenyl-1,2,3,6-tetra hydropyridine (MPTP) or iron chloride, for example, could be due in part to inhibition of oxidative and inflammatory processes [46–49], both of which influence behavior [50].

Weak associations between tea consumption and neurological disorders such as Parkinson's disease have also been suggested [51]. Although the possible mechanisms involved are unclear, tea flavonoids have been shown to protect against iron-induced deficits in striatal neurotransmitter concentrations/turnover [52], as well as to decrease oxidative damage in the substantia nigra [53], both of which are common etiological features associated with Parkinson's disease. However, in these studies tea flavonoids were administered intravenously or by infusion directly into the brain, and hence it is not clear how the outcomes relate to the normal ingestion of tea per se and the consequences of its interactions and biotransformations in the gastrointestinal tract before uptake.

Although the active phenolic component(s) in tea extract affording the protection described are not known, potential candidates include the metabolites of the major components catechin and epigallocatechin gallate, which supplemented alone afford protection against ischemia/reperfusion-induced memory impairment [54] and cell death of hippocampal CA1 neurons [55]. Moreover, Levites and coworkers [49] have also shown that deficits induced by the neurotoxin MPTP, which specifically induce neurodegeneration in dopaminergic neurons similar to that observed in Parkinson's disease, can be ameliorated by green tea extract (0.5 and 1 mg/kg body weight). They also reported that supplementation of epigallocatechin gallate alone provided significant protection, which may account for the neuroprotective properties of the extract itself. However, epigallocatechin gallate was supplemented at higher concentrations (2 and 10 mg/kg body weight) than that present in the extract itself, suggesting possible synergistic interactions of the various components of the tea extract. However, although epigallocatechin gallate had previously been shown to cross the blood-brain barrier [21], the studies described did not determine the localization of the flavonoids in the brain; hence one could only *hypothesize* that their protective actions were mediated directly within the CNS and not from the periphery.

Correlations between moderate wine consumption (3–4 glasses/day) and the incidence of dementia and AD compared with that of nondrinkers [56,57] have also been reported, although some argue the contrary [58]. Reports have also shown that heavy drinkers displayed the poorest results on memory or on intelligence tasks as compared with moderate drinkers or nondrinkers [59]. Orgogozo and colleagues [57] propose several possible protective mechanisms of action regarding moderate wine consumption, including antioxidant and/or anti-inflammatory properties, as well as elevation of plasma apolipoprotein E levels. Although there have been few direct demonstrations that the protective effects associated with wine consumption are due to the flavonoid content, there is some evidence that they are not due to the alcohol content [60].

Attempts to elucidate which possible component(s) contributes neuro-protection have been focused on the stilbene resveratrol. This is found in several types of red wine and is considered one of the substances responsible for the lower incidence of coronary heart diseases among moderate drinkers and has been proposed to exhibit neuroprotective properties. Virgili and Contestabile [61] reported that long-term administration of resveratrol to young adult rats protected them from damage caused by systemic injection of the excitotoxin kainic acid, in the olfactory cortex and the hippocampus. The same treatment, however, is not able to give any significant protection in an ex vivo model of simulated ischemia on hippocampal slices. Although a direct demonstration of the protective effects of wine or a delineation of its mechanisms has not been shown, moderate wine drinking appears to promote some degree of protection against senile dementia or AD.

VI. HERBS AND SPICES

Alternative or complementary medicines are widely used in North America. Consumers in the United States spend an estimated $1.5 billion per year on herbal medicines with projected annual growth of 15%. Germany, the largest market among American or Western European countries, had total sales in 1993 of $1.9 billion for plant-based allopathic medicines (half of these prescribed by physicians) and 5 million prescriptions for ginkgo in 1988. The use of these complementary medicines in dementia therapy varies according to the different cultural traditions. The two most commonly used preparations, *Gingko biloba* and ginseng, are discussed in the following sections. Further neuroprotective herbs and spices, including preparations that contain *Gingko biloba* and ginseng, are highlighted in Table 2.

A. *Ginkgo biloba* (EGb 761)

By far the most extensively studied herbal mixture with respect to brain function is *Ginkgo biloba* (EGb 761) [62–64]. Extracts of the leaves have been used for 5000 years in traditional Chinese medicine for various purposes. Flavonoids such as myricetin, quercetin, and kaempferol are general components in *Ginkgo biloba* extract, which is often standardized to contain 24% ginkgo-flavone glycosides and 6% terpenoids including bilobalide and the ginkgolides A, B, C, M, and J (20-carbon cage molecules with six five-member rings).

A number of epidemiological studies have implicated *Ginkgo biloba* as a potential neuroprotective mixture against age-related and AD-related dementia [65–67]. Although the cause and underlying pathophysiological features of AD are unknown, prominent hypotheses as to the cause center around age-related oxidative injury and inflammatory processes [9] and propose that attenuation of these processes by *Ginkgo biloba* underlies its neuroprotective actions. Further actions promoted by *Gingko biloba* that may contribute to improvement in dementia and other brain aging functions include the reduction in levels of ROS [68,69]; reduction of age-related deficits in oxygen and glucose delivery to the brain by increasing cerebral blood flow [70–75] and protection of membrane fatty acids [76–79], changes in which have been shown to correlate with cognitive performance [80,81], interactions with the muscarinic cholinergic system [82,83] involved in the performance of spatial tasks [84,85], antagonism of platelet activating factor [86–88]; protection of the striatal dopaminergic system [89,90]; and inhibition of monoamine oxidase activity [91–93], hence maintaining levels of monoamines such as NE, 5-HT, and DA, which are known to play essential roles in a variety of brain functions. However, these findings still dictate a significant amount of validation before *Gingko biloba* is fully accepted as a therapy against dementia. This progression is complicated further in light of the fact that the relative content of ginkgolide and bilobalide components and those of the various flavonoids varies across preparations and even seasons [94].

Although these observations highlight possible neuroprotective action by *Gingko biloba*, Oken and colleagues [95] reported in 1998 that after closer examination of the epidemiological data according to certain acceptable scientific criteria, only four studies suggest a potential protective effect of the extract. As such, further validation is required before these herbal remedies are fully accepted as therapies against dementia.

B. Ginseng

Panax ginseng is one of the mostly widely used herbs in traditional Chinese medicine. In addition to controlling functions related to stamina, fatigue, and physical stress, ginseng has been reported to reduce neuronal death and protect against ischemic damage [96–98]. Possible mechanisms of neuroprotection proposed include ability to increase the expression of nerve growth factor [99] and intrinsic antioxidant activity [100,101]. Ginseng has also been investigated as a potential therapy against amyotrophic lateral sclerosis (ALS), a motor neuron disease [102]. Among the observations made in rats given ginseng in drinking water were prolongation in onset of signs of motor impairment and survival. This improvement in memory performance is in agreement with studies describing the beneficial effects of ginseng on memory and learning performance [103–107]. In addition, ginseng's ability to modulate the cholinergic and serotoninergic neurotransmitter systems, damage to which affects spatial working memory [108,109], has also been proposed as a mechanism of neuroprotection [100]. In this regard, behavioral paradigms affected by electroconvulsive shock through modulation of the cholinergic neurotransmitter system are attenuated by ginseng supplementation [110], in particular within brain areas such as the hippocampus, known to be involved in spatial memory tasks [111,112]. It has also been speculated that ginseng acts by enhancing cholinergic systems such as choline acetyltransferase important in the formation of memory [113].

Studies have also shown that behavioral impairments can be forestalled by using individual ginsenosides [97,107,114–122]. The observations made after the application of these individual components appear to support the notion that neuroprotection may result from interactions with neurotransmitter systems. For example, ginsenosides have been shown to increase muscarinic-cholinergic receptor density and levels of acetylcholine in the brain [123]. Unfortunately, despite these observations few epidemiological studies have been performed to complement these findings. Indeed, a comprehensive survey of the literature found only five studies investigating the effects of ginseng on human cognitive performance [124–128], in three of which significant improvement in mental arithmetic and abstraction tests were reported [125–127].

VII. CONCLUSIONS

At the present time, for some neurodegenerative disorders there is very little in the form of treatment and what treatments are available are only effective for a short period and are often associated with debilitating side effects. Current drug therapy does not address the progressive nature of many of these diseases, and ultimately the patient becomes severely disabled and requires nursing/hospital care. In light of this it appears essential that novel strategies with potential to delay the onset or even prevent the manifestation of certain processes believed to contribute to neurological dysfunction be developed. As such renewed attention is being paid to the application of flavonoids commonly found in fruits, vegetables, and beverages such as tea and wine. The findings from these studies highlight their ability to afford neuroprotection, yet evidence of direct action within the brain is lacking. Only a few studies to date have reported that flavonoids localize within the brain. Although further studies are clearly required to support these findings, a more cautious approach must also be taken when attempting to elucidate potential mechanisms of action. Although use of the native flavonoid in in vitro studies has some relevance to the in vivo scenario, the application of the predominant in vivo physiological metabolites that enter the circulation will undoubtedly prove more fruitful in examining possible mechanisms of action.

REFERENCES

1. Ingram DK, Jucker M, Spangler E. Behavioral manifestations of aging. In: Mohr U, Cungworth DL, Capen CC, eds. Pathobiology of the Aging Rat. Washington, DC: ILSI, 1994:149–170.
2. Shukitt-Hale B, Mouzakis G, Joseph JA. Psychomotor and spatial memory performance in aging male Fischer 344 rats. Exp Gerontol 1998; 33:615–624.
3. Bartus RT. Drugs to treat age-related neurodegenerative problems: the final frontier of medical science? J Am Geriat Soc 1990; 38:680–695.
4. Muir JL. Acetylcholine, aging, and Alzheimer's disease. Pharmacol Biochem Behav 1997; 56:687–696.
5. West RL. An application of pre-frontal cortex function theory to cognitive aging. Psychol Bull 1996; 120:272–292.
6. Joseph JA, Hunt WA, Rabin BM, Dalton TK. Possible "accelerated striatal aging" induced by 56Fe heavy-particle irradaition: implications for manned space flights. Radiat Res 1992; 130:88–93.
7. Bickford PC. Motor learning deficits in aged rats are correlated with loss of cerebellar noradrenergic function. Brain Res 1993; 620:133–138.
8. Joseph JA, Villalobos-Molina R, Denisova N, Erat S, Cutler R, Strain JG. Age differences in sensitivity to H_2O_2 or NO-induced reductions in K+-evoked dopamine release from superfused striatal slices: reversals by PBN or Trolox. Free Radic Biol Med 1996; 20:821–830.
9. Christen Y. Oxidative stress and Alzheimer disease. Am J Clin Nutr 2000; 71:621S–629S.
10. Urano S, Sato Y, Otonari T, Makabe S, Susuki S, Ogata M, Endo T. Aging and oxidative stress in neurodegeneration. Biofactors 1998; 7:103–112.
11. Shukitt-Hale B. The effects of aging and oxidative stress on psychomotor and cognitive behavior. Age 1999; 22:9–17.
12. Halliday G, Robinson SR, Shepherd C, Kril J. Alzheimer's disease and inflammation: a review of cellular and therapeutic mechanisms. Clin Exp Pharmacol Physiol 2000; 27:1–8.
13. Combs CK, Johnson DE, Karlo JC, Cannady SB, Landreth GE. Inflammatory mechanisms in Alzheimer's disease: inhibition of beta-amyloid-stimulated proinflammatory responses and neurotoxicity by PPARgamma agonists. J Neurosci 2000; 20:558–567.
14. Benzi G, Moretti A. Age- and peroxidative stress-related modifications of the cerebral enzymatic activities linked to mitochondria and the glutathione system. Free Radic Biol Med 1995; 19:77–101.

15. Egashira T, Takayama F, Yamanaka Y. Effects of bifemelane on muscarinic receptors and choline acetyltransferase in the brains of aged rats following chronic cerebral hypoperfusion induced by permanent occlusion of bilateral carotid arteries. Jpn J Pharmacol 1996; 72:57–65.

16. Joseph JA, Berger RE, Engel BT, Roth GS. Age-related changes in the nigro-striatum: a behavioral and biochemical analysis. J Gerontol 1978; 33: 643–649.

17. Gould TJ, Bickford PC. Age-related deficits in the cerebellar beta-adrenergic signal transduction cascade in Fischer 344 rats. J Pharmacol Exp Ther 1997; 281:965–971.

18. Kornhuber J, Schoppmeyer K, Bendig C, Riederer P. Characterization of [3H] pentazocine binding sites in post-mortem human frontal cortex. J Neural Transm 1996; 103:45–53.

19. Youdim KA, Martin A, Joseph JA. Essential fatty acids and the brain: possible health implications. Int J Dev Neurosci 2000; 18:382–399.

20. Landfield PW, Eldridge JC. The glucocorticoid hypothesis of age-related hippocampal neurodegeneration: role of dysregulated intraneuronal Ca^{2+}. Ann NY Acad Sci 1994; 746:308–321.

21. Suganuma M, Okabe S, Oniyama M, Tada Y, Ito H, Fujiki H. Wide distribution of [3H](-)-epigallocatechin gallate, a cancer preventive tea polyphenol, in mouse tissue. Carcinogenesis 1998; 19:1771–1776.

22. Tsai T-H, Chen Y-F. Determination of unbound hesperetin in rat blood and brain by microdialysis coupled to microbore liquid chromatography. J Food Drug Anal 2000; 8:331–336.

23. Peng HW, Cheng FC, Huang YT, Chen CF, Tsai TH. Determination of naringenin and its glucuronide conjugate in rat plasma and brain tissue by high-performance liquid chromatography. J Chromatogr A 1998; 714:369–374.

24. Schroder-van der Elst JP, van der Heide D, Rokos H, Morreale de Escobar G, Kohrle J. Different tissue distribution, elimination and kinetics of thyroxine and its conformational analog, the synthetic flavonoid Emd 49209 in the rat. Endocrinology 1997; 138:79–84.

25. Schroder-van der Elst JP, van der Heide D, Rokos H, Morreale de Escobar G, Kohrle J. Synthetic flavonoids cross the placenta in the rat and are found in the fetal brain. Am J Physiol 1998; 274:E253–E256.

26. Joseph JA, Strain JG, Jimenez ND, Fisher D. Oxidant injury in PC12 cells—a possible model of calcium deregulation in aging. I. Selectivity of protection against oxidative stress. J Neurochem 1997; 69:1252–1258.

27. Joseph JA, Shukitt-Hale B, McEwen J, Rabin B. Magnesium activation of GTP hydrolysis or incubation in S-Adenosyl-1-methionine reverses 56Fe-induced decrements in oxotremorine-enhancement of K+ evoked striatal dopamine release. Radiat Res 1999; 152:637–641.

28. Joseph JA, Kowatch MA, Maki T, Roth GS. Selective cross activation/inhibition of second messenger systems and the reduction of age-related deficits in the muscarinic control of dopamine release from perfused rat striata. Brain Res 1990; 537:40–48.

29. Joseph JA, Bartus RT, Clody DE, Morgan D, Finch C, Beer B, Sesack S. Psycomotor performance in the senescent rodent: reduction of deficits via striatal dopamine receptor up-regulation. Neurobiol Aging 1983; 4:313–319.

30. Bickford PC, Shukitt-Hale B, Joseph JA. Effects of aging on cerebellar noradrenergic function and motor learning: nutritional interventions. Mech Ageing Dev 1999; 111:141–154.

31. Bickford PC, Gould T, Briederick L, Chadman K, Pollock A, Young D, Shukitt-Hale B, Joseph J. Antioxidant-rich diets improve cerebellar physiology and motor learning in aged rats. Brain Res 2000; 866:211–217.

32. Joseph JA, Shukitt-Hale B, Denisova NA, Bielinski D, Martin A, McEwen JJ, Bickford PC. Reversals of age-related declines in neuronal signal transduction, cognitive and motor behavioral deficits with diets supplemented with blueberry, spinach or strawberry dietary supplementation. J Neurosci 1999; 19:8114–8121.

33. Joseph JA, Shukitt-Hale B, Denisova NA, Prior RL, Cao G, Martin A, Taglialatela G, Bickford PC. Long-term dietary strawberry, spinach, or vitamin E supplementation retards the onset of age-related neuronal signal-transduction and cognitive behavioral deficits. J Neurosci 1998; 18:8047–8055.

34. Youdim KA, Shukitt-Hale B, Martin A, Wang H, Denisova N, Bickford PC, Joseph JA. Short-term dietary supplementation of blueberry polyphenolics: beneficial effects on aging brain performance and peripheral tissue function. Nutr Neurosci 2000; 3:383–397.

35. Micheau J, Riedel G. Protein kinases: which one is the memory molecule? Cell Mol Life Sci 1999; 55:534–548.

36. Serrano PA, Beniston DS, Oxonian MG, Rodriguez WA, Rosenzweig MR, Bennett EL. Differential effects of protein kinase inhibitors and activators on memory formation in the 2 day old chick. Behav Neural Biol 1994; 61:60–72.

37. Colombo PJ, Wetsel W, Gallagher MG. Spatial memory is related to hippocampal subcellular concentrations of calcium-dependent protein kinase C isoforms in young and aged rats. Proc Natl Acad Sci USA 1997; 94:1495–1499.

38. Van der Zee EA, Compaan JC, Bohus B, Luiten PG. Alterations in the immunoreactivity for muscarinic acetylcholine receptors and colocalized PKC gamma in mouse hippocampus induced by spatial discrimination learning. Hippocampus 1995; 5:349–362.

39. Sgambato V, Pages C, Rogard M, Besson MJ, Caboche J. Extracellular signal regulated kinase (ERK) controls immediate early gene induction on corticostriatal stimulation. J Neurosci 1998; 18:8814–8825.

40. Berman DE, Hazvi S, Rosenblum K, Seger R, Dudai Y. Specific and differential activation of mitogen-activated protein kinase cascades by unfamiliar taste in the insular cortex of the behaving rat. J Neurosci 1998; 18:10037–10044.

41. Swank MW, Sweatt JD. Increased histone acetyltransferase and lysine acetyltransferase activity and biphasic activation of the ERK / RSK cascade in insular cortex during novel taste learning. J Neurosci 2001; 21:3383–3391.

42. Selcher JC, Atkins CM, Trzaskos JM, Paylor R, Sweatt JD. A necessity for MAP kinase activation in mammalian spatial learning. Learn Mem 1999; 6:478–490.

43. Schafe GE, Nadel NV, Sullivan GM, Harris A, LeDoux JE. Memory consolidation for contextual and auditory fear conditioning is dependent on protein synthesis, PKA, and MAP kinase. Learn Mem 1999; 6:97–110.

44. Zhen X, Uryu K, Cai G, Johnson GP, Friedman E. Age-associated impairment in brain MAPK signal pathways and the effect of restriction in Fischer 344 rats. J Gerontol A Biol Sci Med Sci 1999; 54:B539–B548.

45. Heiss WD. Therapy of cerebral ischemia. Zr Kardiol 1987; 76:87–98.

46. Hofbauer R, Frass M, Gmeiner B, Handler S, Speiser W, Kapiotis S. The green tea extract epigallocatechin gallate is able to reduce neutrophil transmigration through monolayers of endothelial cells. Wiener Klin Wochenschr 1999; 111:278–282.

47. Lin YL, Lin JK. (-)-Epigallocatechin-3-gallate blocks the induction of nitric oxide synthase by down-regulating lipopolysaccharide-induced activity of transcription factor nuclear factor-kappaB. Mol Pharmacol 1997; 52:465–472.

48. Lin YL, Tsai SH, Lin-Shiau SY, Ho CT, Lin JK. Theaflavin-3,3'-digallate from black tea blocks the nitric oxide synthase by down-regulating the activation of NF-kappaB in macrophages. Eur J Pharmacol 1999; 367:379–388.

49. Levites Y, Weinreb O, Maor G, Youdim MBH, Mandel S. Green tea polyphenol (-) epigallocatechin-3-gallate prevents N-methyl-4-phenyl-1,2,3,6-tetrahydropyridine-induced dopaminergic neurodegeneration. J Neurochem 2001; 78:1073–1082.

50. Hauss-Wegrzyniak B, Vannucchi MG, Wenk GL. Behavioral and ultrastructural changes in by chronic neuroinflammation in young rats. Brain Res 2000; 859: 157–166.

51. Hellenbrand W, Boeing H, Robra BP, Siedler A, Vieregge P, Nischan P, Joerg J, Oertel WH, Schneider E, Ulm G. Diet and Parkinson's disease II: a possible role for the past intake of specific nutrients. Neurology 1996; 47:644–650.

52. Kabuto H, Yokoi I, Mori A. Monoamine metabolites, iron induced seizures, and the anticonvulsant effect of tannins. Neurochem Res 1992; 17:585–590.

53. Lin AM, Chyi BY, Wu LY, Hwang LS, Ho LT. The antioxidative property of green tea against iron-induced oxidative stress in rat brain. Chin J Physiol 1998; 41:189–194.

54. Matsuoka Y, Hasegawa H, Okuda S, Muraki T, Uruno T, Kubota K. Ameliorative effects of tea catechins on active oxygen-related nerve cell injuries. J Pharmacol Exp Therapeut 1995; 274:602–608.

55. Lee S, Suh S, Kim S. Protective effects of the green tea polyphenol (-) epi-gallocatechin gallate against hippocampal neuronal damage after transient global ischemia in gerbils. Neurosci Lett 2000; 287:191–194.

56. Launer LJ, Kalmijn S. Anti-oxidants and cognitive function: a review of clinical and epidemiologic studies. J Neural Transm Suppl 1998; 53:1–8.

57. Orgogozo JM, Dartigues JF, Lafont S, Letenneur L, Commenges D, Salamon R, Renaud S, Breteler MB. Wine consumption and dementia in the elderly: a prospective community study in the Bordeaux area. Rev Neurol 1997;153: 185–192.

58. Leibovici D, Ritchie K, Ledesert B, Touchon J. The effects of wine and tobacco consumption on cognitive performance in the elderly: a longitudinal study of relative risk. Int J Epidemiol 1999; 28:77–81.

59. De Renzi E, Faglioni P, Nichelli P, Pignattari L. Intellectual and memory impairment in moderate and heavy drinkers. Cortex 1984; 20:525–533.

60. Puddey IB, K.D. C, Abdu-Amsha Caccetta R, Beilin LJ. Alcohol, free radicals and antioxidants. Novartis Found Symp 1998; 216:51–62.

61. Virgili M, Contestabile A. Partial neuroprotection of in vivo excitotoxic brain damage by chronic administration of the red wine antioxidant agent, trans-resveratrol in rats. Neurosci Lett 2000; 281:123–126.

62. Diamond BJ, Shifleett SC, Feiwel N, Matheis RJ, Noskin O, Richards JA, Schoenberger NE. Ginkgo biloba extract: mechanisms and clinical indications. Arch Phys Med Rehab 2000; 81:668–678.

63. Clostre F. Ginkgo biloba extract (EGb 761): state of knowledge in the dawn of the year 2000. Ann Pharm Fr 1999; 57:1S8–88.

64. Kidd PM. A review of nutrients and botanicals in the integrative management of cognitive dysfunction. Altern Med Rev 1999; 4:144–161.

65. Simanyi M. Use of special Ginkgo biloba extract for cognitive disorders in the elderly. Wiener Med Wochenschr 1999; 149:231–234.

66. Le Bars PL, Kieser M, Itil KZ. A 26-week analysis of a double-blind, placebo-controlled trial of the ginkgo biloba extract EGb 761 in dementia. Dement Geriat Cogn Disord 2000; 11:230–237.

67. Le Bars PL, Katz MM, Berman N, Itil TM, Freedman AM, Schatzberg AF. A placebo-controlled, double-blind, randomized trial of an extract of Ginkgo biloba for dementia: North American EGb Study Group. JAMA 1997; 278:1327–1332.

68. Oyama Y, Chikahisa L, Ueha T, Kanemaru K, Noda K. Ginkgo biloba extract protects brain neurons against oxidative stress induced by hydrogen peroxide. Brain Res 1996; 712:349–352.

69. Seif-El-Nasr M, El-Fattah AA. Lipid peroxide, phospholipids, glutathione levels and superoxide dismutase activity in rat brain after ischaemia: effect of ginkgo biloba extract. Pharmacol Res 1995; 32:273–278.

70. Eckmann F. Cerebral insufficiency — treatment with Ginkgo-biloba extract: Time of onset of effect in a double-blind study with 60 inpatients. Fortschr der Med 1990; 108:557–560.

71. Gajewski A, Hensch SA. Ginkgo biloba and memory for a maze. Psychol Rep 1999; 84:481–484.

72. Krieglstein J, Beck T, Seibert A. Influence of an extract of Ginkgo biloba on cerebral blood flow and metabolism. Life Sci 1986; 39:2327–2334.

73. Lamm K, Arnold W. The effect of blood flow promoting drugs on cochlear blood flow, perilymphatic pO(2) and auditory function in the normal and noise-damaged hypoxic and ischemic guinea pig inner ear. Hear Res 2000; 141:199–219.

74. Oberpichler H, Beck T, Abdel-Rahman MM, Bielenberg GW, Krieglstein J. Effects of Ginkgo biloba constituents related to protection against brain damage caused by hypoxia. Pharmacol Res Comm 1988; 20:349–368.

75. Rapin JR, Le Poncin Lafitte M. Cerebral glucose consumption: the effect of Ginkgo biloba extract. Presse Med 1986; 15:1494–1497.

76. al-Zuhair H, Abd el-Fattah A, el-Sayed MI. The effect of meclofenoxate with

Ginkgo biloba extract or zinc on lipid peroxide, some free radical scavengers and the cardiovascular system of aged rats. Pharmacol Res 1998; 38:65–72.

77. Klein J, Chatterjee SS, Loffelholz K. Phospholipid breakdown and choline release under hypoxic conditions: inhibition by bilobalide, a constituent of Ginkgo biloba. Brain Res 1997; 755:347–350.

78. Rabin O, Drieu K, Grange E, Chang MC, Rapoport SI, Purdon AD. Effects of EGb 761 on fatty acid reincorporation during reperfusion following ischemia in the brain of the awake gerbil. Mol Chem Neuropathol 1998; 34:79–101.

79. Stoll S, Scheuer K, Pohl O, Muller WE. Ginkgo biloba extract (EGb 761) independently improves changes in passive avoidance learning and brain membrane fluidity in the aging mouse. Pharmacopsychiatry 1996; 29:144–149.

80. Suzuki H, Park SJ, Tamura M, Ando S. Effect of the long-term feeding of dietary lipids on the learning ability, fatty acid composition of brain stem phospholipids and synaptic membrane fluidity in adult mice: a comparison of sardine oil diet with palm oil diet. Mech Ageing Dev 1998; 16:119–128.

81. Scheuer K, Rostock A, Bartsch R, Muller WE. Piracetam improves cognitive performance by restoring neurochemical deficits of the aged rat brain. Pharmacopsychiatry 1999; 32:10–16.

82. Chopin P, Briley M. Effects of four non-cholinergic cognitive enhancers in comparison with tacrine and galanthamine on scopolamine-induced amnesia in rats. Psychopharmacology 1992; 106:26–30.

83. Deberdt W. Interaction between psychological and pharmacological treatment in cognitive impairment. Life Sci 1994; 55:2057–2066.

84. Puumala T, Greijus S, Narinen K, Haapalinna A, Riekkinen P Sr, Sirvio J. Stimulation of alpha-1 adrenergic receptors facilitates spatial learning in rats. Eur Neuropsychopharmacol 1998; 8:17–26.

85. Packard MG, Teather LA. Posttraining estradiol injections enhance memory in ovariectomized rats: cholinergic blockade and synergism. Neurobiol Learn Mem 1997; 68:172–188.

86. Akisu M, Kultursay N, Coker I, Huseyinov A. Platelet-activating factor is an important mediator in hypoxic ischemic brain injury in the newborn rat: flunarizine and Ginkgo biloba extract reduce PAF concentration in the brain. Biol Neonate 1998; 74:439–444.

87. Smith PF, Maclennan K, Darlington CL. The neuroprotective properties of the Ginkgo biloba leaf: a review of the possible relationship to platelet-activating factor (PAF). J Ethnopharmacol 1996; 50:131–139.

88. Spinnewyn B, Blavet N, Clostre F, Bazan N, Braquet P. Involvement of platelet-activating factor (PAF) in cerebral post-ischemic phase in Mongolian gerbils. Prostaglandins 1987; 34:337–349.

89. Ramassamy C, Clostre F, Christen Y, Costentin J. Prevention by a Ginkgo biloba extract (GBE 761) of the dopaminergic neurotoxicity of MPTP. J Pharm Pharmacol 1990; 42:785–789.

90. Ramassamy C, Christen Y, Clostre F, Costentin J. The Ginkgo biloba extract, EGb761, increases synaptosomal uptake of 5-hydroxytryptamine: in-vitro and ex-vivo studies. J Pharm Pharmacol 1992; 44:943–945.

91. White HL, Scates PW, Cooper BR. Extracts of Ginkgo biloba leaves inhibit monoamine oxidase. Life Sci 1996; 58:1315–1321.
92. Wu WR, Zhu XZ. Involvement of monoamine oxidase inhibition in neuroprotective and neurorestorative effects of Ginkgo biloba extract against MPTP-induced nigrostriatal dopaminergic toxicity in C57 mice. Life Sci 1999; 65:157–164.
93. Pardon MC, Joubert C, Perez-Diaz F, Christen Y, Launay JM, Cohen-Salmon C. In vivo regulation of cerebral monoamine oxidase activity in senescent controls and chronically stressed mice by long-term treatment with Ginkgo biloba extract (EGb 761). Mech Ageing Dev 2000; 113:157–168.
94. Sticher O. Quality of Ginkgo preparations. Planta Medica 1993; 59:2–11.
95. Oken BS, Storzbach DM, Kaye JA. The efficacy of Ginkgo biloba on cognitive function in Alzheimer disease. Arch Neurol 1998; 55:1409–1415.
96. Choi SR, Saji H, Iida Y, Magata Y, Yokoyama A. Ginseng pretreatment protects against transient global cerebral ischemia in the rat: measurement of local cerebral glucose utilization by [14C]deoxyglucose autoradiography. Biol Pharma Bull 1996; 19:644–646.
97. Wen TC, Yoshimura H, Matsuda S, Lim JH, Sakanaka M. Ginseng root prevents learning disability and neuronal loss in gerbils with 5-minute forebrain ischemia. Acta Neuropathol 1996; 91:15–22.
98. Zhang YG, Liu TP. Infleuences of ginsenosides Rb1 and Rg1 on reversible focal brain ischemia in rats. Chung Kuo Yao Li Hsueh Pao 1996; 17:44–48.
99. Salim KN, McEwen BS, Chao HM. Ginsenoside Rb1 regulates ChAT, NGF and trkA mRNA expression in the rat brain. Brain Res Mol Brain Res 1997; 47:177–182.
100. Kitts DD, Wijewickreme AN, Hu C. Antioxidant properties of a North American ginseng extract. Mol Cell Biochem 2000; 203:1–10.
101. Keum YS, Park KK, Lee JM, Chun KS, Park JH, Lee SK, Kwon H, Surh YJ. Antioxidant and anti-tumor promoting activities of the methanol extract of heat-processed ginseng. Cancer Lett 2000; 150:41–48.
102. Jiang F, DeSilva S, Turnbull J. Beneficial effect of ginseng root in SOD-1 (G93A) transgenic mice. J Neurol Sci 2000; 180:52–54.
103. Jaenicke B, Kim EJ, Ahn JW, Lee HS. Effect of Panax ginseng extract on passive avoidance retention in old rats. Arch Pharmacol Res 1991; 14:25–29.
104. Lasarova MB, Mosharrof AH, Petkov VD, Markovska VL, Petkov VV. Effect of piracetam and of standardized ginseng extract on the electroconvulsive shock-induced memory disturbances in "step-down" passive avoidance. Acta Physiol Pharmacol Bulg 1987; 13:11–17.
105. Petkov VD, Mosharrof AH. Effects of standardized ginseng extract on learning, memory and physical capabilities. Am J Chin Med 1987; 15:19–29.
106. Petkov VD, Kehayov R, Belcheva S, Konstantinova E, Petkov VV, Getova D, Markovska V. Memory effects of standardized extracts of Panax ginseng (G115), Ginkgo biloba (GK 501) and their combination Gincosan (PHL-00701). Planta Med 1993; 59:106–114.
107. Petkov VD, Cao Y, Todorov I, Lazarova M, Getova D, Stancheva S, Alova L. Behavioral effects of stem-leaves extract from panax ginseng C.A. Meyer. Acta Physiol Pharmacol Bulg 1992; 18:41–48.

108. Lehmann O, Jeltsch H, Lehnardt O, Pain L, Lazarus C, Cassel JC. Combined lesions of cholinergic and serotonergic neurons in the rat brain using 192 IgG-saporin and 5,7-dihydroxytryptamine: neurochemical and behavioural characterization. Eur J Neurosci 2000; 12:67–79.

109. Balse E, Lazarus C, Kelche C, Jeltsch H, Jackisch R, Cassel JC. Intrahippocampal grafts containing cholinergic and serotonergic fetal neurons ameliorate spatial reference but not working memory in rats with fimbria-fornix/cingular bundle lesions. Brain Res Bull 1999; 49:263–272.

110. Mingo NS, Cottrell GA, Mendonca A, Gombos Z, Eubanks JH, Burnham WM. Amygdala-kindled and electroconvulsive seizures alter hippocampal expression of the m1 and m3 muscarinic cholinergic receptor genes. Brain Res 1998; 810:9–15.

111. Vann SD, Brown MW, Erichsen JT, Aggleton JP. Fos imaging reveals differential patterns of hippocampal and parahippocampal subfield activation in rats in response to different spatial memory tests. J Neurosci 2000; 20:2711–2718.

112. Herzog CD, Gandhi C, Bhattacharya P, Walsh TJ. Effects of intraseptal zolpidem and chlordiazepoxide on spatial working memory and high-affinity choline uptake in the hippocampus. Neurobiol Learn Mem 2000; 73:168–179.

113. Stancampiano R, Cocco S, Cugusi C, Sarais L, Fadda F. Serotonin and acetylcholine release response in the rat hippocampus during a spatial memory task. Neuroscience 1999; 89:1135–1143.

114. Chepurnov SA, Chepurnova NE, Kholmanskikh SS, Berdiev RK, Park JK, Son JO. Learning by mice of Y-maze using drinking reinforcement and aversive olfactory stimulus (beneficial effect of ginsenosides). Biull Eksp Biol Med 1996; 122:253–257.

115. Gillis CN. Panax ginseng pharmacology: a nitric oxide link? Biochem Pharmacol 1997; 54:1–8.

116. Benishin CG, Lee R, Wang LC, Liu HJ. Effects of ginsenoside Rb1 on central cholinergic metabolism. Pharmacology 1991; 42:223–229.

117. Ma TC, Yu QH. Effect of 20(S)-ginsenoside-Rg2 and cyproheptadine on two-way active avoidance learning and memory in rats. Arzneimittelforschung 1993; 43:1049–1052.

118. Ma TC, Yu QH, Chen MH. Effects of ginseng stem-leaves saponins on one-way avoidance behavior in rats. Acta Pharmacol Sin 1991; 12:403–406.

119. Liu M. Studies on the anti-aging and nootropic effects of ginsenoside Rg1 and its mechanisms of actions. Prog Physiol Sci 1996; 27:139–142.

120. Liu M, Zhang JT. Effects of ginsenoside Rg1 on c-fos gene expression and cAMP levels in rat hippocampus. Acta Pharmacol Sin 1996; 17:171–174.

121. Lim JH, Wen TC, Matsuda S, Tanaka J, Maeda N, Peng H, Aburaya J, Ishihara K, Sakanaka M. Protection of ischemic hippocampal neurons by ginsenoside Rb1, a main ingredient of ginseng root. Neurosci Res 1997; 28:191–200.

122. Li Z, Guo YY, Wu CF, Li X, Wang JH. Protective effects of pseudoginsenoside-F11 on scopolamine-induced memory impairment in mice and rats. J Pharm Pharmacol 1999; 51:435–440.

123. Zhang JT, Qu ZW, Liu Y, Deng HL. Preliminary study on antiamnestic mechanism of ginsenoside Rg1 and Rb1. Chin Med J 1990; 103:932–938.
124. Smith K, Engels H-J, Martin JA, Wirth JC. Efficacy of a standardised ginseng extract to alter psychological function characteristics at rest and during exercise stress. Med Sci Sport Exerc 1995; 27:S147.
125. D'Angelo L, Grimaldi R, Caravagi M, Marcoli M, Perucca E, Lecchini S, Frigo GM, Crema A. A double-blind placebo-controlled clinical study on the effect of a standardized ginseng extract on psychomotor performance in healthy volunteers. J Ethnopharmacol 1986; 16:15–22.
126. Sorensen H, Sonne J. A double-masked study of the effect of ginseng on cognitive functions. Curr Ther Res 1996; 57:959–968.
127. Winther K, Ranlov C, Rein E, Mehlsen J. Russian root (Siberian Ginseng) improves cognitive functions in middle-aged people whereas Ginkgo biloba seems effective only in the elderly. J Neurol Sci 1997; 150:S90.
128. Thommessen B, Laake K. No identifiable effect of ginseng (Gericomplex) as an adjuvant in the treatment of geriatric patients. Aging (Milano) 1996; 8:417–420.
129. Joseph JA, Shukitt-Hale B, Denisova NA, Bielinski D, Martin A, McEwen JJ, Bickford PC. Reversals of age-related declines in neuronal signal transduction, cognitive, and motor behavioral deficits with blueberry, spinach, or strawberry dietary supplementation. J Neurosci 1999; 19:8114–8121.
130. Sun GY, Xia J, Draczynska-Lusiak B, Simonyi A, Sun AY. Grape polyphenols protect neurodegenerative changes induced by chronic ethanol administration. Neuroreport 1999; 10:93–96.
131. Bagchi D, Garg A, Krohn RL, Bagchi M, Bagchi DJ, Balmoori J, Stohs SJ. Protective effects of grape seed proanthocyanidins and selcted antioxidants against TPA-induced hepatic and brain lipid peroxidation and DNA fragmentation, and peritoneal macrophage activation in mice. Gen Pharmacol 1998; 30:771–776.
132. Hong JT, Ryu SR, Kim HJ, Lee JK, Lee SH, Kim DB, Yun YP, Ryu JH, Lee BM, Kim PY. Neuroprotective effect of green tea extract in experimental ischemia-reperfusion brain injury. Brain Res Bull 2000; 53:743–749.
133. Hong JT, Ryu SR, Kim HJ, Lee JK, Lee SH, Yun YP, Lee BM, Kim PY. Protective effect of green tea extract on ischemia/reperfusion-induced brain injury in Mongolian gerbils. Brain Res 2001; 888:11–18.
134. Inanami O, Asanuma T, Inukai N, Jin T, Shimokawa S, Kasai N, Nakano M, Sato F, Kuwabara M. The suppression of age-related accumulation of lipid peroxides in rat brain by administration of Rooibos tea (Aspalathus linearis). Neurosci Lett 1995; 196:85–88.
135. Komatsu M, Hiramatsu M. The efficacy of an antioxidant cocktail on lipid peroxide level and superoxide dismutase activity in aged rat brain and DNA damage in iron-induced epileptogenic foci. Toxicology 2000; 148:143–148.
136. Hartmann AM, Burleson LE, Holmes AK, Geist CR. Effects of chronic kombucha ingestion on open-field behaviors, longevity, appetitive behaviors, and organs in c57-bl/6 mice: a pilot study. Nutrition 2000; 16:755–761.

137. Hindmarch I, Quinlan PT, Moore KL, Parkin C. The effects of black tea and other beverages on aspects of cognition and psychomotor performance. Psychopharmacology 1998; 139:230–238.

138. Nishiyama N, Zhou Y, Saito H. Beneficial effects of DX-9386, a traditional Chinese prescription, on memory disorder produced by lesioning the amygdala in mice. Biol Pharm Bull 1994; 17:1679–1681.

139. Nishiyama N, Zhou Y, Saito H. Ameliorative effects of chronic treatment using DX-9386, a traditional Chinese prescription, on learning performance and lipid peroxide content in senescence accelerated mouse. Biol Pharm Bull 1994; 17:1481–1484.

140. Nishiyama N, Zhou Y, Takashina K, Saito H. Effects of DX-9386, a traditional Chinese preparation, on passive and active avoidance performances in mice. Biol Pharm Bull 1994; 17:1472–1476.

141. Nishiyama N, Wang YL, Saito H. Beneficial effects of S-113m, a novel herbal prescription, on learning impairment model in mice. Biol Pharm Bull 1995; 18:1498–1503.

142. Nishiyama N, Chu PJ, Saito H. An herbal prescription, S-113m, consisting of biota, ginseng and schizandra, improves learning performance in senescence accelerated mouse. Biol Pharm Bull 1996; 19:388–393.

143. Smriga M, Saito H, Nishiyama N. Hoelen (Poria cocos Wolf) and ginseng (Panax ginseng C. A. Meyer), the ingredients of a Chinese prescription DX-9386, individually promote hippocampal long-term potentiation in vivo. Biol Pharm Bull 1995; 18:518–522.

144. Zhang Y, Saito H, Nishiyama N. Improving effects of DX-9386, a traditional Chinese medicinal prescription, on thymectomy-induced impairment of learning behaviors in mice. Biol Pharm Bull 1994; 17:1199–1205.

145. Zhang Y, Saito H, Nishiyama N, Abe K. Effects of DX-9386, a traditional Chinese medicinal prescription, on long-term potentiation in the dentate gyrus in rats. Biol Pharm Bull 1994; 17:1337–1340.

146. Yagi H, Irino M, Matsushita T, Katoh S, Umezawa M, Tsuboyama T, Hosokawa M, Akiguchi I, Tokunaga R, Takeda T. Spontaneous spongy degeneration of the brain stem in SAM-P/8 mice, a newly developed memory-deficient strain. J Neuropathol Exp Neurol 1989; 48:577–590.

147. Yen TT, Kalasz H, Matkovics B. The effects of Ding lang root extract on the activities of peroxide metabolism enzymes in the rat. In: Matkovics B, Karmazsin L, Kalasz H, eds. Radicals, Ions and Tissue Damage. Budapest: Akademiai Kiado, 1990:293–298.

148. Yen TT. Stimulation of sexual performance in male rats with the root extract of Ding lang (Policias fruticosum L.). Acta Physiol Hung 1990; 75:61–67.

149. Yen TT, Dallo J, Knoll J. The effect of Ding lang and (-) deprenyl on the survival rate of male rats. Acta Physiol Hung 1990; 75:301–302.

150. Yen TT. Improvement of learning ability in mice and rats with the root extract of Ding Lang (Policias fruticosum L.). Acta Physiol Hung 1990; 75:69–76.

151. Xuejiang W, Magara T, Konishi T. Prevention and repair of cerebral ischemia-

reperfusion injury by Chinese herbal medicine, shengmai san, in rats. Free Radic Res 1999; 31:449–455.

152. Sugaya E, Yuyama N, Kajiwara K, Tsuda T, Ohguchi H, Shimizu-Nishikawa K, Kimura M, Sugaya A. Regulation of gene expression by herbal medicines—a new paradigm of gene therapy for multifocal abnormalities of genes. Res Commun Mol Pathol Pharmacol 1999; 106:171–180.

153. Inada K, Yokoi I, Kabuto H, Habu H, Mori A, Ogawa N. Age-related increase in nitric oxide synthase activity in senescence accelerated mouse brain and the effect of long-term administration of superoxide radical scavenger. Mech Ageing Dev 1996; 89:95–102.

154. Hiramatsu M, Edamatsu R, Mori A. Free radicals, lipid peroxidation, SOD activity, neurotransmitters and choline acetyltransferase activity in the aged rat brain. EXS 1992; 62:213–218.

155. Hamada H, Hiramatsu M, Edamatsu R, Mori A. Free radical scavenging action of baicalein. Arch Biochem Biophys 1993; 306:261–266.

156. Hiramatsu M, Haba K, Edamatsu R, Hamada H, Mori A. Increased choline acetyltransferase activity by Chinese herbal medicine Sho-saiko-to-go-keishi-kashakuyaku-to in aged rat brain. Neurochem Res 1989; 14:249–251.

157. Haba K, Ogawa N, Mori A. The effects of sho-saiko-to-go-keishi-ka-shakuyaku-to (TJ-960) on ischemia-induced changes of brain acetylcholine and monoamine levels in gerbils. Neurochem Res 1990; 15:487–493.

158. Chen J. An experimental study on the anti-senility effects of shou xing bu zhi. Chin J Mod Dev Trad Med 1989; 9:226–227, 198.

159. Liu JK, Kabuto H, Hiramatsu M, Mori A. Effects of Guilingji on brain monoamines and their metabolites in mice. Acta Med Okayama 1991; 45:217–222.

160. Liu J, Edamatsu R, Kabuto H, Mori A. Antioxidant action of guilingji in the brain of rats with FeCl3-induced epilepsy. Free Radic Biol Med 1990; 9:451–454.

161. Kabuto H, Asanuma M, Nishibayashi S, Iida M, Ogawa N. Chronic administration of Oren-gedoku-to (TJ15) inhibits ischemia-induced changes in brain indoleamine metabolism and muscarinic receptor binding in the Mongolian gerbil. Neurochem Res 1997; 22:33–36.

162. Fushitani S, Minakuchi K, Tsuchiya K, Takasugi M, Murakami K. Studies on attenuation of post-ischemic brain injury by kampo medicines-inhibitory effects of free radical production. II. J Pharm Soc Jpn 1995; 115:611–617.

163. Xu J, Murakami Y, Matsumoto K, Tohda M, Watanabe H, Zhang S, Yu Q, Shen J. Protective effect of Oren-gedoku-to (Huang-Lian-Jie-Du-Tang) against impairment of learning and memory induced by transient cerebral ischemia in mice. J Ethnopharmacol 2000; 73:405–413.

164. Sasaki K, Sudo T, Kurusu T, Kiuchi T, Yoshizaki F. The mechanism of alteration of monoamine metabolism in brain regions in marble burying behavior-isolated housing mice and effect of oren-gedoku-to on this alteration. J Pharm Soc Jpn 2000; 120:559–567.

165. Kondo Y, Kondo F, Asanuma M, Tanaka K, Ogawa N. Protective effect of oren-

gedoku-to against induction of neuronal death by transient cerebral ischemia in the C57BL/6 mouse. Neurochem Res 2000; 25:205–209.

166. Komatsu M, Ueda Y, Hiramatsu M. Different changes in concentrations of monoamines and their metabolites and amino acids in various brain regions by the herbal medicine/Toki-Shakuyaku-San between female and male senescence-accelerated mice (SAMP8). Neurochem Res 1999; 24:825–831.

167. Toriizuka K, Hou P, Yabe T, Iijima K, Hanawa T, Cyong JC. Effects of Kampo medicine, Toki-shakuyaku-san (Tang-Kuei-Shao-Yao-San), on choline acetyl-transferase activity and norepinephrine contents in brain regions, and mitogenic activity of splenic lymphocytes in ovariectomized mice. J Ethnopharmacol 2000; 71:133–143.

168. Kishikawa M, Sakae M. Herbal medicine and the study of aging in senescence-accelerated mice (SAMP1TA/Ngs). Exp Gerontol 1997; 32:229–242.

169. Itoh T, Murai S, Saito H, Masuda Y. Effects of single and repeated administrations of Toki-shakuyaku-san on the concentrations of brain neurotransmitters in mice. Methods Find Exp Clin Pharmacol 1998; 20:11–17.

170. Nishiyama N, Moriguchi T, Saito H. Beneficial effects of aged garlic extract on learning and memory impairment in the senescence-accelerated mouse. Exp Gerontol 1997; 32:149–160.

171. Moriguchi T, Saito H, Nishiyama N. Anti-ageing effect of aged garlic extract in the inbred brain atrophy mouse model. Clin Exp Pharmacol Physiol 1997; 24:235–242.

172. Moriguchi T, Matsuura H, Itakura Y, Katsuki H, Saito H, Nishiyama N. Allixin, a phytoalexin produced by garlic, and its analogues as novel exogenous substances with neurotrophic activity. Life Sci 1997; 61:1413–1420.

173. Zhang Y, Moriguchi T, Saito H, Nishiyama N. Functional relationship between age-related immunodeficiency and learning deterioration. Eur J Neurosci 1998; 10:3869–3875.

174. Moriguchi T, Takashina K, Chu P-J, Saito H, Nishiyama N. Prolongation of life span and improved learning in the senescence accelerated mouse produced by aged garlic extract. Biol Pharm Bull 1994; 17:1589–1594.

175. Shimada A, Ohta A, Akiguchi I, Takeda T. Inbred SAM-P/10 as a mouse model of spontaneous, inherited brain atrophy. J Neuropathol Exp Neurol 1992; 51:440–450.

176. Shimada A, Hosokawa M, Ohta A, Akiguchi I, Takeda T. Localization of atrophy-prone areas in the aging mouse brain: comparison between the brain atrophy model SAM-P/10 and the normal control SAM-R/1. Neuroscience 1994; 59:859–869.

177. Kaul S, Krishnakantha TP. Influence of retinol deficiency and curcumin/turmeric feeding on tissue microsomal membrane lipid peroxidation and fatty acids in rats. Mol Cell Biochem 1997; 175:43–48.

178. Kaul S, Krishnakanth TP. Effect of retinol deficiency and curcumin or turmeric feeding on brain Na(+)-K+ adenosine triphosphatase activity. Mol Cell Biochem 1994; 137:101–107.

179. Sugiura M, Shoyama Y, Saito H, Abe K. Crocin (crocetin di-gentiobiose ester) prevents the inhibitory effect of ethanol on long-term potentiation in the dentate gyrus in vivo. J Pharmacol Exp Ther 1994; 271:703–707.
180. Abe K, Saito H. Effects of saffron extract and its constituent crocin on learning behaviour and long-term potentiation. Phytother Res 2000; 14:149–152.
181. Suganuma H, Hirano T, Inakuma T. Amelioratory effect of dietary ingestion with red bell pepper on learning impairment in senescence-accelerated mice (SAMP8). J Nutr Sci Vitaminol 1999; 45:143–149.
182. Youdim KA, Spencer JPE, Schroeter H, Rice-Evans C. Flavonoids in neuro-protection. Biol Chem. In press.

9

Flavonoids: Neuroprotective Agents? Modulation of Oxidative Stress–Induced Map Kinase Signal Transduction

Hagen Schroeter
University of Southern California School of Pharmacy
Los Angeles, California, U.S.A.
King's College London
London, England

Jeremy P. E. Spencer
King's College London
London, England

I. INTRODUCTION

The pathological mechanisms of neurodegeneration and the process of aging are increasingly associated with oxidative stress mediated by reactive oxygen species/reactive nitrogen species (ROS/RNS) [1,2]. Several findings substantiate the susceptibility of the central nervous system (CNS) to oxidative stress:

The high density of mitochondria, a major cellular source of ROS, in brain tissue [1,3]

The high rate of oxygen consumption [1,3]

A considerable pool of autoxidizable compounds such as dopamine, L-dopa, and noradrenalin [4]

The presence of locally high levels of potentially excitotoxic glutamate [5]

The high Ca^{2+} traffic across membranes [6]

The lipid composition of the neuronal membrane with its relatively
high content of polyunsaturated fatty acids compared to that of the
nonneuronal cell

Oxidative stress has been shown to contribute to the neuropathological processes
of a number of neurodegenerative disorders [1], including Alzheimer's disease [7],
Parkinson's disease [8], and Huntington's disease [9], as well as being implicated
in neuronal loss associated with age-related cognitive decline [2,10,11], cerebral
ischemia/reperfusion injuries [5,12], seizures, trauma, and neuroinflammation
[13,14]. Several factors might enhance oxidative stress–related neuronal decline
in age-related neurological disorders:

Decrements in energy availability due to reduced mitochondrial
function [15]

Loss of enzyme activities essential in the defense against ROS/RNS or
important in damage removal, especially superoxide dismutase (SOD),
catalase, glutathione peroxidase, and proteasomal functions [16,17]

Decline in the overall antioxidant status due to a low intake of vitamin C,
vitamin E, β-carotene, minerals (e.g., selenium), and other micronutrients based on age-related changes in food intake and digestion
[18–20]

Accumulation of the amyloid β-peptide in Alzheimer's disease and the
deposition of potentially pro-oxidant iron in certain brain regions,
especially the substantia nigra, in Parkinson's disease [21]

Indeed, increased iron levels in the substantia nigra [21], elevation of lipid
peroxidation [22], decline in glutathione concentrations [23], enhanced protein
oxidation [24], and increase in glycation end products [25] are biomarkers
associated with oxidative damage in neurodegeneration and aging.

II. NEURONAL APOPTOSIS: INVOLVEMENT OF THE MITOGEN-ACTIVATED PROTEIN KINASE CASCADE

A. The Mitogen-Activated Protein Kinases

High levels of ROS and RNS can disrupt the normal redox state and shift cells
into the state of oxidative stress, hallmarked by intracellular increases in
products of lipid peroxidation, hydrogen peroxide, and elevated damage to
other biomolecules. It is often stated that the damage to important biomolecules
such as proteins and deoxyribonucleic acid (DNA) as a result of oxidative stress
leads to cell injury or death. Importantly, this oversimplification ignores a
number of stress response mechanisms that cells have developed to coordinate
reactions that ultimately determine the outcome of an oxidative insult. Among

the main stress signaling pathways or central mediators in stress response to oxidative insults are the mitogen-activated protein kinase (MAPK) cascades, the phosphoinositol 3-kinase/Akt pathway (PI3-kinase/Akt), the nuclear factor κB (NF-κB) signaling system, p53 activation, and the heat shock protein response. In general, MAPK pathways transduce extracellular and intracellular stimuli into cellular responses. These responses consist of direct phosphorylation of cytosolic or nuclear target proteins and activation of transcription factors, which consequently modulate gene expression. Typically, MAPK cascades are composed of three proteins, which are subsequently activated by phosphorylation: MAPK kinase kinase (MAPKKK), MAPK kinase (MAPKK), and MAPK [26,27]. These cascades may be initiated by a great variety of stimuli often via small guanosine triphosphate–(GTP)-binding proteins (Ras, Rho, Ral, Rap) [28,29].

Mammals express at least three distinctly regulated groups of MAPKs, which may exist in different isoforms:

1. extracellular signal-regulated kinases (ERK1/2)
2. c-Jun amino-terminal kinases (JNK1/2/3)
3. p38 kinases (p38α/β/γ/δ)

Each MAPKK can be activated by more than one MAPKKK, increasing the complexity and diversity of MAPK signaling. All members of the MAPK family require dual phosphorylation of a threonine and tyrosine residue within the catalytic domain by their respective upstream kinase in order to be activated. Hence ERK, JNK, and p38 contain the specific dual phosphorylation motif Thr-Glu-Tyr, Thr-Pro-Tyr, and Thr-Gly-Try, respectively [27]. Besides the upstream kinases, the activation of MAPKs critically depends on the activity of a special family of dual specificity phosphatases, the MAP kinase phosphatases (MKPs), which inactivate MAPK and therefore play an important role in the dynamics of MAPK signaling.

Active MAPKs function as modulators for differentiation, proliferation, cell death, and survival. Commonly, the activation of ERK1/2 has been linked to cell survival, whereas JNK and p38 [also called the *stress-activated protein kinases* (SAPKs)] have been associated with apoptosis [30]. This perspective is an oversimplification, and the actual roles are highly dependent on the cell type, the state of cell development, the kind of stimulus, and the context of stimulation.

B. Extracellular Signal-Regulated Kinase

Extracellular signal-regulated kinase (ERK) is probably the best-characterized, hence best-understood, member of the MAPK family. The activation of ERK1/2 in neurons triggers a wide variety of possible cellular responses ranging from proliferation and differentiation to cell survival and death.

1. Activation of Extracellular Signal-Regulated Kinase

Classically, ERK is activated in the CNS via growth factors such as the neuronal growth factor (NGF) [31]. The binding of neurotrophins to Trk receptors induces the autophosphorylation of Trk, which subsequently allows the docking of the Shc adaptor protein (Shc). This triggers the phosphorylation of Shc and consequently the recruitment of the Grb2-SOS complex, which leads to the activation of the small G-protein Ras. Ras-GTP initiates the subsequent sequential phosphorylation of Raf1-kinase (a MAPKKK), MEK1/2 (a MAPKK), and finally ERK1/2 (MAPK) [32] (Fig. 1). A novel pathway leading to the activation of ERK1/2 in neuronal cells, which is wholly Ca^{2+}-dependent,

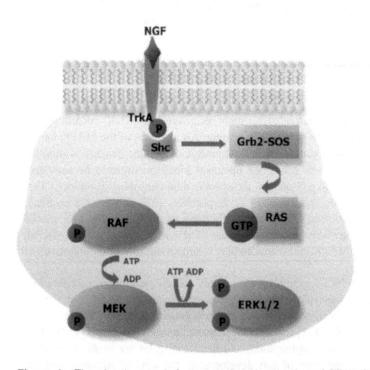

Figure 1 The classic growth factor–mediated activation of ERK1/2. The figure illustrates the binding of the neuronal growth factor (NGF) to its receptor (TrkA). The subsequent autophosphorylation of TrkA allows the binding and of the adapter protein Shc, which in turn triggers the recruitment of Grb2-SOS, subsequently leading to a GDP/GTP exchange and the phosphorylation of Ras. After its activation Ras-GTP initiates the subsequent sequential phosphorylation of Raf1-kinase (MAPKKK), MEK (MAPKK), and finally ERK1/2 (MAPK). ERK, extracellular signal-regulated kinase; GDP/GTP, guanosine diphosphate/guanosine triphosphate; MAPK, mitogen-activated protein kinase.

was recognized in the 1990s and gives insights to cell signaling processes in which calcium homeostasis is a fundamental factor [31,33,34]. These findings implicate the increase in intracellular calcium levels mediated by glutamate receptors with an activation of ERK1/2. This activation of ERK1/2 involves a complex signal transduction involving a pertussis-sensitive G-protein [33] and possibly phosphoinositol-3-kinase [34]. Evidence that Ca^{2+} might be an important regulator of ERK1/2 activation also emerges from the observation that the activation of the neurospecific Ras-GRF1, an activator of the MAPK pathways, is Ca^{2+}/calmodulin-sensitive [31,35,36].

2. Targets for Active Extracellular Signal-Regulated Kinase 1/2

Activation of ERK1/2 can lead to the phosphorylation of a wide array of potential targets in cytosol or nucleus. For example, active ERK1/2 can activate transcription factors [37] and effector kinases—the MAPK-activated protein kinases (MAPKAPKs) such as the mitogen- and stress-activated kinase 1 (MSK1) or the pp90 ribosomal S6 kinases (RSKs) [38]. RSK phosphorylates the Bcl-2 family member BAD, thereby inhibiting its proapoptotic activity [32]. RSK and MSK1 are also potent activators of the cyclic adenosine monophosphate (cAMP) response element binding protein (CREB), a transcription factor for Bcl-2 and therefore an important factor for cell survival [39]. This would suggest that active ERK1/2 plays an important role in prosurvival signal transduction processes. Indeed, different studies clearly indicate that under certain conditions the activation of ERK1/2 is essential to neuronal survival [30,32]. Furthermore, the activation of Ras, an initiator of the ERK1/2 signaling cascade, has been shown also to activate the PI3-kinase/Akt pathway, another important survival pathway in neurons, demonstrating possible interlinks of different survival signals [40]. Interestingly, the protective effects initiated by Ras activation involve the suppression of proapoptotic signals such as the tumor suppressor protein p53 and the Bcl-2 family member Bax and are less potent if either the ERK1/2 or the PI3-kinase/Akt pathway is selectively inhibited [40]. Thus, this describes the outcome of a signal transduction process as the sum of different actions toward the same target and underlines the importance of interlinks among signaling cascades in promoting and enhancing specific signals.

However, the actual role of ERK1/2 seems to be dependent on various parameters since inhibition of ERK1/2 activation during focal ischemia [41], oxidative stress [42], and a model for hippocampal seizures [43] attenuated neuronal death and cellular injury, indicating a proapoptotic role for ERK1/2. In addition, the inhibition of ERK1/2 activation has been demonstrated to protect a mouse neuronal cell line and rat primary cortical neurons from oxidative stress–induced neurotoxicity [44].

C. c-Jun Amino-Terminal Kinase

The stress-activated protein kinases (SAPKs) c-Jun amino-terminal kinase (JNK) and p38 are involved in signal transduction processes that are complex with regard to the large number of different stimuli capable of activating JNK or p38. At least 12 different MAPKKKs involved in the activation of SAPK have been overexpressed, of which the physiological functions, specificities, and regulations remain elusive. JNK exists in mammals in three distinct isoforms, which are differentially expressed in different tissues. The expression pattern of the JNK isoforms may even differ in cells of the same tissue. JNK3 expression is largely restricted to the CNS, whereas JNK1 and JNK2 have a wider pattern of distribution [27,45,46].

1. Activation of c-Jun Amino-Terminal Kinases

JNKs can be activated by a large range of different stimuli, including kainic acid [47], microtubule-disrupting agents [48], double-stranded ribonucleic acid (RNA) [49], viral infections [49], tumor necrosis factors, interleukins [50], osmotic stress [51], and trophic factor withdrawal [30]. JNKs have also been reported to be activated by oxidative stress, particularly that induced by ultraviolet radiation [47], hypoxia [52], ischemia [53], hydrogen peroxide [30,54], peroxynitrite [55], malondialdehyde [56], and 4-hydroxynonenal [57].

In accord with the general scheme of the activation of MAPKs, the JNK signaling cascade can be initiated through the activation of small G-proteins such as Rac/cdc42, a protein belonging to the Rho family of small G-proteins. Rac/cdc42 is activated by guanosine diphosphate/guanosine triphosphate (GDP/GTP) exchange mediated by a variety of guanine nucleotide exchange factors (GEFs) such as SOS, VAV, Tiam, and the neuron-specific Ras-GRF1 in response to a stimulus [45]. Active Rac/cdc42 initiates the subsequent phosphorylation of JNK-specific MAPKKKs such as MEKK2/3,which phosphorylates the MAPKKs, and MKK4/7, which ultimately causes dual phosphorylation of JNKs (MAPK) [27,45,46] (Fig. 2).

Several other MAPKKKs, have been reported to activate the JNK pathway. These include members of the MEKK group (MEKK1–4), the mixed-lineage protein kinases (MLK1–3) [58,59], and members of the apoptosis signaling kinase (ASK) family [60] (Fig. 3). This variety of potential activators of JNKs reflects the broad array of possible stimuli that may lead to JNK phosphorylation. For example, JNKs can be activated via the MAPKKK MEKK1, involving caspases [61]; via the MAPKKK MEKK4, involving the p53-mediated expression of growth arrest and DNA damage–induced protein (GADD45) [62]; or via the tumor necrosis factor α–(TNF-α)-induced activation of the ASK1 [60] (Fig. 3).

As a result of the diversity of stimuli and upstream events that lead to the activation of JNKs, this signaling cascade is complex and not completely understood. Furthermore, the essential Ca^{2+}/calmodulin binding to the neuron-specific

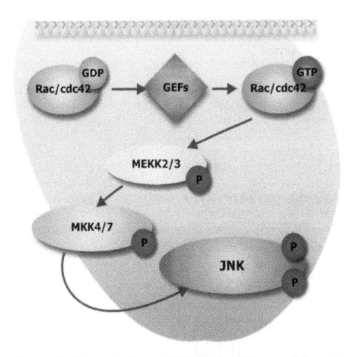

Figure 2 The activation of JNK. Rac/cdc42, a small G-protein, is activated by GDP/GTP exchange mediated by guanine nucleotide exchange factors (GEFs). Active Rac/cdc42 initiates the subsequent phosphorylation of the JNK-specific MAPKKK MEKK2/3, which in turn phosphorylates the MAPKK MKK4/7, ultimately resulting in the dual phosphorylation of JNK (MAPK). JNK, c-Jun amino-terminal kinase; GDP/GTP, guanosine diphosphate/guanosine triphosphate; MAPK, mitogen-activated protein kinase.

Ras-GRF1 [35,36], which may also activate the JNK pathway [63], provides a possible link between intracellular Ca^{2+} homeostasis and activation of JNKs.

The control of such diverse signaling processes is essential for cellular functions, especially in view of the variety of different stimuli involved. It is therefore important to ensure specificity in MAPK activation and function. In the case of JNKs, mechanisms ensuring specificity involve scaffolding proteins, physical interactions between members of a given cascade, and ability of MAPK to regulate indirectly the expression of ligands and inhibitors for receptors that are involved in JNK signaling [45].

2. Targets for Active c-Jun Amino-Terminal Kinase

Active JNKs have a wide range of potential phosphorylation targets in the nucleus as well as in the cytoplasm (Fig. 4). Nuclear substrates for JNKs are transcription

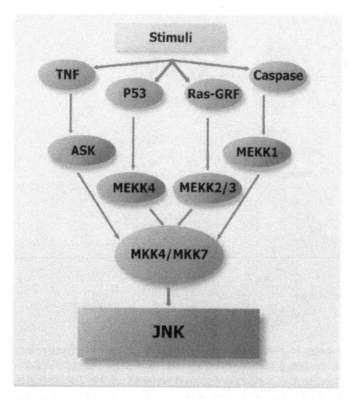

Figure 3 Diverse stimuli are able to activate JNK. This picture exemplifies the various pathways leading to JNK activation, indicating the complexity and diversity of stimuli able to induce JNK-mediated signal transduction. JNK, c-Jun amino-terminal kinase.

factor proteins such as c-Jun [64–66], ATF-2 [67], and ELK-1 [68]. As far as is known, JNKs are the only kinases capable of phosphorylating c-Jun in vivo. c-Jun is part of the activator protein 1 (AP-1) transcription factor, which exists as either a Jun homodimer or as a Jun/Fos heterodimer [69]. The transcriptional activity of the AP-1 complex is increased [69] following phosphorylation of Ser-63 and Ser-73 of c-Jun by JNK [70]. In addition to c-Jun, JNKs also phosphorylate other AP-1 proteins, including JunB and JunD [69,71]. The regulatory effect of JNKs on AP-1 transcription may not be due only to phosphorylation of Jun but may also involve JNK-regulated ubiquitin-mediated degradation of AP-1 proteins [72,73]. JNKs appear to be essential in cytokine- and stress-induced activation of AP-1 but are not required for AP-1 activation in response to other stimuli.

Cytosolic substrates for JNKs include cytoskeletal proteins, the tumor suppressor protein p53, the mitogen-activated kinase activating death domain

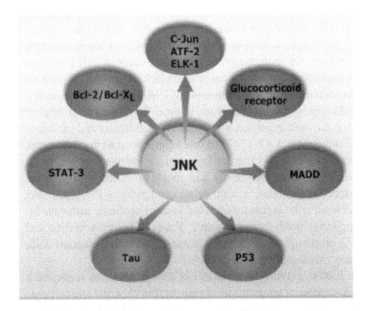

Figure 4 Potential cytosolic and nuclear targets for active JNK. JNK, c-Jun amino-terminal kinase.

protein (MADD), glucocorticoid receptors, Bcl-2 and Bcl-x_L, neurofilaments, tau, and STAT-3, a member of the signal transducer and activators of transcription (STAT) family (Fig. 4).

D. c-Jun Amino-Terminal Kinase Signaling in Neuronal Death

Although JNK activation is often associated with cellular injury and degeneration, it also plays an important part in embryonic morphogenesis, memory formation, and immune defense. The proapoptotic action of JNKs in neurons was initially investigated in neuronal cell death that followed the withdrawal of neurotrophic factors [30]. It was found that JNK activation contributed to the apoptotic response and that JNK-mediated apoptosis was suppressed by activation of survival pathways. Studies in mice utilizing the disruption of the neuronal gene *JNK3* confirmed the role of JNK in stress-induced neuronal apoptosis by demonstrating that *JNK3*-knockout mice were developmentally normal but resistant to excitotoxin-induced neuronal apoptosis [74]. These findings were supported by observations in mice with a mutation in the *c-Jun* gene that altered the JNK phosphorylation sites of c-Jun and led to a resistance to kainic acid–triggered death of hippocampal neurons [75]. Further support arose from data obtained by using dominant-negative c-Jun mutants, which reduced sympathetic neuronal death after NGF withdrawal [76]. A proapoptotic function has also been suggested for JNK activation in NGF-withdrawal-induced neuronal death [77] in a hippocampal model of Huntington's disease [78] and β-amyloid-induced neuronal apoptosis [79]. However, the apoptotic process does not occur in JNK-knockout mice [74] or mice expressing a mutant form of c-Jun lacking the JNK phosphorylation site [75]. In addition, the indirect inhibition of JNK by CEP-1347 protected neuronal PC12 cells and sympathetic neurons in vitro from death after trophic factor withdrawal, β-amyloid exposure, ultraviolet (UV) radiation and oxidative stress [75,79,80]. Furthermore, CEP-1347 protected nigral neurons in vivo against MPTP-induced neuronal death [81] implicating again the involvement of JNK in neurodegeneration. Moreover, JNK activity and apoptosis in cerebellar granule neurons were enhanced after inhibition of the prosurvival pathway PI3-kinase/Akt [82]. Finally JNKs seem to be indirectly involved in other apoptotic pathways by enhancing transcription of death receptors such as Fas-L [83] or by activating and stabilizing the p53 protein [84,85]. Together these data strongly support JNK involvement in apoptosis signaling in neurons.

A review in 2000 by Davis [45] of the role of JNK in apoptosis suggested a mechanism of JNK-dependent apoptosis involving the mitochondria and caspase-3. This hypothesis is based on observations in primary murine embryonic fibroblasts (MEFs) lacking the genes for *Jnk1* and *Jnk2* (Jnk null; no JNK1/2 protein/activity). These *Jnk* null MEFs exhibit profound defects in stress-induced (UV radiation, DNA alkylation, translational inhibition) apoptosis [86]. The defect in the execution of apoptotic cell death was caused by the failure to initiate the JNK-induced cytochrome c release from the mitochondria [86]. This malfunction is significant since it is critical to the subsequent sequential activation of Apaf-1 [87], the initiator caspase caspase-9 [88], and finally the effector caspase caspase-3 [89], all of which are essential in the execution of apoptosis. Tournier and colleagues [86] suggested that the apoptotic response is suppressed in *Jnk* null MEF as a result of the absence of JNK, which is needed to initiate the apoptotic cascade (Fig. 5). However, it is not clear yet by which molecular mechanism JNK mediates the release of cytochrome c from the mitochondria. Although a c-Jun activated transcription seems possible (Fig. 5) [75,76], it is not required for UV-induced apoptosis [86]. Several studies point to the JNK-mediated in vitro phosphorylation/inactivation of the mitochondria-associated antiapoptotic proteins Bcl-2 and Bcl-X$_L$ [90–92] as a possible mechanism, since Bcl-2/Bcl-X$_L$ is known to regulate cytochrome c release (Fig. 5). Others have proposed an as yet unknown adaptor protein as the mediator for JNK actions. Further studies are needed to substantiate this hypothesis and establish the molecular mechanisms.

In summary it is becoming increasingly clear that JNK activation is involved in apoptotic processes in vivo and in vitro either by direct effects of JNK-mediated c-Jun phosphorylation, adverse effects on survival pathways, or indirect effects on cell function via other mediator systems. But apoptosis is not

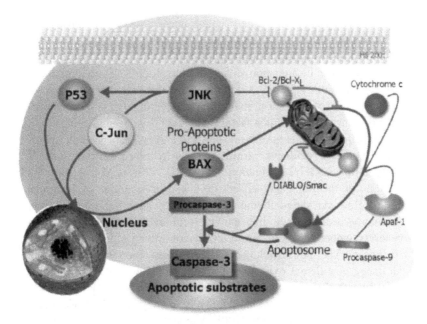

Figure 5 The possible involvement of JNK signaling in the execution of cell death. Once activated, JNK can phosphorylate p53, which in term stabilizes and activates this apoptotic mediator, resulting in the suppression of the antiapoptotic protein Bcl-2 and the induction of the proapoptotic protein BAX. Active JNK may also influence the expression of other proapoptotic molecules via the c-Jun/AP-1-mediated regulation of their expression. The proposed function for active JNK as mediator of the phosphorylation of the mitochondria-associated proteins Bcl-2 and Bcl-X$_L$ alters their antiapoptotic functions and may subsequently result in the release of cytochrome c or other apoptotic factors such as DIABLO/Smac from the mitochondria. The release of cytochrome c in turn promotes the formation of the apoptosome, subsequently leading to the activation of caspase-3, one of the major executors of apoptosis. JNK, c-Jun amino-terminal kinase; AP-1, activator protein 1.

the inevitable outcome of JNK signaling. In fact, apoptotic cell death might rather be the sum of signal transduction processes involving both prosurvival and proapoptotic pathways.

E.   Extracellular Signal-Regulated Kinase1/2 and c-Jun Kinase Signaling and Oxidative Stress

Oxidative stress seems to be a major stimulus for MAPK signaling cascades. However, this picture would appear to be too simplistic since a shift of the overall reductive environment that most cells maintain under physiological conditions toward reduction also leads to the activation of MAPK signal transduction with cell survival or death as a possible consequence. The observation that multiple pathways are sensitive to alteration in intracellular ROS/RNS concentrations indicates that this might represent a common way that cells have evolved to signal a large diversity of different stressful stimuli. Accordingly a large number of redox stress-responsive transcription factors and genes have been identified [93].

Growth factor receptors have been shown to undergo enhanced phosphorylation in response to oxidative insults such as hydrogen peroxide or UV radiation initiating ERK1/2 activation [33,94,95]. This is consistent with demonstrations of the mitogenic effects of low concentrations of hydrogen peroxide [96]. However, the inhibition of ERK1/2 activation has been shown to protect a mouse neuronal cell line and rat primary cortical neurons from oxidative stress–induced neurotoxicity [97], demonstrating a possible involvement in cell death. The JNK signaling cascade has been reported to be activated by a wide range of different oxidants/reductants, including hydrogen peroxide [30,54], lipid peroxidation products [56,57], different types of radiation [47], modulators of intracellular glutathione status [98], peroxynitrite [55], glutamate [99], dithiothreitol, and nitric oxide [100]. Although the links between redox status and MAPK signaling have been known for some time, data demonstrating the molecular basis of such links and identifying the sensors in this redox response are few.

Accumulating evidence indicates that members of the MAPK family or their upstream or downstream partners have such redox-sensitive motives. For example, JNK itself exhibits a redox-sensitive cysteine residue that is not present on ERK or p38. An additional mechanism in JNK redox regulation is its binding to redox-sensitive proteins such as glutathione S-transferases (GSHSTs) [101]. It has been shown that under nonstressed conditions JNK is associated with GSHST, resulting in the inhibition of JNK activity, but that GSHST dissociates from JNK after UV radiation or hydrogen peroxide treatment [101,102]. Forced expression of GSHST decreased JNK activity, increased c-Jun ubiquitination, and reduced c-Jun-mediated transcription [101]. In addition, GSHST expression in NIH 3T3 cells led to the attenuation of hydrogen peroxide–induced JNK activation as well to an increase in ERK1/2 activity [102]. Similarly GSHST and the redox regulatory protein thioredoxin (Trx) bind under nonstressed conditions to the apoptosis signal-regulating kinase ASK1 [103–105], an upstream activator of JNK, and inhibit JNK activity. However, oxidative insults cause the dissociation of the Trx-ASK1 and GSHST complex and the subsequent activation of [103–105].

Ca^{2+}-Homeostasis is another important mediator/regulator of oxidative stress–induced signaling, since both ERK and JNK are sensitive to changes in intracellular Ca^{2+} concentrations [35,36,63,106]. This Ca^{2+} sensitivity of ERK/

JNK signaling might play a role in oxidative stress–induced signaling events, since it has been demonstrated that oxidative insults often influence normal Ca^{2+} homeostasis in cells. With regard to ERK1/2, Samanta and coworkers [107] demonstrated for the first time that hydrogen peroxide–induced activation of ERK1/2 in primary neurons is strictly dependent on extracellular calcium. Ca^{2+}-dependent MAPK activation has also been suggested to play a role in glutamate receptor–mediated neuronal death [34]. The influx of Ca^{2+} into the cytosol from the extracellular space or from intracellular stores after a stressful stimulus can activate the Ca^{2+}/calmodulin kinases, which in turn can stimulate the activation of all three MAPKs [108].

The effect of ROS/RNS on MAPK activation is complex and occurs at multiple levels, further results are needed to elucidate the molecular basis of these interactions.

F. Nitric Oxide and Mitogen-Activated Protein Kinase Signaling

$NO^\bullet$ has been identified as a messenger, promoting Ca^{2+}-dependent neurotransmitter release from synaptic storage vesicles. $NO^\bullet$ modulates exocytosis through cGMP-dependent protein phosphorylation cascades after the classic activation of soluble guanylate cyclase. The high diffusibility of $NO^\bullet$ makes it an ideal retrograde signal for the two forms of synaptic modulation required for learning and memory, namely, long-term potentiation (LTP) in the hippocampus and long-term depression in the cerebellum [109].

There is growing evidence for numerous cGMP-independent mechanisms for $NO^\bullet$-mediated cell signaling. Besides having a role in differentiation and synaptic plasticity, $NO^\bullet$ has been implicated in neuronal apoptosis and, consequently, in neurodegenerative diseases, specifically when $NO^\bullet$ production is increased to toxic levels [1,2,110–112]. In particular, $NO^\bullet$ has been linked to the phenomenon of excitotoxicity involving the overstimulation of the NMDA receptor by glutamate, subsequently triggering a strong intracellular accumulation of Ca^{2+} [113]. This Ca^{2+} overload in neurons leads to a substantial increase in the activity of Ca^{2+}/calmodulin-dependent nitric oxide synthase (neuronal NOS), resulting in a high intracellular concentration of $NO^\bullet$, which has been identified as a mediator of glutamate-induced neuronal death [110]. Furthermore, microglia activation during neuroinflammation is associated with induction of inducible nitric oxide synthase (iNOS) expression, leading to large and sustained $NO^\bullet$ production, which also appears to be causally linked to neuronal apoptosis and neurodegeneration [114,115].

Conversely, there is growing evidence for cGMP-independent $NO^\bullet$-mediated cell signaling toward neuronal survival, and $NO^\bullet$ has been implicated in mechanisms protecting against stress-induced cell injury. The GTP-binding protein Ras is an intermediate in the transduction of signals from membrane receptor tyrosine kinases to MAP kinases [116] (Fig. 6). Ras appears to be a common signaling target of reactive free radicals, including NO$^{\bullet}$, and agents that modulate the cellular redox status, including H_2O_2 and GSH (111). NO$^{\bullet}$ activates Ras by S-nitrosation of a highly conserved cysteine residue, leading to GDP/GTP exchange and downstream signaling [117], such as the activation

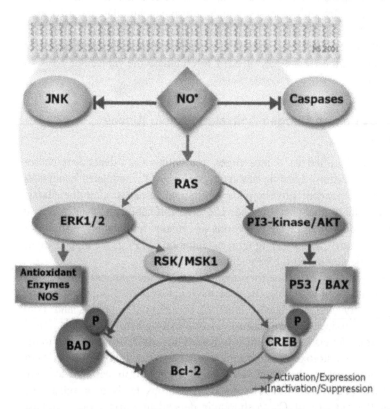

Figure 6 Potential antiapoptotic mechanisms mediated by nitric oxide. The nitrosation of Ras mediated by nitric oxide potentially leads to the activation of ERK1/2 or the PI3-kinase/AKT pathway, resulting in the suppression of proapoptotic proteins such as p53 and BAX or in the upregulation of antioxidant enzymes. Furthermore, Ras activation might mediate the phosphorylation/ activation of the cAMP response element binding protein (CREB), a transcription factor for Bcl-2, and the phosphorylation/inactivation of the proapoptotic protein BAD. In addition, the NO$^{\bullet}$-mediated nitrosation of JNK and caspases may also contribute to the antiapoptotic properties of NO$^{\bullet}$. ERK, extracellular signal-regulated kinase; PI3, phosphoinositol 3-kinase; JNK, c-Jun amino-terminal kinase.

of ERK1/2. ERK1/2 activation, as discussed earlier, may in turn result in the phosphorylation/inactivation of the proapoptotic protein BAD, the expression of antiapoptotic proteins such as Bcl-2, and the suppression of proapoptotic proteins such as BAX (discussed previously) (Fig. 6). It has been proposed that this cysteine residue may represent an important molecular redox trigger, whereby cells can respond to the ambient redox status. In addition, NO$^{\bullet}$ may directly suppress JNK activation via S-nitrosation of a redox-sensitive residue that is not present on ERK or p38 [100,118], thereby also supporting prosurvival mechanisms (Fig. 6).

Ultimately, S-nitrosation may thus regulate several redox-senstive transcription factors, including NF-κB, AP-1, Sp-1, and p53, and modulate levels of active c-*fos* and c-*jun* [119]. Indeed, the expression of some enzymes implicated in antioxidant defenses such as the enzymes for glutathione synthesis [120] and heme oxygenase exhibit ERK1/2 dependency in the regulation of their expression (Fig. 6). Furthermore, the transcription of Cu/ZnSOD is regulated by the transcription factor Elk1 [121] and the promoter for MnSOD expression contain binding sites for Sp1, AP-1, and CREB [122,123], all of which have been linked to regulation by ERK1/2 [32,124] (Fig. 6).

In turn, MAPKs are also involved in the regulation of the gene expression of all three NOS isoenzymes. The expression of iNOS (NOS-2) is regulated by all three MAP kinase pathways in a variety of cell types [125–129]. For example, JNK and ERK1/2 pathways are necessary for lipopolysaccharide (LPS-) and interferon-γ-stimulated iNOS expression in mouse macrophage cells, possibly via α-tumor necrosis factor secretion, whereas p38 inhibited induction [130]. The induction of endothelial NOS (eNOS, NOS-3) by estrogen, fibroblast growth factor, or epidermal growth factor in endothelial cells involves the Ras-ERK pathway [131,132]. eNOS is phosphorylated, and thus activated, by the serine/threonine protein kinase Akt, which is recruited to the cell membrane by PI3-kinase as an antiapoptotic mechanism in the response of endothelial cells to shear stress [133].

III. IMPLICATIONS FOR FLAVONOIDS IN PROTECTION AGAINST NEURODEGENERATION

As indicated in the previous discussion, there is increasing evidence that oxidative stress contributes to the neuropathological mechanisms of disorders that include Alzheimer's [7], Parkinson's [8], and Huntington's diseases [9]; neuronal loss associated with cognitive decline in aging [11]; and cerebral ischemia and seizures [5]. In addition, the molecular mechanisms underlying oxidative stress–induced neuronal damage are emerging and appear to involve an apoptotic mode of death in which ERK1/2 [42,44,107] and JNK [10,45,46] have been strongly implicated. Furthermore, it is becoming clear that products of lipid peroxidation such as 4-hydroxy-2,3-nonenal (4-HNE), lipid hydroperoxides (LOOHs), and oxysterols are important mediators of oxidative stress–mediated apoptosis in the CNS [22,134–136]. Consequently, there is a growing interest in the establishment of therapeutic and dietary strategies to combat oxidative stress–induced damage to the CNS, and attention is turning to the potential neuroprotective effects of dietary antioxidants, especially flavonoids.

IV. FLAVONOIDS: BIOLOGICAL AND CELLULAR PROPERTIES

The major emphasis in recent years has concerned the potent in vitro antioxidant effects of flavonoids described in numerous publications [137–140]. However, flavonoids exhibit various effects on mammalian cells with interesting implications for inflammation [141], cardiovascular disease [142,143], and cancer [144] involving the modulation of redox functions, calcium homeostasis [145], and the activity of various enzyme systems [146] (for more information please refer to Chap. 8). The effects of flavonoids are often pictured as beneficial for cell survival, preventive against oxidative insults, and anticarcinogenic. However, the actions of flavonoids are complex and often seemingly antagonistic or paradoxical. For example, flavonoids have been described as antioxidant agents protecting against oxidative insults to cells and apoptosis [147–149]; other researchers have found flavonoids to be prooxidants and proapoptotic [150–152]. Thus it becomes clear that the effects of flavonoids depend on different factors such as the specific compound used, the cell type, concentrations, experimental design, and the general context in which flavonoids are used. Furthermore, data used to demonstrate biological effects of flavonoids are often observational and fail to explain the molecular/cellular basis of such observations or base the effects solely on the rather unspecific (though important) antioxidant properties of flavonoids.

A. Flavonoids: Neuroprotective Agents In Vivo and In Vitro?

Data on flavonoids in the context of the CNS or CNS-derived cells are starting to accumulate but are not as extensive as for other cells or tissues. Epidemiological and dietary intervention studies in humans and animals in 2000 suggest that flavonoids may play a useful role in preventing neurodegeneration, especially age-related cognitive, motoric, and mood decline, and protect against oxidative stress as well as cerebral ischemia/reperfusion injuries. Studies in humans using flavonoid-containing plant extracts, such as *Ginkgo biloba* extracts, demonstrate positive effects on cognitive function and memory performance in healthy volunteers from all age groups [153–155], as well as in patients with age- or Alzheimer's disease–associated dementia [156–158]. Watanabe and associates [159] reported that *Ginkgo biloba* supplementation in mice had neuromodulatory effects as indicated by a more than three-fold increase of messenger RNA (mRNA) expression for neuronal tyrosine/threonine phosphatases-1, microtubule-associated tau, prolactin, different calcium and chloride channels, as well as transthyretin. Flavonoid-associated [e.g., citrus flavonoids, (epi)catechin, anthocyanidins] antioxidant interventions have also been proposed to be beneficial in hypoxia/ischemia, seizures, Parkinson's disease, increased survival rate in brain cancers, and general age-related neurodegeneration [160–164]. However, since the flavonoids used in these studies were given as food extracts or preparations containing other potential bioactive or antioxidant components rather than as pure flavonoid fractions, the results do not solely reflect the effects of flavonoids. Furthermore, the mechanisms of action of flavonoids remain speculative and may involve the antioxidant properties, the modulation of receptor/protein function, and calcium homeostasis or a combination thereof [165].

In vivo studies in animals demonstrate protective effects of the flavonoids epicatechin and quercetin against ischemia/reperfusion-induced neuronal injury [166,167]. Furthermore, flavonoids showed protective effects against an increase in brain lipid peroxidation after vitamin E deprivation [168], attenuated neuronal damage induced by ethanol administration [169,170], and reduced *O*-ethyl-*S,S*-dipropyl phosphorodithioate–induced neurotoxicity in mice [171]. Interestingly, the oral administration of a catechin-containing antioxidant preparation significantly increased the life span of senescence-accelerated mice [172] and increased SOD activity in the mitochondrial fraction of the striatum and the midbrain, decreased products of lipid peroxidation in cortex and cerebellum, and attenuated the iron(II)-chloride-induced formation of markers of DNA damage in the cortex of aged rats [173]. The authors of this study relate the observed effects mainly to the rather unspecific antioxidant activity of catechin and other compounds contained in the preparation and give no further insights about the mechanisms related to the increase in mitochondrial SOD activity. Other investigators reported an attenuation of age-related decline in cognitive function and behavioral deficits after long-term dietary supplementation with anthocyanidin-rich foods or plant extracts [163,174,175]. The investigators used several neuronal and behavioral parameters including dopamine release, GTP activity, calcium buffering in striatal synaptosomes, rod walking tasks, and water maze performance to elucidate possible mechanisms involved [163,175].

In vitro studies on primary striatal neurons demonstrated the attenuation of oxidized low-density lipoprotein–(Ox-LDL)-induced neurotoxicity by phenolic compounds, indicating overall that flavonoids, in particular epicatechin, kaempferol, and cyanidin, are more effective in attenuating Ox-LDL-induced neurotoxicity than hydroxycinnamates or vitamin C [176]. In subsequent studies the same authors demonstrated that the protective effect of epicatechin against Ox-LDL-induced neurotoxicity is seemingly independent of the classical hydrogen-donating antioxidant properties of epicatechin [177]. This conclusion was based on the following observations:

1. Epicatechin does not prevent the oxLDL-induced oxidative shift in the intracellular redox state [177].
2. Common antioxidants, such as ascorbic acid, only weakly protect against Ox-LDL-induced neurotoxicity when used in 10 times higher concentrations than epicatechin [176].
3. 3'-O-methyl-epicatechin, one of the major in vivo metabolites of epicatechin and a compound with significantly lower antioxidant capacity than epicatechin, equals epicatechin in its ability to prevent neuronal death [177].

The authors indicated that the neuroprotective effects of epicatechin and 3'-O-methyl-epicatechin might be based on their modulation of apoptotic cellular signaling pathways [177] (discussed later). Interestingly, Spencer and colleagues [178] reported that epicatechin and 3'-O-methyl-epicatechin are equally effective in protecting primary cortical neurons against hydrogen peroxide–induced neuronal death, whereas the glucuronidated in vivo metabolite epicatechin-5-β-O-D-glucuronide exerts no significant protection in the same model. This difference in bioactivity was based on the inability of epicatechin-5-β-O-D-glucuronide to enter cells as determined by liquid chromatography mass spectrometry (LC-MS/MS) analysis [178].

Observations in neuronal PC12 cells demonstrated the protective effects of *Ginkgo biloba* extract against hydrogen peroxide–mediated [179] or β-amyloid-induced neuronal death [180,181]. Other authors reported protective effect of *Ginkgo biloba* extract against β-amyloid-mediated neurotoxicity in primary hippocampal neurons [147,182] and implicated antioxidant effects and the modulation of intracellular calcium levels in PC12 cells as possible mechanisms of action [179]. Other lines of research focused on the binding of flavonoids such as apigenin, naringenin, kaempferol, and quercetin-3-O-glucoside to benzodiazepine binding sites [183] of different receptors, including the γ-aminobutyric acid A–(GABA-A)-receptor [184,185] and adenosine receptors [186], and investigated their anxiolytic potential [183] as a possible mechanism of action in the CNS. In addition, the levels of neurotransmitters involved in the pathophysiological mechanisms of mood disorders, such as 5-hydroxytryptamine, noradrenaline, and dopamine, have been found to be modified in cortex, diencephalons, and brainstem of rats after the administration of a flavonoid-containing preparation [187,188]. Bastianetto and coworkers [148] report that antioxidant effects are not the only mechanism of flavonoid-mediated protection against neuronal death and show attenuation of nitric oxide–induced activation of protein kinase C (PKC) to be partially involved in the protection against neurotoxicity. However, the precise mechanisms by which flavonoids exert their neuroprotective actions in vivo and in vitro are presently unknown, and it is currently unclear whether or not these compounds function solely as hydrogen-donating antioxidants or exert their neuroprotective actions independently of such properties.

Furthermore, only a very few investigators have studied the influence of metabolism on the bioactivity of flavonoids [177,178,189–191]; thus most of the effects reported, on the basis of in vitro experiments, cannot be extrapolated to in vivo situations. For example, one of the most intensively investigated flavonoids, quercetin, is often only used as the aglycone in cell culture systems and interesting biological activities such as modulation of the multidrug-resistance protein [192], induction of apoptosis [193], inhibition of the expression of nitric oxide synthase and cyclooxygenase [194], protection against oxidative stress [149], and modulation of the calcium homeostasis [179] have been reported. However, in plants and therefore also in most plant-derived foods and beverages, quercetin is predominantly present in a glycosylated form that completely alters its bioactivity [195–197]. If digested, the glycosides might be taken up or be cleaved and the released quercetin metabolized so that there is almost no free quercetin aglycone in circulation. Thus the bioactivities observed when using the aglycone alone may be very different from the effects of the derivatives of quercetin. It is therefore vital for cell culture experiments to analyze, synthesize, and investigate the in vivo metabolites of the flavonoid under investigation.

V. WHAT ARE THE POTENTIAL MECHANISMS AND TARGETS FOR THE BIOACTIVITY OF FLAVONOIDS?

A. Do Flavonoids Act as Antioxidants In Vivo?

There has been considerable interest in recent years in the cytoprotective and neuroprotective effects of flavonoids, especially in the context of their modes of action as antioxidants. The electron-donating properties of flavonoids are well defined to explain their antioxidant properties in vitro [10–14]. Structurally important features defining the reduction potential of flavonoids are the hydroxylation pattern, a 3',4'-dihydroxy catechol structure in the B-ring, the planarity of the molecule, and the presence of 2,3 unsaturation in conjugation with a 4-oxo-function in the C-ring. Many studies have described the antioxidant efficacy of flavonoids and demonstrated that these polyphenols can inhibit the oxidation of lipids [12,15,16] and other biomolecules such as proteins and DNA [17–19] in vitro. In addition, the ability of flavonoids to act as antioxidants in vitro may be based on their metal-chelating properties [22,23]. However, although flavonoids react rapidly with ROS/RNS in chemical systems in vitro, their reactions in vivo are dependent on the form that is bioavailable to cells and tissues. This becomes especially important if one considers that flavonoids undergo substantial modification when metabolized in mammals, resulting in mainly methylated and/or glucuronidated or sulfated primary metabolites with often lower antioxidant capacity (for further information, see Chap. 14). In addition, the maximal blood plasma concentration of flavonoids and their structurally related metabolites after ingestion of flavonoid-rich food or beverages rarely reaches 5 μM. This compares to an average of 65–100 μM ascorbic acid and relatively high amounts of other antioxidants such as uric acid and vitamin E in the plasma. Furthermore, the slight increases in the antioxidant capacity of blood plasma that follow flavonoid administration to humans or animals are of a very temporary nature and often decline rapidly to reach baseline levels 2–4 h later. On the basis of such considerations the role of flavonoids as important antioxidants in vivo seems unlikely (exception: the gastrointestinal tract).

B. Flavonoids as Modulators of Protein Activities and Intracellular Signaling?

Reports on the bioactivity of flavonoids or flavonoid-rich plant extracts demonstrated that these compounds are able to alter cellular function seemingly independently of their antioxidant potential. These actions include alterations of receptor function [e.g., GABA-A-receptor (183,185)] and enzyme activities [e.g., *mitochondrial calcium* adenosine triphosphatases (ATPases) (198)], influences on the calcium homeostasis [179], and alterations of intracellular cell signaling processes, especially with regard to protein kinase C [148] and mitogen-activated protein kinase (MAPK) signaling. Accumulating evidence suggests that flavonoids interact selectively within MAPK signaling cascades [177,195–197,199,200]. This could have important implications with regard to their possible sites of action in neurons since members of the MAPK family are involved in signaling processes with regard to neuronal survival, regeneration, and death [10,45,46].

In 2001, we demonstrated that pretreatment of primary striatal neurons with low-micromolar concentrations of the flavan-3-ol epicatechin and the flavonol kaempferol potently protected against Ox-LDL-induced neuronal apoptosis [177]. The neuronal death induced by Ox-LDL was characterized by a time- and dose-dependent decrease in MTT reduction and membrane integrity and increase in annexin-V binding, caspase-3-like protease activity, and DNA fragmentation. In addition, it was demonstrated that oxLDL is a rapid activator of ERK1/2 and JNK in striatal neurons, thereby highlighting the potential importance of MAPK cascades as pivotal mediators of oxidative stress signaling in neurons [177]. More importantly, the Ox-LDL-induced activation of ERK1/2 and JNK was strongly inhibited when striatal neurons were pretreated with epicatechin or kaempferol at concentrations that blocked Ox-LDL-induced neuronal death. However, preventing ERK1/2 activation by using the MEK inhibitors PD98059 or U0126 had no effect on neuronal loss caused by Ox-LDL. This finding strongly suggests that the neuroprotective effects of epicatechin and kaempferol are not linked to their potent inhibitory action within the ERK1/2 cascade. The Ox-LDL-induced activation of JNK was also potently attenuated by epicatechin and kaempferol pretreatments. Whereas other authors have reported a direct link between JNK activation and neurotoxicity [79–81], in the absence of a commercially available inhibitor of JNK it remains to be demonstrated that Ox-LDL-induced JNK activation, as observed here, is directly linked to neuronal apoptosis. However, the phosphorylation of the activator protein 1 (AP-1) protein c-Jun on Ser-63 and Ser-73 by JNK causes increased transcriptional activity [70], which has been linked to stress-induced apoptosis. Ox-LDL caused a pronounced phosphorylation of c-Jun, which mirrored the activation of JNK. Significantly, the Ox-LDL-induced activation of c-Jun was abolished in neurons pre-exposed to epicatechin. The complete inhibition of phosphorylation of c-Jun by flavonoids suggests that these polyphenolic compounds may have powerful antiapoptotic actions in neurons. Interestingly, Ox-LDL also induced the cleavage of procaspase-3 and thereby the activation of caspase-3 in striatal neurons. The activation of caspase-3 was completely blocked in flavonoid pretreated neurons, providing compelling evidence in support of a potent antiapoptotic action of flavonoids in these cells. This is further supported by the fact that blocking caspase-3 activity with a selective inhibitor protected neurons against Ox-LDL-induced neurotoxicity [177]. These results are of special interest with regard to the proposed links among JNK activation, impairment of mitochondria function, cytochrome c release, and caspase activation (as discussed previously) [45,86].

Other reports of nonneuronal cells support a potential role for flavonoids as modulators of intracellular signaling, especially with regard to MAPKs. Yoshizumi and associates [201] demonstrated that the flavonol quercetin attenuates the angiotensin II–induced activation of JNK in cultured rat aortic smooth muscle cells. Furthermore, in experiments with RAW 264.7 macrophages Wadsworth and colleagues [202] showed that quercetin inhibits the lipopolysaccharide-(LPS)-induced TNF-α expression by inhibiting the activation of JNK, which subsequently results in a suppression of AP-1-DNA binding. In addition, the flavanol epigallocatechin has been suggested to suppress the proliferation of vascular smooth muscle cells via the inhibition of JNK activation and c-Jun expression [203].

Another interesting target for flavonoids, which is potentially linked to calcium homeostasis, cell signaling processes, and apoptosis is the mitochondrion. Mitochondria play a central role in oxidative stress–induced apoptosis [204] since they contain cytochrome c, thought to be essential in Fas-receptor-independent apoptosis [86,205]. The emerging view is that after mitochondrial depolarization cytochrome c is released, committing the cell to die by either an apoptotic mechanism, involving Apaf-1-mediated caspase-3 activation, or a necrotic process, due to the collapse of the mitochondrial electron transport chain [204]. Thus, blocking the release of cytochrome c might be the key to preventing neuronal death. However, other investigators reported apoptotic processes involving the depolarization of the mitochondria without translocation of cytochrome c to the cytosol [204,206], thus potentially indicating additional mechanisms leading to the recruitment of apoptosis-executing caspases.

Speculatively, the flavonoid-mediated inhibition of apoptosis as demonstrated by Schroeter and coworkers [176,177] and others might occur by a variety of means related to mitochondrial functions, first, by blocking the activation of JNK. This might take place upstream of JNK, influencing one of many MAPKKK activating proteins involved in transducing signals to JNK [or at the level of maintaining the calcium homeostasis, which has been shown to be important in JNK activation]. Second, flavonoids might be able to upregulate the expression of the antiapoptotic proteins Bcl-2 and Bcl-x_L known to inhibit the release of cytochrome c from the mitochondria and to be inactivated through phosphorylation by JNK. Or, third, flavonoids might modulate the mitochondrial permeability transition (MPT) believed to be important in apoptotic cell death by either opening a gateway for cytochrome c release from the mitochondria [204] or activating other mitochondrial-related proapoptotic factors such as DIABLO/smac [207,208]. Inhibitors of MPT pore opening such as cyclosporin A bind to cyclophilin D, which is associated with the adenine nucleotide transporter (ANT), part of the multiprotein complex of the MPT pore [204,209], and inhibit mitochondrial depolarization [206], cytochrome c translocation [210], and cell death [206,210]. Conversely, MPT pore openers such as the ANT-activator atractyloside trigger MPT pore opening, cytochrome c release, and apoptosis [210]. Interestingly, the MPT pore possesses a benzodiazepine binding site [204], and the binding of ligands such as PK11195 [211,212] and antagonists such as flumazenil [213] have been shown to modulate the MPT pore and thus the MPT. Since flavonoids have been reported to bind to benzodiazepine binding sites of GABA and adenosine receptors [183,185], it might be speculated that these compounds exert an effect on the MPT pore and therefore modulate cytochrome c release. Taken together flavonoids might influence the release of proapoptotic cytochrome c from the mitochondrion, thus providing an interesting hypothesis for future investigations.

1. Flavonoids as Modulators of Adenosine Triphosphate Binding Sites: Structure-Activity Relationship

As detailed, a variety of biological effects of flavonoids may be related to their binding to ATP-binding sites of various proteins, including transport ATPases and protein kinases. In fact, quite a few commercially available protein kinase inhibitors have been developed on the basis of the structure of quercetin. These compounds include the widely used PI-3-K inhibitor LY294002 ($IC_{50} \approx 1$–5 μM; the IC_{50} of quercetin is approximately 4–8 μM), the MEK inhibitor

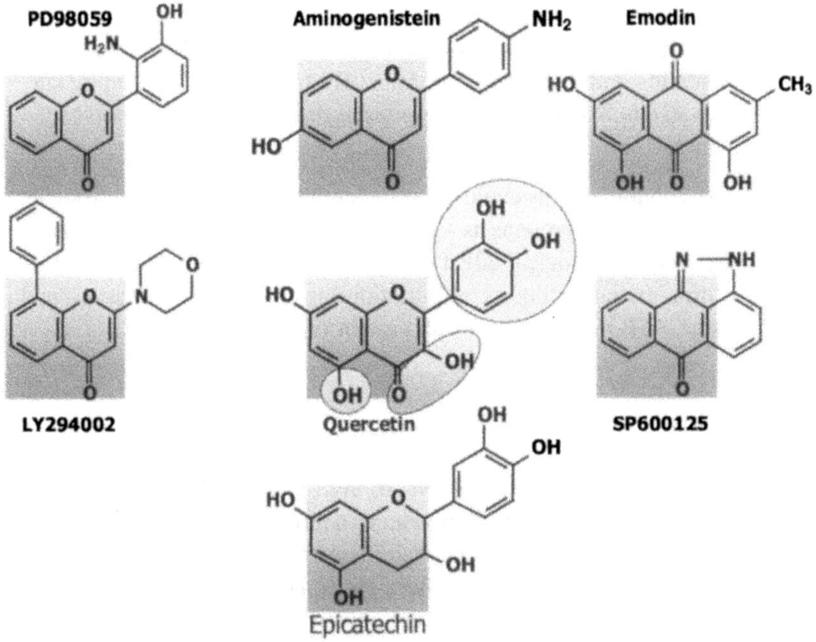

Figure 7 Structural comparison of epicatechin, quercetin, and widely used protein kinase and ATPase inhibitors. Flavonoids might exert their effects by modulating the activity of protein kinases and ATPases on the basis of binding to ATP-binding sites. This picture demonstrates structural similarities of classical kinase inhibitors, quercetin, and epicatechin. The proposed structural requirements for affinity to ATP-binding sites include (with varying degree of importance) A- and C-ring moiety of flavonoids, the 5-hydroxyl group, and the 4-carbonyl function, which are supposed to mimic the adenine moiety of ATP. Although the B-ring moiety may not be directly involved in binding to the ATP-binding domain, it seems to be a determinant of the protein specificity of the inhibitor. ATP, adenosine triphosphate.

PD98059 ($IC_{50} \approx 2\text{--}4 \ \mu M;$), the JNK inhibitor SP600129 ($IC_{50} \approx 90 \ nM$), the tyrosine kinase inhibitors aminogenistein ($IC_{50} \approx 1\text{--}2 \ \mu M$) and emodin ($IC_{50} \approx 5\text{--}19 \ \mu M$), and the CDK2 inhibitor L868276 ($IC_{50} \approx 0.4\text{--}1.6 \ \mu M$). The structural characteristics responsible for interactions with ATP-binding sites on a variety of proteins have been investigated [214–216] (Fig. 7):

1. Rings A and B of the flavonoid structure mimic the adenine ring system of ATP.
2. The 5-hydroxyl group mimics the 6-amino group of ATP and may interact with up to six protein residues of the target protein (the 7-OH group seems not to be involved).
3. The 4-carbonyl function may enhance the inhibitor activity by mimicking the N1 of adenine.
4. The B-ring moiety of the flavonoid structure seems to be important for orientating and directing rings A and B toward the ATP-binding site. Derivatization of the hydroxyl groups on the B-ring has been implicated in modulation of both protein specificity and inhibitor activity.

Thus these findings substantiate the potential interaction of flavonoids with ATP-binding sites on proteins as a potential mechanism for their biological activities. The specific biological effect of a particular flavonoid depends on its bioavailable form, the intracellular concentrations achieved in target tissues, and its affinity to the ATP-binding site of the protein of interest.

VI. CONCLUSIONS

Epidemiological studies and investigations in vivo and in cell culture systems increasingly indicate flavonoids as potential neuroprotective agents. Thus, exploring the mechanisms of the biological effects of flavonoids might help to develop dietary and therapeutic approaches against neurodegeneration.

The classical hydrogen-donating antioxidant properties of flavonoids, which are often used as a possible explanation for their biological activity, may, after all, not be the most important determinant for their effects in vivo. This conclusion is tenable when factors such as in vivo metabolism and the achieved plasma and tissue concentrations of flavonoids/metabolites are taken into account. Therefore, the bioactivity of flavonoids in vivo may rather be supported by their ability to modulate protein functions, intracellular cell signaling, and receptor activities on the basis of interactions, for example, with ATP-binding sites and benzodiazepine-binding domains.

With regard to a potential role for flavonoids against neurodegeneration, the availability of flavonoids and their metabolites to the brain remains to be investigated further. In this context it needs to be considered that perhaps the beneficial effects of flavonoids on memory and motoric functions and against age-related cognitive decline are not exerted directly in the brain but occur in the periphery and are reflected subsequently in functions of the CNS. It might therefore be that factors such as platelet aggregation, vascular tone, hormonal changes, and immune responses modulated by flavonoid intake induce changes in the CNS, which in turn affect neurodegenerative processes.

To understand fully the molecular basis for the in vivo effects of flavonoids and to assess comprehensively their potential as neuroprotective agents, further studies are needed, in particular with regard to metabolism, distribution across the blood-brain barrier, and neuropharmacological properties.

REFERENCES

1. Halliwell B. Reactive oxygen species and the central nervous system. J Neurochem 1992; 59(5):1609–1623.
2. Finkel T, Holbrook NJ. Oxidants, oxidative stress and the biology of ageing. Nature 2000; 408(6809):239–247.
3. Halliwell B, Gutteridge JM. The importance of free radicals and catalytic metal ions in human diseases. Mol Aspects Med 1985; 8(2):89–193.
4. Bindoli A, Rigobello MP, Deeble DJ. Biochemical and toxicological properties of the oxidation products of catecholamines. Free Radic Biol Med 1992;13(4): 391–405.
5. Coyle JT, Puttfarcken P. Oxidative stress, glutamate, and neurodegenerative disorders. Science 1993; 262(5134):689–695.
6. Orrenius S, Nicotera P. The calcium ion and cell death. J Neural Transm Suppl 1994; 43:1–11.
7. Behl C. Alzheimer's disease and oxidative stress: implications for novel therapeutic approaches. Prog Neurobiol 1999; 57(3):301–323.
8. Zhang Y, Dawson VL, Dawson TM. Oxidative stress and genetics in the pathogenesis of Parkinson's disease. Neurobiol Dis 2000; 7(4):240–250.
9. Alexi T, Borlongan CV, Faull RL, Williams CE, Clark RG, Gluckman PD, Hughes PE. Neuroprotective strategies for basal ganglia degeneration: Parkinson's and Huntington's diseases. Prog Neurobiol 2000; 60(5):409–470.
10. Yuan J, Yankner BA. Apoptosis in the nervous system. Nature 2000; 407(6805): 802–809.
11. Quinn J, Kaye J. The neurology of aging. Neurologist 2001; 7:98–112.
12. Oliver CN, Starke-Reed PE, Stadtman ER, Liu GJ, Carney JM, Floyd RA. Oxidative damage to brain proteins, loss of glutamine synthetase activity, and production of free radicals during ischemia/reperfusion-induced injury to gerbil brain. Proc Natl Acad Sci USA 1990; 87(13):5144–5147.
13. McGeer PL, McGeer EG. The inflammatory response system of brain: implications for therapy of Alzheimer and other neurodegenerative diseases. Brain Res Rev 1995; 21(2):195–218.

14. McGeer EG, McGeer PL. The importance of inflammatory mechanisms in Alzheimer disease. Exp Gerontol 1998; 33(5):371–378.

15. Shigenaga MK, Hagen TM, Ames BN. Oxidative damage and mitochondrial decay in aging. Proc Natl Acad Sci USA 1994; 91(23):10771–10778.

16. Tsay HJ, Wang P, Wang SL, Ku HH. Age-associated changes of superoxide dismutase and catalase activities in the rat brain. J Biomed Sci 2000; 7(6):466–474.

17. Keller JN, Huang FF, Markesbery WR. Decreased levels of proteasome activity and proteasome expression in aging spinal cord. Neuroscience 2000; 98(1):149–156.

18. Wakimoto P, Block G. Dietary intake, dietary patterns, and changes with age: an epidemiological perspective. J Gerontol A Biol Sci Med Sci 2001; 56 (spec no 2) (2):65–80.

19. Winkler S, Garg AK, Mekayarajjananonth T, Bakaeen LG, Khan E. Depressed taste and smell in geriatric patients. J Am Dent Assoc 1999; 130(12): 1759–1765.

20. Chau N, Tebi A, Creton C, Belbraouet S, Debry G. Relationship between plasma retinol and infectious diseases in the elderly: a case-control study. Ann Nutr Metab 2000; 44(5–6):256–262.

21. Mohanakumar KP, de Bartolomeis A, Wu RM, Yeh KJ, Sternberger LM, Peng SY, Murphy DL, Chiueh CC. Ferrous-citrate complex and nigral degeneration: evidence for free-radical formation and lipid peroxidation. Ann NY Acad Sci 1994; 738:392–399.

22. Keller JN, Mattson MP. Roles of lipid peroxidation in modulation of cellular signaling pathways, cell dysfunction, and death in the nervous system. Rev Neurosci 1998; 9(2):105–116.

23. Ben-Yoseph O, Boxer PA, Ross BD. Assessment of the role of the glutathione and pentose phosphate pathways in the protection of primary cerebrocortical cultures from oxidative stress. J Neurochem 1996; 66(6):2329–2337.

24. Butterfield DA, Kanski J. Brain protein oxidation in age-related neurodegenerative disorders that are associated with aggregated proteins. Mech Ageing Dev 2001; 122(9):945–962.

25. Vitek MP, Bhattacharya K, Glendening JM, Stopa E, Vlassara H, Bucala R, Manogue K, Cerami A. Advanced glycation end products contribute to amyloidosis in Alzheimer disease. Proc Natl Acad Sci USA 1994; 91(11):4766–4770.

26. Garrington TP, Johnson GL. Organization and regulation of mitogen-activated protein kinase signaling pathways. Curr Opin Cell Biol 1999; 11(2):211–218.

27. Chang L, Karin M. Mammalian MAP kinase signalling cascades. Nature 2001; 410(6824):37–40.

28. Huber LA, Ullrich O, Takai Y, Lutcke A, Dupree P, Olkkonen V, Virta H, de Hoop MJ, Alexandrov K, Peter M. Mapping of Ras-related GTP-binding proteins by GTP overlay following two-dimensional gel electrophoresis. Proc Natl Acad Sci USA 1994; 91(17):7874–7878.

29. Arozarena I, Aaronson DS, Matallanas D, Sanz V, Ajenjo N, Tenbaum SP, Teramoto H, Ighishi T, Zabala JC, Gutkind JS, Crespo P. The Rho family GTPase Cdc42 regulates the activation of Ras/MAP kinase by the exchange factor Ras-GRF. J Biol Chem 2000; 275(34):26441–26448.

30. Xia Z, Dickens M, Raingeaud J, Davis RJ, Greenberg ME. Opposing effects of ERK and JNK-p38 MAP kinases on apoptosis. Science 1995; 270:1326–1331.

31. Rosen LB, Ginty DD, Weber MJ, Greenberg ME. Membrane depolarization and calcium influx stimulate MEK and MAP kinase via activation of Ras. Neuron 1994; 12(6):1207–1221.

32. Bonni A, Brunet A, West AE, Datta SR, Takasu MA, Greenberg ME. Cell survival promoted by the Ras-MAPK signaling pathway by transcription-dependent and -independent mechanisms. Science 1999; 286(5443):1358–1362.

33. Wang X, Martindale JL, Liu Y, Holbrook NJ. The cellular response to oxidative stress: influences of mitogen-activated protein kinase signalling pathways on cell survival. Biochem J 1998; 333(pt 2):291–300.

34. Perkinton MS, Sihra TS, Williams RJ. Ca(2+)-permeable AMPA receptors induce phosphorylation of cAMP response element-binding protein through a phosphatidylinositol 3-kinase-dependent stimulation of the mitogen-activated protein kinase signaling cascade in neurons. J Neurosci 1999; 19(14):5861–5874.

35. Zippel R, Balestrini M, Lomazzi M, Sturani E. Calcium and calmodulin are essential for Ras-GRF1-mediated activation of the Ras pathway by lysophosphatidic acid. Exp Cell Res 2000; 258(2):403–408.

36. Enslen H, Tokumitsu H, Stork PJ, Davis RJ, Soderling TR. Regulation of mitogen-activated protein kinases by a calcium/calmodulin-dependent protein kinase cascade. Proc Natl Acad Sci USA 1996; 93(20):10803–10808.

37. Whitmarsh AJ, Davis RJ. Transcription factor AP-1 regulation by mitogen-activated protein kinase signal transduction pathways. J Mol Med 1996; 74(10): 589–607.

38. Sturgill TW, Ray LB, Erikson E, Maller JL. Insulin-stimulated MAP-2 kinase phosphorylates and activates ribosomal protein S6 kinase II. Nature 1988; 334 (6184):715–718.

39. Deak M, Clifton AD, Lucocq LM, Alessi DR. Mitogen- and stress-activated protein kinase-1 (MSK1) is directly activated by MAPK and SAPK2/p38, and may mediate activation of CREB. EMBO J 1998; 17(15):4426–4441.

40. Mazzoni IE, Said FA, Aloyz R, Miller FD, Kaplan D. Ras regulates sympathetic neuron survival by suppressing the p53-mediated cell death pathway. J Neurosci 1999; 19(22):9716–9727.

41. Alessandrini A, Namura S, Moskowitz MA, Bonventre JV. MEK1 protein kinase inhibition protects against damage resulting from focal cerebral ischemia. Proc Natl Acad Sci USA 1999; 96(22):12866–12869.

42. Stanciu M, Wang Y, Kentor R, Burke N, Watkins S, Kress G, Reynolds I, Klann E, Angiolieri MR, Johnson JW, DeFranco DB. Persistent activation of ERK contributes to glutamate-induced oxidative toxicity in a neuronal cell line and primary cortical neuron cultures. J Biol Chem 2000; 275(16):12200–12206.

43. Murray B, Alessandrini A, Cole AJ, Yee AG, Furshpan EJ. Inhibition of the p44/42 MAP kinase pathway protects hippocampal neurons in a cell-culture model of seizure activity. Proc Natl Acad Sci USA 1998; 95(20):11975–11980.

44. Satoh T, Nakatsuka D, Watanabe Y, Nagata I, Kikuchi H, Namura S. Neuroprotection by MAPK/ERK kinase inhibition with U0126 against oxidative

stress in a mouse neuronal cell line and rat primary cultured cortical neurons. Neurosci Lett 2000; 288(2):163–166.

45. Davis RJ. Signal transduction by the JNK group of MAP kinases. Cell 2000; 109:239–252.

46. Mielke K, Herdegen T. JNK and p38 stress kinases—degenerative effectors of signal-transduction-cascades in the nervous system. Prog Neurobiol 2000; 61 (1):45–60.

47. Ferrer I, Planas AM, Pozas E. Radiation-induced apoptosis in developing rats and kainic acid-induced excitotoxicity in adult rats are associated with distinctive morphological and biochemical c-Jun/AP-1 (N) expression. Neurosci 1997; 80(2): 449–458.

48. Gibson S, Widmann C, Johnson GL. Differential involvement of MEK kinase 1 (MEKK1) in the induction of apoptosis in response to microtubule-targeted drugs versus DNA damaging agents. J Biol Chem 1999; 274(16):10916–10922.

49. Iordanov MS, Wong J, Newton DL, Rybak SM, Bright RK, Flavell RA, Davis RJ, Magun BE. Differential requirement for the stress-activated protein kinase/c-Jun NH(2)-terminal kinase in RNA damage-induced apoptosis in primary and in immortalized fibroblasts. Mol Cell Biol Res Commun 2000; 4(2):122–128.

50. Meier R, Rouse J, Cuenda A, Nebreda AR, Cohen P. Cellular stresses and cytokines activate multiple mitogen-activated-protein kinase kinase homologues in PC12 and KB cells. Eur J Biochem 1996; 236(3):796–805.

51. Meeker R, Fernandes A. Osmotic and glutamate receptor regulation of c-Jun NH(2)-terminal protein kinase in neuroendocrine cells. Am J Physiol Endocrinol Metab 2000; 279(3):E475–E486.

52. Chihab R, Ferry C, Koziel V, Monin P, Daval JL. Sequential activation of activator protein-1-related transcription factors and JNK protein kinases may contribute to apoptotic death induced by transient hypoxia in developing brain neurons. Brain Res Mol Brain Res 1998; 63(1):105–120.

53. Herdegen T, Claret FX, Kallunki T, Martin-Villalba A, Winter C, Hunter T, Karin M. Lasting N-terminal phosphorylation of c-Jun and activation of c-Jun N-terminal kinases after neuronal injury. J Neurosci 1998; 18(14):5124–5135.

54. Mielke K, Brecht S, Dorst A, Herdegen T. Activity and expression of JNK1, p38 and ERK kinases, c-Jun N-terminal phosphorylation, and c-Jun promoter binding in the adult rat brain following kainate-induced seizures. Neurosci 1999; 91(2): 471–483.

55. Go YM, Patel RP, Maland MC, Park H, Beckman JS, Darley-Usmar VM, Jo H. Evidence for peroxynitrite as a signaling molecule in flow-dependent activation of c-Jun NH(2)-terminal kinase. Am J Physiol 1999; 277(4 pt 2):H1647–H1653.

56. Anania FA, Womack L, Jiang M, Saxena NK. Aldehydes potentiate alpha(2)(I) collagen gene activity by JNK in hepatic stellate cells. Free Radic Biol Med 2001; 30(8):846–857.

57. Soh Y, Jeong KS, Lee IJ, Bae MA, Kim YC, Song BJ. Selective activation of the c-Jun N-terminal protein kinase pathway during 4-hydroxynonenal-induced apoptosis of PC12 cells. Mol Pharmacol 2000; 58(3):535–541.

58. Tibbles LA, Ing YL, Kiefer F, Chan J, Iscove N, Woodgett JR, Lassam NJ. MLK-3 activates the SAPK/JNK and p38/RK pathways via SEK1 and MKK3/6. EMBO J 1996; 15(24):7026–7035.

59. Hirai S, Izawa M, Osada S, Spyrou G, Ohno S. Activation of the JNK pathway by distantly related protein kinases, MEKK and MUK. Oncogene 1996; 12(3): 641–650.

60. Hoeflich KP, Yeh WC, Yao Z, Mak TW, Woodgett JR. Mediation of TNF receptor-associated factor effector functions by apoptosis signal-regulating kinase-1 (ASK1). Oncogene 1999; 18(42):5814–5820.

61. Shiah SG, Chuang SE, Kuo ML. Involvement of Asp-Glu-Val-Asp-directed, caspase-mediated mitogen-activated protein kinase kinase 1 Cleavage, c-Jun N-terminal kinase activation, and subsequent Bcl-2 phosphorylation for paclitaxel-induced apoptosis in HL-60 cells. Mol Pharmacol 2001; 59(2):254–262.

62. Takekawa M, Saito H. A family of stress-inducible GADD45-like proteins mediate activation of the stress-responsive MTK1/MEKK4 MAPKKK. Cell 1998; 95(4): 521–530.

63. Kiyono M, Satoh T, Kaziro Y. G protein beta gamma subunit-dependent Racguanine nucleotide exchange activity of Ras-GRF1/CDC25(Mm). Proc Natl Acad Sci USA 1999; 96(9):4826–4831.

64. Adler V, Polotskaya A, Wagner F, Kraft AS. Affinity-purified c-Jun amino-terminal protein kinase requires serine/threonine phosphorylation for activity. J Biol Chem 1992; 267(24):17001–17005.

65. Kallunki T, Su B, Tsigelny I, Sluss HK, Derijard B, Moore G, Davis R, Karin M. JNK2 contains a specificity-determining region responsible for efficient c-Jun binding and phosphorylation. Genes Dev 1994; 8(24):2996–3007.

66. Gupta S, Barrett T, Whitmarsh AJ, Cavanagh J, Sluss HK, Derijard B, Davis RJ. Selective interaction of JNK protein kinase isoforms with transcription factors. EMBO J 1996; 15(11):2760–2770.

67. Kallunki T, Deng T, Hibi M, Karin M. c-Jun can recruit JNK to phospho-rylate dimerization partners via specific docking interactions. Cell 1996; 87(5): 929–939.

68. Cavigelli M, Dolfi F, Claret FX, Karin M. Induction of c-fos expression through JNK-mediated TCF/Elk-1 phosphorylation. EMBO J 1995; 14(23):5957–5964.

69. Karin M, Liu Z, Zandi E. AP-1 function and regulation. Curr Opin Cell Biol 1997; 9(2):240–246.

70. Pulverer BJ, Kyriakis JM, Avruch J, Nikolakaki E, Woodgett JR. Phosphorylation of c-jun mediated by MAP kinases. Nature 1991; 353(6345):670–674.

71. Buschmann T, Yin Z, Bhoumik A, Ronai Z. Amino-terminal-derived JNK fragment alters expression and activity of c-Jun, ATF2, and p53 and increases H2O2-induced cell death. J Biol Chem 2000; 275(22):16590–16596.

72. Fuchs SY, Dolan L, Davis RJ, Ronai Z. Phosphorylation-dependent targeting of c-Jun ubiquitination by Jun N-kinase. Oncogene 1996; 13(7):1531–1535.

73. Fuchs SY, Xie B, Adler V, Fried VA, Davis RJ, Ronai Z. c-Jun NH2-terminal kinases target the ubiquitination of their associated transcription factors. J Biol Chem 1997; 272(51):32163–32168.

74. Yang DD, Kuan CY, Whitmarsh AJ, Rincon M, Zheng TS, Davis RJ, Rakic P, Flavell RA. Absence of excitotoxicity-induced apoptosis in the hippocampus of mice lacking the Jnk3 gene. Nature 1997; 389(6653):865–870.

75. Behrens A, Sibilia M, Wagner EF. Amino-terminal phosphorylation of c-Jun regulates stress-induced apoptosis and cellular proliferation. Nat Genet 1999; 21(3):326–329.

76. Whitfield J, Neame SJ, Paquet L, Bernard O, Ham J. Dominant-negative c-Jun promotes neuronal survival by reducing BIM expression and inhibiting mitochondrial cytochrome c release. Neuron 2001; 29(3):629–643.

77. Eilers A, Whitfield J, Shah B, Spadoni C, Desmond H, Ham J. Direct inhibition of c-Jun N-terminal kinase in sympathetic neurons prevents c-jun promoter activation and NGF withdrawal-induced death. J Neurochem 2001; 76(5):1439–1454.

78. Liu YF. Expression of polyglutamine-expanded Huntingtin activates the SEK1-JNK pathway and induces apoptosis in a hippocampal neuronal cell line. J Biol Chem 1998; 273(44):28873–28877.

79. Troy CM, Rabacchi SA, Xu Z, Maroney AC, Connors TJ, Shelanski ML, Greene LA. Beta-Amyloid-induced neuronal apoptosis requires c-Jun N-terminal kinase activation. J Neurochem 2001; 77(1):157–164.

80. Maroney AC, Finn JP, Bozyczko-Coyne D, O'Kane TM, Neff NT, Tolkovsky AM, Park DS, Yan CY, Troy CM, Greene LA. CEP-1347 (KT7515), an inhibitor of JNK activation, rescues sympathetic neurons and neuronally differentiated PC12 cells from death evoked by three distinct insults. J Neurochem 1999; 73(5):1901–1912.

81. Saporito MS, Thomas BA, Scott RW. MPTP activates c-Jun NH(2)-terminal kinase (JNK) and its upstream regulatory kinase MKK4 in nigrostriatal neurons in vivo. J Neurochem 2000; 75(3):1200–1208.

82. Shimoke K, Yamagishi S, Yamada M, Ikeuchi T, Hatanaka H. Inhibition of phosphatidylinositol 3-kinase activity elevates c-Jun N-terminal kinase activity in apoptosis of cultured cerebellar granule neurons. Brain Res Dev Brain Res 1999; 112(2):245–253.

83. Faris M, Latinis KM, Kempiak SJ, Koretzky GA, Nel A. Stress-induced Fas ligand expression in T cells is mediated through a MEK kinase 1-regulated response element in the Fas ligand promoter. Mol Cell Biol 1998; 18(9):5414–5424.

84. Buschmann T, Potapova O, Bar-Shira A, Ivanov VN, Fuchs SY, Henderson S, Fried VA, Minamoto T, Alarcon-Vargas D, Pincus MR, Gaarde WA, Holbrook NJ, Shiloh Y, Ronai Z. Jun NH2-terminal kinase phosphorylation of p53 on Thr-81 is important for p53 stabilization and transcriptional activities in response to stress. Mol Cell Biol 2001; 21(8):2743–2754.

85. Fuchs SY, Adler V, Pincus MR, Ronai Z. MEKK1/JNK signaling stabilizes and activates p53. Proc Natl Acad Sci USA 1998; 95(18):10541–10546.

86. Tournier C, Hess P, Yang DD, Xu J, Turner TK, Nimnual A, Bar-Sagi D, Jones SN, Flavell RA, Davis RJ. Requirement of JNK for stress-induced activation of the cytochrome c-mediated death pathway. Science 2000; 288(5467):870–874.

87. Yoshida H, Kong YY, Yoshida R, Elia AJ, Hakem A, Hakem R, Penninger JM, Mak TW. Apaf1 is required for mitochondrial pathways of apoptosis and brain development. Cell 1998; 94(6):739–750.

88. Hakem R, Hakem A, Duncan GS, Henderson JT, Woo M, Soengas MS, Elia A, de la Pompa JL, Kagi D, Khoo W, Potter J, Yoshida R, Kaufman SA, Lowe SW, Penninger JM, Mak TW. Differential requirement for caspase 9 in apoptotic pathways in vivo. Cell 1998; 94(3):339–352.

89. Woo M, Hakem R, Soengas MS, Duncan GS, Shahinian A, Kagi D, Hakem A, McCurrach M, Khoo W, Kaufman SA, Senaldi G, Howard T, Lowe SW, Mak TW. Essential contribution of caspase 3/CPP32 to apoptosis and its associated nuclear changes. Genes Dev 1998; 12(6):806–819.

90. Kang CD, Jang JH, Kim KW, Lee HJ, Jeong CS, Kim CM, Kim SH, Chung BS. Activation of c-jun N-terminal kinase/stress-activated protein kinase and the decreased ratio of Bcl-2 to Bax are associated with the auto-oxidized dopamine-induced apoptosis in PC12 cells. Neurosci Lett 1998; 256(1):37–40.

91. Fan MY, Goodwin M, Vu T, Brantley-Finley C, Gaarde WA, Chambers TC. Vinblastine-induced phosphorylation of Bcl-2 and Bcl-X-L is mediated by JNK and occurs in parallel with inactivation of the Raf-1/MEK/ERK cascade. J Biol Chem 2000; 275:29980–29985.

92. Choi WS, Yoon SY, Chang, II, Choi EJ, Rhim H, Jin BK, Oh TH, Krajewski S, Reed JC, Oh YJ. Correlation between structure of Bcl-2 and its inhibitory function of JNK and caspase activity in dopaminergic neuronal apoptosis. J Neurochem 2000; 74(4):1621–1626.

93. Allen RG, Tresini M. Oxidative stress and gene regulation. Free Radic Biol Med 2000; 28(3):463–499.

94. Lander HM, Ogiste JS, Pearce SF, Levi R, Novogrodsky A. Nitric oxide-stimulated guanine nucleotide exchange on p21ras. J Biol Chem 1995; 270(13): 7017–7020.

95. Lander HM, Ogiste JS, Teng KK, Novogrodsky A. p21ras as a common signaling target of reactive free radicals and cellular redox stress. J Biol Chem 1995; 270(36):21195–21198.

96. Burdon RH. Superoxide and hydrogen peroxide in relation to mammalian cell proliferation. Free Radic Biol Med 1995; 18(4):775–794.

97. Satoh T, Nakatsuka D, Watanabe Y, Nagata I, Kikuchi H, Namura S. Neuroprotection by MAPK/ERK kinase inhibition with U0126 against oxidative stress in a mouse neuronal cell line and rat primary cultured cortical neurons. Neurosci Lett 2000; 288(2):163–166.

98. Oguro T, Hayashi M, Nakajo S, Numazawa S, Yoshida T. The expression of heme oxygenase-1 gene responded to oxidative stress produced by phorone, a glutathione depletor, in the rat liver: the relevance to activation of c-jun n-terminal kinase. J Pharmacol Exp Ther 1998; 287(2):733–778.

99. Schwarzschild MA, Cole RL, Hyman SE. Glutamate, but not dopamine, stimulates stress-activated protein kinase and AP-1-mediated transcription in striatal neurons. J Neurosci 1997; 17(10):3455–3466.

100. Park HS, Huh SH, Kim MS, Lee SH, Choi EJ. Nitric oxide negatively regulates c-Jun N-terminal kinase/stress-activated protein kinase by means of S-nitrosylation. Proc Natl Acad Sci USA 2000; 97(26):14382–14387.

101. Adler V, Yin Z, Fuchs SY, Benezra M, Rosario L, Tew KD, Pincus MR, Sardana M, Henderson CJ, Wolf CR, Davis RJ, Ronai Z. Regulation of JNK signaling by GSTp. EMBO J 1999; 18(5):1321–1334.

102. Yin Z, Ivanov VN, Habelhah H, Tew K, Ronai Z. Glutathione S-transferase p elicits protection against H2O2-induced cell death via coordinated regulation of stress kinases. Cancer Res 2000; 60(15):4053–4057.

103. Saitoh M, Nishitoh H, Fujii M, Takeda K, Tobiume K, Sawada Y, Kawabata M, Miyazono K, Ichijo H. Mammalian thioredoxin is a direct inhibitor of apoptosis signal-regulating kinase (ASK) 1. EMBO J 1998; 17(9):2596–2606.

104. Yin ZM, Liu AH, Jiang Y. Glutathione S-transferase pi protects serum depletion-induced cell death by inhibiting ASK1-MKK7-JNK pathway in the 293 cells. Acta Biochim Biophys Sin 2001; 33:185–190.

105. Cho SG, Lee YH, Park HS, Ryoo K, Kang KW, Park J, Eom SJ, Kim MJ, Chang TS, Choi SY, Shim J, Kim Y, Dong MS, Lee MJ, Kim SG, Ichijo H, Choi EJ. Glutathione S-transferase mu modulates the stress-activated signals by suppressing apoptosis signal-regulating kinase 1. J Biol Chem 2001; 276(16): 12749–12755.

106. Nimnual AS, Yatsula BA, Bar-Sagi D. Coupling of Ras and Rac guanosine triphosphatases through the Ras exchanger Sos. Science 1998; 279(5350): 560–563.

107. Samanta S, Perkinton MS, Morgan M, Williams RJ. Hydrogen peroxide enhances signal-responsive arachidonic acid release from neurons: role of mitogen-activated protein kinase. J Neurochem 1998; 70(5):2082–2090.

108. Vanhoutte P, Barnier JV, Guibert B, Pages C, Besson MJ, Hipskind RA, Caboche J. Glutamate induces phosphorylation of Elk-1 and CREB, along with c-fos activation, via an extracellular signal-regulated kinase-dependent pathway in brain slices. Mol Cell Biol 1999; 19(1):136–146.

109. Garthwaite J, Boulton CL. Nitric oxide signaling in the central nervous system. Annu Rev Physiol 1995; 57:683–706.

110. Dawson VL, Dawson TM, London ED, Bredt DS, Snyder SH. Nitric oxide mediates glutamate neurotoxicity in primary cortical cultures. Proc Natl Acad Sci USA 1991; 88:6368–6371.

111. Lipton SA, Choi Y-B, Pan Z-H, Lei SZ, Chen H-SV, Sucher NJ, Loscalzo J, Singel DJ, Stamler JS. A redox-based mechanism for the neuroprotective and neuro-destructive effects of nitric oxide and related nitroso-compounds. Nature 1993; 364:626–632.

112. Tatton WG, Olanow CW. Apoptosis in neurodegenerative diseases: the role of mitochondria. Biochim Biophys Acta 1999; 1410:195–213.

113. Nicholls DG, Ward MW. Mitochondrial membrane potential and neuronal glutamate excitotoxicity: mortality and millivolts. Trends Neurosci 2000; 23:166–174.

114. Liberatore GT, Jackson-Lewis V, Vukosavic S, Mandir AS, Vila M, McAuliffe WG, Dawson VL, Dawson TM, Przedborski S. Inducible nitric oxide synthase stimulates dopaminergic neurodegeneration in the MPTP model of Parkinson disease. Nat Med 1999; 5(12):1403–1409.

115. Kim WG, Mohney RP, Wilson B, Jeohn GH, Liu B, Hong JS. Regional difference in susceptibility to lipopolysaccharide-induced neurotoxicity in the rat brain: role of microglia. J Neurosci 2000; 20(16):6309–6316.

116. Schaeffer HF, Weber MJ. Mitogen-activated protein kinase: specific messages from ubiquitous messengers. Mol Cell Biol 1999; 19:2435–2444.

117. Lander HM, Hajjar DP, Hempstead BL, Mirza UA, Chait BT, Campbell S, Quilliam LA. A molecular redox switch on p21(ras). Structural basis for the nitric oxide-p21(ras) interaction. J Biol Chem 1997; 272(7):4323–4326.

118. So HS, Park RK, Kim MS, Lee SR, Jung BH, Chung SY, Jun CD, Chung HT. Nitric oxide inhibits c-Jun N-terminal kinase 2 (JNK2) via S-nitrosylation. Biochem Biophys Res Commun 1998; 247(3):809–813.

119. Marshall HE, Merchant K, Stamler JS. Nitrosation and oxidation in the regulation of gene expression. FASEB J 2000; 14:1889–1900.

120. Zipper LM, Mulcahy RT. Inhibition of ERK and p38 MAP kinases inhibits binding of Nrf2 and induction of GCS genes. Biochem Biophys Res Commun 2000; 278(2):484–492.

121. Chang MS, Yoo HY, Rho HM. Positive and negative regulatory elements in the upstream region of the rat Cu/Zn-superoxide dismutase gene. Biochem J 1999; 339(pt 2):335–341.

122. Das KC, Lewis-Molock Y, White CW. Activation of NF-kappa B and elevation of MnSOD gene expression by thiol reducing agents in lung adenocarcinoma (A549) cells. Am J Physiol 1995; 269(5 pt 1):L588–L602.

123. Seo SJ, Kim HT, Cho G, Rho HM, Jung G. Sp1 and C/EBP-related factor regulate the transcription of human Cu/Zn SOD gene. Gene 1996; 178(1–2):177–185.

124. Sgambato V, Pages C, Rogard M, Besson MJ, Caboche J. Extracellular signal-regulated kinase (ERK) controls immediate early gene induction on corticostriatal stimulation. J Neurosci 1998; 18(21):8814–8825.

125. Kan H, Xie Z, Finkel MS. Norepinephrine-stimulated MAP kinase activity enhances cytokine-induced NO production by rat cardiac myocytes. Am J Physiol 1999; 276(1 pt 2):H47–H52.

126. Singh K, Balligand JL, Fischer TA, Smith TW, Kelly RA. Regulation of cytokine-inducible nitric oxide synthase in cardiac myocytes and microvascular endothelial cells. Role of extracellular signal-regulated kinases 1 and 2 (ERK1/ERK2) and STAT1 alpha. J Biol Chem 1996; 271(2):1111–1117.

127. Chan ED, Winston BW, Uh ST, Wynes MW, Rose DM, Riches DW. Evaluation of the role of mitogen-activated protein kinases in the expression of inducible nitric oxide synthase by IFN-gamma and TNF-alpha in mouse macrophages. J Immunol 1999; 162:415–422.

128. Da Silva J, Pierrat B, Mary JL, Lesslauer W. Blockade of p38 mitogen-activated protein kinase pathways inhibits inducible nitric-oxide synthase expression in mouse astrocytes. J Biol Chem 1997; 272:28373–28380.

129. Bhat NR, Zhang P, Bhat AN. Cytokine induction of inducible nitric oxide synthase in an oligodendrocyte cell line: role of p38 mitogen-activated protein kinase activation. J Neurochem 1999; 72(2):472–478.

130. Chan ED, Riches DW. IFN-gamma + LPS induction of iNOS is modulated by ERK, JNK/SAPK, and p38(mapk) in a mouse macrophage cell line. Am J Physiol Cell Physiol 2001; 280(3):C441–C450.

131. Chen Z, Yuhanna IS, Galcheva-Gargova Z, Karas RH, Mendelsohn ME, Shaul PW. Estrogen receptor alpha mediates the nongenomic activation of endothelial nitric oxide synthase by estrogen. J Clin Invest 1999; 103:401–406.

132. Zheng J, Bird IM, Melsaether AN, Magness RR. Activation of the mitogen-activated protein kinase cascade is necessary but not sufficient for basic fibroblast growth factor- and epidermal growth factor-stimulated expression of endothelial nitric oxide synthase in ovine fetoplacental artery endothelial cells. Endocrinology 1999; 140:1399–1407.

133. Dimmeler S, Fleming I, Fisslthaler B, Hermann C, Busse R, Zeiher AM. Activation of nitric oxide synthase in endothelial cells by Akt-dependent phosphorylation. Nature 1999; 399:601–605.

134. Camandola S, Poli G, Mattson MP. The lipid peroxidation product 4-hydroxy-2,3-nonenal increase AP-1-binding activity through caspase activation in neurons. J Neurochem 2000; 74(1):159–168.

135. Kumari MV, Yoneda T, Hiramatsu M. Effect of "beta CATECHIN" on the life span of senescence accelerated mice (SAM-P8 strain). Biochem Mol Biol Int 1997; 41(5):1005–1011.

136. Mark RJ, Keller JN, Kruman I, Mattson MP. Basic FGF attenuates amyloid beta-peptide-induced oxidative stress, mitochondrial dysfunction, and impairment of Na+/K+-ATPase activity in hippocampal neurons. Brain Res 1997; 756(1–2): 205–214.

137. Sichel G, Corsaro C, Scalia M, Di Bilio AJ, Bonomo RP. In vitro scavenger activity of some flavonoids and melanins against O$_2$-(.). Free Radic Biol Med 1991; 11(1):1–8.

138. Bors W, Michel C, Schikora S. Interaction of flavonoids with ascorbate and determination of their univalent redox potentials: a pulse radiolysis study. Free Radic Biol Med 1995; 19(1):45–52.

139. Salah N, Miller NJ, Paganga G, Tijburg L, Bolwell GP, Rice-Evans C. Polyphenolic flavanols as scavengers of aqueous phase radicals and as chain-breaking antioxidants. Arch Biochem Biophys 1995; 322(2):339–346.

140. Rice-Evans CA, Miller NJ, Paganga G. Structure-antioxidant activity relationships of flavonoids and phenolic acids. Free Radic Biol Med 1996; 20(7):933–956.

141. Middleton E Jr, Kandaswami C. Effects of flavonoids on immune and inflammatory cell functions. Biochem Pharmacol 1992; 43(6):1167–1179.

142. Hertog MG, Feskens EJ, Hollman PC, Katan MB, Kromhout D. Dietary antioxidant flavonoids and risk of coronary heart disease: the Zutphen Elderly Study. Lancet 1993; 342(8878):1007–1011.

143. Hirvonen T, Pietinen P, Virtanen M, Ovaskainen ML, Hakkinen S, Albanes D, Virtamo J. Intake of flavonols and flavones and risk of coronary heart disease in male smokers. Epidemiology 2001; 12(1):62–67.

144. Edenharder R, von Petersdorff I, Rauscher R. Antimutagenic effects of flavonoids, chalcones and structurally related compounds on the activity of 2-amino-3-

methylimidazo[4,5-f]quinoline (IQ) and other heterocyclic amine mutagens from cooked food. Mutat Res 1993; 287(2):261–274.

145. Yule DI, Kim ET, Williams JA. Tyrosine kinase inhibitors attenuate "capacitative" Ca2+ influx in rat pancreatic acinar cells. Biochem Biophys Res Commun 1994; 202(3):1697–1704.

146. Middleton E Jr, Kandaswami C, Theoharides TC. The effects of plant flavonoids on mammalian cells: implications for inflammation, heart disease, and cancer. Pharmacol Rev 2000; 52(4):673–751.

147. Bastianetto S, Ramassamy C, Dore S, Christen Y, Poirier J, Quirion R. The Ginkgo biloba extract (EGb 761) protects hippocampal neurons against cell death induced by beta-amyloid. Eur J Neurosci 2000; 12(6):1882–1890.

148. Bastianetto S, Zheng WH, Quirion R. The Ginkgo biloba extract (EGb 761) protects and rescues hippocampal cells against nitric oxide-induced toxicity: involvement of its flavonoid constituents and protein kinase C. J Neurochem 2000; 74(6):2268–2277.

149. Ishige K, Schubert D, Sagara Y. Flavonoids protect neuronal cells from oxidative stress by three distinct mechanisms. Free Radic Biol Med 2001; 30(4): 433–446.

150. Yamashita N, Kawanishi S. Distinct mechanisms of DNA damage in apoptosis induced by quercetin and luteolin. Free Radic Res 2000; 33(5):623–633.

151. Chan FL, Choi HL, Chen ZY, Chan PS, Huang Y. Induction of apoptosis in prostate cancer cell lines by a flavonoid, baicalin. Cancer Lett 2000; 160(2):219–228.

152. Wenzel U, Kuntz S, Brendel MD, Daniel H. Dietary flavone is a potent apoptosis inducer in human colon carcinoma cells. Cancer Res 2000; 60(14):3823–3831.

153. Wesnes KA, Ward T, McGinty A, Petrini O. The memory enhancing effects of a Ginkgo biloba/Panax ginseng combination in healthy middle-aged volunteers. Psychopharmacologia 2000; 152(4):353–361.

154. Kennedy DO, Scholey AB, Wesnes KA. The dose-dependent cognitive effects of acute administration of Ginkgo biloba to healthy young volunteers. Psychopharmacologia 2000; 151(4):416–423.

155. Cockle SM, Kimbe S, Hindmarch I. The effects of Gingko biloba extract (LI 1370) supplementation on activities of daily living in free living older volunteers: a questionnare survey. Hum Psychopharm Clin 2000; 15:227–235.

156. Maurer K, Ihl R, Dierks T, Frolich L. Clinical efficacy of Gingko biloba special extract Egb 761 in dementia of the Alzheimer type. Phytomedicine 1998; 5: 417–424.

157. Oken BS, Storzbach DM, Kaye JA. The efficacy of Ginkgo biloba on cognitive function in Alzheimer disease. Arch Neurol 1998; 55(11):1409–1415.

158. van Dongen MCJM, van Rossum E, Kessels AGH, Sielhorst HJG, Knipschild PG. The efficacy of ginko for elderly people with dementia and age-associated memory impairment: new results of a randomised clinical trial. J Am Geriatr Soc 2000; 48:1183–1194.

159. Watanabe CM, Wolffram S, Ader P, Rimbach G, Packer L, Maguire JJ, Schultz PG, Gohil K. The in vivo neuromodulatory effects of the herbal medicine Ginkgo biloba. Proc Natl Acad Sci USA 2001; 98(12):6577–6580.

160. Gramaglia A, Loi GF, Mongioj V, Baronzio GF. Increased survival in brain metastatic patients treated with stereotactic radiotherapy, omega three fatty acids and bioflavonoids. Anticancer Res 1999; 19(6C):5583–5586.

161. Smith PF, Maclennan K, Darlington CL. The neuroprotective properties of the Ginkgo biloba leaf: a review of the possible relationship to platelet-activating factor (PAF). J Ethnopharmacol 1996; 50(3):131–139.

162. de Rijk MC, Breteler MM, den Breeijen JH, Launer LJ, Grobbee DE, van der Meche FG, Hofman A. Dietary antioxidants and Parkinson disease: The Rotterdam Study. Arch Neurol 1997; 54(6):762–765.

163. Joseph JA, Shukitt-Hale B, Denisova NA, Prior RL, Cao G, Martin A, Taglialatela G, Bickford PC. Long-term dietary strawberry, spinach, or vitamin E supplementation retards the onset of age-related neuronal signal-transduction and cognitive behavioral deficits. J Neurosci 1998; 18(19):8047–8055.

164. Logani S, Chen MC, Tran T, Le T, Raffa RB. Actions of Ginkgo biloba related to potential utility for the treatment of conditions involving cerebral hypoxia. Life Sci 2000; 67(12):1389–1396.

165. Joseph JA, Denisova N, Fisher D, Bickford P, Prior R, Cao G. Age-related neurodegeneration and oxidative stress: putative nutritional intervention. Neurol Clin 1998; 16(3):747–755.

166. Inanami O, Watanabe Y, Syuto B, Nakano M, Tsuji M, Kuwabara M. Oral administration of (-)catechin protects against ischemia-reperfusion-induced neuronal death in the gerbil. Free Radic Res 1998; 29:359–365.

167. Shutenko Z, Henry Y, Pinard E, Seylaz J, Potier P, Berthet F, Girard P, Sercombe R. Influence of the antioxidant quercetin in vivo on the level of nitric oxide determined by electron paramagnetic resonance in rat brain during global ischemia and reperfusion. Biochem Pharmacol 1999; 57(2):199–208.

168. Yamagishi M, Osakab N, Takizawa T, Osawa T. Cacao liquor polyphenols reduce oxidative stress without maintaining alpha-tocopherol levels in rats fed a vitamin E-deficient diet. Lipids 2001; 36(1):67–71.

169. Sun GY, Xia J, Draczynska-Lusiak B, Simonyi A, Sun AY. Grape polyphenols protect neurodegenerative changes induced by chronic ethanol administration. Neuroreport 1999; 10(1):93–96.

170. La Grande L, Wang M, Watkins R, Ortiz D, Sanchez ME, Konst J, Lee C, Reyes E. Protective effects of the flavonoid mixture, sylimarin, on fetal rat brain and liver. J Ethnopharmacol 1995; 65:53–61.

171. Ray SD, Wong V, Rinkovsky A, Bagchi M, Raje RR, Bagchi D. Unique organoprotective properties of a novel IH636 grape seed proanthocyanidin extract on cadmium chloride-induced nephrotoxicity, dimethylnitrosamin (DMN)-induced splenotoxicity and MOCAP-induced neurotoxicity in mice. Res Commun Mol Pathol Pharmcol 2000; 107:105–128.

172. Kruman I, Bruce-Keller AJ, Bredesen D, Waeg G, Mattson MP. Evidence that 4-hydroxynonenal mediates oxidative stress-induced neuronal apoptosis. J Neurosci 1997; 17(13):5089–5100.

173. Komatsu M, Hiramatsu M. The efficacy of an antioxidant cocktail on lipid peroxide level and superoxide dismutase activity in aged rat brain and

DNA damage in iron-induced epileptogenic foci. Toxicology 2000; 148(2–3): 143–148.

174. Winter JC. The effects of an extract of Ginkgo biloba, EGb 761, on cognitive behavior and longevity in the rat. Physiol Behav 1998; 63(3):425–433.

175. Joseph JA, Shukitt-Hale B, Denisova NA, Bielinski D, Martin A, McEwen JJ, Bickford PC. Reversals of age-related declines in neuronal signal transduction, cognitive, and motor behavioral deficits with blueberry, spinach, or strawberry dietary supplementation. J Neurosci 1999; 19(18):8114–8121.

176. Schroeter H, Williams RJ, Matin R, Iversen L, Rice-Evans CA. Phenolic antioxidants attenuate neuronal cell death following uptake of oxidized low-density lipoprotein. Free Radic Biol Med 2000; 29(12):1222–1233.

177. Schroeter H, Spencer JPE, Rice-Evans C, Williams RJ. Flavonoids protect neurones from oxidized low-density lipoprotein-induced apoptosis involving JNK, c-jun and caspase-3. Biochem J 2001; 358:547–557.

178. Spencer JP, Schroeter H, Crossthwaithe AJ, Kuhnle G, Williams RJ, Rice-Evans C. Contrasting influences of glucuronidation and O-methylation of epicatechin on hydrogen peroxide-induced cell death in neurons and fibroblasts. Free Radic Biol Med 2001; 31(9):1139–1146.

179. Wang H, Joseph JA. Structure-activity relationships of quercetin in antagonizing hydrogen peroxide-induced calcium dysregulation in PC12 cells. Free Radic Biol Med 1999; 27(5–6):683–694.

180. Shin-Ya K, Kunigami T, Kim JS, Seto H. Protective effect of catechin against beta-amyloid toxicity in hippocampal neurons and PC12 cells. J Neurochem 1997; 69:S42–S52.

181. Yao Z, Drieu K, Papadopoulos V. The Ginkgo biloba extract EGb 761 rescues the PC12 neuronal cells from beta-amyloid-induced cell death by inhibiting the formation of beta-amyloid-derived diffusible neurotoxic ligands. Brain Res 2001; 889(1–2):181–190.

182. Dore S, Bastianetto S, Kar S, Quirion R. Protective and rescuing abilities of IGF-I and some putative free radical scavengers against beta-amyloid-inducing toxicity in neurons. Ann NY Acad Sci 1999; 890:356–364.

183. Paladini AC, Marder M, Viola H, Wolfman C, Wasowski C, Medina JH. Flavonoids and the central nervous system: from forgotten factors to potent anxiolytic compounds. J Pharm Pharmacol 1999; 51(5):519–526.

184. Dekermendjian K, Kahnberg P, Witt MR, Sterner O, Nielsen M, Liljefors T. Structure-activity relationships and molecular modeling analysis of flavonoids binding to the benzodiazepine site of the rat brain GABA(A) receptor complex. J Med Chem 1999; 42(21):4343–4350.

185. Medina JH, Viola H, Wolfman C, Marder M, Wasowski C, Calvo D, Paladini AC. Overview—flavonoids: a new family of benzodiazepine receptor ligands. Neurochem Res 1997; 22(4):419–425.

186. Moro S, van Rhee AM, Sanders LH, Jacobson KA. Flavonoid derivatives as adenosine receptor antagonists: a comparison of the hypothetical receptor binding site based on a comparative molecular field analysis model. J Med Chem 1998; 41(1):46–52.

187. Calapai G, Crupi A, Firenzuoli F, Costantino G, Inferrera G, Campo GM, Caputi AP. Effects of Hypericum perforatum on levels of 5-hydroxytryptamine, noradrenaline and dopamine in the cortex, diencephalon and brainstem of the rat. J Pharm Pharmacol 1999; 51(6):723–728.

188. Calapai G, Crupi A, Firenzuoli F, Inferrera G, Squadrito F, Parisi A, De Sarro G, Caputi A. Serotonin, norepinephrine and dopamine involvement in the antidepressant action of hypericum perforatum. Pharmacopsychiatry 2001; 34(2):45–49.

189. Moon J, Tsushida T, Nakahara K, Terao J. Identification of quercetin 3-O-beta-D-glucuronide as an antioxidative metabolite in rat plasma after oral administration of quercetin. Free Radic Biol Med 2001; 30(11):1274–1285.

190. Koga T, Meydani M. Effect of plasma metabolites of (+)-catechin and quercetin on monocyte adhesion to human aortic endothelial cells. Am J Clin Nutr 2001; 73(5):941–948.

191. Spencer JP, Schroeter H, Kuhnle G, Srai SK, Tyrrell RM, Hahn U, Rice-Evans C. Epicatechin and its in vivo metabolite, 3'-O-methyl epicatechin, protect human fibroblasts from oxidative-stress-induced cell death involving caspase-3 activation. Biochem J 2001; 354(pt 3):493–500.

192. Leslie EM, Mao Q, Oleschuk CJ, Deeley RG, Cole SP. Modulation of multidrug resistance protein 1 (MRP1/ABCC1) transport and atpase activities by interaction with dietary flavonoids. Mol Pharmacol 2001; 59(5):1171–1180.

193. Rong Y, Yang EB, Zhang K, Mack P. Quercetin-induced apoptosis in the monoblastoid cell line U937 in vitro and the regulation of heat shock proteins expression. Anticancer Res 2000; 20(6B):4339–4345.

194. Raso GM, Meli R, Di Carlo G, Pacilio M, Di Carlo R. Inhibition of inducible nitric oxide synthase and cyclooxygenase-2 expression by flavonoids in macrophage J774A.1. Life Sci 2001; 68(8):921–931.

195. Saija A, Scalese M, Lanza M, Marzullo D, Bonina F, Castelli F. Flavonoids as antioxidant agents: importance of their interaction with biomembranes. Free Radic Biol Med 1995; 19(4):481–486.

196. Morand C, Manach C, Crespy V, Remesy C. Quercetin 3-O-beta-glucoside is better absorbed than other quercetin forms and is not present in rat plasma. Free Radic Res 2000; 33(5):667–676.

197. Cesquini M, Tenor AC, Torsoni MA, Stoppa GR, Pereira AL, Ogo SH. Quercetin diminshes the binding of hemoglobin to the red blood cell membrane. J AntiAging Med 2001; 4:55–63.

198. Di Pietro A, Godinot C, Bouillant ML, Gautheron DC. Pig heart mitochondrial ATPase: properties of purified and membrane-bound enzyme. Effects of flavonoids. Biochimie 1975; 57(8):959–967.

199. Kobuchi H, Roy S, Sen CK, Nguyen HG, Packer L. Quercetin inhibits inducible ICAM-1 expression in human endothelial cells through the JNK pathway. Am J Physiol 1999; 277(3 pt 1):C403–C411.

200. Kong AN, Yu R, Chen C, Mandlekar S, Primiano T. Signal transduction events elicited by natural products: role of MAPK and caspase pathways in homeostatic response and induction of apoptosis. Arch Pharm Res 2000; 23(1):1–16.

201. Yoshizumi M, Tsuchiya K, Kirima K, Kyaw M, Suzaki Y, Tamaki T. Quercetin inhibits Shc- and phosphatidylinositol 3-kinase-mediated c-Jun N-terminal kinase activation by angiotensin II in cultured rat aortic smooth muscle cells. Mol Pharmacol 2001; 60(4):656–665.

202. Wadsworth TL, McDonald TL, Koop DR. Effects of Ginkgo biloba extract (EGb 761) and quercetin on lipopolysaccharide-induced signaling pathways involved in the release of tumor necrosis factor-alpha. Biochem Pharmacol 2001; 62(7): 963–974.

203. Lu LH, Lee SS, Huang HC. Epigallocatechin suppression of proliferation of vascular smooth muscle cells: correlation with c-jun and JNK. Br J Pharmacol 1998; 124(6):1227–1237.

204. Green DR, Reed JC. Mitochondria and apoptosis. Science 1998; 281:1309–1312.

205. Li P, Nijhawan D, Budihardjo I, Srinivasula SM, Ahmad M, Alnemri ES, Wang X. Cytochrome c and dATP-dependent formation of Apaf-1/caspase-9 complex initiates an apoptotic protease cascade. Cell 1997; 91(4):479–489.

206. Kingham PJ, Pocock JM. Microglial apoptosis induced by chromogranin A is mediated by mitochondrial depolarisation and the permeability transition but not by cytochrome c release. J Neurochem 2000; 74(4):1452–1462.

207. Srinivasula SM, Hegde R, Saleh A, Datta P, Shiozaki E, Chai J, Lee RA, Robbins PD, Fernandes-Alnemri T, Shi Y, Alnemri ES. A conserved XIAP-interaction motif in caspase-9 and Smac/DIABLO regulates caspase activity and apoptosis. Nature 2001; 410(6824):112–116.

208. Goyal L. Cell death inhibition: keeping caspases in check. Cell 2001; 104(6): 805–808.

209. Tafani M, Schneider TG, Pastorino JG, Farber JL. Cytochrome c-dependent activation of caspase-3 by tumor necrosis factor requires induction of the mitochondrial permeability transition. Am J Pathol 2000; 156(6):2111–2121.

210. Zamzami N, Marchetti P, Castedo M, Hirsch T, Susin SA, Masse B, Kroemer G. Inhibitors of permeability transition interfere with the disruption of the mitochondrial transmembrane potential during apoptosis. FEBS Lett 1996; 384(1): 53–57.

211. Donahue RJ, Razmara M, Hoek JB, Knudsen TB. Direct influence of the p53 tumor suppressor on mitochondrial biogenesis and function. FASEB J 2001; 15(3):635–644.

212. Chelli B, Falleni A, Salvetti F, Gremigni V, Lucacchini A, Martini C. Peripheral-type benzodiazepine receptor ligands: mitochondrial permeability transition induction in rat cardiac tissue. Biochem Pharmacol 2001; 61(6): 695–705.

213. Galeffi F, Sinnar S, Schwartz-Bloom RD. Diazepam promotes ATP recovery and prevents cytochrome c release in hippocampal slices after in vitro ischemia. J Neurochem 2000; 75(3):1242–1249.

214. Walker EH, Pacold ME, Perisic O, Stephens L, Hawkins PT, Wymann MP, Williams RL. Structural determinants of phosphoinositide 3-kinase inhibition by wortmannin, LY294002, quercetin, myricetin, and staurosporine. Mol Cell 2000; 6(4):909–919.

215. Conseil G, Baubichon-Cortay H, Dayan G, Jault JM, Barron D, Di Pietro A. Flavonoids: a class of modulators with bifunctional interactions at vicinal ATP- and steroid-binding sites on mouse P-glycoprotein. Proc Natl Acad Sci USA 1998; 95(17):9831–9836.
216. De Azevedo WF Jr, Mueller-Dieckmann HJ, Schulze-Gahmen U, Worland PJ, Sausville E, Kim SH. Structural basis for specificity and potency of a flavonoid inhibitor of human CDK2, a cell cycle kinase. Proc Natl Acad Sci USA 1996; 93(7):2735–2740.

10

Mitochondrial Actions of Flavonoids and Isoflavonoids

Clinton S. Boyd and Enrique Cadenas
University of Southern California School of Pharmacy
Los Angeles, California, U.S.A.

I. INTRODUCTION

Flavonoids are a group of naturally occurring polyphenolic compounds, ubiquitously distributed in plants (1). More than 4000 flavonoids have been identified and can be divided into the following subgroups of monomeric structures: flavones, flavonols, flavanols, flavanones, and anthocyanidins. Isoflavonoids, including the isoflavone genistein, are structural congeners of flavonoids. The high diversity of individual polyphenols within each of these subgroups derives mainly from the number and configuration of hydroxyl groups, and the nature and extent of alkylation. Numerous in vitro, ex vivo, and in vivo studies have established that flavonoids and isoflavonoids exert a wide range of biological activities, including both cytoprotective and cytotoxic effects [1–8]. The biochemical targets that manifest the effects of these polyphenolic compounds are distributed throughout the subcellular compartments. This review focuses on one such compartment, the mitochondrion.

Mitochondria play pivotal roles in both the life and the death of the cell. First, mitochondrial integrity and the efficiency of oxidative phosphorylation are essential for cellular bioenergetics and viability [9]. Second, mitochondria appear to play a central role in apoptosis [10,11]. Specifically, mitochondria-dependent apoptotic pathways are responsive to "intrinsic" proapoptotic stimuli such as oxidative stress. The latter is of particular interest in that the oxyradical generation by the mitochondrial respiratory chain is the main contributor to the hydrogen peroxide (H_2O_2) steady-state concentration in most cells [12]. Mitochondria can thus influence the cellular redox state, to which many redox-active biochemical species and pathways are sensitive. Many of the biological effects of flavonoids and isoflavonoids have been attributed to the antioxidant capacity of these compounds, as determined in vitro, including their reducing properties (defined by their redox potential) and their ability to scavenge radicals and chelate metals [1]. Thus, as a first approach, it can be surmised that flavonoids and isoflavonoids may modulate mitochondrial function through redox mechanisms. In fact, early in vitro and ex vivo studies investigated structure-activity relationships to correlate the redox potential of these compounds with their ability to inhibit mitochondrial bioenergetics. These studies were extensively reviewed in the first volume of Flavonoids in Health and Disease [2] and are summarized and extended in the current review.

However, evidence is accumulating from recent studies that flavonoids and isoflavonoids may have more specific effects on biochemical pathways, seemingly independent of their hydrogen-donating properties or of their ability to alter the intracellular redox state. This new and exciting possibility is investigated in the current chapter with respect to mitochondrial function at the level of (1) mitochondrial bioenergetics, (2) oxyradical generation and mitochondrial antioxidant systems, and (3) mitochondria-dependent apoptosis. In particular, the finding that mitogen-activated protein kinase (MAPK) signaling pathways can be modulated by flavonoids [7,13,14] has important implications for the potential regulation of mitochondria-dependent apoptosis by such polyphenols. The latter is considered with respect to the MAPK-dependent regulation of antioxidant enzymes and the Bcl-2 family of proteins, which are known to regulate cytochrome c release from mitochondria [11,15].

II. THE EFFECT OF FLAVONOIDS AND ISOFLAVONES ON MITOCHONDRIAL RESPIRATION AND OXIDATIVE PHOSPHORYLATION

Mitochondria are described as the "powerhouses" of the cell, supplying the adenosine triphosphate (ATP) demand for the multiple energy-dependent cellular processes via oxidative phosphorylation. Thus the health of cells is dependent on the functional state of these organelles. This is of particular biological significance considering the high energy demand of cancer cells and may explain the cytotoxic or antineoplastic action of many flavonoids [5,6]. For example, potent inhibitors of mitochondrial respiration such as rotenone, an isoflavone derivative, and antimycin A result in cell death of tumors. For this reason, mitochondrial respiration and oxidative phosphorylation represent a likely starting point for an investigation into the effects of bioflavonoids on mitochondrial function.

A. Inhibition of the Mitochondrial Electron Transport Chain

The mitochondrial electron transport chain (ETC) or respiratory chain comprises a series of membrane-bound redox-active intermediates including flavoproteins, quinones, cytochromes, and iron-sulfur clusters [9] (Fig. 1). The latter facilitate the thermodynamically controlled transfer of electrons from conjugate redox pairs of low redox potential ($E^{0'}$) [e.g., -320 mV for reduced nicotinamide-adenine dinucleotide/oxidized nicotinamide-adenine dinucleotide (NADH/NAD$^+$)] to the final electron donor, O_2 (with a high redox potential of $+820$

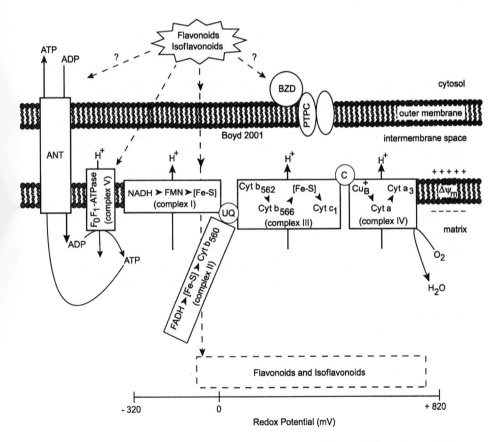

Figure 1 Mitochondrial bioenergetics, redox potentials, and the effect of flavonoids and isoflavonoids.

mV for the $^1/_2O_2/H_2O$ pair). Iron-sulfur clusters generally have lower redox potentials than cytochromes. Thus iron-sulfur clusters tend to be more concentrated at the electron donor (respiratory substrate) end of the ETC, whereas cytochromes are more prominent at the electron acceptor (O_2) end. Accordingly, the cytochromes and iron-sulfur clusters are arranged into discrete complexes, forming a sequential array of redox potentials (Fig. 1). These include NADH-ubiquinone oxidoreductase (complex I), succinate-ubiquinone oxidoreductase (complex II), ubiquinol-cytochrome c oxidoreductase (complex III), and cyto-chrome c oxidase (complex IV). Approximate redox potentials for these complexes are -300, -220, $+230$, and $+550$ mV, respectively. Ubiquinone (coenzyme Q; $E_{0'}$ = $+0.04$ mV) is an important electron carrier, common to both complexes I and II and required to direct electron flow from the oxidation of respiratory substrates to complexes III and IV. Another important shuttle is the hemoprotein cytochrome c ($E_{0'}$ = $+250$ mV), located in the mitochondrial intermembrane space. Cytochrome c receives electrons from complex III and is then oxidized by cytochrome c oxidase, resulting in the efficient four-electron reduction of O_2 to H_2O_2, the terminal step in respiration [9].

As a whole, flavonoids and isoflavones cover almost the entire spectrum of redox potentials inherent to the ETC [2,16,17]. Therefore, on electro-chemical grounds, it is feasible that these polyphenolic compounds can interact with mitochondrial redox centers and divert or impair electron flux through the ETC (Fig. 1). There is also the potential for complex-specific interactions depending on their redox potential. Extensive in vitro and ex vivo structure-activity studies were performed to determine the effect of a plethora of flavonoids and isoflavones on mitochondrial electron transport and to correlate potency with their redox potential [16,18–21]. These and other studies were extensively reviewed in the first volume of this series [2]. The reader is referred to the first volume for more detailed information; only the key findings are summarized here. All the polyphenolic compounds tested were either inhibitory or ineffective toward complex I and complex II, defined as NADH-oxidase and succino-oxidase activity, respectively [2]. None was found to exert a stimulatory effect. From the onset, it should be noted that the multitude and complexity of the flavonoid and isoflavone structures allow only the formulation of generalized structure-activity rules.

1. Complex I (Reduced Nicotinamide-Adenine Dinucleotide-Oxidase)

NADH-oxidase activity was based on the O_2 uptake in the presence of complex I substrates (e.g., NADH) and is dependent on complexes I, III, and IV. Hydroxy-lation of the flavonoid B-ring was an important determinant of inhibitory potency toward NADH-oxidase [2]. The absence of B-ring hydroxyl substituents or their methylation or glycosylation decreased inhibitory potency. The conjugation of the $C_{2,3}$ double bond (e.g., comparison of flavonols and dihydroflavonols) and the presence of the 4-keto group were also very important structural features. A study with a series of 3,5,7-trihydroxyflavones revealed that inhibitory potency correlated with the number and configuration of the B-ring hydroxyl groups (i.e., their redox potential). In fact, the redox potentials of the latter 3,5,7-trihydroxyflavones falls within the range covered by NADH-ubiquinone reductase (-340 to $+100$ mV) [16,17]. Other studies involving a wider group of flavonoids confirmed the importance of B-ring hydroxyl group configuration but found a weaker correlation with the number of hydroxyl groups [2]. There were also strong indications that inhibitory potency is also dependent on other structural features unrelated to redox potential. In summary, flavonols were the most potent inhibitors of NADH-oxidase, and the isoflavone congener may also be more potent than the corresponding flavone-like structure. The latter is not surprising considering that rotenone, an iso-flavone derivative, is a very potent and specific complex I inhibitor, routinely used in mitochondrial research.

2. Complex II (Succino-oxidase)

Succino-oxidase activity was based on the O_2 uptake in the presence of complex II substrates (e.g., succinate) and is dependent on complexes II, III, and IV. Overall succino-oxidase activity was less sensitive to inhibition by flavonoids and isoflavones than NADH-oxidase activity [2]. Only anthocyanidins showed a greater potency toward succino-oxidase than NADH-oxidase. In the case of succino-oxidase, there was a clearer relationship between both the number and configuration of B-ring hydroxyl groups and inhibitory potency, as predicted by studies with model phenolic groups [22,23]. For example, the order of potency was as follows: chalcone > flavone $\geq$ flavonol > dihydroflavonol > anthocyanidin [2]. Catechins were inactive. Within the flavones, a strong correlation existed between the inhibitory potency and the number of hydroxyl groups, such that $4 > 6 > 5 > 3 > 2 > 1 = 0$. Indeed, the inhibitory potency of a series of 3,5,7-trihydroxyflavones toward succino-oxidase correlated with their $E_{1/2}$ potentials, which fall within the range of redox potentials inherent to succinate-ubiquinone reductase (-260 to $+120$ mV) [16,17]. Furthermore, the redox potentials of flavonoids determine their tendency to auto-oxidize, oxidize, with the subsequent generation of o-semiquinones and oxygen-centered radicals. In fact, auto-oxidizable flavonoids have $E_{1/2}$ values that are comparable to the one-electron reduction potential range for superoxide anion ($O_2 \cdot ^-$) generation by quinone in the NADH-ubiquinone reductase domain of the respiratory chain (-70 to $+30$ mV) [16]. Thus, some flavonoids may damage complexes I and II through a redox-cycling mechanism involving reactive quinoid and oxygen species.

In summary, the free in vitro studies discussed indicate that flavonoids do show some specificity in inhibitory potency toward mitochondrial complex I or complex II activity. However, complex I did appear to be more sensitive to a larger number of flavonoids, most likely because inhibitory potency is not only related to redox potential. The chalcone butein was the most potent inhibitor of both complexes I and II, with similar potency toward each complex. Polyphenolic compounds such as quercetin, flavone, and genistein were more potent against complex I, whereas anthocyanidins were more potent against complex II. The catechins were essentially inactive against both ETC complexes. Similarly, (+)-epicatechin and ($\pm$)-catechin were found to have no effect on β-NADH oxidation by rat liver submitochondrial particles, and taxifolin and (-)-epicatechin gallate caused minor inhibition [24]. In contrast, pycnogenol, the "French maritime pine bark extract" (PBE), competitively inhibited O_2 consumption and β-NADH oxidation by whole rat liver mitochondria and submitochondrial particles. The decrease in electron transport activity was attributed to an inhibition of complexes I and II and a direct inhibition of complex III independent of an upstream effect on complexes I and II. This indicates that natural plant extracts contain a multitude of other polyphenolic phytochemicals and bioflavonoids, besides the catechins, which are likely to be potent inhibitors of all components of the mitochondrial ETC.

B.   Inhibition of Mitochondrial Proton F_0F_1-Adenosine Triphosphatase

The F_0F_1–adenosine triphosphatase (ATPase) (or ATP synthase; complex V), located in the mitochondrial inner membrane, catalyzes the synthesis of ATP from adenosine diphosphate (ADP) (Fig. 1). Electron flux through the ETC during respiration is associated with the vectorial transport of protons (H^+) from the matrix to the intermembrane space. Complexes I, III, and IV are the "proton pump," setting up the electrochemical gradient: a proton (pH) gradient and a transmembrane potential ($\Delta\Psi_m$). Proton reentry into the matrix through the F_0 channel of complex V provides the energy to drive ATP synthesis by the F_1 component. Mitochondrial ATP is exchanged for ADP in the cytoplasm by the adenine nucleotide translocase (ANT), also located in the mitochondrial inner membrane (Fig. 1).

Certain flavonoids are capable of inhibiting a variety of ATPases, including the oligomycin-sensitive mitochondrial F_0F_1-ATPase (complex V), Ca^{2+}, $Mg2^+$-ATPase, and Na^+, K^+-ATPase (2,25). For example, the effect of quercetin on ATPases and the implication for the regulation of glycolysis in tumor cells have been reviewed [26]. Early ex vivo and in vitro structure-activity studies on bovine F_1-ATPase from whole mitochondria [18], submitochondrial particles [27], or the purified enzyme [28] indicated that flavonols (e.g., morin and quercetin) were the most potent inhibitors of the flavonoids tested. This suggested the importance of the 3-hydroxyl group. In fact, very few flavonoids were found to have inhibitory affects on F_1-ATPase. A more recent study confirmed the inhibitory effect of flavonoids on mitochondrial F_1-ATPase from rat brain and liver and extended the study to include other classes of polyphenolic phytochemicals [25]. In particular, resveratrol and genistein were among the most potent inhibitors, displaying noncompetitive kinetics. (+)-Catechin, (+)-epicatechin, (-)-epicatechin, and (-)-epigallocatechin were ineffective. Overall mitochondrial F_1-ATPase was relatively resistant to flavonoids and isoflavones in comparison to complexes I and II. A reason for this may be that none of the studies found a correlation between inhibitory potency and the number or configuration of the B-ring hydroxyl groups.

The preceding findings suggest that the redox potential is not a determining factor and that other mechanisms must be responsible for the observed inhibition. One possibility is that flavonoids having structural similarities with the adenosine moiety may compete for the binding site of ADP/ATP of the F_0F_1-ATPase, as has been reported for a variety of ATP-dependent enzymes and steroid-binding sites [29]. This would be consistent with the ability of certain flavonoids to inhibit a number of different ATPase enzymes. Furthermore, the ANT also possesses ADP-and ATP-binding domains [30]. Flavonoid hydroxy-ethylrutosides and procyanidolic oligomers were found to increase the respiratory control ratio (RCR) of isolated rat liver mitochondria by stimulating ADP-dependent state 3 respiration [31]. The mechanism involved a promotion of ADP uptake, which was sensitive to atractyloside, a selective inhibitor of ANT activity. Hydroxyethylrutosides and procyanidolic oligomers are classified as venotropic drugs, used in the treatment of chronic venous deficiency or arteriopathy. Thus, an increase in ANT activity could explain the ability of these compounds to protect cultured human endothelial cells from the hypoxia-induced decrease in ATP content [31]. It is unclear how these compounds would facilitate ADP uptake, but it may involve a switching of ANT to a conformation that favors ADP transport over ATP transport.

C. Respiratory Control, Coupling of Mitochondria, and the Permeability Transition

As discussed, ATP synthesis is driven by the "proton motive force" of the electrochemical gradient set up during the electron flow through the respiratory chain. For this reason, the two processes of oxidation and phosphorylation are described as being coupled (oxidative phosphorylation). In normally functioning, tightly coupled mitochondria oxidation proceeds primarily when ATP is synthesized from ADP. The dependence of respiration on ADP levels is defined as respiratory control, a key property of coupled mitochondria. The respiratory control ratio (RCR) is calculated as the ratio of the respiration rate in the presence of ADP (i.e., state 3 respiration) and the rate of resting respiration in the absence of ADP (i.e., state 4 respiration). Agents that uncouple oxidation from phosphorylation (known as *uncouplers*) remove respiratory control and thus allow state 4 respiration to proceed at maximal rates, but without the conservation of energy in the form of ATP. Uncoupling is thus reflected by a decrease in the RCR.

Dissipation of the electrochemical gradient (pH gradient and $\Delta\Psi_m$) results in rapid uncoupling. Classical uncouplers include protonophores, such as 2,4-dinitrophenol and cyanide m-chlorophenylhydrazone, and the K^+ ionophore valinomycin. An in vitro model system investigated the ability of flavonoids to uncouple artificial vesicles reconstituted with cytochrome oxidase [32]. Flavonoids uncoupled the vesicles by affecting both the transmembrane potential difference and the transmembrane pH difference. Flavones were slightly more effective than flavanones, whereas the flavonol quercetin exhibited hardly any uncoupling activity. With respect to biological systems, it has been well demonstrated that in plants flavonoids act as uncouplers, possibly serving a role as metabolic regulators [2]. Furthermore, the cytotoxicity of flavonoids toward cultured rat hepatocytes correlated strongly with their ability to dissipate the mitochondrial transmembrane potential [6]. In isolated rat liver mitochondria, quercetin, pinobanksin, pinocembrin, and derivatives were reported to stimulate succinate-driven state 4 respiration, as would be consistent with an uncoupling effect [33].

By definition, the uncoupling effect of certain flavonoids should be independent of their inhibitory effects on mitochondrial respiration or F_0F_1-ATPase, suggesting an additional mode of action of flavonoids against mitochondrial function. A collapse of the transmembrane potential is likely under conditions in which the permeability barrier created by the mitochondrial inner membrane is compromised (as occurs in the presence of ionophores). Calcium, phosphate, oxidative stress, adenine nucleotide depletion, and membrane depolarization can induce such a nonspecific increase in the permeability of the inner membrane, in an event called the *mitochondrial permeability transition* (MPT) [30,34]. The MPT can be selectively inhibited by cyclosporin A and is believed to involve the assembly of a multiprotein complex to form a nonspecific pore that spans the inner and outer mitochondrial membranes. The latter assembly is referred to as the *permeability transition pore complex* (PTPC) (Fig. 1). Its exact composition is unknown, but appears to comprise cyclophilin D, ANT, the voltage-dependent anion channel (porin), and a benzodiazepine-binding site [10,30,34].

Quercetin and other flavonoids were reported to prevent the MPT induced by mefenamic acid or Ca^{2+} plus phosphate in isolated rat liver mitochondria [33]. The flavonoids were also found to prevent the oxidation of mitochondrial protein sulfhydryl groups associated with the MPT. The authors proposed that the protective effect of quercetin most likely involved the scavenging of free radicals or an inhibition of respiration. As an alternative mechanism, it is interesting to note that flavonoids can act as agonists or inverse agonists to the benzodiazepine-binding site of the rat brain γ-amino butyric acid A (GABA-A) receptor complex [35]. Finally, the role of the MPT in mitochondria-dependent apoptotic pathways and the impact of flavonoid treatment are discussed in the following section.

III. OXYRADICAL PRODUCTION BY MITOCHONDRIA AND THE ANTIOXIDANT EFFECTS OF FLAVONOIDS

Mitochondria are a main source of H_2O_2 production, contributing 40–90% to the cellular $[H_2O_2]_{ss}$, depending on the tissue [12]. Superoxide anion ($O_2 \cdot^-$), the stoichiometric precursor of H_2O_2, is formed predominantly by ubisemiquinone auto-oxidation during electron flux through the respiratory chain (Fig. 2). Secondarily, $O_2 \cdot^-$ can be generated as a by-product of NADH-dehydrogenase (complex I) activity. Once generated, $O_2 \cdot^-$ is vectorially released from the inner membrane into the matrix, where it undergoes disproportionation to H_2O_2 by Mn-superoxide dismutase (Mn-SOD). H_2O_2 that escapes the matrix glutathione peroxidase activity can freely diffuse from the mitochondria to the cytosol, contributing to the cellular $[H_2O_2]_{ss}$ and redox state. Furthermore, monoamine oxidase, located in the outer mitochondrial membrane, catalyzes the oxidative deamination of monoamines, such as dopamine, with the concomitant production of H_2O_2 (Fig. 2). The latter route of H_2O_2 production could surpass that by the inner mitochondrial membrane [12]. Thus $O_2 \cdot^-$ is continuously generated by mitochondrial respiration and accounts for $\sim 2\%$ of oxygen uptake by the organelle under physiological conditions [12]. A further burst of $O_2 \cdot^-$ generation can be observed under pathophysiological conditions, for example, on the release of cytochrome c from the mitochondria in response to certain proapoptotic stimuli [36,37] (Fig. 2).

The antioxidant effects of flavonoids in vitro have been extensively described and are related to their ability to act as reducing agents (defined by their redox potential), free radical scavengers, and metal chelators [1,38–40]. Besides the intramitochondrial $[H_2O_2]_{ss}$, mitochondria are rich in redox-active species, including, inter alia, flavoproteins, pyridine nucleotides, reduced glutathione (GSH), and transition metals. It is anticipated that flavonoids, by virtue of their antioxidant properties, may influence the intramitochondrial redox state. Several in vitro studies have demonstrated that flavonoids can protect against lipid peroxidation induced in mitochondria by both nonenzymatic and enzymatic sources [41–43]. Nonenzymatic agents include the Fe^{2+}-ascorbic acid system and the alkyl radical generating system, whereas enzymatic mechanisms

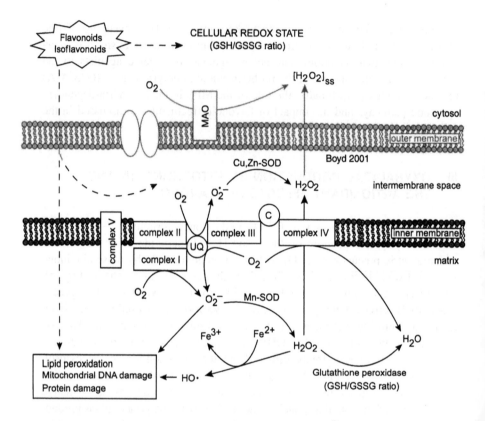

Figure 2 Oxyradical production by mitochondria and the role of antioxidant enzymes: potential targets for flavonoids and isoflavonoids.

include complex I substrate (NADH-dependent) oxidation during mitochondrial respiration. As an example of the former approach, an early in vitro investigation with isolated rat brain mitochondria [41] conducted an extensive structure-activity study of the effects of flavonoids on lipid peroxidation induced by the Fe^{2+}-ascorbic acid system. In general, although some flavonoids actually promoted lipid peroxidation, the antiperoxidative action of the majority of the compounds could be ranked according to those features known to confer the most potent antioxidant properties to flavonoids: (1) the number and configuration of hydroxylated substitutions on rings A and B, especially the 3', 4'-dihydroxy catechol structure in ring B and a free 3-hydroxyl substitution; (2) the planarity of molecule; (3) conjugation of the $C_{2,3}$ double bond; and (4) the 4-keto moiety in the C-ring [1, 17]. However, a more recent study found that methylation of certain flavonoids (e.g., quercetin, pinobanksin, and pinocembrin) improved their ability to inhibit ADP/Fe^{2+}-induced lipid peroxidation in isolated rat liver mitochondria (33). This finding is consistent with the fact that increased hydrophobicity would facilitate access of the flavonoid into the membranes, the site of lipid peroxidation [44]. Thus, the antioxidant potential of the flavonoid depends not only on its structure, but also on the type and source of the radical and the localization of the oxidized molecular target (e.g., mitochondrial matrix vs. membrane compartments).

The ability of certain flavonoids to reduce lipid peroxidation in the enzymatic, NADH-dependent model [42,43] most likely involved a dual effect: the ability of the flavonoids to scavenge radicals and a direct inhibition of complex I activity. As discussed earlier, complex I activity is sensitive to most flavonoids in vitro; the inhibitory potency is only partially dependent on the redox potential of the compounds. Furthermore, although methylation of a flavonoid such as quercetin promoted its antiperoxidative effect, it reduced the inhibitory potency of the flavonoids in relation to complex I and II respiration [33]. Therefore, it is important to bear in mind that the protective vs. toxic effect of a flavonoid may be the net result of its ability to inhibit respiratory complexes, scavenge radicals, and influence other mitochondrial functions.

Flavonoids may also influence antioxidant systems within the mitochondria, which normally serve to protect mitochondria against the oxyradicals generated during respiration. Such defenses could include glutathione peroxidase and two SOD isoforms, which differ in their subcellular distribution. Cu, Zn-SOD is found in the cytosol, whereas Mn-SO is abundant in the mitochondrial matrix. In 2001, two studies altered the traditional view of SOD and mitochondrial function. First, a Cu, Zn-SOD enzyme has been identified in the mitochondrial intermembrane space [45]. The latter enzyme has properties similar to those of the cytosolic form of the enzyme, and both appear to be encoded by the same gene. Second, evidence has been presented that a percentage of mitochondrial $O_2 \bullet^-$ production is vectorially released from the cytosolic side of the inner membrane space directly into the intermembrane space [46]. Thus some $O_2 \bullet^-$ can escape the matrix Mn-SOD, potentially to be disproportionated to H_2O_2 by the intermembrane Cu, Zn-SOD. A significant point of this finding is that H_2O_2 produced in this manner would escape glutathione peroxidase in the mitochondrial matrix.

The transcription of the Cu, Zn-SOD gene is regulated by the transcription factor Elk1 (Ets domain protein-1) (47), whereas the promoter for Mn-SOD expression contains consensus sequences for Sp-1, Ap-1 (activator protein-1), and CREB (c-AMP response element binding protein) [48,49]. The regulation of the latter transcription factors has been linked to both the ERK and JNK groups of MAPKs [50]. In turn, as discussed later, flavonoids have been shown to influence MAPK (mitogen-activated protein kinase) signaling pathways, including ERK and JNK (c-Jun amino-terminal kinase) [7,13,14], suggesting that flavonoids may modulate the expression of both cytosolic and mitochondrial SOD enzymes

(Figs. 2 and 3). Some validity is given to the latter possibility by two in vivo studies. First, the oral administration of a catechin-containing preparation to aged rats increased SOD activity in the mitochondrial fraction of the striatum and midbrain [51]. Second, the ability of (-)-catechin, orally administered to gerbils, to protect dose-dependently against neuronal death induced by transient ischemia-reperfusion injury was associated with an increase in the "superoxide-scavenging" ability of the brain tissue [52].

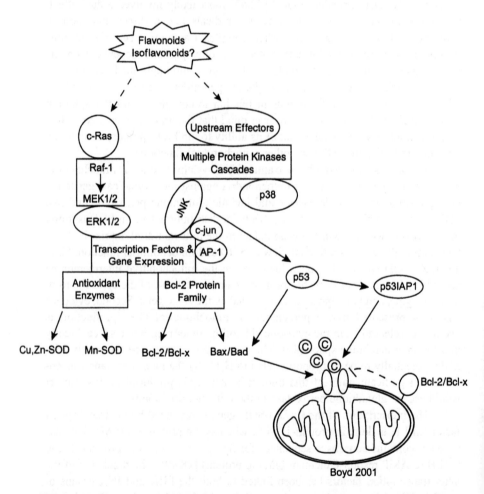

Figure 3 The role of MAPK signaling pathways in mitochondria-dependent apoptosis: antioxidant enzymes, the Bcl-2 protein family, and the influence of flavonoids and isoflavonoids. MAPK, mitogen-activated protein kinase.

IV. APOPTOSIS AND FLAVONOIDS

Apoptosis, or programmed cell death, is a concerted, active process involving the coordination of multiple signaling cascades [53]. Discrete signaling cascades are activated in response to specific proapoptotic stimuli, although there may be a degree of cross-talk between the apoptotic pathways. Key determinants of apoptosis include the nature of proapoptotic stimulus, cell type, cellular redox status, and preexisting activation state of the pro- versus antiapoptotic biochemical machinery.

The predominant effect of flavonoid and isoflavonoid supplementation in ex vivo cell culture models appears to be one of promoting apoptosis [54–57]. This is repeatedly observed in studies with transformed cancer cells, leading to the descriptions "cytoprotective" and/or "chemopreventive" [6,58]. Two polyphenolic compounds that have been extensively studied in anticancer research are quercetin and genistein, a flavonoid and isoflavone, respectively. However, ex vivo studies with primary cultured cells in 2000 and 2001 showed that some flavonoids can prevent apoptosis promoted by agents that induce oxidative stress [7,8,59]. The outcome of flavonoid treatment is expected to show a complex dependence on a number of factors, including the type of flavonoid, its concentration, the type of cell (e.g., transformed versus nontransformed), the mechanisms of action of the flavonoid, the nature of the proapoptotic stimulus, and the specific apoptotic signaling pathway that is activated.

There is a growing recognition that mitochondria play a central role in apoptosis [10,11]. Mitochondria-dependent apoptotic pathways are responsive to "intrinsic" proapoptotic stimuli that cause perturbations in the intracellular environment, particularly oxidative stress. In addition, other apoptotic signaling pathways appear to converge onto the mitochondria. This raises the question of whether flavonoids may modulate events upstream or downstream of the mitochondria-dependent apoptotic cascade, in addition to their potential effect on mitochondrial respiration, oxidative phosphorylation, and oxyradical generation (as discussed earlier).

A. Flavonoids and the Mitochondria-Dependent Apoptotic Pathway

The hemoprotein cytochrome c, located in the intermembrane space, serves a dual role in mitochondrial function [37]. First, it is essential for electron transfer from complex III to cytochrome oxidase (complex IV) during mitochondrial respiration. Second, cytochrome c is one of a number of apoptogenic or apoptosis-inducing factors (AIFs) involved in mitochondria-dependent apoptosis [10,11]. In response to certain proapoptotic stimuli, cytochrome c dissociates from the respiratory chain along the inner membrane and is released from the intermembrane space into the cytosol (Fig. 4).

Cytochrome c–dependent mitochondrial apoptosis involves a number of concerted steps leading to the formation of the apoptosome and the downstream activation of caspase-3 [10,11]. The predominant mechanism is believed to involve the following events (Fig. 4): (1) a transient collapse of the mitochondrial transmembrane potential; (2) opening of the permeability transition pore (PTP); (3) release of cytochrome c and procaspase-9; (4) recruitment of Apaf-1 (apoptosis protease activating factor-1) and dATP and assembly of the mitochondrial apoptosome; (5) activation of procaspase-9 by the apoptosome; and (6) the caspase-9-catalyzed activation of caspase-3, the molecular executor of apoptosis. Amplification through the caspase cascade leads to the downstream

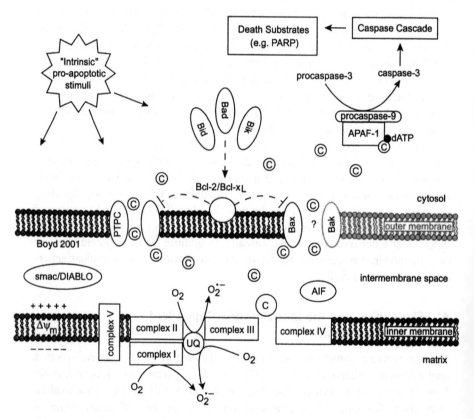

Figure 4 Mitochondria-dependent apoptosis: Cytochrome-c–dependent caspase-3 activation and the role of the Bcl-2 protein family.

proteolysis of numerous "death" substrates [e.g., poly-(ADP-ribose) polymerase (PARP)] and subsequent cellular damage. It should be noted, however, that some dispute exists over the temporal relationship between the dissipation of the mitochondrial transmembrane potential and the induction of the permeability transition. Furthermore, it is unclear whether these two events are actually necessary for the release of cytochrome c (and other apoptogenic factors), and thus mitochondria-mediated apoptosis [37].

The "Bcl-2 family" of oncoproteins is known to play an important role in apoptosis through their ability to regulate cytochrome c release from mitochondria [11,15]. The antiapoptotic proteins Bcl-2 and Bcl-x_L prevent cytochrome c release, whereas the proapoptotic family members (e.g., Bad, Bid, Bik, Bax) facilitate cytochrome c efflux or block the protective effects of Bcl-2 and Bcl-x_L. The mechanism involved is unclear; the "Bcl-2 family" proteins may interact directly with the MTP (mitochondrial permeability transition) protein complex (the PTPC) or form independent ionic pores in the outer mitochondrial membrane (Fig. 3). Nonetheless, cytochrome c–dependent caspase-3 activation and changes in the expression or phosphorylation state of "Bcl-2 family" proteins are taken as indicative of mitochondria-dependent apoptotic pathways. It is important to remember that other apoptogenic proteins are also present in the mitochondrial intermembrane space, including smac/DIABLO and flavoprotein (AIF) [10,11,60]. The "release stimuli" for the latter factors, which are currently being elucidated, may also involve the permeability transition or the "Bcl-2 family" proteins [37].

In 2001 a hypothesis was proposed that the redox state of cytochrome c, after release into the cytoplasm, acts a fail-safe mechanism in the regulation of apoptosis [61]. It is argued that cytochrome c induces apoptosis only if it is present in the cytoplasm in the oxidized state. It is predicted that the redox state of this hemoprotein will be sensitive to the cellular redox state such that cytosolic GSH levels will maintain the protein in a reduced state. An increase in pro-oxidants (e.g., reactive oxygen species) or a decrease in GSH will switch cytochrome c to the oxidized state, with the induction of apoptosis. Cytochrome c released during apoptosis results in an impairment of the electron flux through the respiratory chain with a subsequent burst of $O_2^{\bullet-}$ generation by the mitochondria [36,37] (Fig. 3). It could be proposed that this "burst" facilitates the oxidation of cytochrome c. In this regard, the finding that flavonoids such as taxifolin, catechin, epicatechin, and, in particular, epicatechin gallate can directly reduce cytochrome c in vitro is of interest [24]. Therefore, flavonoids may protect cells by promoting the reduction of cytochrome c directly or by preventing its oxidation through modification of the cellular redox state. Specifically, the latter may involve scavenging the apoptotic "burst" of reactive oxygen species produced by the mitochondria. In addition, it will be shown that cell culture studies are pointing to a role for flavonoids and isoflavones upstream of cytochrome c release, i.e., prior to its oxidation in the cytosol.

The natural isoflavone genistein is a potent inhibitor of tyrosine kinase and induces cell cycle arrest and apoptosis in a number of tumor cell lines. An early (1993) study found that the induction of G2/M cell cycle arrest by genistein in MC-7 (human breast), Jurkat (human T-cell leukemia), and L-929 (mouse transformed fibroblasts) tumor cells was consistently associated with an increase in mitochondrial number and/or activity [54]. A more recent (2000) study with the MRC-5 cell line (human lung fibroblasts) reported a similar finding [56]. Treatment with genistein or oxidative stress, induced by H_2O_2 or buthionine sulfoximine (an inhibitor of intracellular GSH synthesis), resulted in cell cycle arrest and an increase in mitochondrial mass and deoxyribonucleic acid (DNA) content. Furthermore, studies have provided strong evidence that genistein-induced apoptosis is mediated by the mitochondria-dependent apoptotic pathway. For example, in RPE-J cells, genistein treatment caused a reduction in the mitochondrial transmembrane potential, an opening of the permeability transition pore in a caspase-independent manner, and the release of cytochrome c [57]. Cytochrome c release was associated with the downstream activation of caspase-3 and subsequent nuclear condensation and DNA fragmentation. Interestingly, the ANT inhibitor bongkrekic acid, a blocker of the MPT, prevented the apoptotic effects of genistein described. The overexpression of Bcl-x_L protein was found to be responsible for the resistance of human prostate cancer PC-3 cells to staurosporine-induced apoptosis [62]. Although the pro apoptotic Bax and BAD proteins were successfully incorporated into the mitochondrial membranes, there was no activation of mitochondria-dependent apoptosis. Treatment with genistein restored the sensitivity to staurosporine by downregulating Bcl-x_L protein expression, thereby promoting the MPT [62]. Similarly, the flavonoid quercetin was found to restore sensitivity to CD95-(Apo1/Fas)-mediated apoptosis in HPB-ALL cells, which typically show a resistance to the latter proapoptotic stimulus [55]. Quercetin treatment promoted the dissipation of the mitochondrial trans-membrane potential, caspase-3 activation, and DNA fragmentation. Importantly, quercetin induced apoptosis even though treatment was associated with a reduction in the level of intracellular reactive oxygen species [55].

The activation of mitochondria-mediated apoptosis in transformed cells appears to be a common feature of flavonoids, too. For example, in a structure-activity study with human leukemia HL-60 cells [58], the ability of low micro-molar concentrations of flavonoids to induce mitochondria-mediated apoptosis was as follows: apigenin > quercetin > myricetin > kaempferol. For all the flavonoids tested, the key biochemical events included elevated level of reactive oxygen species, dissipation of the mitochondrial transmembrane potential, release of cytochrome c, procaspase-9 processing, caspase-3 activation, and proteolytic cleavage of PARP. Another study on the same leukemia cell line found that theaflavins, (-)-epigallocatechin-3-gallate, and penta-O-galloyl-β-d-glucose [which is structurally related to (-)-epigallocatechin-3-gallate] could induce apoptosis in an identical manner to that described [63]. Similarly, green tea catechins, including (-)-epigallocatechin-3-gallate, (-)-epigallocatechin, and (-)-epicatechin-3-gallate, were reported to induce mitochondria-dependent apoptosis in human prostate cancer DU145 cells [64]. Observed events included an increase in reactive oxygen species production and mitochondrial depolarization, but treatment did not alter the expression of Bcl-2, Bcl-x_L, or Bad. The order of potency was as follows: (-)-epicatechin-3-gallate > (-)-epigallocatechin-3-gallate > (-)-epigallocatechin > (-)-epicatechin, which was inactive (64). Finally, flavone was a potent inducer of apoptosis in HT-29 human colon carcinoma cells without an effect on nontransformed murine colonocytes [65]. The latter effect of flavone was associated with a change in the messenger ribonucleic acid (mRNA) levels of cell cycle-and apoptosis-related genes, including an upregulation of Bak mRNA without an effect on BAX, and a downregulation of NF-κB and Bcl-x_L mRNA levels. To reiterate, the proapoptotic effects of flavonoids and isoflavones against cancer cells in ex vivo models may explain their cancer chemopreventive activity [6,58]. One shortcoming of the studies discussed is that, in nearly all cases, there was a lack of comparison of the effects of the flavonoids/isoflavones in transformed cells versus the effects of their non-transformed counterparts.

Recent studies with primary cultures, particularly neurons, are finding a protective effect for a variety of polyphenolic compounds, including flavonoids, against a number of proapoptotic agents. For example, the death of cultured rat hippocampal cells, induced by nitric oxide–related species, could be prevented by pretreatment with the stilbene resveratrol or the flavonoids quercetin and (+)-catechin [4]. Similarly, the flavonoid content of *Ginkgo biloba* extract (EGb 761) was found to protect cultured rat hippocampal cells from the proapoptotic effects of nitric oxide–related species and β-amyloid-derived peptides [66,67]. Of importance is the observation that the antiapoptotic effects of the flavonoids, but not resveratrol, appeared to be related to their ability to block the activation of protein kinase C, rather than just their antioxidant potential [4,66]. Many flavonoids were found to protect against glutamate-induced excitotoxicity in cultured rat primary neurons or the mouse hippocampal cell line HT-22, a model system of oxidative stress [3]. Three distinct mechanisms of protection were observed: an increase in the intracellular GSH levels, a reduction of reactive oxygen species, and the prevention of the NMDA (n-methyl-D-aspartate) receptor-mediated Ca^{2+} influx. Interestingly, the neuroprotective flavonoids appeared to operate specifically by one of the possible mechanisms [3]. Of importance here is the fact that mitochondria-mediated pathways are central to neuronal apoptosis, especially with respect to the hippocampal glutamate excitotoxicity model or oxidative stress.

For example, the induction of apoptosis in cultured primary striatal neurons by oxidized low-density lipoprotein (Ox-LDL) was prevented by pretreatment with low micromolar concentrations of epicatechin and kaempferol [7]. Although the flavonoid pretreatments did not prevent an increase in intracellular oxidative stress, there was a potent inhibition of MAPK signaling and caspase-3 activation. Similarly, epicatechin protected against H_2O_2-induced apoptosis in cultured human fibroblasts [8]. Epicatechin reduced caspase-3-like protease activity and protected mitochondrial function from the damage induced by the oxidative stress. Importantly, in both studies $3'$-O-methyl-epicatechin, the in vivo metabolite of epicatechin, displayed the same potency as epicatechin in attenuating oxidative stress–mediated cell death [7,8]. This suggests that the antiapoptotic effects of these flavonoids are not simply related to their hydrogen-donating properties.

An ex vivo study with rat hepatocytes found a strong correlation between the cytotoxicity of a flavonoid and its ability to induce an early collapse of the mitochondrial transmembrane potential [6]. In contrast, nontoxic flavonoids had no effect on the transmembrane potential. As discussed previously, the ability of flavonoids and isoflavones to inhibit mitochondrial respiration and adenosine triphosphate (ATP) synthesis or uncouple oxidative phosphorylation represents a potential mechanism whereby they can trigger dissipation of the transmembrane potential and thus cytochrome c–dependent apoptosis. Thus, it is interesting to note that epicatechin, which had little or no proapoptotic effect in cancer cells [64] and exerted antiapoptotic effects in primary culture models [7,8,59], had essentially no inhibitory effect against complex I, II, or III activity and F_0F_1-ATPase activity [24,25]. However, the preceding discussion has also indicated that the pro- or antiapoptotic effects of flavonoids may be explained by more specific mechanisms, potentially independently of their antioxidant capacity. These include effects on protein kinase cascades and gene expression, including MAPK and the Bcl-2 family of proteins.

B.   Mitogen-Activated Protein Kinases, Mitochondria, and Flavonoids

Mitogen-activated protein kinases (MAPKs) are integral components of cellular phosphorylation/dephosphorylation signalling cascades [50,68]. In response to extracellular and intracellular signals, the activation of MAPKs is initiated by upstream effector molecules and specific protein kinase cascades (e.g., tyrosine kinases, MAPK kinase kinases, and MAPK kinases) and deactivated by MAPK phosphatases. Mammalian MAPKs can be divided into at least 3 groups, with identifiable isoenzymes: (1) extracellular signal-related kinases (ERK1/2); (2) c-Jun N-terminal kinases (JNK1/2/3); (3) p38 kinases (p38α/β/γ/δ) [50,68] (Fig. 4). Distinct signal transduction pathways activate the latter groups, although some overlap may occur, depending on the stimuli. The activation of ERK1/2 by growth factors involves a well-defined pathway, including the G-protein effector molecule c-Ras (p21^{ras}), Raf-1, and MEK1/2. Both JNK and p38 respond to proinflammatory cytokines and environmental stress, but their activation may occur via common or parallel pathways. The activation of JNK and p38 by a variety of environmental stressors, including ultraviolet (UV) irradiation, osmotic shock, oxidative stress, proinflammatory cytokines, and trophic factor withdrawal, has led to their alternative designation as stress-activated protein kinases (SAPKs). Likely effectors for SAPKs include the G-proteins Rac1 and Cdc42, and in some cases, c-Ras. Downstream targets of MAPK include numerous cytosolic proteins, nuclear proteins, and transcription factors. Thus, ultimately, the MAPKs can modulate gene and protein expression involved in cell differentiation, proliferation, survival, and death [69]. For example, it has already been discussed that both JNK and ERK can regulate the gene expression of the antioxidant enzymes Cu, Zn-SOD, and Mn-SOD [47–49] (Fig. 4). Importantly, the JNKs are the only kinases known to phosphorylate c-Jun in vivo, thereby activating this transcription factor, which is a component of the activator protein 1 (AP-1) transcription factor complex. Typically ERK1/2 activation is associated with cell proliferation and survival, whereas the SAPKs are predominantly associated with apoptotic events [70,71].

Activated MAPKs have been found to modulate mitochondria-mediated apoptotic pathways, JNK in particular [71]. In fact, it appears that MAPK signaling may represent a potential cross-link between the different apoptotic pathways. JNK activation is not required for death-receptor-mediated apoptosis but is required for caspase-9 activation by the mitochondrial pathway, induced by a variety of proapoptotic stimuli [72,73]. JNK activation and c-Jun phosphorylation were found to be necessary to promote cytochrome c release from mitochondria, with the sequential assembly of the apoptosome and caspase-3 activation. The molecular mechanism of this effect was unclear, but it appeared to involve regulation of the expression and phosphorylation state of the Bcl-2 protein family and their recruitment to the mitochondrial outer membrane (Fig. 4). In a number of apoptotic models, JNK activation was associated with a downregulation of the antiapoptotic Bcl-2 and Bcl-x$_L$ and upregulation of the proapoptotic Bax and Bad [74–77]. Two cell culture studies provided very strong evidence that JNK activation resulted in the phosphorylation of Bcl-2 and Bcl-x$_L$ ex vivo, with the induction of apoptosis [76,77]. In other words, Bcl-2 and Bcl-x$_L$ appear to be substrates of JNK, and phosphorylation results in their inactivation, thereby abolishing their ability to prevent cytochrome c release.

JNK can also stabilize the proapoptotic p53 through phosphorylation [78] (Fig. 4). Once phosphorylated, p53 can impair mitochondria function and induce the permeability transition through its mediator p53-regulated apoptosis-inducing protein 1 (p53AIP1), a mitochondrial protein [79]. Furthermore, the expression of the proapoptotic Bax is regulated by p53 [80]. Alternatively, p53 might activate JNK via the MAPK kinase MEKK4, involving the p53-mediated expression of the growth arrest and DNA damage–induced protein (GADD45) [81]. Also, caspase-3 activation may, under certain conditions, activate JNK downstream via MAPK kinase MEKK1 [82]. Finally the burst of $O_2^{\bullet-}$ generation by mitochondria associated with cytochrome c release [36,37] could result in the activation of JNK, which is known to be sensitive to H_2O_2 [70,71]. Activated JNK could, in turn, promote further cytochrome c release (as discussed), thereby amplifying mitochondria-dependent apoptosis. These observations strengthen the point that apoptotic pathways can be interdependent, and JNK activation potentially play a central proapoptotic role.

In contrast, several lines of evidence suggest that the activation of the Ras-Raf-ERK1/2 pathway may oppose JNK-mediated effects with respect to mito-chondria-dependent apoptosis (Fig. 4). ERK1/2 can phosphorylate and activate mitogen-and stress-activated kinase 1 (MSK1) and pp90 ribosomal S6 kinase (RSK) [83,84]. RSK can phosphorylate BAD, thereby inhibiting its proapop-totic effects [83]. Furthermore, both RSK and MSK1 are potent activators of cyclic adenosine monophosphate (cAMP) element binding protein (CREB), a transcription factor for Blc-2, and therefore important for cell survival [84]. The upregulation or overexpression of Bcl-2 has been reported to be associated with the inactivation of JNK [75,85]. Thus, ERK1/2 activation could ultimately lead to JNK inactivation. Finally, the activation of c-Ras has been shown to result in a suppression of Bax expression, via a mechanism that involved activation of both the ERK1/2 signaling cascade and the phosphatidylinositol-3'-kinase (PI-3)/Akt pathway [80].

Oxidative stress is a major stimulus for MAPK signaling cascades, as evidenced by the number of redox-responsive transcription factors and antioxidant genes that are downstream targets of these enzymes [69]. The G-protein c-Ras is a common signaling target for reactive oxygen species and nitric oxide–related species with important consequences for ERK1/2 signaling [86–89]. JNK is also activated by a number of agents that alter the redox state, including H_2O_2 [71,87], peroxynitrite [90], and nitric oxide–related species [86–88]. Similarly to ERK1/2, effector molecules upstream of JNK (e.g., c-Ras, Rac1 or Cdc42) appear to be redox-sensitive. In addition, JNK can be directly affected by oxidative stress, an effect that may be related to the redox-sensitive cysteine present in JNK, but not ERK or p38 [91,92]. Furthermore, redox-sensitive proteins such as glutathione-S-transferase and thioredoxin can directly interact with JNK or apoptosis-signal regulating kinase (ASK1), an upstream activator of JNK [93–95]. Therefore, certain flavonoids may influence MAPK pathways through an effect on the cellular redox potential, depending on their intrinsic antioxidant capacity.

In fact, strong evidence is accumulating from different models that flavonoids can influence MAPK signaling pathways, and this influence may represent a specific mechanism whereby flavonoids can modulate apoptosis. At low concentrations, epicatechin gallate and epigallocatechin gallate activated ERK2, JNK1, and p38 in mammalian cell lines [14]. MAPK activation was associated with an increase in the gene expression of c-Fos, c-Jun, and antioxidant and detoxifying enzymes, such as glutathione-S-transferase. However, at higher concentrations, these flavonoids activated the caspase pathway and apoptosis, that activation may reflect a predominance of the JNK1 pathway over ERK. In contrast, quercetin (in low micromolar concentrations) inhibited the activation of the JNK pathway by proapoptotic stimuli, resulting in a downregulation of AP-1 and ICAM-1 (intracellular adhesion molecule-1) expression in human endothelial cells [13]. Similarly, a study with cultured primary striatal neurons [7] found that pretreatment with low micromolar concentrations of epicatechin or 3-O-methylepicatechin strongly inhibited the Ca^{2+}-dependent activation of ERK1/2 and JNK by oxidized low-density lipoprotein (oxLDL). Inhibition of JNK activation by these compounds (and another flavonoid, kaempferol) abolished c-Jun phosphorylation. Inhibition of JNK activation, but not ERK1/2 activation, was associated with cell survival and the neuroprotective effects of the flavonoids. Importantly, the flavonoids also reduced procaspase-3 cleavage (i.e., caspase-3 activation) and caspase-3-like protease activity induced by oxLDL [7]. The latter is of particular relevance as the mitochondria-dependent pathways typically mediate oxidative stress–induced apoptosis. Thus, the results of this study strongly suggest a relationship between JNK and mitochondria that can be modified by flavonoid treatment.

In summary, the effect of a flavonoid on MAPK signaling is likely to depend on the concentration and type of flavonoid, the preexisting degree of activation of the main MAPK pathways, and the cellular environment, such as the redox state to which MAPKs are sensitive. It is important to note that epicatechin and its in vivo metabolite 3-O-methylepicatechin were equally potent in their ability to prevent JNK activation and apoptosis in response to an oxidative stress [7]. This finding suggests that the latter effect of these flavonoids is not related to their hydrogen-donating abilities. Flavonoids may influence MAPK signaling pathways by a direct inhibitory or stimulatory effect on the enzyme itself or by modulation of signal transduction events upstream of the relevant MAPK (Fig. 4). Ultimately, an effect on MAPK signaling could represent a specific mechanism by which flavonoids and isoflavones influence mitochondrial function.

V. CONCLUSIONS

This review has demonstrated that flavonoids can influence mitochondrial function, either at the level of respiration and oxidative phosphorylation or of the mitochondrial apoptosome. These biochemical effects were elucidated in vitro with isolated mitochondria or cell culture systems. However, the biological significance of these effects to mitochondria in vivo needs to be clarified and depends on the bioavailability of flavonoids [96].

After absorption, dietary flavonoids are metabolized in the small intestine or through xenobiotic metabolic systems in the liver. Main metabolic routes include methylation, glucuronidation, and sulfation [97–100]. In turn, tissue distribution (including the ability to cross the blood-brain barrier), subcellular localization, and access to mitochondria depend on the extent and type of metabolism. The in vitro structure-activity studies discussed clearly showed that metabolism of the hydroxyl groups (e.g., glycosylation, glucuronidation, or methylation) reduced the antioxidant capacity of flavonoids, their inhibitory potency toward complex II and F_0F_1-ATPase activity, and their proapoptotic potential. Alternatively, methylation may serve to enhance flavonoid stability, membrane penetration, and access to subcellular compartments (e.g., mitochondria). In other words, the in vivo biochemical properties of flavonoids cannot necessarily be predicted from their in vitro properties. Future studies should (1) identify the biologically relevant metabolic derivatives of dietary flavonoids; (2) compare the biochemical effects of these derivatives to those of their aglycones in vitro to predict likely mitochondrial effects; and (3) determine the in vivo mitochondrial effects of those dietary flavonoids, which are known to be taken up by cells and thus are likely to have access to mitochondria. Finally, an effort should be made to identify specific and selective molecular targets of flavonoids with respect to mitochondria function, e.g., MAPK signaling, and to rationalize their importance as potential therapeutic agents.

REFERENCES

1. Rice-Evans CA, Miller NJ, Paganga G. Structure-antioxidant activity relationship of flavonoids and phenolic acids. Free Radic Biol Med 1996; 20:933–956.
2. Hodnick WF, Pardini RS. Inhibition of mitochondrial function by flavonoids. In: Rice-Evans CA, Packer L, eds. Flavonoids in Health and Disease. Vol. 1. New York: Marcel Dekker, 1998:179–197.
3. Ishige K, Schubert D, Sagara Y. Flavonoids protect neuronal cells from oxidative stress by three distinct mechanisms. Free Radic Biol Med 2001; 30:433–446.
4. Bastianetto S, Zheng WH, Quirion R. Neuroprotective abilities of resveratrol and other red wine constituents against nitric oxide-related toxicity in cultured hippocampal neurons. Br J Pharmacol 2000; 131:711–720.
5. Colic M, Pavelic K. Molecular mechanisms of anticancer activity of natural dietetic products. J Mol Med 2000; 78:333–336.
6. Galati G, Teng S, Moridani MY, Chan TS, O'Brien PJ. Cancer chemoprevention and apoptosis mechanisms induced by dietary polyphenolics. Drug Metab Drug Interact 2000; 17:311–349.

7. Schroeter H, Spencer JPE, Rice-Evans C, Williams RJ. Flavonoids protect neurones from oxidized low-density lipoprotein-induced apoptosis involving JNK, c-jun and caspase-3. Biochem J 2001; 358:547–557.

8. Spencer JPE, Schroeter H, Kuhnle G, Srai SK, Tyrrell RM, Hahn U, Rice-Evans C. Epicatechin and its *in vivo* metabolite, 3'-O-methyl epicatechin, protect human fibroblasts from oxidative-stress-induced cell death involving caspase-3 activation. Biochem J 2001; 354:493–500.

9. Hatefi Y. The mitochondrial electron transport and oxidative phosphorylation system. Annu Rev Biochem 1985; 54:1015–1069.

10. Green DR, Reed JC. Mitochondria and apoptosis. Science 1998; 281:1309–1312.

11. Adrain C, Martin SJ. The mitochondrial apoptosome: a killer unleashed by the cytochrome seas. Trends Biochem Sci 2001; 26:390–397.

12. Boveris A, Cadenas E. Mitochondrial production of hydrogen peroxide regulation by nitric oxide and the role of ubisemiquinone. IUBMB Life 2000; 50:245–250.

13. Kobuchi H, Roy S, Sen CK, Nguyen HG, Packer L. Quercetin inhibits inducible ICAM-1 expression in human endothelial cells through the JNK pathway. Am J Physiol Cell Physiol 1999; 277:C403–C411.

14. Kong A-NT, Yu R, Chen C, Mandlekar S, Primiano T. Signal transduction events elicited by natural products: role of MAPK and caspase pathways in homeostatic response and induction of apoptosis. Arch Pharm Res 2000; 23:1–16.

15. Kluck RM, Bossy-Wetzel E, Green DR, Newmeyer DD. The release of cytochrome c from mitochondria: a primary site for Bcl-2 regulation of apoptosis. Science 1997; 275:1132–1136.

16. Hodnick WF, Milosavljević EB, Nelson JH, Pardini RS. Electrochemistry of flavonoids: relationships between redox potentials, inhibition of mitochondrial respiration, and production of oxygen radicals by flavonoids. Biochem Pharmacol 1988; 37:2607–2611.

17. Van Acker SABE, van den Berg D-J, Tromp MNJL, Griffioen DH, van Bennekom WP, van der Vijgh WJF, Bast A. Structural aspects of antioxidant activity of flavonoids. Free Radic Biol Med 1996; 20:331–342.

18. Bohmont C, Aaronson LM, Mann K, Pardini RS. Inhibition of mitochondrial NADH oxidase, succinoxidase, and ATPase by naturally occurring flavonoids. J Nat Prod 1987; 50:427–433.

19. Hodnick WF, Kung FS, Roettger WJ, Bohmont CW, Pardini RS. Inhibition of mitochondrial respiration and production of toxic oxygen radicals by flavonoids: a structure-activity study. Biochem Pharmacol 1986; 35:2345–2357.

20. Hodnick WF, Bohmont CW, Capps C, Pardini RS. Inhibition of mitochondrial NADH-oxidase (NADH-Coenzyme Q oxidoreductase) enzyme system by flavonoids: a structure-activity study. Biochem Pharmacol 1987; 36:2873–2874.

21. Hodnick WF, Duval DL, Pardini RS. Inhibition of mitochondrial respiration and cyanide-stimulated generation of reactive oxygen species by selected flavonoids. Biochem Pharmacol 1994; 47:573–580.

22. Cheng SC, Pardini RS. Structure-inhibition relationships of various phenolic compounds towards mitochondrial respiration. Pharmacol Res Commun 1978; 10:897–910.

23. Cheng SC, Pardini RS. Inhibition of mitochondrial respiration by model phenolic compounds. Biochem Pharmacol 1979; 28:1661–1667.

24. Moini H, Arroyo A, Vaya J, Packer L. Bioflavonoid effects on the mitochondrial respiratory electron transport chain and cytochrome c redox state. Redox Rep 1999; 4:35–41.

25. Zheng J, Ramirez VD. Inhibition of mitochondrial proton F0F1-ATPase/ATP synthase by polyphenolic phytochemicals. Br J Pharmacol 2000; 130:1115–1123.

26. Racker E. Effect of quercetin on ATP-driven pumps and glycolysis. Prog Clin Bio Res 1986; 213:257–271.

27. Lang DR, Racker E. Effects of quercetin and F1 inhibitor on mitochondrial ATPase and energy-linked reactions in submitochondrial particles. Biochim Biophys Acta 1974; 333:180–186.

28. Di Pietro A, Godinot C, Bouillant ML, Gautheron DC. Pig heart mitochondrial ATPase: properties of purified and membrane-bound enzyme: effects of flavonoids. Biochimie 1975; 57:959–967.

29. Conseil G, Baubichon-Cortay H, Dayan G, Jault JM, Barron D, Di Pietro A. Flavonoids: a class of modulators with bifunctional interactions at vicinal ATP-and steroid-binding sites on mouse P-glycoprotein. Proc Natl Acad Sci USA 1998; 95:9831–9836.

30. Halestrap AP. The mitochondrial permeability transition: its molecular mechanism and role in reperfusion injury. Biochem Soc Symp 1999; 66:181–203.

31. Janssens D, Delaive E, Houbion A, Eliaers F, Remacle J, Michiels C. Effect of venotropic drugs on the respiratory activity of isolated mitochondria and in endothelial cells. Br J Pharmacol 2000; 130:1513–1524.

32. van Dijk C, Driessen AJ, Recourt K. The uncoupling efficiency and affinity of flavonoids for vesicles. Biochem Pharmacol 2000; 60:1593–1600.

33. Santos AC, Uyemura SA, Lopes JL, Bazon JN, Mingatto FE, Curti C. Effect of naturally occurring flavonoids on lipid peroxidation and membrane permeability transition in mitochondria. Free Radic Biol Med 1998; 24:1455–1461.

34. Crompton M, Virji S, Doyle V, Johnson N, Ward JM. The mitochondrial permeability transition pore. Biochem Soc Symp 1999; 66:167–179.

35. Dekermendjian K, Kahnberg P, Witt MR, Sterner O, Nielsen M, Liljefors T. Structure-activity relationship and molecular modelling analysis of flavonoids binding to the benzodiazepine site of the rat brain GABA(A) receptor complex. J Med Chem 1999; 42:4343–4350.

36. Cai J, Jones DP. Superoxide in apoptosis: mitochondrial generation triggered by cytochrome c loss. J Biol Chem 1998; 273:11401–11404.

37. Cai J, Yang J, Jones DP. Mitochondrial control of apoptosis: the role of cytochrome c. Biochim Biophys Acta 1998; 1366:139–149.

38. Bors W, Heller W, Michel C. The chemistry of flavonoids. In: Rice-Evans CA, Packer L, eds. Flavonoids in Health and Disease. Vol. 1. New York: Marcel Dekker, 1998:111–136.

39. Jovanovic SV, Steenken S, Simic MG, Hara Y. Antioxidant properties of flavonoids: Reduction potentials and electron transfer reactions of flavonoid

radicals. In: Rice-Evans, CA, Packer L, eds. Flavonoids in Health and Disease. Vol. 1. New York: Marcel Dekker, 1998:137–161.

40. Morel I, Cillard P, Cillard J. Flavonoid-metal interactions in biological systems. In: Rice-Evans CA, Packer L, eds. Flavonoids in Health and Disease. Vol. 1. New York: Marcel Dekker, 1998:163–177.

41. Ratty AK, Das NP. Effects of flavonoids on nonenzymatic lipid peroxidation: structure-activity relationship. Biochem Med Metab Biol 1988; 39:69–79.

42. Gao Z, Huang K, Yang X, Xu H. Free radical scavenging and antioxidant activities of flavonoids extracted from the radix of Scutellaria baicalensis Georgi Biochim Biophys Acta 1999; 1472:643–650.

43. Haraguchi H, Yoshida N, Ishikawa H, Tamura Y, Mizutani K, Kinoshita T. Protection of mitochondrial functions against oxidative stresses by isoflavans from Glycyrrhiza glabra. J Pharm Pharmacol 2000; 52:219–223.

44. Arora A, Nair MG, Strasburg GM. Structure-activity relationships for antioxidant activities of a series of flavonoids in a liposomal system. Free Radic Biol Med 1998; 24:1355–1363.

45. Okado-Matsumoto A, Fridovich I. Subcellular distribution of superoxide dismutases (SOD) in rat liver- Cu,Zn-SOD in mitochondria. J Biol Chem 2001; 276:38388–38393.

46. Han D, Williams E, Cadenas E. Mitochondrial respiratory chain-dependent generation of superoxide anion and its release into the intermembrane space. Biochem J 2001; 353:411–416.

47. Chang MS, Yoo HY, Rho HM. Positive and negative regulatory elements in the upstream region of the rat Cu/Zn-superoxide dismutase gene. Biochem J 1999; 339:335–341.

48. Das KC, Lewis-Molock Y, White CW. Activation of NF-kappa B and elevation of MnSOD gene expression by thiol reducing agents in lung adenocarcinoma (A549) cells. Am J Physiol 1995; 269:L588–L602.

49. Seo SJ, Kim HT, Cho G, Rho HM, Jung G. Sp1 and C/EBP-related factor regulate the transcription of human Cu/Zn SOD gene. Gene 1996; 178:177–185.

50. Chang L, Karin M. Mammalian MAP kinase signalling cascades. Nature 2001; 410:37–40.

51. Komatsu M, Hiramatsu M. The efficacy of an antioxidant cocktail on lipid peroxide level and superoxide dismutase activity in aged rat brain and DNA damage in iron-induced epileptogenic foci. Toxicology 2000; 148:143–148.

52. Inanami O, Watanabe Y, Syuto B, Nakano M, Tsuji M, Kuwabara M. Oral administration of (-)catechin protects against ischemia-reperfusion-induced neuronal death in the gerbil. Free Rad Res 1998; 29:359–365.

53. Hengartner MO. The biochemistry of apoptosis. Nature 2000; 407:770–775.

54. Pagliacci MC, Spinozzi F, Migliorati G, Fumi G, Smacchia M, Grignani F, Riccardi C, Nicoletti I. Genistein inhibits tumour cell growth *in vitro* but enhances mitochondrial reduction of tetrazolium salts: a further pitfall in the use of the MTT assay for evaluating cell growth and survival. Eur J Cancer 1993; 29A: 1573–1577.

55. Russo M, Palumbo R, Tedesco I, Mazzarella G, Russo P, Iacomino G, Russo GL. Quercetin and anti-CD95(Fas/Apo1) enhance apoptosis in HPB-ALL cell line. FEBS Lett 1999; 462:322–328.

56. Lee HC, Yin PH, Lu CY, Chi CW, Wei YH. Increase of mitochondria and mitochondrial DNA in response to oxidative stress in human cells. Biochem J 2000; 348:425–432.

57. Yoon HS, Moon SC, Kim ND, Park BS, Jeong MH, Yoo YH. Genistein induces apoptosis of RPE-J cells by opening mitochondrial PTP. Biochem Biophys Res Commun 2000; 276:151–156.

58. Wang IK, Lin-Shiau SY, Lin JK. Induction of apoptosis by apigenin and related flavonoids through cytochrome c release and activation of caspase-9 and caspase-3 in leukaemia HL-60 cells. Eur J Cancer 1999; 35:1517–1525.

59. Schroeter H, Williams RJ, Matin R, Iversen L, Rice-Evans CA. Phenolic antioxidants attenuate neuronal cell death following uptake of oxidized low-density lipoprotein. Free Radic Biol Med 2000; 29:1222–1233.

60. Susin SA, Lorenzo HK, Zamzami N, Marzo I, Snow BE, Brothers GM, Mangion J, Jacotot E, Costantini P, Loeffler M, Larochette N, Goodlett DR, Aebersold R, Siderovski DP, Penninger JM, Kroemer G. Molecular characterization of mitochondrial apoptosis-inducing factor. Nature 1999; 397:441–446.

61. Hancock JT, Desikan R, Neill SJ. Does the redox status of cytochrome c act as a fail-safe mechanism in the regulation of programmed cell death? Free Radic Biol Med 2001; 31:697–703.

62. Li X, Marani M, Mannucci R, Kinsey B, Andriani F, Nicoletti I, Denner L, Marcelli M. Overexpression of BCL-X$_L$ underlies the molecular basis for resistance to staurosporine-induced apoptosis in PC-3 cells. Cancer Res 2001; 61:1699–1706.

63. Pan MH, Lin JH, Lin-Shiau SY, Lin JK. Induction of apoptosis by penta-O-galloyl-beta-D-glucose through activation of caspase-3 in human leukemia HL-60 cells. Eur J Pharmacol 1999; 381:171–183.

64. Chung LY, Cheung TC, Kong SK, Fung KP, Choy YM, Chan ZY, Kwok TT. Induction of apoptosis by green tea catechins in human prostate cancer DU145 cells. Life Sci 2001; 68:1207–1214.

65. Wenzel U, Kuntz S, Brendel MD, Daniel H. Dietary flavone is a potent apoptosis inducer in human colon carcinoma cells. Cancer Res 2000; 60:3823–3831.

66. Bastianetto S, Zheng W-H, Quirion R. The Ginkgo biloba extract (EGb 761) protects and rescues hippocampal cells against nitric oxide-induced toxicity: involvement of its flavonoid constituents and protein kinase C. J Neurochem 2000; 74:2268–2277.

67. Bastianetto S, Ramassamy C, Dore S, Christen Y, Poirier J, Quirion R. The Ginkgo biloba extract (EGb 761) protects hippocampal neurons against cell death induced by beta-amyloid. Eur J Neurosci 2000; 12:1882–1890.

68. Schaeffer HF, Weber MJ. Mitogen-activated protein kinase: specific messages from ubiquitous messengers. Mol Cell Biol 1999; 19:2435–2444.

69. Allen RG, Tresini M. Oxidative stress and gene regulation. Free Radic Biol Med 2000; 28:463–499.

70. Xia Z, Dickens M, Raingeaud J, Davis RJ, Greenberg ME. Opposing effects of ERK and JNK-p38 MAP kinases on apoptosis. Science 1995; 270:1326–1331.
71. Davis RJ. Signal transduction by the JNK group of MAP kinases. Cell 2000; 109:239–252.
72. Tournier C, Hess P, Yang DD, Xu J, Turner TK, Nimnual A, Bar-Sagi D, Jones SN, Flavell RA, Davis RJ. Requirement of JNK for stress-induced activation of the cytochrome c-mediated death pathway. Science 2000; 288: 870–874.
73. Whitfield J, Neame SJ, Paquet L, Bernard O, Ham J. Dominant-negative c-Jun promotes neuronal survival by reducing BIM expression and inhibiting mito-chondrial cytochrome c release. Neuron 2001; 29:629–643.
74. Kang CD, Jang JH, Kim KW, Lee HJ, Jeong CS, Kim CM, Kim SH, Chung BS. Activation of c-jun N-terminal kinase/stress-activated protein kinase and the decreased ratio of Bcl-2 to Bax are associated with the auto-oxidized dopamine-induced apoptosis in PC12 cells. Neurosci Lett 1998; 256:37–40.
75. Choi WS, Yoon SY, Chang, II, Choi EJ, Rhim H, Jin BK, Oh TH, Krajewski S, Reed JC, Oh YJ. Correlation between structure of Bcl-2 and its inhibitory function of JNK and caspase activity in dopaminergic neuronal apoptosis. J Neurochem 2000; 74:1621–1626.
76. Yamamoto K, Ichijo H, Korsmeyer SJ. BCL-2 is phosphorylated and inactivated by an ASK1/Jun N-terminal protein kinase pathway normally activated at G(2)/M. Mol Cell Biol 1999; 19:8469–8478.
77. Fan MY, Goodwin M, Vu T, Brantley-Finley C, Gaarde WA, Chambers TC. Vinblastine-induced phosphorylation of Bcl-2 and Bcl-X-L is mediated by JNK and occurs in parallel with inactivation of the Raf-1/MEK/ERK cascade. J Biol Chem 2000; 275:29980–29985.
78. Buschmann T, Potapova O, Bar-Shira A, Ivanov VN, Fuchs SY, Henderson S, Fried VA, Minamoto T, Alarcon-Vargas D, Pincus MR, Gaarde WA, Holbrook NJ, Shiloh Y, Ronai Z. Jun NH2-terminal kinase phosphorylation of p53 on Thr-81 is important for p53 stabilization and transcriptional activities in response to stress. Mol Cell Biol 2001; 21:2743–2754.
79. Oda K, Arakawa H, Tanaka T, Matsuda K, Tanikawa C, Mori T, Nishimori H, Tamai K, Tokino T, Nakamura Y, Taya Y. p53AIP1, a potential mediator of p53-dependent apoptosis, and its regulation by Ser-46-phosphorylated p53. Cell 2000; 102:849–862.
80. Mazzoni IE, Said FA, Aloyz R, Miller FD, Kaplan D. Ras regulates sympathetic neuron survival by suppressing the p53-mediated cell death pathway. J Neurosci 1999; 19:9716–9727.
81. Takekawa M, Saito H. A family of stress-inducible GADD45-like proteins mediate activation of the stress-responsive MTK1/MEKK4 MAPKKK. Cell 1998; 95:521–530.
82. Shiah SG, Chuang SE, Kuo ML. Involvement of Asp-Glu-Val-Asp-directed, caspase-mediated mitogen-activated protein kinase kinase 1 cleavage, c-Jun N-terminal kinase activation, and subsequent Bcl-2 phosphorylation for paclitaxel-induced apoptosis in HL-60 cells. Mol Pharmacol 2001; 59:254–262.

83. Bonni A, Brunet A, West AE, Datta SR, Takasu MA, Greenberg ME. Cell survival promoted by the Ras-MAPK signaling pathway by transcription-dependent and independent mechanisms. Science 1999; 286:1358–1362.

84. Deak M, Clifton AD, Lucocq LM, Alessi DR. Mitogen-and stress-activated protein kinase-1 (MSK1) is directly activated by MAPK and SAPK2/p38, and may mediate activation of CREB. EMBO J 1998; 17:4426–4441.

85. Park DS, Stefanis L, Yan CY, Farinelli SE, Greene LA. Ordering the cell death pathway: differential effects of BCL2, an interleukin-1-converting enzyme family protease inhibitor, and other survival agents on JNK activation in serum/nerve growth factor-deprived PC12 cells. J Biol Chem1996; 271: 21898–21905.

86. Lander HM, Ogiste JS, Pearce SF, Levi R, Novogrodsky A. Nitric oxide-stimulated guanine nucleotide exchange on p21^{ras}. J Biol Chem 1995; 270:7017–7020.

87. Lander HM, Ogiste JS, Teng KK, Novogrodsky A. p21^{ras} as a common signaling target of reactive free radicals and cellular redox stress. J Biol Chem 1995; 270:21195–21198.

88. Lander HM, Jacovina AT, Davis RJ, Tauras JM. Differential activation of mitogen-activated protein kinases by nitric oxide-related species. J Biol Chem 1996; 271:19705–19709.

89. Lander HM, Hajjar DP, Hempstead BL, Mirza UA, Chait BT, Campbell S, Quilliam LA. A molecular redox switch on p21^{ras}. J Biol Chem 1997; 272: 4323–4326.

90. Go YM, Patel RP, Maland MC, Park H, Beckman JS, Darley-Usmar VM, Jo H. Evidence for peroxynitrite as a signaling molecule in flow-dependent activation of c-Jun NH$_2$-terminal kinase. Am J Physiol 1999; 277:H1647–H1653.

91. So HS, Park RK, Kim MS, Lee SR, Jung BH, Chung SY, Jun CD, Chung HT. Nitric oxide inhibits c-Jun N-terminal kinase 2 (JNK2) via S-nitrosylation. Biochem Biophys Res Commun 1998; 247:809–813.

92. Park HS, Huh SH, Kim MS, Lee SH, Choi EJ. Nitric oxide negatively regulates c-Jun N-terminal kinase/stress-activated protein kinase by means of S-nitrosylation. P Natl Acad Sci USA 2000; 97:14382–14387.

93. Saitoh M, Nishitoh H, Fujii M, Takeda K, Tobiume K, Sawada Y, Kawabata M, Miyazono K, Ichijo H. Mammalian thioredoxin is a direct inhibitor of apoptosis signal-regulating kinase (ASK) 1. EMBO J 1998; 17:2596–2606.

94. Adler V, Yin Z, Fuchs SY, Benezra M, Rosario L, Tew KD, Pincus MR, Sardana M, Henderson CJ, Wolf CR, Davis RJ, Ronai Z. Regulation of JNK signaling by GSTp. EMBO J 1999; 18:1321–1334.

95. Cho SG, Lee YH, Park HS, Ryoo K, Kang KW, Park J, Eom SJ, Kim MJ, Chang TS, Choi SY, Shim J, Kim Y, Dong MS, Lee MJ, Kim SG, Ichijo H, Choi EJ. Glutathione S-transferase mu modulates the stress-activated signals by suppressing apoptosis signal-regulating kinase 1. J Biol Chem 2001; 276:12749–12755.

96. Rice-Evans C, Spencer JP, Schroeter H, Rechner AR. Bioavailability of flavonoids and potential bioactive forms *in vivo*. Drug Metab Drug Interact 2000; 17:291–310.

97. Spencer JP, Chowrimootoo G, Choudhury R, Debnam ES, Srai SK, Rice-Evans C. The small intestine can both absorb and glucuronidate luminal flavonoids. FEBS Lett 1999; 458:224–230.

98. Kuhnle G, Spencer JP, Chowrimootoo G, Schroeter H, Debnam ES, Srai SK, Rice-Evans C, Hahn U. Resveratrol is absorbed in the small intestine as resveratrol glucuronide. Biochem Biophys Res Commun 2000; 272:212–217.

99. Kuhnle G, Spencer JP, Schroeter H, Shenoy B, Debnam ES, Srai SK, Rice-Evans C, Hahn U. Epicatechin and catechin are O-methylated and glucuronidated in the small intestine. Biochem Biophys Res Commun 2000; 277:507–512.

100. Okushio K, Suzuki M, Matsumoto N, Nanjo F, Hara Y. Identification of (-)-epicatechin metabolites and their metabolic fate in the rat. Drug Metab Dispos 1999; 27:309–316.

11

Gene Modulation of HaCaT Cells Induced by Pine Bark Extract

Bertrand Henri Rihn
Teaching Hospital of Brabois
Vandoeuvre, France

Claude Saliou
Johnson & Johnson Consumer Products Worldwide
Skillman, New Jersey, U.S.A.

Pine barks have been used for centuries as herbal remedies. During the 15th century pine decoctions were used for wound healing, as noted by Minner in 1497 in *Thesaurus Medicaminum*. Maritime Indians were the first reported to have used pine bark (decoction of white pine bark) to treat Jacques Cartier's crew from scurvy during the winter of 1535. Maritime Indians applied various barks to treat wounds and skin sores [1]. The French maritime pine (*Pinus maritima*) bark extract (PBE) is a mixture particularly rich in oligomeric procyanidins and other bioflavonoids such as taxifolin, catechin, and epicatechin. The French maritime pine grows on the weather-beaten sand dunes of the Bay of Biscay in the southwestern corner of France. Pine bark extracts have become popular again as dietary supplements. Beside their antioxidant functions, their other biological properties are now being characterized. Genes are the Rosetta stone of human health and disease. Thus gene expression analysis or genomics studies, using recently developed complementary deoxyribonucleic acid (cDNA) arrays, help tremendously to identify markers of disease, therapeutic targets, and potential pharmacological activities [2]. Using this approach, the effects of PBE on the gene expression profile of the human keratinocyte (HaCaT) cell line were investigated.

I. INTRODUCTION

Numerous flavonoids have been reported to affect cellular signaling processes, from kinases to transcription factors. In fact, oligomeric procyanidins from pine bark were found to prevent the activation of the proinflammatory transcription factor NF-κ-B on ultraviolet (UV) exposure [3]. The same procyanidins are also potent modulators of nitric oxide metabolism [4,5].

To understand the protective mechanism afforded by PBE-supplemented cells, their basal gene expression profile was determined and compared to that of nonsupplemented cells [6]. As expected, only a small proportion (83 genes) of the 588 genes analyzed were detected in either group. However, of these 83 genes, 39 genes showed an expression significantly (more than two-fold) increased or decreased (Table 1).

Interestingly, a group of overexpressed genes is involved in stress response. Thioredoxin peroxidase 2 plays an important role in eliminating peroxides generated during cellular metabolism and signaling cascades. This effect should be related to the increase of gluthatione level in macrophage by PBE observed by Rimbach and colleagues [7]. The ultraviolet (UV) excision repair protein HHR23B is involved in nucleotide excision repair of deoxyribonucleic acid (DNA) damage. Last, the heat shock protein HSP70 is generally expressed in response to a stress. However, antioxidants such as curcumin have been suggested to induce its expression, thereby increasing cell resistance to stress [8]. The expression of the inhibitor for helix-loop-helix protein ID-3 is known to be dysregulated in keratinocyte cell lines such as HaCaT [9]. The increased dynein light chain 1 (dlc1) expression points to the regulation of intracellular trafficking through cytoskeleton interaction.

II. PROTECTIVE EFFECT OF PINE BARK EXTRACT
AGAINST ULTRAVIOLET B

Using a similar approach, it is noteworthy that several UV-upregulated genes are normalized after PBE pretreatment (Table 2). For instance, nucleoside diphosphate kinase B has a major role in the synthesis of nucleoside triphosphates [other than adenosine triphosphate (ATP)]. In addition, nucleoside diphosphate kinase B interacts with nuclease hypersensitive element (NHE) upstream of the c-*myc* gene to transactivate c-*myc* [10]. The messenger level of thymosin β_{10} was increased 2.8-fold by UV and was normalized when cells were PBE-supplemented. This protein plays an important role in the organization of the cytoskeleton by binding to and sequestering actin monomers (g-actin) and, as a result, inhibits actin polymerization. The heat shock 27-kd protein (HSP27) is known to be associated with the cytoskeleton components α- and β-tubulin and the microtubules.

Table 1 Relative Expression of the Most and Least Expressed Genes[a]

GenBank	Genes overexpressed in PBE-supplemented cells	P/C
X69111	Inhibitor for helix-loop-helix protein ID-3	10.3
X67643	Heat shock 70-kd protein 1	5.7
M11886	HLA class 1 histocompatibility antigen C-4 α chain	4.9
D21090	UV excision repair protein HHR 23B	4.7
U32944	Cytoplasmic dynein light chain 1	4.6
X67951	Thioredoxin peroxidase 2	3.9
M36429	Transducin β_2	3.7
X00351	β-Actin	2.0
X53587	β_4-Integrin	1.7
K00558	α-Tubulin	1.6
	Housekeeping genes	
X56932	23 kd highly basic protein	1.1
	Genes overexpressed in control cells	
M22489	Bone morphogenetic protein 2A	0.08
X01060	CD71, transferin receptor protein	0.07
X59798	Cyclin D1 (G/S specific)	0.07
D13866	α-Catenin	0.07
M74088	Adenomatous polyposis coli protein	0.05
X06234	Calgranulin A	0.04
X06233	Calgranulin A	0.04
M28372	Cellular nucleic acid binding protein	0.04

[a] HaCaT confluent cells were grown without and with 25 µg/mL of PBE (Horphag Research Ltd, Guernsey, UK). Twenty-four hours later the messengers were oligo-dT purified and [^{32}P]-dATP labeled. Approximately 1.10^6 CPM of each probe was used for filter hybridization. After stringent washing, blots were exposed for phosphorimaging (BAS2000 BioImager). The cDNA dots were analyzed, normalized to radioactivity per square millimeter, and the RNA abundance was quantified as compared. P/C, ratio of relative abundancy of a given mRNA from PBE-supplemented cells (P) or its equivalent from control cells (C); PBE, pine boric extract; dATP, IP-3, CPM, HLA, human leukocyte antigen; cDNA, complementary deoxyribonucleic acid; UV, ultraviolet, HaCaT, human keratinocyte; mRNA, messenger ribonucleic acid.

Furthermore, HSP27, as do other chaperones, translocates to the nucleus during heat shock. As for thymosine β_{10}, HSP27 expression was normalized in PBE-supplemented cells. Consistent with the present results, other antioxidants have been reported to prevent the upregulation of HSP27 [11]. Interestingly, anchorage dependence of butyrate-treated HT29 colon carcinoma cells was correlated with a lower level of S19 messenger ribonucleic acid (mRNA) expression

Table 2 PBE Supplementation Reduces Ultraviolet B–induced Genes in HaCaT Cells[a]

Protein	Genbank	CTRL	UV	PBE	PBE + UV
60S Ribosomal protein L6	X69391	1.24	1.85	0.83	1.25
40S Ribosomal protein S19	M81757	0.30	0.52	nd	0.34
Y Box binding protein 1	M83234	0.28	0.54	0.28	0.19
G3PDH	X01677	0.95	1.03	0.60	0.60
Heat shock 27 kd	X54079	0.94	6.46	1.16	1.07
Nucleoside diphosphate kinase B	L16785	0.11	0.26	0.16	0.07
Thymosine β_{10}	M92381	0.60	1.65	nd	0.61
23-kd Highly basic protein	X56932	1.00	1.00	1.00	1.00

[a] HaCaT cells were supplemented with PBE for 24 h before their exposure to sham (control cells) or 150 mJ/cm^2 UV (treated cells). mRNA were extracted 4 h after sham or UV exposure. The gene expression of (1) control, (2) PBE-supplemented, (3) UV-treated, and (4) PBE-supplemented and UV-treated was determined by high-density filter arrays (Atlas human cDNA expression arrays, Clontech Laboratories, Inc.). One microgram of oligo-dT purified mRNA was [^{32}P]-dATP-labeled. Approximately 0.5.10^6 CPM of each probe was used for filter hybridization. After stringent washing, blots were exposed for phosphorimaging (BAS2000 BioImager). The cDNA dots were analyzed, normalized to radioactivity/per square millimeter by subtracting the nearest equivalent, and nonspecific dot considered as background, and the RNA abundance was quantified as compared to the expression of 23-kd highly basic protein (X56932). HaCaT, human keratinocyte; PBE, pine bark extract; UV, ultraviolet; cDNA, complementary deoxyribonucleic acid; mRNA, messenger ribonucleic acid; [^{32}P]-dATP, CPM, CTRL.

when compared with untreated cell [12]. Indeed, in our experiment, the level of S19 mRNA expression was normalized by PBE treatment.

III. CONCLUSIONS

Physiological and biochemical effects of PBE are, to date, not well studied: PBE has been shown to inhibit platelet aggregation [13] and therefore proposed in venous insufficiency [14]. In 1999 some biochemical targets of PBE action were proposed, e.g., inhibition of the respiratory electron transport chain [15]. PBE also dose-dependently inhibited the activities of xanthine oxidase, xanthine dehydrogenase, horseradish peroxidase, and lipoxygenase [16]. Nardili and associates [17] demonstrated a clear effect of PBE on the protein kinases A and C at 20 µg/mL in vitro. Clinical trials with further evaluation on psoriatic lesions are also required. There seems to be sufficient evidence on use of PBE in skin lesions, and PBE seems well tolerated, as indicated by trials in 2000 on venous insufficiency [14,18]. In addition, as dosage of ferrulic acid, a major component of PBE, has become available [19], pharmacocinetic studies are eventually possible.

ACKNOWLEDGMENTS

B.H.R. is indebted to M.C. Bottin and S. Mohr for excellent technical assistance.

REFERENCES

1. Chandler FR, Freeman L, Hooper SN. Herbal remedies of the Maritime Indians. J Ethnopharmacol 1979; 1:49–68.
2. Gerhold D, Rushmore T, Caskey CT. DNA chips: promising toys have become powerful tools. Trends Biochem Sci 1999; 24:168–173.
3. Saliou C, Rimbach G, Moini H, McLaughlin L, Hosseini S, Lee J, Watson RR, Packer L. Solar ultraviolet-induced erythema in human skin and nuclear factor-kappa-B-dependent gene expression in keratinocytes are modulated by a French maritime pine bark extract. Free Radic Biol Med 2001; 30:154–160.
4. Packer L, Rimbach G, Virgili F. Antioxidant activity and biologic properties of a procyanidin-rich extract from pine (*Pinus maritima*) bark, pycnogenol. Free Radic Biol Med 1999; 27:704–724.
5. Park YC, Rimbach G, Saliou C, Valacchi G, Packer L. Activity of monomeric, dimeric, and trimeric flavonoids on NO production, TNF-alpha secretion, and NF-kappa B-dependent gene expression in RAW 264.7 macrophages. FEBS Lett 2000; 465:93–97.
6. Rihn B, Saliou C, Bottin MC, Keith G, Packer L. From ancient remedies to modern therapeutics: pine bark uses in skin disorders revisited. Phytother Res 2001; 15:76–78.
7. Rimbach G, Virgili F, Park YC, Packer L. Effect of procyanidins from *Pinus maritima* on glutathione levels in endothelial cells challenged by 3-morpholino-sydnonimine or activated macrophages. Redox Rep 1999; 4:171–177.
8. Sood A, Mathew R, Trachtman H. Cytoprotective effect of curcumin in human proximal tubule epithelial cells exposed to shiga toxin. Biochem Biophys Res Commun 2001; 283:36–41
9. Langlands K, Down GA, Kealey T. ID proteins are dynamically expressed in normal epidermis and dysrgulated in squamous cell carcinoma. Cancer Res 2000; 60:5929–5933.

10. Postel EH, Berberich SJ, Flint SJ, Ferrone CA. Human c-myc transcription factor PuF identified as nm23-H2 nucleoside diphosphate kinase, a candidate suppressor of tumor metastasis. Science 1993; 261:428–429.

11. Gorman AM, Heavey B, Creagh E, Cotter TG, Samali A. Antioxidant-mediated inhibition of the heat shock response leads to apoptosis. FEBS Lett 1999; 445:98–102.

12. Kondoh N, Schweinfest CW, Henderson KW, Papas TS. Differential expression of S19 ribosomal protein, laminin-binding protein, and human lymphocyte antigen class I messenger RNAs associated with colon carcinoma progression and differentiation. Cancer Res 1992; 52:791–796.

13. Araghi-Niknam M, Hosseini S, Larson D, Rohdewald P, Watson RR. Pine bark extract reduces platelet aggregation. Integr Med 2000; 2:73–77.

14. Arcangeli P. Pycnogenol in chronic venous insufficiency. Fitoterapia 2000;71:236–244.

15. Moini H, Guo Q, Packer L. Enzyme inhibition and protein-binding action of the procyanidin-rich French maritime pine bark extract, pycnogenol: effect on xanthine oxidase. J Agric Food Chem 2000; 48:5630–9.

16. Moini H, Arroyo A, Vaya J, Packer L. Bioflavonoid effects on the mitochondrial respiratory electron transport chain and cytochrome c redox state. Redox Rep 1999; 4:35–41.

17. Nardini M, Scaccini C, Packer L, Virgili F. In vitro inhibition of the activity of phosphorylase kinase, protein kinase C and protein kinase A by caffeic acid and a procyanidin-rich pine bark (*Pinus maritima*) extract. Biochim Biophys Acta 2000; 1474:219–225.

18. Petrassi C, Mastromarino A, Spartera C. Pycnogenol in chronic venous insufficiency. Phytomedicine 2000; 7:383–383.

19. Virgili F, Pagana G, Bourne L, Rimbach G, Natella F, Rice-Evans C, Packer L. Ferulic acid excretion as a marker of consumption of a French maritime pine (*Pinus maritima*) bark extract. Free Radic Biol Med 2000; 28:1249–1256.

12

Cytoprotective and Cytotoxic Effects of Flavonoids

Jeremy P. E. Spencer and Catherine A. Rice-Evans
King's College London
London, England

Hagen Schroeter
University of Southern California School of Pharmacy
Los Angeles, California, U.S.A.
King's College London
London, England

I. INTRODUCTION

Flavonoids have been ascribed a wide range of beneficial properties related to human health, including cancer [1–6]; cardiovascular diseases [7], including coronary heart disease [2,4,8] and atherosclerosis [9]; inflammation [4,10]; and other diseases in which an increase in oxidative stress has been implicated [11–14]. Their ability to act as classical electron- (or hydrogen-) donating antioxidants in vitro has been intensively reported [15–17] and used to explain their protective effects against oxidative stress. Structurally important features that define this antioxidant activity are the hydroxylation pattern, in particular a $3',4'$-dihydroxy catechol structure in the B-ring and the presence of 2,3 unsaturation in conjugation with a 4-oxo-function in the C-ring [17]. The antioxidant efficacy of flavonoids has been described for protection against oxidative damage to a variety of cellular biomolecules. For example, flavonoids inhibit the oxidation of low-density lipoprotein [18–22] and deoxyribonucleic acid (DNA) [23–26] in vitro. In addition, flavonoids are effective scavengers of reactive nitrogen species in the form of peroxynitrite [27–30] and limit dopamine oxidation mediated by peroxynitrite in a structure-dependent way involving oxidation or nitration of the flavonoid ring system [31]. Furthermore, their antioxidant properties have also been attributed to their abilities to chelate transition metal ions [32–35] and their potential to quench singlet oxygen [36,37].

The ability of flavonoids to act as effective antioxidants in vivo is dependent on the extent of their conjugation and biotransformation on absorption. Flavonoids are substrates for uridine diphosphate–(UDP)-glucuronosyl transferases, catechol-O-methyl transferases, and sulfotransferases in the small intestine and liver [38–43] and can also be degraded to secondary phenolic acid metabolites in the colon [38–40] (Chaps. 14 and 17). Most importantly it is now clear that those flavonoids with the most pronounced antioxidant activity, as demonstrated in vitro, are those that are metabolized to the greatest extent in vivo [41]. This phase I/II metabolism effectively lowers the antioxidant activity of the flavonoid either by affecting the B-ring catechol by O-methylation or by glucuronidating the A-ring, thereby lowering hydrogen-donating ability and significantly increasing the rate of renal excretion. These observations have important implications for the

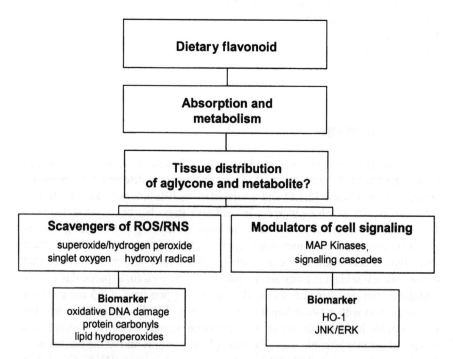

Figure 1 Flavonoids and their in vivo metabolite forms may act by direct reaction with oxidizing species in the body, thereby reducing the accumulation of end products of oxidative damage in the cell, or by modulation of intracellular signaling events.

determination of their mechanism of protection against oxidative stress in vivo and have provided evidence that their mode of protection in vivo may involve processes that are independent of their classic antioxidant effects in scavenging of reactive oxygen and reactive nitrogen species (Fig. 1).

This chapter examines the large body of evidence that demonstrates the potential of flavonoids to be both cytoprotective against oxidative *stress* (Sec. II) and cytotoxic (Sec. III) in vitro. Many of these initial cytoprotection studies characterized the flavonoids as direct scavengers of free radical species, thereby reducing the levels of oxidative damage to cell biomolecules (Sec. VI) and attenuating the overall loss of cell function. More recent studies have also highlighted that they may also act by mechanisms independent of classical H-donating antioxidant reactions, such as specific interactions within intracellular signalling pathways (Sec. IV). With all of these studies, however, caution must be used in interpreting the results since in many instances, with the exception of topical skin applications or effects in the oral cavity and gastrointestinal tract, one must question whether the flavonoids applied are present in the circulation in that form and thus exposed to cells in vivo. With this in mind, the effects of the more physiologically relevant in vivo flavonoid metabolites on cells are also reviewed (Sec. V).

II. PROTECTION AGAINST OXIDATIVE STRESS–INDUCED CELL DAMAGE

The precise mechanisms by which flavonoids may protect different cell populations from oxidative insults are currently unknown. However, potential mechanisms that involve their classical antioxidant properties, interactions with mitochondria, modulation of intracellular signaling cascades, and stimulation of adaptive responses have been proposed. Flavonoids, such as quercetin, have been shown to be cytotoxic in many cell systems by mechanisms involving the production of oxygen radicals through an auto-oxidation process dependent on pH and the presence of oxygen. For example, quercetin has been observed to initiate an adaptive response in low doses with the effect of protecting the cells against higher doses of quercetin and other compounds, such as hydrogen peroxide and mitomycin C [44]. However, most studies have not highlighted an exact mechanism for the protective effects observed against oxidative cell injury; often they have suggested that classical antioxidant processes are involved. Other possible mechanisms might include an ability to increase intracellular glutathione (GSH), the prevention of Ca^{2+} influx in the presence of high levels of reactive oxygen species (ROS) [45], and direct interactions within cell signaling cascades and with adenosine triphosphate (ATP) binding sites (discussed in Sec. III).

A. Protection Against Peroxide and Other Reactive Oxygen Species

A number of studies have investigated the ability of flavonoids to protect against neuronal death induced by oxidative stress [46–50] (Chap. 9). For example, flavonoids such as epicatechin and quercetin have been shown to reduce the neurotoxicity induced by oxidized low-density lipoprotein [50–53]. There has been an equivalent amount of interest in the ability of flavonoids to protect against oxidative injury, particularly that induced by hydrogen peroxide, in a variety of cell systems. For example, the flavanol epicatechin has been shown to evoke strong protection against cell damage, caspase-3 activation, and binding of annexin V-CY3.18 (an early marker of apoptosis) induced by hydrogen peroxide in dermal fibroblasts [54,55] and primary cortical neurons [55]. The protective effects of epicatechin were apparent at low micromolar concentrations, similar to those reported to be detectable in human plasma after oral ingestion of epicatechin [56] or green tea [57,58], and were independent of direct interactions between the flavonoid and the peroxide. These findings are consistent with those of other investigators who have shown that the other main flavanol, catechin, evokes significant protection against fibroblast (3T3 Swiss) damage induced by high levels of reactive oxygen species (ROS) generated extracellularly by xanthine-xanthine oxidase [59] and against hydrogen peroxide–induced damage to cultured rat hepatocytes (BL-9) [60]. The protective effect of catechins has also been studied in cultured human umbilical vein endothelial cells exposed to toxicity induced by linoleic acid hydroperoxide (LOOH) [61]. Catechins interacted with LOOH present in the medium or near the surface of membranes, but not with LOOH incorporated into cellular membranes. In addition, the interaction of catechin with α-tocopherol may provide synergistic protection against the cytotoxicity of LOOH [61].

The flavonol quercetin has been the subject of much interest in terms of its beneficial properties against oxidative stress [62] and has been shown to exert a potent protective actions against hydrogen peroxide–induced apoptosis of rat thymocytes [63] and cell death in rat hepatocytes (BL-9) [60], as well as reducing oxidative stress and cell damage in liver tumor cells induced by AAPH (2,2-azobis(2-aminopropane) dihydrochloride) [64]. Quercetin and another flavonol, kaempferol, as well as catechin and the flavone taxifolin, have been observed to suppress the cytotoxicity of $O_2^{\bullet-}$ and H_2O_2 to Chinese hamster V79 cells, as assessed by the ability of the flavonoids to prevent the decrease in the number of cell colonies induced by the oxidants [65]. Furthermore, quercetin has been shown to protect cutaneous tissue-associated cells (human skin fibroblasts, keratinocytes, and endothelial cells) from oxidative injury induced by buthionine sulfoximine (BSO), an inhibitor of GSH synthesis [66].

Baicalein (5,6,7-trihydroxy-2-phenyl-4H-1-benzopyran-4-one), a naturally occurring flavone, has also been observed to prevent human dermal fibroblast cell damage induced by reactive oxygen species such as hydrogen peroxide, tert-butyl hydroperoxide, and superoxide anions in a concentration-dependent manner [67]. Interestingly, the protection evoked by baicalein was more effective than that of the iron chelator deferoxamine, the hydroxyl radical scavengers dimethyl sulfoxide and ethanol, the lipid peroxidation chain blocker α-tocopherol, and the xanthine oxidase inhibitor allopurinol, indicating that the protective effects may involve a more complex mechanism than simple direct oxidant scavenging. Furthermore, *Scutellaria baicalensis Georgi*, which also contains baicalein as well as the other flavones, wogonin, wogonoside, and baicalin, has been found to be effective in protecting human neuroblastoma SH-SY5Y cells against hydrogen peroxide–induced cell damage and lipid peroxidation [68,69].

The flavone glycoside diosmin and its aglycone form, diosmetin, are reported to be effective inhibitors of cell membrane damage in primary cultures of rat hepatocytes induced by erythromycin estolate and oxidative stress caused by tert-butylhydroperoxide (tBH) [70]. Diosmetin, but not diosmin, protected against tBH toxicity, and this protection was related to a decreased extent of lipid peroxidation and reduced loss of glutathione. Metal complexes (Fe^{2+}, Fe^{3+}, Cu^{2+}, Zn^{2+}) of the quercetin glycoside, rutin, and epicatechin have been found to be more potent than the parent flavonoids in protecting red blood cells against asbestos-induced injury [71], suggesting that flavonoid metal complexes may be an additional effective therapy for the inflammatory response associated with the inhalation of asbestos fibers.

These studies highlight the abilities of flavonoids, in particular quercetin and epicatechin, to reduce oxidative stress and cell death induced by ROS in different cell systems. Although it is likely that flavonoids can reduce oxidative stress–induced damage to cells by direct interactions with the oxidants in the extracellular environment, a few studies suggest that other mechanisms are involved, as protection is also observed when the oxidant and flavonoid are not present together in the culture medium. In this case, intracellular interactions between oxidant and flavonoid may explain the protection observed. However, as will be outlined later, other nonantioxidant mechanisms have also been proposed.

B. Protection Against Ultraviolet A/Ultraviolet B–Induced Cell Damage

Ultraviolet (UV) radiation–induced oxidative stress in skin cells [72,73], as well as infiltration of leukocytes and depletion of antigen-presenting cells [74], plays an important role in the induction of immune suppression [74,75], DNA damage [76] and photocarcinogenesis [76,77]. In particular, ultraviolet B (UV-B) exposure may result in oxidative stress and skin injury that has been associated with a variety of skin diseases including photoaging [78], inflammation [74,75,79–81], and cancer [75,76]. Many investigations have shown that topical treatment with [82–85] or oral consumption [86] of green tea polyphenols inhibits UV-B-induced skin tumorigenesis in different animal models. The most chemopreventive constituent in green tea, proposed to be responsible for the biochemical or pharmacological effects observed in these studies, is (−)−epigallocatechin-3-gallate, although treatment of the skin with green tea polyphenols in general has been shown to modulate the biochemical pathways involved in inflammatory responses, cell proliferation, and responses of chemical tumor promoters, as well as ultraviolet light–induced inflammatory markers of skin inflammation [82,84,85].

The protective effects of green tea treatment on human skin either topically or consumed orally against UV light–induced inflammatory or carcinogenic responses are not well understood. Pretreatment of cultured fish cells with quercetin inhibited both the photoinduction of cyclobutane pyrimidine dimer photolyase and enhancement of cellular attachment in a similar manner to that of H7 (1-(5-isoquinoline solfunil)-2 methlypiperazine), a strong inhibitor for protein kinase C [87], suggesting that interactions in cell signaling processes might be involved. Epigallocatechin gallate (EGCG) has been shown to exert preventive effects against photocarcinogenesis and phototoxicity in mouse models. The topical application of EGCG to human skin before UV irradiation (four times minimal erythema dose) caused decreases in UV-induced production of hydrogen peroxide (68–90%) and nitric oxide (30–100%) in both the epidermis and the dermis in a time-dependent manner [85]. EGCG pretreatment also inhibited UV-induced infiltration of inflammatory leukocytes, particularly CD11b(+) cells (a surface marker of monocytes/macrophages and neutrophils), into the skin, reduced UV-induced epidermal lipid peroxidation, was found to restore the UV-induced decrease in GSH level, and protected against the decrease in glutathione peroxidase activity. The topical application of EGCG to C3H/HeN mice before a single dose of UV-B (90 mJ/cm^2) exposure inhibited UV-B-induced infiltration of leukocytes, specifically the CD11b+ cell type, and myeloperoxidase activity, a marker of tissue infiltration of leukocytes [82]. EGCG treatment was also found to prevent UV-B-induced depletion in the number of antigen-presenting cells when immunohistochemically detected as class II major histocompatibility complex (MHC+) la+ cells. These data suggest that topical applications of EGCG to skin may afford some protection against UV-B-induced immunosuppression, photoaging, inflammatory dermatoses, and photocarcinogenesis.

III. PRO-OXIDANT EFFECTS OF FLAVONOIDS IN CELL SYSTEMS

The huge interest in flavonoids and isoflavones, fueled by their powerful antioxidant and estrogenic effects, respectively, has led to their proposed use as anticarcinogens and cardioprotective agents, prompting a dramatic increase in their consumption as dietary supplements. However, there are a number of studies that highlight the potentially toxic effects of excessive flavonoid intake [3,88–91]. For example, at higher doses (high micromolar to low millimolar) in cell studies, flavonoids may act as cytotoxic agents [91]; as mutagens [92]; as pro-oxidants that generate free radicals [90,93,94]; as inducers of apoptosis [3,95]; and as inhibitors of key enzymes involved in hormone metabolism. Flavonoids containing a free hydroxyl group at position 3 of the C ring, a free hydroxyl group at position 7 of the A-ring, and a B ring with a catechol (quercetin) or pyrogallol structure (epigallocatechin or epigallocatechin gallate), or a structure that after metabolic activation is transformed into a catechol or a pyrogallol, have been suggested to be the most genotoxic to eukaryotic cells. However, these cytotoxic effects are concentration-dependent as flavonols such as quercetin and myricetin are damaging to cells at much lower concentrations than flavanols such as epicatechin or non-catechol-containing flavanones such as hesperetin. This finding is in agreement with an induction of revertants in *Salmonella typhimurium* TA98 and the induction of chromosomal aberrations in V79 cells by flavonols such as quercetin [92]. Quercetin, in particular, although possessing several biological activities that can be useful in protection against oxidative stress [60,63], has been discussed as a compound that is potentially cytotoxic and/or genotoxic to normal or cancer cells [96–104].

A. Generation of Reactive Oxygen Species by Flavonoids in Cell Cultures Systems

Recently there has been interest in the ability of some flavonoids to generate cytotoxic amounts of hydrogen peroxide or express a pro-oxidant nature in some cell culture models. In cell culture, chemical reactions between media and flavonoid are often ignored, in particular the potential "artifactual" generation of ROS and hydrogen peroxide by reducing agents. It has been shown that addition of high concentrations of strong reducing agents (high micromolar to millimolar), such as flavonoids, to cell culture media can lead to generation of substantial amounts of H_2O_2 (10–100 µM) [93]. This effect has led to concern that some of the reported effects of flavonoids such as quercetin and epigallo-catechin gallate on cells in culture may be due to H_2O_2 generation, produced as a result of the interaction of these compounds with cell culture media. This compound-dependent generation may be viewed both negatively in that peroxide is a known inducer of both apoptosis and necrosis in cells [50,55] and positively in that small amounts of generated peroxide may result in an adaptive response in cells that would act to protect them against a subsequent oxidative insult. It should be noted, however, that artifactual generation of peroxide is negligible when flavonoids are exposed at physiologically relevant low concentrations in media containing transferrin [50,55].

As mentioned, the ability of flavonoids to generate peroxide seems to be dependent on the presence of a pyrogallol or catechol structure; the former flavonoids generate more peroxide than the latter [in decreasing ability to generate peroxide: myricetin > baicalein > quercetin > (-)-epicatechin > (+)-catechin > fisetin = 7,8-dihydroxy flavone] [94] (Fig. 2). The amount of peroxide generated by myricetin (pyrogallol-type flavonoid) was proportional to its concentration and was dependent on the amount of dissolved oxygen in the bufier and was inhibited by the addition of superoxide dismutase [94]. These results suggest that under specific culture conditions, where oxygen saturation is high, some flavonoids may generate small amounts of hydrogen peroxide by donation of electrons from their pyrogallol or catechol structures to oxygen, to form the superoxide anion radical and subsequently peroxide.

Other investigations have provided evidence that flavonoids may also induce DNA damage in cultured cells [105,106]. For example, morin and naringenin [106] and kaempferol [107] induce a concentration-dependent peroxidation of nuclear membrane lipids and DNA strand breaks in isolated rat liver nuclei [106]. Quercetin and myricetin have been shown both to inhibit growth of and cause dose-dependent increases in DNA strand breakage in Caco-2, HepG2, HeLa cells, and normal human lymphocytes [24]. None of the flavonoids was observed to induce DNA base oxidation above normal cellular levels, although all caused a depletion of reduced glutathione, which, in the case of quercetin, occurred before cell death. These studies demonstrate the pro-oxidant activities of polyphenolic flavonoids, which are generally considered as antioxidants and anticarcinogens, and may suggest their possible dual role in mutagenesis and carcinogenesis.

B. Pro-Oxidant Effects and Anticarcinogenesis

The ability of flavonoids to act as pro-oxidants in vitro and in cell systems has also been suggested to be a potential mechanism by which they may act as anti-carcinogenic compounds in vivo as a result of their abilities to promote death of cancer cells. Quercetin has been thoroughly investigated for its abilities to express antiproliferative effects [100,103,108,109] and induce death predominantly by an apoptotic mechanism in cancer cell lines [95,97–99,102–105,110–116]. For example, the exposure of quercetin and the isoflavone genistein to the colonic

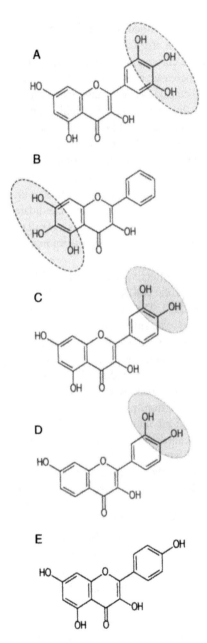

Figure 2 Structures of some common flavonols: (A) myricetin, (B) baicalein, (C) quercetin, (D) fisetin, and (E) kaempferol. Light shading highlights the presence of a pyrogallol structure, whereas dark shading highlights the presence of a catechol function. The ability to generate hydrogen peroxide decreases from (A) to (E).

cancer cell lines, Caco-2, HT-29, and rat nontransformed intestinal crypt cells, IEC-6 induced strong chromatin condensation, a marker of apoptosis [109]. Furthermore, quercetin has been found to induce chromatin and nuclear fragmentation in human leukemia HL-60 cells [115], increase caspase-3 activation in the malignant cell line HPB-ALL [113] and HL-60 cells [104], and cause activation of caspase-9 and release of cytochrome c in HL-60 cells [104]. The inhibition of cancer cell proliferation by quercetin may proceed via an inhibition of epidermal growth factor (EGF) receptor phosphorylation, an integral part of the proliferation mechanism in cultures of colonic tumor cells [114], or via inhibition of cell cycle progression through transient M phase accumulation and subsequent G2 arrest, as was observed in MCF-7 human breast cancer cells [111]. The many studies with quercetin in this area would suggest that this flavonol is a promising candidate chemotherapy in vivo. However, for bioavailability data suggest that as a result of the extensive metabolism of quercetin in the small intestine and liver the potential for action in vivo may be limited [41].

Other flavonoids have also been shown to be effective in cancer cell killing. The flavone apigenin is capable of inducing apoptotic cell death characterized by DNA fragmentation and activation of caspases [95]. The involvement of hydrogen peroxide in the mechanism for cytotoxicity was ruled out in this study as catalase failed to eliminate the cytotoxic effects. A synthetic flavone acetic acid (FAA) that has been reported to have antitumor activity against a variety of transplanted tumors in mice inhibits the proliferation of endothelial cells by a superoxide-dependent mechanism and induces apoptosis by a nitric oxide– and superoxide-independent mechanism [117]. In addition, pycnogenol, a preparation derived from pine bark, which contains high amounts of procyanidins [13], selectively induces death in human mammary cancer cells (derived from human fibrocystic mammary tissue) but not in normal human mammary MCF-10 cells [118], indicating that flavonoids may be of use for the selective killing to tumor cells in vivo. Wine polyphenols, which include epicatechin, have been shown to have a direct inhibitory effect on the proliferation of human prostate cancer cell lines that was found to be mediated by the production of NO [119]. Furthermore, two dietary flavonoids isolated from the leaves of *Morus alba* (Moraceae), quercetin-3-*O*-β-D-glucopyranoside and quercetin-3,7-di-*O*-β-D-glucopyranoside, were found to possess a significant inhibitory effect on the growth of the human promyelocytic leukemia cells (HL-60) [120]. Genistein and quercetin have been found to inhibit the proliferation and migration of ras-oncogene-driven tumor cells, rat breast adenocarcinoma, and human bladder carcinoma cell lines [121]. Baicalin and baicalein have been shown to induce apoptosis in the androgen-positive and -negative human prostatic carcinoma cell lines LNCaP and JCA-1 [116].

Resveratrol (3,4',5-trihydroxy-*trans*-stilbene) (Fig. 3), a common phytoalexin found in grape skins, peanuts, and red wine, has been speculated to

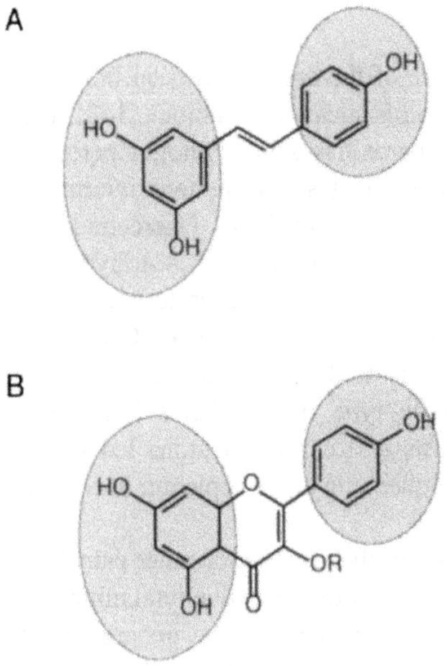

A

B

Figure 3 Structure of resveratrol. Highlighted areas show its structural similarity with the flavonoid kaempferol in both the A-ring and the B-ring.

act as an antioxidant, promote nitric oxide production, inhibit platelet aggregation, increase high-density lipoprotein cholesterol, and thereby serve as a cardioprotective agent [122]. It has structural similarity to certain flavonoids such as kaempferol in that its structure mimics both the A-ring and the B-ring of the flavonoid (Fig. 3). In 2001, it was proposed that resveratrol can function as a potent cancer chemopreventive agent, and there has been a great deal of experimental effort directed toward defining this effect [122,123]. Resveratrol has been shown to be a potent inducer of apoptosis in many different cancer cell models [124–132] and is also capable of producing strong growth inhibitory effects on cancer cells, causing both cell cycle arrest [132–134] and inhibition of cell proliferation [135–138]. However, in 2002 resveratrol was found to have no growth-inhibitory effects on 4T1 breast cancer in vivo [139], even though it is a potent inhibitor of 4T1 breast cancer cells in vitro [139]. This lack of activity in vivo may be due to the extensive metabolism of resveratrol that occurs on absorption to produce glucuronidated [140,141] and sulfated [136,142] forms that may possess different bioactivity against cancer cells. At present there are no data to suggest that these metabolites are capable of cancer chemopreventive effects similar to those of resveratrol.

C.   Depletion of Cellular Thiols

A possible mechanism of the cytotoxic nature of flavonoids has been linked to their potential intracellular interactions with thiols such as glutathione [143,144]. Catechin oxidation in the presence of cellular peroxidases and hydrogen peroxide leads to the generation of monoglutathione conjugates (via quinone intermediates), which has the effect of lowering cellular GSH levels [144]. Quercetin [145] and other flavonoids such as taxifolin, luteolin, fisetin, and 3,3′,4′-trihydroxy-flavone have also been found to react with GSH to generate mono-and diglutathionyl adducts [146–148]. Interestingly, the conjugation of GSH with catechin occurred on the B-ring [144], in contrast to the A-ring of quercetin, as a result of the formation of quinone methide–type intermediates [145,147]. Quercetin exposure to both dermal fibroblasts and cortical neurons led to the rapid formation of glutathionyl conjugates of quercetin and a depletion in cellular thiols (Spencer et al., in preparation).

Similar chemical reactions are involved in the reaction of other catechols such as the catecholamines, dopamine, and L-dihydroxyphenylalanine (L-DOPA) with cysteine or GSH [149–152] and can lead to the generation of mitochondrial toxins with relevance to Parkinson's disease [153–155]. As well as possible cytotoxic effects of lowering cellular thiol levels or binding to cysteine residues at the active site of specific enzymes, flavonoids could act beneficially by acting to limit the formation of the potentially cytotoxic catecholamine-thiol adducts, in a manner similar to that observed for dihydro-lipoic acid [151]. Because of the structural similarity of the flavonoid B-ring and their ability to donate electrons efficiently to form quinones, it is conceivable that specific flavonoids may be of use to prevent such neurotoxic compounds as 5-S-cysteinyl dopamine from forming in vivo.

IV.   ANTIOXIDANT EFFECTS INDEPENDENT OF CLASSICAL REDUCING PROPERTIES

Many studies have demonstrated that flavonoids have mechanisms of action independent of their conventional hydrogen-donating free radical scavenging properties by modulating enzyme activities, interfering with pathways of intermediary metabolism, downregulating the expression of adhesion molecules, acting at various sites within signal transduction cascades, and mimicking substrates for various binding sites [4,156–159]. Apigenin has been identified as a flavonoid that effectively blocks intercellular adhesion molecule 1 (ICAM-1) upregulation and leukocyte adhesion in response to cytokines in vitro. Tumor necrosis factor–(TNF)-induced ICAM-1 upregulation in vivo effectively is blocked by apigenin through a mechanism that is unrelated to free radical scavenging or leukocyte function [156]. Furthermore, the adhesion of leukocytes to cytokine-treated endothelial cells was blocked in cells cotreated with apigenin [157]. Epigallocatechin strongly reduces ultraviolet-A-induced heme oxygenase 1 activation in skin-derived fibroblasts, however, it was also observed to activate collagenase and cyclooxygenase expression (635), indicating that the effect of this green tea polyphenol on cellular stress responses is complex and may involve direct effects on signal transduction as well as changes that may be associated with its antioxidant activity.

The potential role flavonoids play in intracellular signaling events triggered in response to oxidative stress is becoming increasingly clear, especially with regard to the influence they have on protein kinases, phosphatidylinositol-3 kinase, and nuclear factor-κB (NF-κB) [160,161]. Among the identified signal transducers affected by flavonoids are phosphoinositide 3-kinase (PI-3K) and protein kinase C (PKC), which are key players in many cellular responses, including cell multiplication, apoptosis, and transformation. The strong inhibitory effects of quercetin on both NF-κB binding activity and oxidative deoxyribonucleic acid (DNA) damage suggest that its antioxidant activity may outweigh its oxidative potential in the cellular environment, which might contribute to quercetin's reported anticarcinogenic and anti-inflammatory effects. Quercetin has also been reported to be an inhibitor of phospholipase A2 (PLA2) and lipoxygenase and effectively inhibits the oxidative stress–induced expression of heat shock protein 68 (Hsp68) [162]. Accumulating evidence also links flavonoids to interactions within mitogen-activated protein kinase (MAPK) signaling cascades [50,163–166]. For example, quercetin downregulates both phorbol 12-myristate 13-acetate (PMA) and tumor necrosis factor-α (TNF-α) expression in human endothelial cell line ECV304 (ECV) via inhibiting both activator protein 1 (AP-1) activation and the Jun Kinase (JNK) pathway [163]. In addition, quercetin has been observed to exert a significant inhibitory effect on 4-hydroxy-2-nonenal– (HNE)-induced JNK activation [167]. It has been proposed that low concentrations of flavonoids, such as quercetin, may activate the MAPK pathway [extracellular signal-regulated kinase (ERK2), JNK1, p38], leading to expression of survival genes (c-Fos, c-Jun) and defensive genes (phase II detoxifying enzymes; glutathione-S-transferase, quinone reductase), resulting in survival and protective mechanisms (homeostasis response). However, increasing the concentrations of these compounds additionally activates the caspase pathway, leading to apoptosis [164]. The *Gingko biloba* extract, EGb 761, which is rich in flavonoids, has been observed to suppresses c-fos messenger ribonucleic acid (mRNA) expression, followed by AP-1 DNA activation and in Jurkat T cells, suggesting that the step in the signal transduction pathway for AP-1 activation that is inhibited by EGb 761 is upstream to c-fos mRNA expression [168].

Recent interest has been directed to the use of flavonoids and flavonoid derivatives as benzodiazepine receptor (BDZ-R) ligands [159]. Benzodiazepines are the most widely prescribed class of psychoactive drugs in current therapeutic use but have unwanted side effects such as sedation, myorelaxation, ataxia, amnesia, and ethanol and barbiturate potentiation, and tolerance, and it is postulated that a new family of ligands based on flavonoids might prevent these side effects. First isolated from plants and used as tranquilizers in folkloric medicine, some natural flavonoids have been shown to possess a selective and relatively mild affinity for BDZ-Rs and have a pharmacological profile compatible with a partial agonistic action. In a logical extension of this discovery, of various synthetic derivatives of those compounds, such as 6,3'-dinitroflavone, have been found to have a very potent anxiolytic effect not associated with myorelaxant, amnestic, or sedative actions [159]. Because of their selective pharmacological profile and low intrinsic efficacy at the BDZ-Rs, flavonoid derivatives, such as those described, could represent an improved therapeutic tool in the treatment of anxiety.

Catechins and specific flavonols have been reported to interact with proteins such as the mitochondrial adenosine triphosphatase (ATPase) [169], calcium plasma membrane ATPase [170], protein kinase A [171–173], and protein kinase C [160,174–176] through binding to the ATP binding site [177]. Quercetin was found to cause a 50% inhibition of calcium transport at a concentration of 15 μM; morin and rutin had similar effects at concentrations approximately 10-fold higher [170]. The order of inhibitory potency of the flavonoids on the Ca^{2+}-transport ATPase from synaptosomal vesicles could be linked to their solubility in the membrane lipid phase, since, in addition, quercetin exhibited strong inhibition of phosphorylation of membrane proteins by ATP in synaptosomal vesicles, whereas rutin and morin had no significant effect [170]. Furthermore, certain flavonoids have been shown to be very potent and highly selective inhibitors of cyclin-dependent kinases (CDKs), which regulate the initiation, progression, and completion of the cell cycle, and are thus critical for cell growth [177]. As tumor development is closely associated with genetic alteration and deregulation of CDKs and their regulators, these flavonoid inhibitors of CDKs may act as useful anticancer therapeutics in the future.

Epicatechin, epigallocatechin, epicatechin gallate, and epigallocatechin gallate have been shown to reduce the ability of macrophages to bind oxidized erythrocytes, suggesting that the activity of lectinlike receptors of macrophages for oxidized erythrocytes may be regulated by oxidative mechanisms [178]. The same green tea catechins have also been shown to affect pro- and anti-inflammatory cytokines produced by human leukocytes in vitro. The production of interleukin-1β was decreased and that of interleukin 10 enhanced in leukocytes treated with EGC, ECG, and EGCG, but there was no effect on the production of interleukin-6 or tumor necrosis factor-α [179]. Although these effects suggest anti-inflammatory properties of the tea-derived catechins, they were observed at concentrations that were unlikely to be achievable in plasma in vivo and are therefore unlikely to contribute to the protective effects of tea-derived flavonoids in inflammatory diseases. The antioxidative activity of the tea polyphenols theaflavin, theaflavin-3-gallate, theaflavin-3,3'-digallate, (-)-epigallocatechin-3-gallate, and gallic acid has been shown to be due not only to their ability to scavenge ROS and superoxide in HL-60 cells but also to their ability to block xanthine oxidase and related oxidative signal transducers [180].

It has also been suggested that the protective mechanism of flavonoids may include the inhibition of enzymatic functions other than oxidases, e.g., the inhibition of lipoxygenase and thus prevention of the formation of leukotrienes [181–185]. Analysis of 12 aglycone flavonoids showed that inhibitory potency and selectivity against 5-lipoxygenase are conferred by the presence of a catechol group in ring B as part of a 3,4-dihydroxycinnamoyl structure [186]. In contrast, "cross-over" of inhibitory selectivity is observed in compounds containing few hydroxyl substituents (with None in ring B), which are selective against cyclo-oxygenase. Other studies indicate that the flavonoids such as taxifolin, eriodic-tyol, hesperetin, and luteolin can inhibit myeloperoxidase (MPO) release from activated human neutrophils, and taxifolin and eriodictyol also strongly inhibit MPO activity [187]. Furthermore, flavonoids that are methylated at a single OH group in the B-ring are only inhibitory when they react with activated neutrophils in the presence of myeloperoxidase. Early investigations demonstrated that quercetin, apigenin, and taxifolin (dihydroquercetin) influence anti–immuno-globulin E–(IgE)-induced histamine release and hydrogen peroxide generation in basophil-containing leukocyte suspensions [188,189] and may impair the oxidative burst of neutrophils to an extent dependent on their hydrophobicity [190,191]. For example, quercetin inhibited the generation of superoxide anions by neutrophils [191].

Flavonoids have also been investigated for their ability to influence the normal functioning of mitochondria and in particular the mitochondrial respira-tory chain complexes [192–195]. The flavonoids robinetin, rhamnetin, eupatorin, baicalein, 7,8-dihydroxyflavone, and norwogonin all inhibited beef heart mito-chondrial succinoxidase and NADH-oxidase activities and, in every case, the concentration required to inhibit low-density lipoprotein oxidation by 50% (IC_{50}) observed for the NADH-oxidase enzyme system was lower than for succinox-idase activity, demonstrating a primary site of inhibition in the complex I reduced nicotinamide-adenine dinucleotide–(NADH–coenzyme Q reductase) portion of the respiratory chain [192,193]. In addition, the flavonoids with adjacent tri-hydroxyl, *para*-dihydroxyl, or ortho-dihydroxl (catechol) groups exhibited a substantial rate of auto-oxidation, which was accelerated by the addition of cyanide, demonstrating that the CN-/flavonoid interaction can generate super-oxide nonenzymatically.

Luteolin and quercetin have been shown to be potent antileishmanial agents and may therefore be useful in the chemotherapy of leishmaniasis, which is a major concern in developing countries. Both luteolin and quercetin inhibited the growth of *Leishmania donovani* promastigotes and amastigotes in vitro, inhibited DNA synthesis in promastigotes, and promoted topoisomerase II–mediated linearization of kDNA minicircles [196]. Quercetin has nonspecific effects on normal human T cells; however, luteolin was nontoxic, suggesting it could be a strong candidate for antileishmanial drug design. Another study suggested that the skin cancer chemopreventive effects of silymarin (a flavonoid drug from *Silybum marianum* used in liver disease) are mediated via impairment of EGFR signaling that ultimately leads to perturbation in cell cycle progression, resulting in the inhibition of proliferation and induction of growth arrest [197].

V. IN VIVO FLAVONOID METABOLITES

At present, studies investigating the bioactivity of biologically relevant flavonoid metabolites, which include glucuronides and O-methylated forms, are few [54,55]. However, a number of investigations from 1998 to 2001 studied the ability of the biologically relevant metabolites of quercetin to provide protection against lipid peroxidation, in particular the oxidation of low-density lipoprotein (LDL) [198–201]. Both quercetin and quercetin 3-O-β-D-glucuronide were found to protect against peroxynitrite-induced consumption of endogenous lycopene, β-carotene, and α-tocopherol, suggesting that dietary quercetin, and one of its major in vivo forms, may be capable of inhibiting peroxynitrite-induced oxidative modification of LDL in the circulation [198]. Other studies have also helped to strengthen the proposal that this conjugated quercetin metabolite retains the ability to protect cellular and subcellular membranes from peroxidative attack by reactive oxygen species and peroxidative enzymes [199,201].

There are also a few studies that have studied the influence of metabolism on the bioactivity of flavonoids in cell models [50,54,55,202,203]. The ability of glucuronidated epicatechin (epicatechin-7-and epicatechin-5-O-β-D-glucuronides) to protect neurons and fibroblasts against oxidative stress–induced cell death has been investigated [55]. A mixture of epicatechin-5-O-β-D-and epicatechin-7-O-β-D-glucuronides exhibits no significant protection against peroxide-induced loss in neuronal or fibroblast viability and fails to prevent peroxide-induced caspase-3 activation in both cell models [55] (Fig. 4). This lack of activity may be based on the increased polarity of the glucuronide and its reduced ability to partition, which would limit its access to cells. This conclusion is consistent with the observation that uptake of epicatechin glucuronide into both cortical neurons and dermal fibroblasts was not detectable [55]. Another possibility is that the A-ring is the structurally important feature for cell recognition or biological action within cells and in vivo, and presence of a bulky glucuronide on the A-ring limits its activity. However, in vivo, the

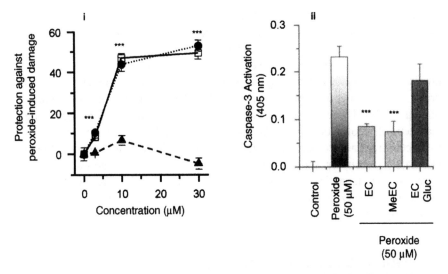

Figure 4   Plot 1: Protection against peroxide-induced cell damage by epicatechin (□), 3′-O-methyl epicatechin (●), and epicatechin glucuronide (▲). Plot 2: Increased activity of caspase-3-like protease activity induced by hydrogen peroxide (50 μM) and inhibition by pretreatment with epicatechin, 3′-O-methyl epicatechin, and epicatechin glucuronides.

possibility exists that epicatechin and other phenolic glucuronides may be cleaved under local conditions of inflammation and the free aglycone released to express cellular effects. Indeed, β-glucuronidases are present in a number of tissues within the body [204] and may be released by certain cells. For example, histamine causes rapid exocytosis of β-glucuronidase from lung macrophages [205] and luteolin monoglucuronide has been shown to be cleaved to free luteolin by β-glucuronidase released from neutrophils stimulated with ionomycin [206,207].

In contrast, epicatechin and 3′-O-methyl epicatechin have been shown to elicit strong cytoprotective effects in fibroblasts and neurons [54,55] (Fig. 4) and are associated with both cell types [55]. In addition, 3′-O-methyl-epicatechin was as effective as epicatechin in protecting neurons against oxidized LDL–(Ox-LDL)-induced activation of JNK, c-jun, and procaspase-3 [50]. The ability of the O-methylated metabolite to exert protection similar to that of the native epicatechin is interesting as H-donating potential of the two compounds is very different, as the O-methylated compound has a reduced capacity to donate electrons from its B-ring [54]. This evidence would suggest that flavonoids may function to protect cells against death induced by oxidants by a mechanism independent of their classic antioxidant properties, a concept that is beginning to receive much attention.

However, in vivo demethylation may occur intracellularly by the action of cytochrome P_{450} enzymes that cleave the O-methylated metabolite to epicatechin.

The in vivo metabolites of eriocitrin (a flavonoid glycoside present in citrus)—eriodictyol, homoeriodictyol, methylated eriodictyol, 3,4-dihydroxyhydrocinnamic acid—and their conjugates were found in plasma and renal excreted urine after oral ingestion of eriocitrin [208]. After administration of eriocitrin, plasma exhibited elevated resistance to lipid peroxidation, therefore suggesting that metabolites of eriocitrin may be capable of antioxidant action in vivo. 3-O-β-D-Glucuronide has also been suggested to act as an effective antioxidant in blood plasma low-density lipoprotein, because this conjugated metabolite was found to have a substantial antioxidant effect on copper ion–induced oxidation of human plasma low-density lipoprotein as well as 1,1-diphenyl-2-picrylhydrazyl radical-scavenging activity [203].

VI. REDUCTION IN CELL BIOMARKERS OF OXIDATIVE DAMAGE

As well as studies examining the role flavonoids play in reducing cellular injury and death induced by oxidative stress, there are many studies that have examined their ability to reduce biomarkers of oxidative damage to cells. Exposure of cells to an oxidative environment rapidly results in the increase over time of damaged products of DNA [209], lipid [210], and protein [211], and the amount and type of oxidative lesions induced are dependent on the oxidizing species [212]. These oxidative products of cellular biomolecules are often used as biomarkers of damage to cells by ROS and are also associated with the viability of the cell in general. Increases in damaged products may in turn lead to mitochondrial dysfunction and stimulation of signaling cascades, which may lead to induction of repair of such damage, on the one hand, or to the death by necrotic or apoptotic mechanisms, on the other hand (Fig. 5). The ultimate fate of the cell is dependent on the level of damage, but it is generally assumed that a reduction in these biomarkers of oxidative damage is beneficial to the cell. Flavonoids have been examined for their abilities to attenuate the oxidant-induced increase in damaged products; the section highlights some of the studies in which they seem to have beneficial effects.

A. Cellular Lipid Oxidation

Flavonoid pretreatment of cells may act to prevent the peroxidation of cellular membrane lipids. For example, the preincubation of cultured retinal cells with eriodictyol, luteolin, quercetin, and taxifolin evoked protection against Fe^{2+} ascorbate-induced lipid peroxidation and increases in intracellular oxidative stress

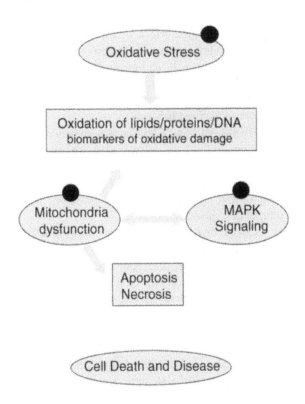

● Possible sites of flavonoid action

Figure 5 The potential sites of action of flavonoids in the prevention of cell damage. Oxidative stress results in the oxidation of lipids, proteins, and deoxyribonucleic acid (DNA), which is in turn sensed by signaling pathways, such as the MAPK signaling cascade and mitochondria. These events interlink to cause induction of apoptosis/necrosis in response to the initial oxidative insult. Flavonoids may act by direct scavenging of the oxidizing species or intermediate, by direct protein interactions with signaling molecules, or by interaction with mitochondria. MAPK, mitogen-activated protein kinase.

[213]. Addition of quercetin to culture cell medium enhanced the rate of growth of baby hamster kidney (BHK-21) cells and also diminished levels of lipid peroxidation breakdown products [214]. Furthermore, quercetin and three other flavonoids, hesperetin, naringenin, rutin, were found to be effective inhibitors of lipid oxidation in two different in vitro experimental models: (1) $Fe^{(2+)}$-induced linoleate peroxidation, by detection of conjugated dienes, and (2) auto-oxidation

of rat cerebral membranes, by use of thiobarbaturic acid for assay of free malondialdehyde production [215]. The ability to protect was observed to differ in each system: rutin > hesperetin > quercetin > naringenin in system (1); quercetin > rutin > hesperetin > naringenin in system (2), indicating the their inhibitory nature is related not only to their structural characteristics but also to their ability to interact with and penetrate the lipid bilayers.

The flavonoids, 7-monohydroxyethylrutoside, fisetin, and naringenin also significantly inhibit lipid peroxidation in rat liver microsomes and may be capable of replacing α-tocopherol as a chain-breaking antioxidant [216]. Catechin-rich extracts isolated from evening primrose seeds have also been shown to attenuate the oxidation of L-α-phosphatidylcholine liposomes and oxidation of leukemic L1210 murine cell membranes [217].

The *Ginkgo biloba* extract, EGb761, and its active constituents have been shown to be effective in attenuating lipid peroxidation, changes in sulfydryl group binding sites on the membrane proteins, and apoptosis induced by hydroxyl radicals in rat cerebellar granule cells [218]. Specifically, the total flavonoid component of EGb761 and a mixture of flavonoids and terpenes protected cells from oxidative damage and apoptosis; however, the total terpene fraction of EGb761 was not effective. The flavonoids baicalein and baicalin effectively inhibit lipid peroxidation of rat brain cortex mitochondria induced by Fe^{2+}-ascorbic acid, AAPH, or reduced nicotinamide-adenine dinucleotide phosphate (NADPH). Wogonin and wogonoside also show significant effects on NADPH-induced lipid peroxidation [69].

One study has reported that tea flavonoids and other polyphenols (theaflavin digallate, theaflavin, epigallocatechin gallate, epigallocatechin, and gallic acid) can inhibit arterial wall cell-mediated LDL oxidation [219]. Examination of the mechanism of action indicated that compounds such as theaflavin digallate may act by decreasing superoxide production in macrophages and by chelating iron ions.

In contrast to many studies that identify beneficial effects of dietary flavonoids against cellular lipid oxidation, the action of flavonoids on bovine leukemia virus–transformed lamb fibroblasts (line FLK) and HL-60 cells was accompanied by lipid peroxidation [90]. Their toxicity was partly prevented by iron chelator desferrioxamine and antioxidant N,N'-diphenyl-p-phenylene di-amine, a result that pointed to the involvement of oxidative stress in their cytotoxicity. Interestingly, the toxicity of quercetin was partly prevented by nontoxic concentrations of other flavonoids examined, thus suggesting potential neutralization of quercetin cytotoxicity by intake of flavonoid mixtures. In another study, supplementation of rat hepatocyte cultures with the flavonoid myricetin led to the formation of phenoxyl radical intermediates, as detected in intact cells by electron paramagnetic resonance (EPR) spectroscopy [220]. These phenoxyl radicals corresponded to one-electron oxidation products of myricetin and have been suggested to be potential mediators of cellular toxicity. However, myricetin was found to be able to inhibit lipid peroxidation induced by iron in hepatocyte culture, suggesting that the intermediate generation of phenoxyl radicals might also contribute to the antioxidant mechanism of myricetin.

B. Deoxyribonucleic Acid Oxidation

Exposure of cells to oxidative stress may lead to the generation of (1) oxidized DNA base lesions, such as 8-OH-guanine, and/or (2) single and double strand breaks, in both nuclear and mitochondrial DNA [212]. For example, exposure of human bronchial epithelial cells to H_2O_2 causes rapid increases in purine and pyrimidine base oxidation products, DNA strand breakage, and eventual loss of cellular viability [221,222]. Flavonoids have been implicated as effective inhibitors of DNA damage induced by a variety of oxidative [23,25,223] and nitrative stresses [26]. As with other cell studies involving flavonoids, the most extensively studied are the flavonols, in particular quercetin. Quercetin and rutin have been observed to protect Caco-2 and Hep G2 cells against hydrogen peroxide–induced DNA damage [25] as measured by the comet assay [25,224–226].

In agreement with these studies, quercetin (at concentrations above 10 μM) and myricetin (>100 μM) decreased oxidant-induced DNA strand breakage and oxidized pyrimidine bases in human lymphocytes exposed to hydrogen peroxide [23]. Another flavonol, myricetin, has been shown to protect against oxidative base DNA modification induced by ferric nitrilotriacetate (Fe-NTA) in primary rat hepatocyte cultures [227]. This reduction in DNA damage may result from an induction of the DNA repair capacity of hepatocytes after exposure to myricetin, which was suggested to be due to an activation of DNA excision repair enzymes, in particular those that remove the more mutagenic purine oxidation products from DNA of hepatocytes [227,228]. Furthermore, quercetin and myricetin protect Caco-2 cells against DNA strand breakage induced by hydrogen peroxide, although rutin and kaempferol were not effective in this cell system [225]. In agreement with this, quercetin and rutin have also been found to reduce DNA single strand breaks in Caco-2 cells stimulated by tert-butylhydroperoxide (tert-BOOH) and menadione exposure [223,229], and quercetin inhibits both oxidative DNA damage and NF-κB binding activity in HepG2 cells exposed to hydrogen peroxide [161].

Data from an in vivo study also suggest that quercetin or its metabolites may protect against DNA damage. Freshly collected lymphocytes from diabetic patients treated for 2 weeks on a high flavonol diet (mostly quercetin) were more resistant to hydrogen peroxide–induced DNA damage than those from patients on the unsupplemented diet [230]. Another in vivo investigation examined the effects of green tea catechins on N-nitrosobis(2-oxopropyl)amine- nitrosobis(2-oxopropyl)amine-(BOP)-induced oxidative stress in pancreas and liver. Hamsters injected with BOP showed increases in the concentration of lipid peroxides and the amount of 8-OHdG (8-hydroxy-deoxyguanosine), which were significantly depressed in those supplemented with a 0.1% solution of green tea catechins as drinking water [231]. Furthermore, pretreatment of cultured human lung cells with the green tea flavonoid EGCG provided significant protection against the induction of DNA strand breakage and genetic damage induced by tobacco-specific nitrosamines [4-(N-methyl-N-n-trosamino)-1-(3-pyridyl)-1-butanone, a metabolite of nicotine] and stimulated human phagocytes [232].

One possible mechanism for the observed protection against DNA base oxidation by flavonoid might reside in their potent transition metal ion binding abilities [32,33,35], as it is known that the generation of base modification products by H_2O_2 is dependent on the presence of DNA bound transition metal ions [233,234]. However, in contrast to the protective effects observed for flavonoids against oxidative insults, the intracellular steady state of oxidized DNA bases, formed as a result of normal ongoing damage and repair in human colon cells, was not altered by exposure to anthocyanins or anthocyanidins [235]. This finding would suggest that flavonoid may act directly to scavenge the damaging oxidant but does not alter the normal cellular level of DNA damage.

VII. CONCLUSION

It is clear that flavonoids elicit a number of different effects on a wide variety of different cell types ranging from potent protective activities against oxidative stress in some cells to strong pro-oxidant natures in others. At present there is no clear pattern that emerges that helps define a mechanism of protection or pro-oxidant nature, although it appears that the effects observed may be more complex than the view that was held until recently that they acted only as potent classical antioxidants. New data are emerging, including some studies that have used the biologically relevant metabolite forms of flavonoids, that suggest that their bioactivity is also dependent on their ability to interact with redox-sensitive intracellular signaling cascades and possibly with mitochondria (Fig. 6). An overview of all the work since the 1970s would suggest that the mechanism of action of these dietary agents in vivo is likely to involve both radical scavenging properties and interactions within signaling cascades. Clearly, to determine the precise mechanism, more studies using the flavonoid metabolite forms in appropriate cell systems are required. The days of direct addition of dietary forms have passed, as the in vivo relevance of this type of study will always be questioned in the light of the absorption and metabolism pathways. How

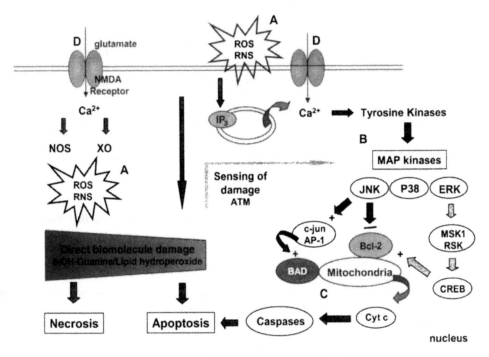

Figure 6 Possible cellular sites of action of flavonoids leading to protection against cell damage and death by either apoptosis or necrosis: (A) Direct scavenging of the radical or oxidizing species resulting in a reduction in oxidative damage biomarkers; (B) specific interactions within intracellular signaling cascades, for example, the MAPK pathway; reduction in oxidative stress–induced phosphylation/activation of JNK and c-jun leading to eventual induction of apoptosis; (C) interactions with mitochondria either at the level of ATP binding sites or in respiratory chain complexes; (D) cell receptor interactions. MAPK, mitogen-activated protein kinase; ATP, adenosine triphosphate; JNK, Jun kinase.

flavonoids act to prevent disease and improve the quality of older life is still a relative mystery, and further research is required before detailed information on mechanism is obtained.

ACKNOWLEDGMENTS

We acknowledge the support of both the Biotechnology and Biological Sciences Research Council and a European Union Fifth Framework RTD Programme Grant (grant no. QLK4-1999-01590).

REFERENCES

1. Colic M, Pavelic K. Molecular mechanisms of anticancer activity of natural dietetic products. J Mol Med 2000; 78:333–336.
2. Eastwood MA. Interaction of dietary antioxidants in vivo: how fruit and vegetables prevent disease? Q J Med 1999; 92:527–530.
3. Galati G, Teng S, Moridani MY, Chan TS, O'Brien PJ. Cancer chemoprevention and apoptosis mechanisms induced by dietary polyphenolics. Drug Metabol Drug Interact 2000; 17:311–349.
4. Middleton EJ, Kandaswami C, Theoharides TC. The effects of plant flavonoids on mammalian cells: implications for inflammation, heart disease, and cancer. Pharmacol Rev 2000; 52:673–751.
5. Rice-Evans C. Implications of the mechanisms of action of tea polyphenols as antioxidants in vitro for chemoprevention in humans. Proc Soc Exp Biol Med 1999; 220:262–266.
6. Dragsted LO, Strube M, Leth T. Dietary levels of plant phenols and other non-nutritive components: could they prevent cancer? Eur J Cancer Prev 1997; 6:522–528.
7. Riemersma RA, Rice-Evans CA, Tyrrell RM, Clifford MN, Lean ME. Tea flavonoids and cardiovascular health. Q J Med 2001; 94:277–282.
8. Giugliano D. Dietary antioxidants for cardiovascular prevention. Nutr Metab Cardiovasc Dis 2000; 10:38–44.
9. Wedworth SM, Lynch S. Dietary flavonoids in atherosclerosis prevention. Ann Pharmacother 1995; 29:627–628.
10. Manthey JA. Biological properties of flavonoids pertaining to inflammation. Microcirculation 2000; 7:S29–S34
11. Diplock AT, Charleux JL, Crozier-Willi G, Kok FJ, Rice-Evans C, Roberfroid M, et al. Functional food science and defence against reactive oxidative species. Br J Nutr 1998; 80 (suppl 1):S77–S112.
12. Harborne JB, Williams CA. Advances in flavonoid research since 1992. Phytochemistry 2000; 55:481–504.
13. Packer L, Rimbach G, Virgili F. Antioxidant activity and biologic properties of a procyanidin-rich extract from pine (Pinus maritima) bark, pycnogenol. Free Radic Biol Med 1999; 27:704–724.
14. Surh Y. Molecular mechanisms of chemopreventive effects of selected dietary and medicinal phenolic substances. Mutat Res 1999; 428:305–327.
15. Sekher PA, Chan TS, O'Brien PJ, Rice-Evans CA. Flavonoid B-ring chemistry and antioxidant activity: fast reaction kinetics. Biochem Biophys Res Commun 2001; 282:1161–1168.
16. Rice-Evans C. Plant polyphenols: free radical scavengers or chain-breaking antioxidants? Biochem Soc Symp 1995; 61:103–116.
17. Rice-Evans CA, Miller NJ, Paganga G. Structure-antioxidant activity relationships of flavonoids and phenolic acids. Free Radic Biol Med 1996; 20:933–956.
18. Green ES, Cooper CE, Davies MJ, Rice-Evans C. Antioxidant drugs and the

inhibition of low-density lipoprotein oxidation. Biochem Soc Trans 1993; 21:362–366.

19. Heller FR, Descamps O, Hondekijn JC. LDL oxidation: therapeutic perspectives. Atherosclerosis 1998; 137 (suppl):S25–S31

20. Brown JE, Rice-Evans CA. Luteolin-rich artichoke extract protects low density lipoprotein from oxidation in vitro. Free Radic Res 1998; 29:247–255.

21. Fuhrman B, Aviram M. flavonoids protect LDL from oxidation and attenuate atherosclerosis. Curr Opin Lipidol 2001; 12:41–48.

22. Hayek T, Fuhrman B, Vaya J, Rosenblat M, Belinky P, Coleman R, et al. Reduced progression of atherosclerosis in apolipoprotein E-deficient mice following consumption of red wine, or its polyphenols quercetin or catechin, is associated with reduced susceptibility of LDL to oxidation and aggregation. Arterioscler Thromb Vasc Biol 1997; 17:2744–2752.

23. Duthie SJ, Collins AR, Duthie GG, Dobson VL. Quercetin and myricetin protect against hydrogen peroxide-induced DNA damage (strand breaks and oxidised pyrimidines) in human lymphocytes. Mutat Res 1997; 393:223–231.

24. Duthie SJ, Johnson W, Dobson VL. The effect of dietary flavonoids on DNA damage (strand breaks and oxidised pyrimdines) and growth in human cells. Mutat Res Genet Toxicol Environ Mutagen 1997; 390:141–151.

25. Aherne SA, O'Brien NM. Protection by the flavonoids myricetin, quercetin, and rutin against hydrogen peroxide-induced DNA damage in Caco-2 and Hep G2 cells. Nutr Cancer 1999; 34:160–166.

26. Zhao K, Whiteman M, Spencer JPE, Halliwell B. DNA damage by nitrite and peroxynitrite: protection by dietary phenols. Methods Enzymol 2001; 335:296–307.

27. Haenen GR, Paquay JB, Korthouwer RE, Bast A. Peroxynitrite scavenging by flavonoids. Biochem Biophys Res Commun 1997; 236:591–593.

28. Heijnen CG, Haenen GR, van Acker FA, van D, V, Bast A. Flavonoids as peroxynitrite scavengers: the role of the hydroxyl groups. Toxicol In Vitro 2001; 15:3–6.

29. Arteel GE, Schroeder P, Sies H. Reactions of peroxynitrite with cocoa procyanidin oligomers. J Nutr 2000; 130:2100S–2104S.

30. Arteel GE, Sies H. Protection against peroxynitrite by cocoa polyphenol oligomers. FEBS Lett 1999; 462:167–170.

31. Kerry N, Rice-Evans C. Inhibition of peroxynitrite-mediated oxidation of dopamine by flavonoid and phenolic antioxidants and their structural relationships. J Neurochem 1999; 73:247–253.

32. Morel I, Lescoat G, Cillard P, Cillard J. Role of flavonoids and iron chelation in antioxidant action. Methods Enzymol 1994; 234:437–443.

33. Brown JE, Khodr H, Hider RC, Rice-Evans CA. Structural dependence of flavonoid interactions with Cu2+ ions: implications for their antioxidant properties. Biochem J 1998; 330(pt 3):1173–1178.

34. Sestili P, Guidarelli A, Dacha M, Cantoni O. Quercetin prevents DNA single strand breakage and cytotoxicity caused by tert-butylhydroperoxide: free radical scavenging versus iron chelating mechanism. Free Radic Biol Med 1998; 25:196–200.

35. Cheng IF, Breen K. On the ability of four flavonoids, baicilein, luteolin,

naringenin, and quercetin, to suppress the Fenton reaction of the iron-ATP complex. Biometals 2000; 13:77–83.

36. Tournaire C, Croux S, Maurette MT, Beck I, Hocquaux M, Braun AM. Antioxidant activity of flavonoids: efficiency of singlet oxygen (1 delta g) quenching. J Photochem Photobiol B 1993; 19:205–215.

37. Devasagayam TP, Subramanian M, Singh BB, Ramanathan R, Das NP. Protection of plasmid pBR322 DNA by flavonoids against single-stranded breaks induced by singlet molecular oxygen. J Photochem Photobiol B 1995; 30:97–103.

38. Scheline RR. Metabolism of oxygen heterocyclic compounds. In: CRC Handbook of Mammalian Metabolism of Plant Compounds. Boca Raton, FL: CRC Press, 1999:243–295.

39. Rice-Evans C, Spencer JPE, Schroeter H, Rechner AR. Bioavailability of flavonoids and potential bioactive forms in vivo. Drug Metabol Drug Interact 2000; 17:291–310.

40. Spencer JPE, Schroeter H, Rechner A, Rice-Evans C. Bioavailability of flavan-3-ols and procyanidins: gastrointestinal tract influences and their relevance to bioactive forms *in vivo*. Antiox Redox Signal 2001; 3:1023–1040.

41. Spencer JPE, Chowrimootoo G, Choudhury R, Debnam ES, Srai SK, Rice-Evans C. The small intestine can both absorb and glucuronidate luminal flavonoids. FEBS Lett 1999; 458:224–230.

42. Scalbert A, Williamson G. Dietary intake and bioavailability of polyphenols. J Nutr 2000; 130:2073S–2085S.

43. Rice-Evans C. Flavonoid antioxidants. Curr Med Chem 2001; 8:797–807.

44. Oliveira NG, Rodrigues AS, Chaveca T, Rueff J. Induction of an adaptive response to quercetin, mitomycin C and hydrogen peroxide by low doses of quercetin in V79 Chinese hamster cells. Mutagenesis 1997; 12:457–462.

45. Ishige K, Schubert D, Sagara Y. flavonoids protect neuronal cells from oxidative stress by three distinct mechanisms. Free Radic Biol Med 2001; 30:433–446.

46. Kim H, Kim YS, Kim SY, Suk K. The plant flavonoid wogonin suppresses death of activated C6 rat glial cells by inhibiting nitric oxide production. Neurosci Lett 2001; 309:67–71.

47. Levites Y, Weinreb O, Maor G, Youdim MB, Mandel S. Green tea polyphenol (-)-epigallocatechin-3-gallate prevents N-methyl-4-phenyl-1,2,3,6-tetrahydropyridine-induced dopaminergic neurodegeneration. J Neurochem 2001; 78:1073–1082.

48. Chen C, Wei T, Gao Z, Zhao B, Hou J, Xu H, et al. Different effects of the constituents of EGb761 on apoptosis in rat cerebellar granule cells induced by hydroxyl radicals. Biochem Mol Biol Int 1999; 47:397–405.

49. Yao Z, Drieu K, Papadopoulos V. The Ginkgo biloba extract EGb 761 rescues the PC12 neuronal cells from beta-amyloid-induced cell death by inhibiting the formation of beta-amyloid-derived diffusible neurotoxic ligands. Brain Res 2001; 889:181–190.

50. Schroeter H, Spencer JPE, Rice-Evans C, Williams RJ. flavonoids protect neurons from oxidized low-density-lipoprotein-induced apoptosis involving c-Jun N-terminal kinase (JNK), c-Jun and caspase-3. Biochem J 2001; 358:547–557.

51. Negre-Salvayre A, Salvayre R. Quercetin prevents the cytotoxicity of oxidized LDL on lymphoid cell lines. Free Radic Biol Med 1992; 12:101–106.
52. Negre-Salvayre A, Alomar Y, Troly M, Salvayre R. Ultraviolet-treated lipoproteins as a model system for the study of the biological effects of lipid peroxides on cultured cells. III. The protective effect of antioxidants (probucol, catechin, vitamin E) against the cytotoxicity of oxidized LDL occurs in two different ways. Biochim Biophys Acta 1991; 1096:291–300.
53. Negre-Salvayre A, Mabile L, Delchambre J, Salvayre R. alpha-Tocopherol, ascorbic acid, and rutin inhibit synergistically the copper-promoted LDL oxidation and the cytotoxicity of oxidized LDL to cultured endothelial cells. Biol Trace Elem Res 1995; 47:81–91.
54. Spencer JPE, Schroeter H, Kuhnle G, Srai SK, Tyrrell RM, Hahn U, Rice-Evans C. Epicatechin and its in vivo metabolite, 3′-O-methyl epicatechin, protect human fibroblasts from oxidative-stress-induced cell death involving caspase-3 activation. Biochem J 2001; 354:493–500.
55. Spencer JPE, Schroeter H, Crossthwaithe AJ, Kuhnle G, Williams RJ, Rice-Evans C. Contrasting influences of glucuronidation and O-methylation of epicatechin on hydrogen peroxide-induced cell death in neurons and fibroblasts. Free Radic Biol Med 2001; 31:1139–1146.
56. Ullrich SE. Modulation of immunity by ultraviolet radiation: key effects on antigen presentation. J Invest Dermatol 1995; 105:30S–36S.
57. Pietta PG, Simonetti P, Gardana C, Brusamolino A, Morazzoni P, Bombardelli E. Catechin metabolites after intake of green tea infusions. Biofactors 1998; 8:111–118.
58. Kivits GAA, vanderSman FJP, Tijburg LBM. Analysis of catechins from green and black tea in humans: a specific and sensitive colorimetric assay of total catechins in biological fluids. Int J Food Sci Nutr 1997; 48:387–392.
59. Subirade I, Fernandez Y, Periquet A, Mitjavila S. Catechin protection of 3T3 Swiss fibroblasts in culture under oxidative stress. Biol Trace Elem Res 1995; 47:313–319.
60. Nagata H, Takekoshi S, Takagi T, Honma T, Watanabe K. Antioxidative action of flavonoids, quercetin and catechin, mediated by the activation of glutathione peroxidase. Tokai J Exp Clin Med 1999; 24:1–11.
61. Kaneko T, Matsuo M, Baba N. Inhibition of linoleic acid hydroperoxide-induced toxicity in cultured human umbilical vein endothelial cells by catechins. Chem Biol Interact 1998; 114:109–119.
62. Formica JV, Regelson W. Review of the biology of quercetin and related bioflavonoids. Food Chem Toxicol 1995; 33:1061–1080.
63. Oyama Y, Noguchi S, Nakata M, Okada Y, Yamazaki Y, Funai M, et al. Exposure of rat thymocytes to hydrogen peroxide increases annexin V binding to membranes: inhibitory actions of deferoxamine and quercetin. Eur J Pharmacol 1999; 384:47–52.
64. Plumb GW, Dupont MS, Williamson G. Modulation of AAPH-induced oxidative stress in cell culture by flavonoids. Biochem Soc Trans 1997; 25:S560
65. Nakayama T, Yamada M, Osawa T, Kawakishi S. Suppression of active oxygen-induced cytotoxicity by flavonoids. Biochem Pharmacol 1993; 45:265–267.

66. Skaper SD, Fabris M, Ferrari V, Dalle CM, Leon A. Quercetin protects cutaneous tissue-associated cell types including sensory neurons from oxidative stress induced by glutathione depletion: cooperative effects of ascorbic acid. Free Radic Biol Med 1997; 22:669–678.

67. Gao D, Tawa R, Masaki H, Okano Y, Sakurai H. Protective effects of baicalein against cell damage by reactive oxygen species. Chem Pharm Bull (Tokyo) 1998; 46:1383–1387.

68. Gao Z, Huang K, Xu H. Protective effects of flavonoids in the roots of Scutellaria baicalensis Georgi against hydrogen peroxide-induced oxidative stress in HS-SY5Y cells. Pharmacol Res 2001; 43:173–178.

69. Gao Z, Huang K, Yang X, Xu H. Free radical scavenging and antioxidant activities of flavonoids extracted from the radix of Scutellaria baicalensis Georgi. Biochim Biophys Acta 1999; 1472:643–650.

70. Villa P, Cova D, De Francesco L, Guaitani A, Palladini G, Perego R. Protective effect of diosmetin on in vitro cell membrane damage and oxidative stress in cultured rat hepatocytes. Toxicology 1992; 73:179–189.

71. Kostyuk VA, Potapovich AI, Vladykovskaya EN, Korkina LG, Afanas'ev IB. Influence of metal ions on flavonoid protection against asbestos-induced cell injury. Arch Biochem Biophys 2001; 385:129–137.

72. Krutmann J, Grewe M. Involvement of cytokines, DNA damage, and reactive oxygen intermediates in ultraviolet radiation-induced modulation of intercellular adhesion molecule-1 expression. J Invest Dermatol 1995; 105:67S–70S.

73. Tyrrell RM. Ultraviolet radiation and free radical damage to skin. Biochem Soc Symp 1995; 61:47–53.

74. Garssen J, Vandebriel RJ, van Loveren H. Molecular aspects of UVB-induced immunosuppression. Arch Toxicol Suppl 1997; 19:97–109.

75. Hart PH, Grimbaldeston MA, Finlay-Jones JJ. Sunlight, immunosuppression and skin cancer: role of histamine and mast cells. Clin Exp Pharmacol Physiol 2001; 28:1–8.

76. de Gruijl FR, van Kranen HJ, Mullenders LH. UV-induced DNA damage, repair, mutations and oncogenic pathways in skin cancer. J Photochem Photobiol B 2001; 63:19–27.

77. Sarasin A. The molecular pathways of ultraviolet-induced carcinogenesis. Mutat Res 1999; 428:5–10.

78. Godar DE. Light and death: photons and apoptosis. J Invest Dermatol Symp Proc 1999; 4:17–23.

79. Takashima A, Bergstresser PR. Impact of UVB radiation on the epidermal cytokine network. Photochem Photobiol 1996; 63:397–400.

80. Kondo S, Sauder DN. Keratinocyte-derived cytokines and UVB-induced immunosuppression. J Dermatol 1995; 22:888–893.

81. Takashima A. UVB-dependent modulation of epidermal cytokine network: roles in UVB-induced depletion of Langerhans cells and dendritic epidermal T cells. J Dermatol 1995; 22:876–887.

82. Katiyar SK, Mukhtar H. Green tea polyphenol (-)-epigallocatechin-3-gallate treatment to mouse skin prevents UVB-induced infiltration of leukocytes,

depletion of antigen-presenting cells, and oxidative stress. J Leukoc Biol 2001; 69:719–726.

83. Katiyar SK, Elmets CA. Green tea polyphenolic antioxidants and skin photo-protection (review). Int J Oncol 2001; 18:1307–1313.

84. Katiyar SK, Matsui MS, Elmets CA, Mukhtar H. Polyphenolic antioxidant (-)-epigallocatechin-3-gallate from green tea reduces UVB-induced inflammatory responses and infiltration of leukocytes in human skin. Photochem Photobiol 1999; 69:148–153.

85. Katiyar SK, Afaq F, Perez A, Mukhtar H. Green tea polyphenol (-)-epigallocatechin-3-gallate treatment of human skin inhibits ultraviolet radiation-induced oxidative stress. Carcinogenesis 2001; 22:287–294.

86. Srivastava PJ, Chandra S, Arif AJ, Singh C, Panday V. Metal chelators/antioxidants: approaches to protect erythrocytic oxidative stress injury during Plasmodium berghei infection in Mastomys coucha. Pharmacol Res 1999; 40:239–241.

87. Mitani H, Uchida N, Shima A. Induction of cyclobutane pyrimidine dimer photolyase in cultured fish cells by UVA and blue light. Photochem Photobiol 1996; 64:943–948.

88. Skibola CF, Smith MT. Potential health impacts of excessive flavonoid intake. Free Radic Biol Med 2000; 29:375–383.

89. Sergediene E, Jonsson K, Szymusiak H, Tyrakowska B, Rietjens IM, Cenas N. Prooxidant toxicity of polyphenolic antioxidants to HL-60 cells: description of quantitative structure-activity relationships. FEBS Lett 1999; 462:392–396.

90. Dickancaite E, Nemeikaite A, Kalvelyte A, Cenas N. Prooxidant character of flavonoid cytotoxicity: structure-activity relationships. Biochem Mol Biol Int 1998; 45:923–930.

91. Agullo G, Gamet-Payrastre L, Fernandez Y, Anciaux N, Demigne C, Remesy C. Comparative effects of flavonoids on the growth, viability and metabolism of a colonic adenocarcinoma cell line (HT29 cells). Cancer Lett 1996; 105:61–70.

92. Silva ID, Gaspar J, daCosta GG, Rodrigues AS, Laires A, Rueff J. Chemical features of flavonols affecting their genotoxicity. Potential implications in their use as therapeutical agents. Chem Biol Interact 2000; 124:29–51.

93. Halliwell B, Clement MV, Ramalingam J, Long LH. Hydrogen peroxide: ubiquitous in cell culture and in vivo? IUBMB Life 2000; 50:251–257.

94. Miura YH, Tomita I, Watanabe T, Hirayama T, Fukui S. Active oxygens generation by flavonoids. Biol Pharm Bull 1998; 21:93–96.

95. Sakagami H, Jiang Y, Kusama K, Atsumi T, Ueha T, Toguchi M, et al. Induction of apoptosis by flavones, flavonols (3-hydroxyflavones) and isoprenoid-substituted flavonoids in human oral tumor cell lines. Anticancer Res 2000; 20:271–277.

96. Weber LP, Kiparissis Y, Hwang GS, Niimi AJ, Janz DM, Metcalfe CD. Increased cellular apoptosis after chronic aqueous exposure to nonylphenol and quercetin in adult medaka (Oryzias latipes). Comp Biochem Physiol C Toxicol Pharmacol 2002; 131:51–59.

97. Pawlikowska-Pawlega B, Jakubowicz-Gil J, Rzymowska J, Gawron A. The effect of quercetin on apoptosis and necrosis induction in human colon adenocarcinoma cell line LS180. Folia Histochem Cytobiol 2002; 39:217–218.

98. Rong Y, Yang EB, Zhang K, Mack P. Quercetin-induced apoptosis in the mono-blastoid cell line U937 in vitro and the regulation of heat shock proteins expression. Anticancer Res 2000; 20:4339–4345.

99. Wei YQ, Zhao X, Kariya Y, Fukata H, Teshigawara K, Uchida A. Induction of apoptosis by quercetin: involvement of heat shock protein. Cancer Res 1994; 54:4952–4957.

100. Csokay B, Prajda N, Weber G, Olah E. Molecular mechanisms in the antiproliferative action of quercetin. Life Sci 1997; 60:2157–2163.

101. O'Brien NM, Woods JA, Aherne SA, O'Callaghan YC. Cytotoxicity, genotoxicity and oxidative reactions in cell-culture models: modulatory effects of phytochem-icals. Biochem Soc Trans 2000; 28:22–26.

102. Rzymowska J, Gawron A, Pawlikowska-Pawlega B, Jakubowicz-Gil J, Wojcier-owski J. The effect of quercetin on induction of apoptosis. Folia Histochem Cytobiol 1999; 37:125–126.

103. Iwashita K, Kobori M, Yamaki K, Tsushida T. Flavonoids inhibit cell growth and induce apoptosis in B16 melanoma 4A5 cells. Biosci Biotech Biochem 2000; 64:1813–1820.

104. Wang IK, Lin-Shiau SY, Lin JK. Induction of apoptosis by apigenin and related flavonoids through cytochrome c release and activation of caspase-9 and caspase-3 in leukaemia HL-60 cells. Eur J Cancer 1999; 35:1517–1525.

105. Yamashita N, Kawanishi S. Distinct mechanisms of DNA damage in apoptosis induced by quercetin and luteolin. Free Radic Res 2000; 33:623–633.

106. Sahu SC, Gray GC. Lipid peroxidation and DNA damage induced by morin and naringenin in isolated rat liver nuclei. Food Chem Toxicol 1997; 35:443–447.

107. Sahu SC, Gray GC. Kaempferol-induced nuclear DNA damage and lipid peroxidation. Cancer Lett 1994; 85:159–164.

108. Knowles LM, Zigrossi DA, Tauber RA, Hightower C, Milner JA. flavonoids suppress androgen-independent human prostate tumor proliferation. Nutr Cancer 2000; 38:116–122.

109. Kuo SM. Antiproliferative potency of structurally distinct dietary flavonoids on human colon cancer cells. Cancer Lett 1996; 110:41–48.

110. Bhatia N, Agarwal C, Agarwal R. Differential responses of skin cancer-chemopreventive agents silibinin, quercetin, and epigallocatechin 3-gallate on mitogenic signaling and cell cycle regulators in human epidermoid carcinoma A431 cells. Nutr Cancer 2002; 39:292–299.

111. Choi JA, Kim JY, Lee JY, Kang CM, Kwon HJ, Yoo YD, et al. Induction of cell cycle arrest and apoptosis in human breast cancer cells by quercetin. Int J Oncol 2001; 19:837–844.

112. Asea A, Ara G, Teicher BA, Stevenson MA, Calderwood SK. Effects of the flavonoid drug quercetin on the response of human prostate tumours to hyperthermia in vitro and in vivo. Int J Hyperthermia 2001; 17:347–356.

113. Russo M, Palumbo R, Tedesco I, Mazzarella G, Russo P, Iacomino G, et al. Quercetin and anti-CD95(Fas/Apo1) enhance apoptosis in HPB-ALL cell line. FEBS Lett 1999; 462:322–328.

114. Richter M, Ebermann R, Marian B. Quercetin-induced apoptosis in colorectal tumor cells: possible role of EGF receptor signaling. Nutr Cancer 1999; 34:88–99.

115. Xiao D, Zhu SP, Gu ZL. Quercetin induced apoptosis in human leukemia HL-60 cells. Zhongguo Yao Li Xue Bao 1997; 18:280–283.

116. Chen S, Ruan Q, Bedner E, Deptala A, Wang X, Hsieh TC, et al. Effects of the flavonoid baicalin and its metabolite baicalein on androgen receptor expression, cell cycle progression and apoptosis of prostate cancer cell lines. Cell Prolif 2001; 34:293–304.

117. Harris SR, Panaro NJ, Thorgeirsson UP. Oxidative stress contributes to the anti-proliferative effects of flavone acetic acid on endothelial cells. Anticancer Res 2000; 20:2249–2254.

118. Huynh HT, Teel RW. Selective induction of apoptosis in human mammary cancer cells (MCF-7) by pycnogenol. Anticancer Res 2000; 20:2417–2420.

119. Kampa M, Hatzoglou A, Notas G, Damianaki A, Bakogeorgou E, Gemetzi C, et al. Wine antioxidant polyphenols inhibit the proliferation of human prostate cancer cell lines. Nutr Cancer 2000; 37:223–233.

120. Kim SY, Gao JJ, Kang HK. Two flavonoids from the leaves of Morus alba induce differentiation of the human promyelocytic leukemia (HL-60) cell line. Biol Pharm Bull 2000; 23:451–455.

121. Lu HQ, Niggemann B, Zanker KS. Suppression of the proliferation and migration of oncogenic ras-dependent cell lines, cultured in a three-dimensional collagen matrix, by flavonoid-structured molecules. J Cancer Res Clin Oncol 1996; 122: 335–342.

122. Bhat KPL, Kosmeder JW, Pezzuto JM. Biological effects of resveratrol. Antioxid Redox Signal 2001; 3:1041–1064.

123. Gusman J, Malonne H, Atassi G. A reappraisal of the potential chemo-preventive and chemotherapeutic properties of resveratrol. Carcinogenesis 2001; 22:1111–1117.

124. She QB, Bode AM, Ma WY, Chen NY, Dong Z. Resveratrol-induced activation of p53 and apoptosis is mediated by extracellular-signal-regulated protein kinases and p38 kinase. Cancer Res 2001; 61:1604–1610.

125. Park JW, Choi YJ, Suh SI, Baek WK, Suh MH, Jin IN, et al. Bcl-2 overexpression attenuates resveratrol-induced apoptosis in U937 cells by inhibition of caspase-3 activity. Carcinogenesis 2001; 22:1633–1639.

126. Mahyar-Roemer M, Katsen A, Mestres P, Roemer K. Resveratrol induces colon tumor cell apoptosis independently of p53 and precede by epithelial differ-entiation, mitochondrial proliferation and membrane potential collapse. Int J Cancer 2001; 94:615–622.

127. Dorrie J, Gerauer H, Wachter Y, Zunino SJ. Resveratrol induces extensive apoptosis by depolarizing mitochondrial membranes and activating caspase-9 in acute lymphoblastic leukemia cells. Cancer Res 2001; 61:4731–4739.

128. Ahmad N, Adhami VM, Afaq F, Feyes DK, Mukhtar H. Resveratrol causes WAF-

1/p21-mediated G(1)-phase arrest of cell cycle and induction of apoptosis in human epidermoid carcinoma A431 cells. Clin Cancer Res 2001; 7:1466–1473.

129. Shih A, Davis FB, Lin HY, Davis PJ. Resveratrol induces apoptosis in thyroid cancer cell lines via a MAPK-and p53-dependent mechanism. J Clin Endocrinol Metab 2002; 87:1223–1232.

130. She QB, Huang C, Zhang Y, Dong Z. Involvement of c-jun NH(2)-terminal kinases in resveratrol-induced activation of p53 and apoptosis. Mol Carcinog 2002; 33:244–250.

131. Lee SH, Ryu SY, Kim HB, Kim MY, Chun YJ. Induction of apoptosis by 3,4'-dimethoxy-5-hydroxystilbene in human promyeloid leukemic HL-60 cells. Planta Med 2002; 68:123–127.

132. Joe AK, Liu H, Suzui M, Vural ME, Xiao D, Weinstein IB. Resveratrol induces growth inhibition, S-phase arrest, apoptosis, and changes in biomarker expression in several human cancer cell lines. Clin Cancer Res 2002; 8:893–903.

133. Waffo-Teguo P, Hawthorne ME, Cuendet M, Merillon JM, Kinghorn AD, Pezzuto JM, et al. Potential cancer-chemopreventive activities of wine stilbenoids and flavans extracted from grape (Vitis vinifera) cell cultures. Nutr Cancer 2002; 40:173–179.

134. Wolter F, Akoglu B, Clausnitzer A, Stein J. Downregulation of the cyclin D1/Cdk4 complex occurs during resveratrol-induced cell cycle arrest in colon cancer cell lines. J Nutr 2001; 131:2197–2203.

135. Sgambato A, Ardito R, Faraglia B, Boninsegna A, Wolf FI, Cittadini A. Resveratrol, a natural phenolic compound, inhibits cell proliferation and prevents oxidative DNA damage. Mutat Res 2001; 496:171–180.

136. Schneider Y, Vincent F, Duranton B, Badolo L, Gosse F, Bergmann C, et al. Antiproliferative effect of resveratrol, a natural component of grapes and wine, on human colonic cancer cells. Cancer Lett 2000; 158:85–91.

137. Holian O, Wahid S, Atten MJ, Attar BM. Inhibition of gastric cancer cell proliferation by resveratrol: role of nitric oxide. Am J Physiol Gastrointest Liver Physiol 2002; 282:G809–G816

138. Brakenhielm E, Cao R, Cao Y. Suppression of angiogenesis, tumor growth, and wound healing by resveratrol, a natural compound in red wine and grapes. FASEB J 2001; 15:1798–1800.

139. Bove K, Lincoln DW, Tsan MF. Effect of resveratrol on growth of 4T1 breast cancer cells in vitro and in vivo. Biochem Biophys Res Commun 2002; 291:1001–1005.

140. de Santi C, Pietrabissa A, Mosca F, Pacifici GM. Glucuronidation of resveratrol, a natural product present in grape and wine, in the human liver. Xenobiotica 2000; 30:1047–1054.

141. Kuhnle G, Spencer JPE, Chowrimootoo G, Schroeter H, Debnam ES, Srai SK, et al. Resveratrol is absorbed in the small intestine as resveratrol glucuronide. Biochem Biophys Res Commun 2000; 272:212–217.

142. de Santi C, Pietrabissa A, Spisni R, Mosca F, Pacifici GM. Sulphation of resveratrol, a natural compound present in wine, and its inhibition by natural flavonoids. Xenobiotica 2000; 30:857–866.

143. Moridani MY, Scobie H, Jamshidzadeh A, Salehi P, O'Brien PJ. Caffeic acid, chlorogenic acid, and dihydrocaffeic acid metabolism: glutathione conjugate formation. Drug Metab Dispos 2001; 29:1432–1439.

144. Moridani MY, Scobie H, Salehi P, O'Brien PJ. Catechin metabolism: glutathione conjugate formation catalyzed by tyrosinase, peroxidase, and cytochrome p450. Chem Res Toxicol 2001; 14:841–848.

145. Awad HM, Boersma MG, Vervoort J, Rietjens IM. Peroxidase-catalyzed formation of quercetin quinone methide-glutathione adducts. Arch Biochem Biophys 2000; 378:224–233.

146. Awad HM, Boersma MG, Boeren S, van Bladeren PJ, Vervoort J, Rietjens IM. The regioselectivity of glutathione adduct formation with flavonoid quinone/quinone methides is pH-dependent. Chem Res Toxicol 2002; 15:343–351.

147. Rietjens IM, Awad HM, Boersma MG, van Iersel ML, Vervoort J, van Bladeren PJ. Structure activity relationships for the chemical behaviour and toxicity of electrophilic quinones/quinone methides. Adv Exp Med Biol 2002; 500:11–21.

148. Awad HM, Boersma MG, Boeren S, van Bladeren PJ, Vervoort J, Rietjens IM. Structure-activity study on the quinone/quinone methide chemistry of flavonoids. Chem Res Toxicol 2001; 14:398–408.

149. Shen XM, Zhang F, Dryhurst G. Oxidation of dopamine in the presence of cysteine: characterization of new toxic products. Chem Res Toxicol 1997; 10:147–155.

150. Shen XM, Dryhurst G. Further insights into the influence of L-cysteine on the oxidation chemistry of dopamine: reaction pathways of potential relevance to Parkinson's disease. Chem Res Toxicol 1996; 9:751–763.

151. Spencer JPE, Jenner P, Daniel SE, Lees AJ, Marsden DC, Halliwell B. Conjugates of catecholamines with cysteine and GSH in Parkinson's disease: possible mechanisms of formation involving reactive oxygen species. J Neurochem 1998; 71:2112–2122.

152. Fremont L, Belguendouz L, Delpal S. Antioxidant activity of resveratrol and alcohol-free wine polyphenols related to LDL oxidation and polyunsaturated fatty acids. Life Sci 1999; 64:2511–2521.

153. Li H, Dryhurst G. Irreversible inhibition of mitochondrial complex I by 7-(2-aminoethyl)-3,4-dihydro-5-hydroxy-2H-1,4-benzothiazine-3-carboxylic acid (DHBT-1): a putative nigral endotoxin of relevance to Parkinson's disease. J Neurochem 1997; 69:1530–1541.

154. Li H, Shen XM, Dryhurst G. Brain mitochondria catalyze the oxidation of 7-(2-aminoethyl)-3,4-dihydro-5-hydroxy-2H-1,4-benzothiazine-3-carboxylic acid (DHBT-1) to intermediates that irreversibly inhibit complex I and scavenge glutathione, potential relevance to the pathogenesis of Parkinson's disease. J Neurochem 1998; 71:2049–2062.

155. Zhang F, Dryhurst G. Effects of l-cysteine on the oxidation chemistry of dopamine—new reaction pathways of potential relevance to idiopathic parkinsons disease. J Med Chem 1994; 37:1084–1098.

156. Panes J, Gerritsen ME, Anderson DC, Miyasaka M, Granger DN. Apigenin inhibits tumor necrosis factor-induced intercellular adhesion molecule-1 upregulation in vivo. Microcirculation 1996; 3:279–286.

157. Gerritsen ME, Carley WW, Ranges GE, Shen CP, Phan SA, Ligon GF, Percy CA. Flavonoids inhibit cytokine-induced endothelial cell adhesion protein gene expression. Am J Pathol 1995; 147:278–292.

158. Soriani M, Rice-Evans C, Tyrrell RM. Modulation of the UVA activation of haem oxygenase, collagenase and cyclooxygenase gene expression by epigallocatechin in human skin cells. FEBS Lett 1998; 439:253–257.

159. Medina JH, Viola H, Wolfman C, Marder M, Wasowski C, Calvo D, Paladini AC. Overview—flavonoids: a new family of benzodiazepine receptor ligands. Neurochem Res 1997; 22:419–425.

160. Gamet-Payrastre L, Manenti S, Gratacap MP, Tulliez J, Chap H, Payrastre B. Flavonoids and the inhibition of PKC and PI 3-kinase. Gen Pharmacol 1999; 32:279–286.

161. Musonda CA, Chipman JK. Quercetin inhibits hydrogen peroxide (H_2O_2)-induced NF-kappaB DNA binding activity and DNA damage in HepG2 cells. Carcinogenesis 1998; 19:1583–1589.

162. Gosslau A, Rensing L. Induction of Hsp68 by oxidative stress involves the lipoxygenase pathway in C6 rat glioma cells. Brain Res 2000; 864:114–123.

163. Kobuchi H, Roy S, Sen CK, Nguyen HG, Packer L. Quercetin inhibits inducible ICAM-1 expression in human endothelial cells through the JNK pathway. Am J Physiol 1999; 277:C403–C411

164. Kong AN, Yu R, Chen C, Mandlekar S, Primiano T. Signal transduction events elicited by natural products: role of MAPK and caspase pathways in homeostatic response and induction of apoptosis. Arch Pharm Res 2000; 23:1–16.

165. Nakahata N, Kyo R, Kutsuwa M, Ohizumi Y. Inhibition of mitogen-activated protein kinase cascade by baicalein, a flavonoid of natural origin. Nippon Yakurigaku Zasshi 1999; 114 (Suppl 1):215P–219P.

166. Lo YY, Wong JM, Cruz TF. Reactive oxygen species mediate cytokine activation of c-Jun NH2-terminal kinases. J Biol Chem 1996; 271:15703–15707.

167. Uchida K, Shiraishi M, Naito Y, Torii Y, Nakamura Y, Osawa T. Activation of stress signaling pathways by the end product of lipid peroxidation: 4-hydroxy-2-nonenal is a potential inducer of intracellular peroxide production. J Biol Chem 1999; 274:2234–2242.

168. Mizuno M, Droy-Lefaix MT, Packer L. Gingko biloba extract EGb 761 is a suppresser of AP-1 transcription factor stimulated by phorbol 12-myristate 13-acetate. Biochem Mol Biol Int 1996; 39:395–401.

169. Di Pietro A, Godinot C, Bouillant ML, Gautheron DC. Pig heart mitochondrial ATPase: properties of purified and membrane-bound enzyme: effects of flavonoids. Biochimie 1975; 57:959–967.

170. Barzilai A, Rahamimoff H. Inhibition of Ca2+-transport ATPase from synaptosomal vesicles by flavonoids. Biochim Biophys Acta 1983; 730:245–254.

171. Revuelta MP, Cantabrana B, Hidalgo A. Mechanisms involved in kaempferol-induced relaxation in rat uterine smooth muscle. Life Sci 2000; 67:251–259.

172. Revuelta MP, Hidalgo A, Cantabrana B. Involvement of cAMP and beta-

adrenoceptors in the relaxing effect elicited by flavonoids on rat uterine smooth muscle. J Auton Pharmacol 1999; 19:353–358.

173. Revuelta MP, Cantabrana B, Hidalgo A. Depolarization-dependent effect of flavonoids in rat uterine smooth muscle contraction elicited by CaCl$_2$. Gen Pharmacol 1997; 29:847–857.

174. Rosenblat M, Belinky P, Vaya J, Levy R, Hayek T, Coleman R, et al. Macrophage enrichment with the isoflavan glabridin inhibits NADPH oxidase-induced cell-mediated oxidation of low density lipoprotein. A possible role for protein kinase C. J Biol Chem 1999; 274:13790–13799.

175. Ursini F, Maiorino M, Morazzoni P, Roveri A, Pifferi G. A novel antioxidant flavonoid (IdB 1031) affecting molecular mechanisms of cellular activation. Free Radic Biol Med 1994; 16:547–553.

176. Kantengwa S, Polla BS. flavonoids, but not protein kinase C inhibitors, prevent stress protein synthesis during erythrophagocytosis. Biochem Biophys Res Commun 1991; 180:308–314.

177. Fischer PM, Lane DP. Inhibitors of cyclin-dependent kinases as anti-cancer therapeutics. Curr Med Chem 2000; 7:1213–1245.

178. Beppu M, Watanabe T, Yokota A, Ohmori S, Kikugawa K. Water-soluble antioxidants inhibit macrophage recognition of oxidized erythrocytes. Biol Pharm Bull 2001; 24:575–578.

179. Crouvezier S, Powell B, Keir D, Yaqoob P. The effects of phenolic components of tea on the production of pro- and anti-inflammatory cytokines by human leukocytes in vitro. Cytokine 2001; 13:280–286.

180. Lin JK, Chen PC, Ho CT, Lin-Shiau SY. Inhibition of xanthine oxidase and suppression of intracellular reactive oxygen species in HL-60 cells by theaflavin-3,3'-digallate, (-)-epigallocatechin-3-gallate, and propyl gallate. J Agric Food Chem 2000; 48:2736–2743.

181. de Groot H, Rauen U. Tissue injury by reactive oxygen species and the protective effects of flavonoids. Fundam Clin Pharmacol 1998; 12:249–255.

182. Dehmlow C, Erhard J, de Groot H. Inhibition of Kupffer cell functions as an explanation for the hepatoprotective properties of silibinin. Hepatology 1996; 23:749–754.

183. Dehmlow C, Murawski N, de Groot H. Scavenging of reactive oxygen species and inhibition of arachidonic acid metabolism by silibinin in human cells. Life Sci 1996; 58:1591–1600.

184. Deak G, Muzes G, Lang I, Nekam K, Gonzalez-Cabello R, Gergely P, et al. Effects of two bioflavonoids on certain cellular immune reactions in vitro. Acta Physiol Hung 1990; 76:113–121.

185. Pignol B, Etienne A, Crastes dP, Deby C, Mencia-Huerta JM, Braquet P. Role of flavonoids in the oxygen-free radical modulation of the immune response. Prog Clin Biol Res 1988; 280:173–182.

186. Moroney MA, Alcaraz MJ, Forder RA, Carey F, Hoult JR. Selectivity of neutrophil 5-lipoxygenase and cyclo-oxygenase inhibition by an anti-inflammatory flavonoid glycoside and related aglycone flavonoids. J Pharm Pharmacol 1988; 40:787–792.

187. 'T HB, Ip VCT Van Dijk H, Labadie RP. How flavonoids inhibit the generation of luminol-dependent chemiluminescence by activated human neutrophils. Chem Biol Interact 1990; 73:323–335.
188. Ogasawara H, Fujitani T, Drzewiecki G, Middleton EJ. The role of hydrogen peroxide in basophil histamine release and the effect of selected flavonoids. J Allergy Clin Immunol 1986; 78:321–328.
189. Middleton EJ. Effect of flavonoids on basophil histamine release and other secretory systems. Prog Clin Biol Res 1986; 213:493–506.
190. Pagonis C, Tauber Al, Pavlotsky N, Simons ER. Flavonoid impairment of neutrophil response. Biochem Pharmacol 1986; 35:237–245.
191. Busse WW, Kopp DE, Middleton EJ. Flavonoid modulation of human neutrophil function. J Allergy Clin Immunol 1984; 73:801–809.
192. Hodnick WF, Duval DL, Pardini RS. Inhibition of mitochondrial respiration and cyanide-stimulated generation of reactive oxygen species by selected flavonoids. Biochem Pharmacol 1994; 47:573–580.
193. Hodnick WF, Milosavljevic EB, Nelson JH, Pardini RS. Electrochemistry of flavonoids: relationships between redox potentials, inhibition of mitochondrial respiration, and production of oxygen radicals by flavonoids. Biochem Pharmacol 1988; 37:2607–2611.
194. Hodnick WF, Kung FS, Roettger WJ, Bohmont CW, Pardini RS. Inhibition of mitochondrial respiration and production of toxic oxygen radicals by flavonoids: a structure-activity study. Biochem Pharmacol 35:2345–2357.
195. Hodnick WF, Roettger WJ, Kung FS, Bohmont CW, Pardini RS. Inhibition of mitochondrial respiration and production of superoxide and hydrogen peroxide by flavonoids: a structure activity study. Prog Clin Biol Res 1986; 213:249–252.
196. Mittra B, Saha A, Chowdhury AR, Pal C, Mandal S, Mukhopadhyay S, et al. Luteolin, an abundant dietary component is a potent anti-leishmanial agent that acts by inducing topoisomerase II-mediated kinetoplast DNA cleavage leading to apoptosis. Mol Med 2000; 6:527–541.
197. Ahmad N, Gali H, Javed S, Agarwal R. Skin cancer chemopreventive effects of a flavonoid antioxidant silymarin are mediated via impairment of receptor tyrosine kinase signaling and perturbation in cell cycle progression. Biochem Biophys Res Commun 1998; 247:294–301.
198. Terao J, Yamaguchi S, Shirai M, Miyoshi M, Moon JH, Oshima S, et al. Protection by quercetin and quercetin 3-O-beta-D-glucuronide of peroxynitrite-induced antioxidant consumption in human plasma low-density lipoprotein. Free Radic Res 2001; 35:925–931.
199. Shirai M, Moon JH, Tsushida T, Terao J. Inhibitory effect of a quercetin metabolite, quercetin 3-O-beta-D-glucuronide, on lipid peroxidation in liposomal membranes. J Agric Food Chem 2001; 49:5602–5608.
200. Yamamoto N, Moon JH, Tsushida T, Nagao A, Terao J. Inhibitory effect of quercetin metabolites and their related derivatives on copper ion-induced lipid peroxidation in human low-density lipoprotein. Arch Biochem Biophys 1999; 372:347–354.

201. da Silva EL, Piskula MK, Yamamoto N, Moon JH, Terao J. Quercetin metabolites inhibit copper ion-induced lipid peroxidation in rat plasma. FEBS Lett 1998; 430:405–408.
202. Koga T, Meydani M. Effect of plasma metabolites of (+)-catechin and quercetin on monocyte adhesion to human aortic endothelial cells. Am J Clin Nutr 2001; 73:941–948.
203. Moon J, Tsushida T, Nakahara K, Terao J. Identification of quercetin 3-O-beta-D-glucuronide as an antioxidative metabolite in rat plasma after oral administration of quercetin. Free Radic Biol Med 2001; 30:1274–1285.
204. Tukey RH, Strassburg CP. Human UDP-glucuronosyltransferases: metabolism, expression, and disease. Annu Rev Pharmacol Toxicol 2000; 40:581–616.
205. Triggiani M, Gentile M, Secondo A, Granata F, Oriente A, Taglialatela M, et al. Histamine induces exocytosis and IL-6 production from human lung macrophages through interaction with H_1 receptors. J Immunol 2001; 166:4083–4091.
206. Shimoi K, Saka N, Nozawa R, Sato M, Amano I, Nakayama T, et al. Deglucuronidation of a flavonoid, luteolin monoglucuronide, during inflammation. Drug Metab Dispos 2001; 29:1521–1524.
207. Shimoi K, Saka N, Kaji K, Nozawa R, Kinae N. Metabolic fate of luteolin and its functional activity at focal site. Biofactors 2000; 12:181–186.
208. Miyake Y, Shimoi K, Kumazawa S, Yamamoto K, Kinae N, Osawa T. Identification and antioxidant activity of flavonoid metabolites in plasma and urine of eriocitrin-treated rats. J Agric Food Chem 2000; 48:3217–3224.
209. Halliwell B, Aruoma OI. Dna damage by oxygen-derived species—its mechanism and measurement in mammalian systems. FEBS Lett 1991; 281:9–19.
210. Halliwell B, Chirico S. Lipid-peroxidation—its mechanism, measurement, and significance. Am J Clin Nutr 1993; 57:S715–S725
211. Berlett BS, Stadtman ER. Protein oxidation in aging, disease, and oxidative stress. J Biol Chem 1997; 272:20313–20316.
212. Halliwell B, Gutteridge MC. Free Radicals in Biology and Medicine. 3d ed. Oxford: Clarendon Press, 1999.
213. Areias FM, Rego AC, Oliveira CR, Seabra RM. Antioxidant effect of flavonoids after ascorbate/Fe(2+)-induced oxidative stress in cultured retinal cells. Biochem Pharmacol 2001; 62:111–118.
214. Alliangana DM. Effects of beta-carotene, flavonoid quercitin and quinacrine on cell proliferation and lipid peroxidation breakdown products in BHK-21 cells. East Afr Med J 1996; 73:752–757.
215. Saija A, Scalese M, Lanza M, Marzullo D, Bonina F, Castelli F. Flavonoids as antioxidant agents: importance of their interaction with biomembranes. Free Radic Biol Med 1995; 19:481–486.
216. van Acker FA, Schouten O, Haenen GR, van D, V, Bast A. Flavonoids can replace alpha-tocopherol as an antioxidant. FEBS Lett 2000; 473:145–148.
217. Balasinska B, Troszynska A. Total antioxidative activity of evening primrose (*Oenothera paradoxa*) cake extract measured in vitro by liposome model and murine L1210 cells. J Agric Food Chem 1998; 46:3558–3563.
218. Xin W, Wei T, Chen C, Ni Y, Zhao B, Hou J. Mechanisms of apoptosis in rat

cerebellar granule cells induced by hydroxyl radicals and the effects of EGb761 and its constituents. Toxicology 2000; 148:103–110.

219. Yoshida H, Ishikawa T, Hosoai H, Suzukawa M, Ayaori M, Hisada T, et al. Inhibitory effect of tea flavonoids on the ability of cells to oxidize low density lipoprotein. Biochem Pharmacol 1999; 58:1695–1703.

220. Morel I, Abalea V, Sergent O, Cillard P, Cillard J. Involvement of phenoxyl radical intermediates in lipid antioxidant action of myricetin in iron-treated rat hepatocyte culture. Biochem Pharmacol 1998; 55:1399–1404.

221. Spencer JPE, Jenner A, Chimel K, Aruoma OI, Cross CE, Wu R, et al. DNA strand breakage and base modification induced by hydrogen-peroxide treatment of human respiratory-tract epithelial-cells. FEBS Lett 1995; 374:233–236.

222. Spencer JPE, Jenner A, Aruoma OI, Cross CE, Wu R, Halliwell B. Oxidative DNA damage in human respiratory tract epithelial cells: time course in relation to DNA strand breakage. Biochem Biophys Res Commun 1996; 224:17–22.

223. Aherne SA, O'Brien NM. Mechanism of protection by the flavonoids, quercetin and rutin, against tertbutylhydroperoxide-and menadione-induced DNA single strand breaks in Caco-2 cells. Free Radic Biol Med 2000; 29:507–514.

224. O'Brien NM, Woods JA, Aherne SA, O'Callaghan YC. Cytotoxicity, genotoxicity and oxidative reactions in cell-culture models: modulatory effects of phytochemicals. Biochem Soc Trans 2000; 28:22–26.

225. Duthie SJ, Dobson VL. Dietary flavonoids protect human colonocyte DNA from oxidative attack in vitro. Eur J Nutr 1999; 38:28–34.

226. Noroozi M, Angerson WJ, Lean ME. Effects of flavonoids and vitamin C on oxidative DNA damage to human lymphocytes. Am J Clin Nutr 1998; 67:1210–1218.

227. Abalea V, Cillard J, Dubos MP, Sergent O, Cillard P, Morel I. Repair of iron-induced DNA oxidation by the flavonoid myricetin in primary rat hepatocyte cultures. Free Radic Biol Med 1999; 26:1457–1466.

228. Morel I, Abalea V, Cillard P, Cillard J. Repair of oxidized DNA by the flavonoid myricetin. Methods Enzymol 2001; 335:308–316.

229. Sestili P, Guidarelli A, Dacha M, Cantoni O. Quercetin prevents DNA single strand breakage and cytotoxicity caused by tert-butylhydroperoxide: free radical scavenging versus iron chelating mechanism. Free Radic Biol Med 1998; 25:196–200.

230. Lean ME, Noroozi M, Kelly I, Burns J, Talwar D, Sattar N, et al. Dietary flavonols protect diabetic human lymphocytes against oxidative damage to DNA. Diabetes 1999; 48:176–181.

231. Takabayashi F, Harada N, Tahara S, Kaneko T, Hara Y. Effect of green tea catechins on the amount of 8-hydroxydeoxyguanosine (8-OHdG) in pancreatic and hepatic DNA after a single administration of N-nitrosobis(2-oxopropyl)amine (BOP). Pancreas 1997; 15:109–112.

232. Weitberg AB, Corvese D. The effect of epigallocatechin galleate and sarcophytol A on DNA strand breakage induced by tobacco-specific nitrosamines and stimulated human phagocytes. J Exp Clin Cancer Res 1999; 18:433–437.

233. Dizdaroglu M, Rao G, Halliwell B, Gajewski E. Damage to the DNA bases in

mammalian chromatin by hydrogen-peroxide in the presence of ferric and cupric ions. Arch Biochem Biophys 1991; 285:317–324.

234. Aruoma OI, Halliwell B, Gajewski E, Dizdaroglu M. Damage to the bases in DNA induced by hydrogen-peroxide and ferric ion chelates. J Biol Chem 1989; 264:20509–20512.

235. Pool-Zobel BL, Bub A, Schroder N, Rechkemmer G. Anthocyanins are potent antioxidants in model systems but do not reduce endogenous oxidative DNA damage in human colon cells. Eur J Nutr 1999; 38:227–234.

13

Understanding the Bioavailability of Flavonoids Through Studies in Caco-2 Cells

Thomas Walle, Richard A. Walgren, U. Kristina Walle, Alema Galijatovic, and Jaya b. Vaidyanathan
Medical University of South Carolina
Charleston, South Carolina, U.S.A.

I. INTRODUCTION

A large and steadily growing body of information, most from in vitro studies [1], supports the beneficial role of flavonoids in the prevention of human disease. A prerequisite for this proposition is that the flavonoids need to be bioavailable to exert these beneficial effects in vivo. Yet, to date, the question of whether these dietary compounds are indeed bioavailable has only limited answers [2,3]. Attempts to ascertain the bioavailability of the flavonoids have been limited by the complex behavior of these molecules in in vivo conditions. It is now known that the bioavailability is restricted by such factors as limited absorption [4], extensive metabolism [5], degradation due to the intestinal microflora [6,7] as well as intestinal enzymes [8,9], and, for some flavonoids, limited chemical stability [10,11].

To gain a better understanding of the various factors affecting the bioavailability of flavonoids, there have been two general approaches: (1) isolated rodent intestinal preparations and (2) human intestinal cell culture using Caco-2 cells. The first of these approaches has many proponents and has yielded much information about mechanisms of transport as well as metabolism [12–18]. As these experiments are done with rat or other rodent tissues, the relevance of the information obtained with this system to the human may, however, sometimes be questionable.

The second approach, initiated rather recently, has also led to new information about the mechanisms of both transport and metabolism. This approach has the attractive advantage of using cultured human cells. Although highly relevant to the clinical situation, the Caco-2 cell line has disadvantages as well.

The objectives of this chapter are to examine the utility of the Caco-2 cell monolayer as a model for studies of the oral bioavailability of dietary flavonoids in humans and to outline the information that has been obtained with this system.

II. Caco-2 CELL MONOLAYER

As a result of the complexity of the in vivo environment, we have adopted a systematic and reductionist approach to understand the underlying factors affecting the bioavailability and mechanisms governing the absorption of flavonoids. Our studies have employed a model of the principal barrier of intestinal absorption, the monolayer of columnar epithelial cells or enterocytes lining the intestinal lumen. Our selected model of the intestinal epithelium is the human Caco-2 cell line. Caco-2 cells are well-differentiated human intestinal cells that have been extensively characterized [19–22]. Caco-2 cells were originally isolated from a human colonic adenocarcinoma [23]. Despite their origin, these cells more closely resemble enterocytes than colonocytes both morphologically and biochemically. Differentiated Caco-2 monolayers express transport systems for sugars, amino acids, bile acids, and dipeptides and express many of the brush border membrane enzymes, including aminopeptidase, alkaline phosphatase, sucrase, dipeptidyl aminopeptidase, and γ-glutamyl transpeptidase.

Examination of colonocyte-and enterocyte-specific markers has demonstrated that fully differentiated Caco-2 cells more closely resemble enterocytes. For example, Caco-2 cells produce surfactantlike particles, a secreted membrane produced in the colon and small intestine. Examination of the composition of the Caco-2 surfactantlike particle [24] has shown that in early differentiation these cells produce particles with a mixed enteric and colonic composition containing the colonic markers tissue-unspecific alkaline phosphatase (colonic), surfactant protein A (colonic), and α_1-antitrypsin (enteric). Three to six days after reaching confluence (i.e., with further differentiation), there are a loss of colonic marker expression and an increase in enteric marker expression, including a transition from production of the tissue-unspecific alkaline phosphatase to production of the intestinal alkaline phosphatase.

In addition to the resemblance of Caco-2 monolayers to enterocytes, Caco-2 monolayers are a well-validated and accepted preclinical model for predicting human intestinal absorption of drug molecules. Apparent permeability values (see Sec. III. B) obtained in Caco-2 cells have been shown to be predictive of human in vivo permeability [25,26] and have been strongly correlated with data on the oral absorption in humans [27–29]. In a 2001 study, it was also demonstrated that the Caco-2 cell is a good model for the human intestinal ATP-binding cassette (ABC) transporters, which are of great importance in regulation and absorption of many foreign compounds [30].

III. TRANSPORT EXPERIMENTS WITH Caco-2 MONOLAYERS

A. Cell Culture

The experimental setup is depicted in Figure 1. Caco-2 cells are grown as monolayers in Eagle's minimum essential medium with Earle's salts, 10% fetal calf serum, 1% nonessential amino acids, penicillin (100 U/mL), and streptomycin (0.1 mg/mL) in a humidified 37 °C incubator with 5% carbon dioxide. Stock cultures are split 1:12 when just confluent, using trypsin with ethylene diamine tetra acetic acid (EDTA). For transport studies, the Caco-2 cells are seeded at a density of 100,000 cells per 1-cm^2 tissue culture insert containing 0.4-μm pore size permeable polycarbonate membranes (Transwells, Corning Costar Corp., Cambridge, MA). The inserts are placed in 12-well tissue culture plates (Fig. 1). The volume of cell culture medium within the inserts is 0.5 mL (apical or mucosal side), and the surrounding wells contain 1.5 mL (basolateral or serosal side). The medium on both sides of the cell layer is changed every 2 days. The integrity of the cell monolayers is evaluated by measuring the transepithelial electrical resistance (TEER) values with a volt/ohmmeter (Millicell-ERS, Millipore Corp., Bedford, MA). Only cell inserts with resistance values exceeding 400 Ω cm^2 are used for transport experiments. The transport of [^{14}C]mannitol, a marker of paracellular transport, is also measured in the inserts. The cell monolayers are considered "tight" or well formed when the mannitol transport is less than ≈ 0.3% of the dose per hour, corresponding to an apparent permeability coefficient (P_{app}) of 0.5 × 10^{-6} cm/s (discussed later). The inserts are used for experiments at 18–30 days after seeding. Before experiments, the cells are washed twice for 30 min with

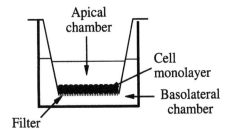

Figure 1 Caco-2 cell monolayer in tissue culture insert (Transwell).

warm Hanks' balanced salt solution (HBSS) buffered to pH 7.4 with 25-mM HEPES. The buffer is then replaced with fresh HBSS buffer on one side of the cell layer and flavonoid in HBSS buffer on the other side. Loading the flavonoid on the apical side with sampling from the basolateral side mimics absorption, whereas loading on the basolateral side with sampling from the apical side mimics efflux. [14C]Mannitol is added to the apical side of all inserts, and a basolateral sample is assayed by liquid scintillation spectrometry at the end of the experiment. In general the transport is linear for at least 3 h. However, when metabolism occurs, as is common for flavonoids, it has been our experience that it is best to do the transport experiments for no longer than 1 h.

B. Calculations

The apparent permeability coefficients (P_{app}), expressed in centimeters per second [27], are calculated as $\Delta Q/\Delta t \times 1/60 \times 1/A \times 1/C_0$, where $\Delta Q/\Delta t$ is the permeability rate (micrograms per minute), A is the surface area of the membrane (square centimeters), and C_0 is the initial concentration in the donor chamber (micrograms per milliliter).

IV. ABSORPTION OF FLAVONOID GLUCOSIDES

Most dietary flavonoids are consumed as glycosides. For example, the most prevalent dietary flavonoid, quercetin, is present in the diet in a number of

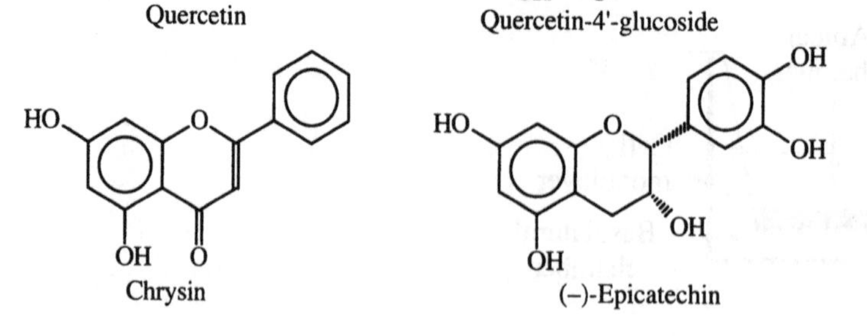

Figure 2 Chemical structures of flavonoids.

glycosylated forms, the most abundant of which is quercetin $4'$-β-O-glucoside (QG) (Fig. 2). The great diversity of dietary flavonoids has posed significant challenges to measuring the bioavailability of these compounds. It was originally thought that the glycosides were too polar to pass through the lipid membranes of the enterocytes. It was thus presumed that the absorption of flavonoids required hydrolysis by the intestinal bacterial microflora to release the aglycone [31]. This was never demonstrated directly but served as a reasonable hypothesis. In 1995, Hollman and associates put forward the hypothesis that the glucosides of flavonoids, e.g., QG, were absorbed intact via the sodium-dependent glucose transporter 1 (SGLT1) [32]. Again, this was never demonstrated but remained as another reasonable hypothesis.

In 1998, Walgren and colleagues demonstrated for the first time, using the human Caco-2 cell monolayer as a model, that whereas quercetin was reasonably well absorbed, QG was not absorbed but rather was efficiently effluxed across the enterocytes [4]. This surprising finding led to further in-depth studies of the mechanisms involved in flavonoid transport using Caco-2 cells and resulted in two key observations. The first was that QG was in fact absorbed across the apical (mucosal) membrane via the SGLT1 transporter and accumulated within the cell. This was demonstrated by using appropriate tools for SGLT1 inhibition in the Caco-2 cells and was rigorously confirmed in SGLT1-transfected CHO cells [33]. The second finding was that although QG accumulated within the enterocyte, it was not further translocated across the basolateral membrane. This condition was due, at least in part, to the fact that it was transported out of the cell across the apical membrane by a transporter opposing SGLT1. Through kinetic studies as well as use of a selective transport inhibitor, MK-571, this was strongly suggested to be the multidrug resistance–associated protein 2 (MRP2) transporter [34]. Only when using a high concentration of this inhibitor together with high concentrations of QG was it possible to override this efflux and achieve some transcellular absorption of QG [34]. These scenarios are summarized in Fig. 3.

It would appear that the conclusions based on these Caco-2 cell studies do indeed apply to the human in vivo condition, as there are still a lack of evidence for QG absorption, and only evidence for absorption of metabolites of quercetin [9,35,36]. The Caco-2 cell studies have also pointed to an additional mechanism governing the absorption of flavonoids, i.e., hydrolysis of QG to quercetin [33], which could then be absorbed (Fig. 3). The importance of this pathway has been well supported by a study in ileostomy patients in which the administration of QG resulted in complete hydrolysis to quercetin, presumably within the small intestine, followed by effective absorption of quercetin [9]. The nature of the glucosidase(s) involved and their exact location have been addressed [8,37] but require much further study.

Observations with the isoflavonoid genistin (genistein-7-glucoside) support our findings with QG. As for QG, this glucoside was not absorbed by the Caco-2

Lumen

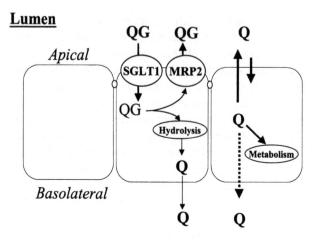

Figure 3 Fluxes and fates of quercetin (Q) and quercetin-4′-glucoside (QG) in the human enterocyte monolayer model Caco-2.

cells but instead effluxed. The MRP2 transport inhibitor reduced this efflux by 87% [38]. Also, as for QG, genistin was hydrolyzed to its aglycone in this preparation.

The extent to which the findings with QG and genistin apply to other flavonoid glucosides or flavonoids conjugated with other sugar moieties is not yet known. Further experiments in this area may benefit from using the Caco-2 cell as a model.

V. ABSORPTION OF FLAVONOID AGLYCONES

Considering the preceding findings, it becomes obvious that the absorption of flavonoid aglycones is of great importance. For quercetin, shown in Fig. 3, the absorption appeared to be transcellular, although efflux was greater than absorption [4]. This observation was not pursued further in the Caco-2 cell system, because of very limited stability of this flavonoid in cell culture systems [11]. We thus selected a chemically stable flavonoid for these studies, i.e., chrysin (see Fig. 2), which is a somewhat more lipid-soluble analog of quercetin with only two hydroxyl groups. For chrysin, we expected to see a high rate of transcellular absorption. However, this was not the case [5]. Instead, the transcellular absorption of this flavonoid seemed to be limited by efficient metabolism in the Caco-2 cell monolayer. In further studies with chrysin, we could conclude that glucuronidation via the uridine diphosphate–(UDP)-glucuronosyltransferases (UGTs) and sulfation via the sulfotransferases (SULTs) were highly efficient in the Caco-2 cells [39] and, presumably, also in the normal human intestine. The bioavailability of chrysin was thus highly impaired by glucuronidation and sulfation, and the metabolites formed were efficiently effluxed to the apical side, presumably by the apical membrane transporter MRP2 [5]. This was the first time that enterocyte metabolism/transport was demonstrated to be the key determinant of the bioavailability of flavonoid aglycones (Fig. 4).

These studies with chrysin were extended into a clinical investigation in which seven normal volunteers received a single oral dose of chrysin. Data obtained from this study yielded a very low estimated bioavailability of 0.003–0.02% [40]. This was in agreement with the results of the Caco-2 cell study and, extrapolating from the in vitro studies, was likely due to extensive presystemic glucuronidation and sulfation. Interestingly, when Caco-2 cells were pretreated with chrysin for several days, using concentrations that may be anticipated in a clinical setting, one of the metabolic pathways, i.e., glucuronidation, was induced (Fig. 4) [41]. The results were an upregulation of UGT1A1 in the Caco-2 cells [42] and considerably more efficient glucuronidation and elimination of chrysin. Whether this also applies to the in vivo condition in the human intestine remains to be demonstrated.

Epicatechin is one of the flavonoids in green tea and is present in tea leaves as the aglycone. It has been the subject of additional transport studies in our laboratory. Much to our surprise, we were unable to detect any apical to basolateral absorption of this compound across the Caco-2 cell monolayers [43]. Although this result may in part be due to limited detection sensitivity, it is doubtful that a rate of transport below our level of detection would be

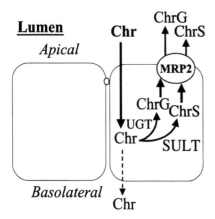

Figure 4 Fluxes and fates of chrysin (Chr) in the human enterocyte monolayer model Caco-2. G, glucuronide conjugate; S, sulfate conjugate; UGT, uridine diphosphate-glucuronosyltransferase; SULT, sulfotransferase.

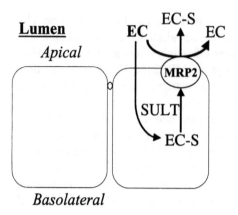

Figure 5 Fluxes and fates of epicatechin (EC) in the human enterocyte mono-layer model Caco-2. S, sulfate conjugate; SULT, sulfotransferase.

significant. On the other hand, epicatechin was effectively effluxed with a P_{app} value of 1.3×10^{-6} cm/s (compared to mannitol transport in these cells of 0.3×10^{-6} cm/s). This observed unidirectional transport once again suggested the presence of an efflux transporter. Although verapamil, a P-glycoprotein trans-porter inhibitor, had no effect on this efflux, MK-571, an MRP2 transporter inhibitor, inhibited the efflux by 50%. Of further importance was the observation that MK-571, when added together with epicatechin on the apical side of the monolayer, produced low but measurable absorption of epicatechin. In addition, as with chrysin, a sulfate conjugate of epicatechin was formed by the Caco-2 cells. This conjugate showed highly efficient efflux to the apical side by the MRP2 transporter. These scenarios are summarized in Figure 5. These observations are supported by clinical data, which have demonstrated that epicatechin and other tea flavonoids have very low oral bioavailability in humans [44]. It remains to be seen whether the interaction of epicatechin with MRP2 will hold true in vivo studies.

VI. STUDIES OF ADDITIONAL FLAVONOIDS

A number of additional flavonoids have been studied by using the Caco-2 cell model. These include the simplest form of the flavone class of flavonoids, the highly lipophilic unsubstituted flavone, which has been shown to diffuse readily across the enterocyte monolayer [45]. It also includes the highly polar hesperidin glycosides, which are suggested to be transported at a low rate via the para-cellular pathway [46]. Another citrus flavonoid, 7-geranyloxycoumarin, has been shown to have a low transcellular permeation rate but was also shown to accumulate in the Caco-2 cells [47]. Nobiletin, a lipophilic polymethoxylated citrus flavonoid, showed high accumulation in Caco-2 cells, in contrast to the hydrophilic luteolin [48]. Proanthocyanidins and in particular their polymeric forms have demonstrated low transport rates in Caco-2 cells [49].

VII. POTENTIAL PROBLEMS WITH Caco-2 CELLS

One potential difficulty of work with the Caco-2 cells is related to the tightness of the monolayer. This is particularly important for compounds with very low net transport rates. In the original development of this transport model, it was recommended to measure the TEER value both before and after a transport experiment. A TEER value $\geq 300 \, \Omega \cdot cm^2$ in general meant a tight monolayer. However, this may not necessarily be true. As have some other laboratories, we have selected to test [^{14}C]-mannitol transport as well. The relationship between P_{app} for mannitol and the corresponding TEER values over a period of a year is shown in Figure 6. On the basis of this information, we have in our laboratory classified a tight monolayer to have a TEER value $\geq 400 \, \Omega \cdot cm^2$, which produces a P_{app} for mannitol $<0.5 \times 10^{-6}$ cm/s.

On rare occasions, we have observed changes in the Caco-2 cell transport of certain compounds. At these times the TEER values have been "normal," but higher mannitol P_{app} values have been observed, changing from 0.2–0.5 to

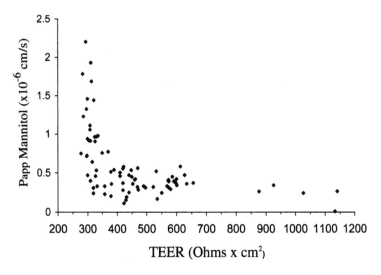

Figure 6 Mannitol flux versus TEER values in confluent Caco-2 cells grown in Transwells. TEER, transepithelial electrical resistance.

$0.7-1.8 \times 10^{-6}$ cm/s. Of great concern was the observation that with these cells certain flavonoids that normally demonstrate a complete lack of absorption (e.g., QG and epicatechin) now clearly demonstrated absorption (P_{app} values $\approx 0.6 \times 10^{-6}$ cm/s). Although this finding may have been due to changes in transporter expression or cell morphological characteristics, we do not know the exact reason for this change in the Caco-2 cell transport. However, after use of a new batch of frozen cells into culture, this problem was overcome.

VIII. CONCLUSIONS

Transport studies with Caco-2 monolayers in conjunction with molecularly specific analytical methodologies provided exciting opportunities to examine systematically the steps involved in the transport of compounds across the principal transport barrier of the intestine, the enterocyte. The ability to simplify the number of confounding variables (e.g., the absence of intestinal microflora and pancreatic enzymes and lack of peristaltic mixing) is both a strength and a limitation of the system. Using this system, we have learned that the degree of absorption of the flavonoids is not due only to glycosylation or nonglycosylation of the molecule and degree of lipophilicity. Rather, absorption of the flavonoids is much more complicated and dependent on myriad interactions with numerous transporters and enzymes contained in or on the surface of the membranes and within the cytosol of enterocytes. To date several proteins have been shown to play key roles, including (1) SGLT1, (2) MRP2, (3) P-glycoprotein, (4) UGTs, and (5) SULTs. At present it is not known how the expression levels of these proteins compare for this cell culture system and in vivo conditions; nor are we fully cognizant of the potential interindividual variability of these proteins. Further defining these differences in expression will become key in understanding how these proteins regulate the bioavailability of the flavonoids. At this time it is also uncertain how other dietary constituents impact the absorption of flavonoids or, possibly of greater importance, how flavonoids impact the absorption of other ingested compounds (i.e., nutrients and pharmaceutical agents). It is clear, though, that the Caco-2 cell culture system can be a powerful tool in helping us address these questions and that the resulting information will increase our understanding of the methods through which cellular transport processes influence the disposition and action of natural products and may aid in the further development of new therapeutic agents.

ACKNOWLEDGMENTS

This research was supported by the National Institutes of Health grant GM55561 and the USDA grant CSREES 00-35200-9071.

REFERENCES

1. Middleton EJ, Kandaswami C, Theoharides TC. The effects of plant flavonoids on mammalian cells: implications for inflammation, heart disease, and cancer. Pharmacol Rev 2000; 52:673–751.
2. Graefe EU, Derendorf H, Veit M. Pharmacokinetics and bioavailability of the flavonol quercetin in humans. Int J Clin Pharmacol Ther 1999; 37:219–233.
3. Rice-Evans C, Spencer JPE, Schroeter H, Rechner AR. Bioavailability of flavonoids and potential bioactive forms in vivo. Drug Metab Drug Interact 2000; 17:291–310.
4. Walgren RA, Walle UK, Walle T. Transport of quercetin and its glucosides across human intestinal epithelial Caco-2 cells. Biochem Pharmacol 1998; 55:1721–1727.
5. Walle UK, Galijatovic A, Walle T. Transport of the flavonoid chrysin and its conjugated metabolites by the human intestinal cell line Caco-2. Biochem Pharmacol 1999; 58:431–438.
6. Bokkenheuser VD, Shackleton CHL, Winter J. Hydrolysis of dietary flavonoid glycosides by strains of intestinal *Bacteroides* from humans. Biochem J 1987; 248:953–956.
7. Walle T, Walle UK, Halushka PV. Carbon dioxide is the major metabolite of quercetin in humans. J Nutr 2001; 131:2648–2652.
8. Day AJ, DuPont MS, Ridley S, Rhodes M, Rhodes MJC, Morgan MRA, Williamson G. Deglycosylation of flavonoid and isoflavonoid glycosides by human small intestine and liver β-glucosidase activity. FEBS Lett 1998; 436:71–75.
9. Walle T, Otake Y, Walle UK, Wilson FA. Quercetin glucosides are completely hydrolyzed in ileostomy patients before absorption. J Nutr 2000; 130:2658–2661.
10. Nordström CG. Autoxidation of quercetin in aqueous solution: an elucidation of the autoxidation reaction. Suomen Kemistilehti 1968; B41:351–353.
11. Boulton DW, Walle UK, Walle T. Fate of the flavonoid quercetin in human cell lines: chemical instability and metabolism. J Pharm Pharmacol 1999; 51:353–359.
12. Inoue H, Yokota H, Taniyama H, Kuwahara A, Ogawa H, Kato S, Yuasa A: 1-Naphthyl β-D-glucuronide formed intraluminally in rat small intestine mucosa and absorbed into the colon. Life Sci 1999; 65:1579–1588.
13. Kuhnle G, Spencer JPE, Schroeter H, Shenoy B, Debnam ES, Srai SKS, Rice-Evans C, Hahn U. Epicatechin and catechin are O-methylated and glucuronidated in the small intestine. Biochem Biophys Res Commun 2000; 277:507–512.
14. Gee JM, DuPont MS, Day AJ, Plumb GW, Williamson G, Johnson IT. Intestinal transport of quercetin glycosides in rats involves both deglycosylation and interaction with the hexose transport pathway. J Nutr 2000; 130:2765–2771.
15. Donovan JL, Crespy V, Manach C, Morand C, Besson C, Scalbert A, Remesy C. Catechin is metabolized by both the small intestine and liver of rats. J Nutr 2001; 131:1753–1757.
16. Andlauer W, Stumpf C, Fürst P. Intestinal absorption of rutin in free and conjugated forms. Biochem Pharmacol 2001; 62:369–374.
17. Spencer JP, Schroeter H, Shenoy B, Srai SK, Debnam ES, Rice-Evans C. Epicatechin is the primary bioavailable form of procyanidin dimers B2 and B5 after transfer across the small intestine. Biochem Biophys Res Commun 2001; 285:588–593.
18. Crespy V, Morand C, Besson C, Manach C, Démigné C, Rémésy C. Comparison

of the intestinal absorption of quercetin, phloretin and their glucosides in rats. J Nutr 2001; 131:2109–2114.

19. Hidalgo IJ, Raub TJ, Borchardt RT. Characterization of the human colon carcinoma cell line (Caco-2) as a model system for intestinal epithelial permeability. Gastroenterology 1989; 96:736–749.

20. Meunier V, Bourrié M, Berger Y, Fabre G. The human intestinal epithelial cell line Caco-2: pharmacological and pharmacokinetic applications. Cell Biol Toxicol 1995; 11:187–194.

21. Gan L-SL, Thakker DR. Applications of the Caco-2 model in the design and development of orally active drugs: elucidation of biochemical and physical barriers posed by the intestinal epithelium. Adv Drug Delivery Rev 1997; 23:77–98.

22. Artursson P, Borchardt RT. Intstinal drug absorption and metabolism in cell cultures: Caco-2 and beyond. Pharm Res 1997; 14:1655–1658.

23. Fogh J, Fogh JM, Orfeo T. One hundred and twenty-seven cultured human tumor cell lines producing tumors in nude mice. J Natl Cancer Inst 1977; 59:221–225.

24. Engle MJ, Goetz GS, Alpers DH. Caco-2 cells express a combination of colonocyte and enterocyte phenotypes. J Cell Physiol 1998; 174:362–369.

25. Lennernäs H. Human jejunal effective permeability and its correlation with preclinical absorption models. J Pharm Pharmacol 1997; 49:627–638.

26. Lennernäs H. Human intestinal permeability. J Pharm Sci 1998; 87:403–410.

27. Artursson P. Epithelial transport of drugs in cell culture. I. A model for studying the passive diffusion of drugs over intestinal absorbtive (Caco-2) cells. J Pharm Sci 1990; 79:476–482.

28. Artursson P, Karlsson J. Correlation between oral drug absorption in humans and apparent drug permeability coefficients in human intestinal epithelial (Caco-2) cells. Biochem Biophys Res Commun 1991; 175:880–885.

29. Yee S. In vitro permeability across Caco-2 cells (colonic) can predict in vivo (small intestinal) absorption in man — fact or myth. Pharm Res 1997; 14:763–766.

30. Taipalensuu J, Törnblom H, Lindberg G, Einarsson C, Sjöqvist F, Melhus H, Garberg P, Sjöström B, Lundgren B, Artursson P. Correlation of gene expression of ten drug efflux proteins of the ATP-binding cassette transporter family in normal human jejunum and in human intestinal epithelial Caco-2 cell monolayers. J Pharmacol Exp Ther 2001; 299:164–170.

31. Griffiths LA, Barrow A. Metabolism of flavonoid compounds in germ-free rats. Biochem J 1972; 130:1161–1162.

32. Hollman PCH, de Vries JHM, van Leeuwen SD, Mengelers MJB, Katan MB. Absorption of dietary quercetin glycosides and quercetin in healthy ileostomy volunteers. Am J Clin Nutr 1995; 62:1276–1282.

33. Walgren RA, Lin J-T, Kinne RK-H, Walle T. Cellular uptake of dietary flavonoid quercetin 4′-β-glucoside by sodium-dependent glucose transporter SGLT1. J Pharmacol Exp Ther 2000; 294:837–843.

34. Walgren RA, Karnaky KJ Jr, Lindenmayer GE, Walle T. Efflux of dietary flavonoid quercetin 4′-β-glucoside across human intestinal Caco-2 cell monolayers by apical multidrug resistance-associated protein-2. J Pharmacol Exp Ther 2000; 294:830–836.

35. Morand C, Crespy V, Manach C, Besson C, Demigné C, Rémésy C. Plasma

metabolites of quercetin and their antioxidant properties. Am J Physiol 1998; 275:R212–R219.

36. Graefe EU, Wittig J, Mueller S, Riethling A-K, Uehleke B, Drewelow B, Pforte H, Jacobasch G, Derendorf H, Veit M. Pharmacokinetics and bioavailability of quercetin glycosides in humans. J Clin Pharmacol 2001; 41:492–499.

37. Day AJ, Cañada FJ, Diaz JC, Kroon PA, Mclauchlan R, Faulds CB, Plumb GW, Morgan MRA, Williamson G. Dietary flavonoid and isoflavone glycosides are hydrolysed by the lactase site of lactase phlorizin hydrolase. FEBS Lett 2000; 468:166–170.

38. Walle UK, French KL, Walgren RA, Walle T. Transport of genistein-7-glucoside by human intestinal Caco-2 cells: potential role for MRP2. Res Comm Mol Pathol Pharmacol 1999; 103:45–56.

39. Galijatovic A, Otake Y, Walle UK, Walle T. Extensive metabolism of the flavonoid chrysin by human Caco-2 and Hep G2 cells. Xenobiotica 1999; 29:1241–1256.

40. Walle T, Otake Y, Brubaker JA, Walle UK, Halushka PV. Disposition and metabolism of the flavonoid chrysin in normal volunteers. Br J Clin Pharmacol 2001; 51:143–146.

41. Galijatovic A, Walle UK, Walle T. Induction of UDP-glucuronosyltransferase by the flavonoids chrysin and quercetin in Caco-2 cells. Pharm Res 2000; 17:21–26.

42. Galijatovic A, Otake Y, Walle UK, Walle T. Induction of UDP-glucuronosyltransferase UGT1A1 by the flavonoid chrysin in Caco-2 cells — Potential role in carcinogen bioinactivation. Pharm Res 2001; 18:374–379.

43. Vaidyanathan JB, Walle T. Transport and metabolism of the tea flavonoid (−)-epicatechin by the human intestinal cell line Caco-2. Pharm Res 2001; 18:1420–1425.

44. Warden BA, Smith LS, Beecher GR, Balentine DA, Clevidence BA. Catechins are bioavailable in men and women drinking black tea throughout the day. J Nutr 2001; 131:1731–1737.

45. Kuo S-M. Transepithelial transport and accumulation of flavone in human intestinal Caco-2 cells. Life Sci 1998; 63:2323–2331.

46. Kim M, Kometani T, Okada S, Shimuzu M. Permeation of hesperidin glycosides across Caco-2 monolayers via the paracellular pathway. Biosci Biotech Biochem 1999; 63:2183–2188.

47. Murakami A, Wada K, Ueda N, Sasaki K, Haga M, Kuki W, Takahashi Y, Yonei H, Koshimizu K, Ohigashi H. In vitro absorption and metabolism of a citrus chemopreventive agent, auraptene, and its modifying effects on xenobiotic enzyme activities in mouse livers. Nutr Cancer 2000; 36:191–199.

48. Murakami A, Kuwahara S, Takahashi Y, Ito C, Furukawa H, Ju-ichi M, Koshimizu K, Ohigashi H. In vitro absorption and metabolism of nobiletin, a chemopreventive polymethoxyflavonoid in citrus fruits. Biosci Biotech Biochem 2001; 65:194–197.

49. Déprez S, Mila I, Huneau J-F, Tomé D, Scalbert A. Transport of proanthocyanidin dimer, trimer and polymer across monolayers of human intestinal epithelial Caco-2 cells. Antiox Redox Signal 2001; 3:957–967.

14

Metabolism in the Small Intestine and Gastrointestinal Tract

Jeremy P. E. Spencer and Catherine A. Rice-Evans
King's College London
London, England

Surjit Kaila Singh Srai
Royal Free University College Medical School
London, England

I. INTRODUCTION

Flavonoids have been the center of huge research interest over the last decade [1–4]. They are the most abundant polyphenols in the human diet and are divided into six main classes based on the degree of oxidation of the C-ring, the hydroxylation pattern of the ring structure, and the substitution in the 3-position: flavanols (e.g., epicatechin), flavonols (e.g., quercetin), flavones (e.g., luteolin), flavanones (e.g., naringenin), isoflavones (e.g., genistein) and anthocyanidins (e.g., cyanidin) [3] (Fig. 1). A large number of in vitro studies have characterized them as powerful antioxidants against both reactive oxygen and reactive nitrogen species [3,5–15]. Flavonoids with the highest antioxidant potential in vitro contain a B-ring catechol group that readily donates a hydrogen (electron) to stabilize a radical species [3]. Until recently, the ability of flavonoids to act as classical H-donating antioxidants was believed to underlie many of their reported health effects [16–22]. However, the extent of their antioxidant potential in vivo is dependent on the absorption, metabolism, distribution, and excretion of these compounds within the body after ingestion and the reducing properties of the resulting metabolites. An understanding of the processes involved in the absorption and distribution of polyphenols is essential for

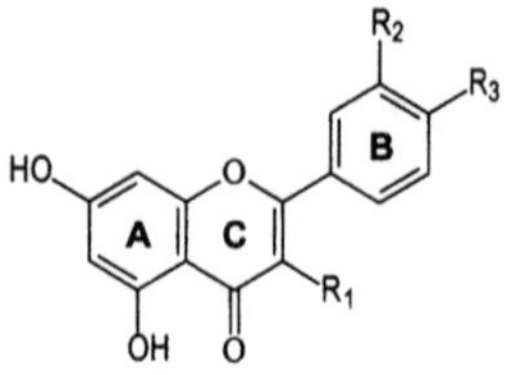

Flavonols

	R1	R2	R3
Quercetin	OH	H	H
Kaempferol	OH	H	H
Flavones			
Luteolin	H	OH	OH
Apigenin	H	H	OH

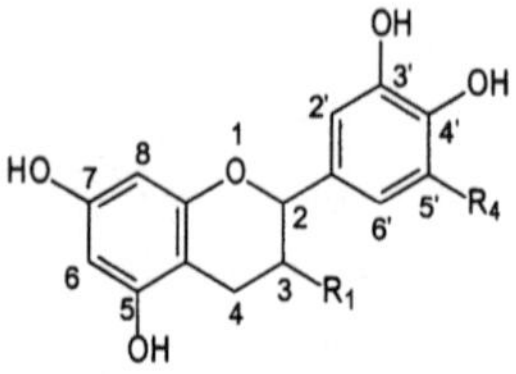

Flavanols

	R1	R2
Catechin	OH	H
Epicatechin	OH	H
EGC	OH	OH
ECG	gallate	H
EGCG	gallate	OH

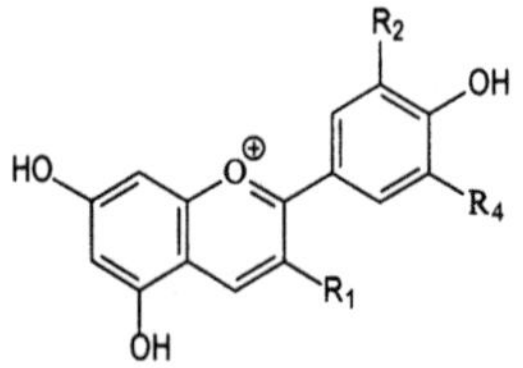

Anthocyanidins

	R2	R4
Cyanidin	OH	H
Malvidin	OCH$_3$	OCH$_3$
Delphinidin	OH	OH

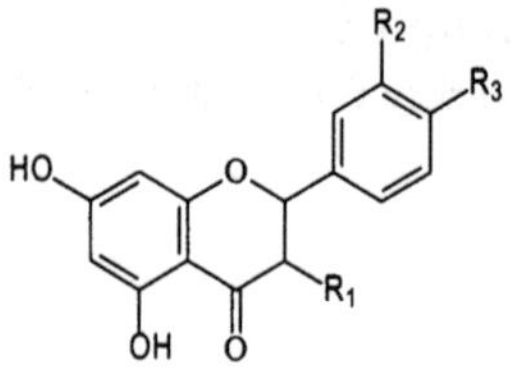

Flavanones

	R1	R2	R3
Taxifolin	OH	OH	OH
Naringenin	H	H	OH
Hesperetin	OH	OCH$_3$	H

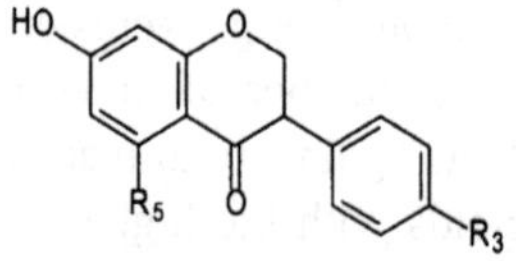

Isoflavones

	R3	R5
Genistein	OH	OH
Daidzein	OH	H

Figure 1 The structures of the five main classes of flavonoids. The major differences between the individual groups reside in the hydroxylation pattern of the ring structure, the degree of saturation of the C-ring, and the substitution in the 3-position.

determining their bioactivities in vivo and their significance. Since the late 1990s, much information has accumulated on the biotransformation of flavonoids in the small intestine and gastrointestinal tract [1,2,17,23–28], as well as the hepatic metabolism [29–32]. This chapter highlights the main sites of biotransformation of flavonoids within the gastrointestinal tract, the major metabolites generated in the small and large intestine, and the implications of this modification in determining how flavonoids may act in vivo.

II. MODIFICATION OF FLAVONOIDS IN THE MOUTH AND STOMACH

Few studies have investigated the ability of saliva and gastric juice to alter the flavonoid structure. Saliva has been found to have little effect on the stability of green tea catechins [33], however, degalloylation of flavanol gallate esters, such as epigallocatechin gallate, in human saliva has been observed [34]. Incubation of procyanidin oligomers (dimer-hexamer) in human saliva for up to 30 min does not result in modification of the compounds [28], suggesting that these compounds remain intact on entering the stomach. The quercetin rutinoside rutin has been shown to be hydrolysed by cell-free extracts of human salivary cultures [35,36]. In contrast, the quercetin-3-rhamnoside quercitrin is not susceptible to hydrolysis, suggesting that only rutin-glycosidase-elaborating organisms occur in saliva [36]. Furthermore, oral streptococci isolated from the mouth of normal individuals have been found to hydrolyse rutin to quercetin [37]. The streptococcal rutinase was found to be cytosolic and constitutive and to have a pH optimum of 6.5, and its liberation of quercetin from rutin in the mouth was hypothesized to be involved in intraoral cancer. The interaction of flavanols and procyanidins with salivary proteins has shown that (+)− catechin has a higher affinity for proline-rich proteins than (-)-epicatechin, and C(4)-C(8)-linked procyanidin dimers bind more strongly to them than their C(4)-C(6) counterparts [38]. Polyphenol-protein binding in the form of adsorption with high-molecular-weight salivary proteins, bacterial cells, and mucous materials may be one explanation for the observed decrease in quercetin mutagenicity after incubation with saliva [39].

Procyanidin oligomers ranging from a dimer to a decamer (isolated from *Theobroma cacao*) have been observed to be unstable under conditions of low pH similar to that present in the gastric juice of the stomach [40]. On incubation of the procyanidins with simulated gastric juice, oligomers rapidly decompose essentially to epicatechin monomeric and dimeric units but also to other oligomeric units [40]. Procyanidins may decompose in mild acidic environments as they are readily cleaved to form flavan-3-ol and quinone methide, which is in equilibrium with a carbocation in stronger acidic conditions (Fig. 2) [41]. The carbocation is converted to an anthocyanin on heating in alcoholic solutions, and the quinone

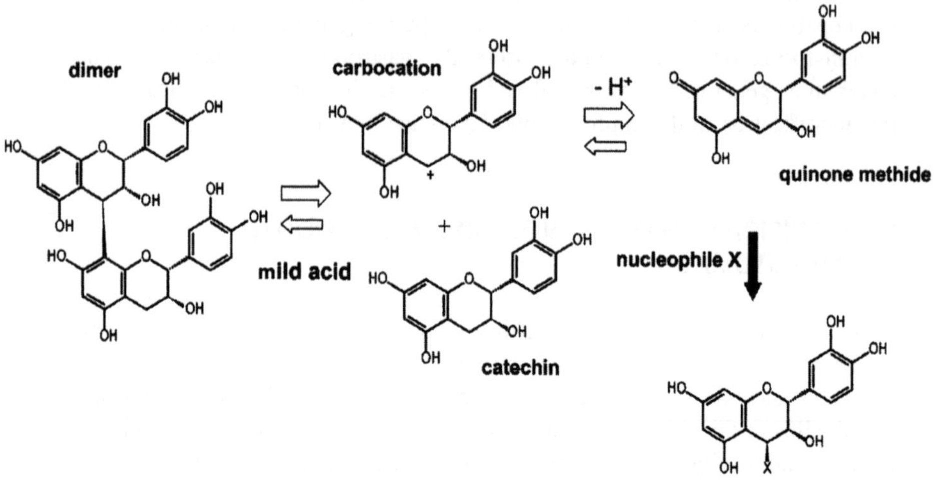

Figure 2 Decomposition of procyanidins in mild acidic environments (dimer as example). Procyanidins are readily cleaved in mild acid solutions to form flavan-3-ol and quinone methide, which is in equilibrium with a carbocation in stronger acidic conditions. The carbocation is converted to an anthocyanin on heating in alcoholic solutions. The quinone methide may be captured with a suitable nucleophile (X).

methide may be captured with a nucleophile. During incubation in acid, other oligomeric units, such as trimer and tetramer, are also formed and degraded, and in some instances a time-dependent formation of larger oligomers occurs. Thus, absorption of flavanols and procyanidins, for example, after consumption of chocolate or cocoa, is likely to be influenced by preabsorption events in the gastric lumen within the residence time. However, consideration needs to be given to the food matrix, which may influence the pH environment of the procyanidins and their subsequent decomposition. Monomeric flavonoid glycosides have been observed to be stable in the acidic environment of the stomach and are not observed to undergo nonenzymatic deglycosylation [42]. Interestingly, the flavonoids ponciretin, hesperetin, naringenin, and diosmetin and phenolic acids generated from flavonoids by human intestinal microflora (discussed later) have been observed to be effective inhibitors of the growth of *Helicobacter pylori*, a bacterium known to cause problems in the stomach of some patients [43].

III. METABOLISM AND CONJUGATION IN THE SMALL INTESTINE

Generally flavonoids are present in plants conjugated to sugars, and therefore it is these glycosides that are ingested in the diet and enter the gastrointestinal tract. The exception to this rule are the flavan-3-ols, such as the catechins and procyanidins, which are almost always present in the diet in the nonglycosylated form [3]. There are many factors that influence the extent and rate of absorption of ingested compounds by the small intestine [44]. These include physicochemical factors such as molecular size, lipophilicity, solubility, and pKa and biological factors including gastric and intestinal transit time, lumen pH, membrane permeability, and first-pass metabolism [45,46].

A. Effects of Intestinal Juice

On transfer from the stomach to the jejunum (the top two-fifths of the small intestine) the pH rises from about 2.0 to 7.0. It is well known that polyphenolic compounds such as those with catechol structures can oxidize in neutral and alkaline pH environments. Epigallocatechin gallate (EGCG) has been observed to oxidize rapidly in authentic intestinal fluid (measured pH of 8.5) with the amount of EGCG decreasing 81.6% in only 5 min [47], whereas a similar incubation in murine plasma (pH 7.4) resulted in only a 29.3% decrease in amount. However, oxidation of EGCG resulted in the formation of dimerized products that were observed to possess greater superoxide radical scavenging activity then EGCG itself and have powerful iron chelating properties [47]. Another factor to consider may be the relative abilities of these polyphenols to bind to proteins in the food matrices in question. In complex food matrices, the pH is likely to be buffered for long periods, and, therefore, oxidation of the flavanols may only occur to a limited extent during intestinal transit. In addition, it has been suggested that ascorbate significantly increases the stability of flavanols incubated in intestinal fluid [48], and therefore the presence of ascorbate in vivo may stabilize the polyphenols in the neutral or alkaline environment of the small intestine.

B. Jejunal and Ileal Conjugation and Metabolism

Many studies have indicated that significant transfer of ingested flavonoids occurs from the lumen of the small intestine to the mesenteric circulation and that extensive metabolism and conjugation of the flavonoid occur during this transfer [2,49–56]. Isolated preparations of rat small intestine [57] have been utilized to study absorption and metabolism in the small intestine and can provide information on events occurring in both the jejunum and the ileum [50,52,56,58–60]. This model (Fig. 3) allows study of the intestinal transfer of dietary

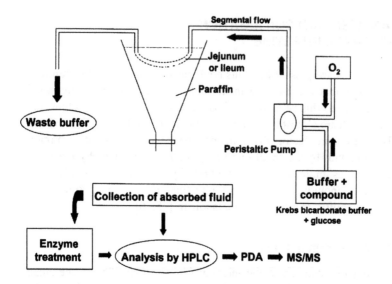

Figure 3 Diagrammatic representation of the isolated rat small intestine perfusion model. Buffers and compound are perfused for up to 90 min at a rate of 8 mL/min at 37 °C by using a peristaltic pump. Serosal fluid was collected from the bottom of the paraffin chamber and analyzed by HPLC and mass spectrometry. HPLC, high-performance liquid chromatography.

polyphenols (and their glycosides) and can be used to assess the rate of absorption from the lumen. The solute under study appears on the serosal surface in the same form as if it were transferred to the mesenteric circulation, and therefore, enterocyte metabolism of flavanols and procyanidins, as well as their rate of transfer across specific gut regions, may be studied [52]. Tissue viability is assessed by measurement of glucose transfer [57], and viability is confirmed by showing that fluid transfer continues at a constant rate for the 90-min collection period and that glucose concentration in the absorbed fluid is more than double that initially present in the perfused buffer [52].

Absorption studies utilizing this model, with a wide range of flavonoids and their glycosides and hydroxycinnamates, show that there was in almost all cases extensive metabolism of the polyphenol in the enterocyte during transfer from the luminal to the serosal side [52] (Table 1, jejunum; Table 2, ileum). The flavonoid glycosides, luteolin-7-glucoside, kaempferol-3-glucoside, and querce-tin-3-glucoside, were cleaved by rat jejunal or ileal mucosa, suggesting the presence of β-glucosidase action before efflux into the serosal fluid. The major products transferred across the small intestinal epithelium were glucuronides of the parent aglycone or of the hydrolysed glycoside, although O-methylated metabolites were also observed [52]. With an identical model, the major

Table 1 Summary of the Absorption and Metabolism of Flavonoids in the Isolated Rat Jejunum Model

Perfused compound (jejunum)	Total % absorbed (90 min)	% of total absorbed		Total %	% total absorbed		Total % of metabolite
		Perfused compound	Aglycone of perfused compound		Glucuronides	Other[a]	
Quercetin	37.8	2.4	*	2.4	91.3	6.3	97.6
Quercetin-3-glucoside	31.0	69.1	8.0	77.1	22.8	0.0	22.8
Rutin	10.3	100.0	0.0	100.0	0.0	0.0	0.0
Kaempferol	45.4	0.6	*	0.6	96.8	2.5	99.3
Kaempferol-3-glucoside	23.2	0.0	0.0	0.0	97.2	2.8	100.0
Luteolin	36.2	3.0	*	3.0	93.7	3.3	97.0
Luteolin-7-glucoside	27.9	0.0	11.5	11.5	85.5	3.0	88.5
Hesperetin	8.9	82.2	0.0	82.2	17.8	0.0	17.8
Naringenin	27.2	0.0	0.0	0.0	100.0	0.0	100.0
Naringenin-7-glucoside	35.0	0.0	0.0	0.0	100.0	0.0	100.0
Catechin	13.1	4.6	*	4.6	65.3	30.1	95.4
Epicatechin	11.6	5.2	*	5.2	61.7	33.1	94.8
Resveratrol	6.0	1.0	0.0	1.0	99.0	0.0	99.0

[a]O-methylated glucuronide and O-methylated metabolites.

Table 2 Summary of the Absorption and Metabolism of Flavonoids in the Isolated Rat Ileum Model

Perfused Compound (ileum)	Total % absorbed (90 min)	% of total absorbed			% of total absorbed		Total % of metabolite
		Perfused compound	Aglycone of perfused compound	Total %	Glucuronides	Other[a]	
Quercetin	19.3	17.1	*	17.1	82.9	0.0	82.9
Quercetin-3-glucoside	8.8	31.8	36.4	68.2	31.8	0.0	31.8
Rutin	0.6	100.0	0.0	100.0	NS	NS	NS
Kaempferol	59.1	32.7	*	32.7	67.3	0.0	67.3
Kaempferol-3-glucoside	11.4	0.0	82.6	82.6	18.4	0.0	18.4
Luteolin	25.6	12.9	*	12.9	87.1	0.0	87.1
Luteolin-7-glucoside	9.2	0.0	47.8	47.8	52.2	0.0	52.2
Hesperetin	11.9	98.3	0.0	98.3	0.7	0.0	0.7
Naringenin	–	–	–	–	–	–	–
Naringenin-7-glucoside	–	–	–	–	–	–	–
Catechin	66.16	69.3	*	69.3	22.9	7.8	30.7
Epicatechin	55.90	66.3	*	66.3	27.4	6.3	33.7
Resveratrol	2.3	0.5	*	0.5	99.5	0.0	99.5

[a] O-methylated glucuronide and O-methylated metabolites.

metabolite of both naringenin and naringenin-7-glucoside was identified as naringenin glucuronide, thus supporting the notion that glucuronidation as well as hydrolysis can occur at the intestinal epithelium [61]. These perfusion studies are consistent with those of Shimoi and associates [62], who studied transfer of flavonoids from the mucosa to the serosal side of a rat everted intestine model, finding that both hydrolysis of luteolin-7-glucoside to luteolin and glucuronidation of luteolin occur during transfer across the small intestine. The extent of glucuronidation in these experiments seemed dependent on the flavonoid structure, in that the flavonoids with a substituted hydroxyl group on the B-ring (i.e., hesperetin) were less predisposed to glucuronidation, whereas the flavonoids containing a 3′,4′-ortho-dihydroxy (or catechol) B-ring were transferred predominantly as glucuronides [52] (Tables 1 and 2). For example, the jejunal transfer of quercetin resulted in its being almost totally glucuronidated (97.6% of total transferred), whereas the absorption of hesperetin resulted in a much lower level of glucuronidation (17.8%) (Table 1). Monophenolic B-ring flavonoids were also extensively glucuronidated, in particular naringenin, which was only detected in serosal fluid glucuronidated. Similar patterns of metabolism were observed in the ileum. However, in general glucuronidation occurred to a lesser extent (Table 2), in line with studies that have recorded lower levels of phase I and II enzymes present in the ileum compared to the jejunum. Glucuronidation of these flavonoids was observed to occur predominantly at the 5- and 7-positions on the A-ring, a process that would be expected to have little influence on the resulting antioxidant potential of the metabolite. Indeed, 1999 studies identified the 5-O-β-glucuronide of catechin and epicatechin excreted in the urine of rats post ingestion, which does not interfere with their antioxidant properties (as assessed by their ability to scavenge superoxide) [32,63]. Interestingly, two uridine diphosphate–(UDP)-glucuronosyltransferase (UGT) isoforms, UGT1A8 and UGT1A10, of human intestinal mucosa, which are absent in liver, have been identified by reverse transcriptase-polymerase chain reaction [49].

Similar studies with resveratrol led to the detection of the glucuronide of resveratrol almost exclusively in the serosal fluid [58] (Table 1), and another flavonoid, diosmetin, was also found to be rapidly glucuronidated in the rat [64]. Whereas the major metabolites observed on the serosal side after perfusion of the jejunum with catechin or epicatechin were always glucuronidated, there were also high levels of both O-methylated and O-methylated-glucuronide forms [28,59] (Tables 1 and 2). 3′-O- and 4′-O-methylated derivatives of the flavanols were detected at high levels in the serosal fluid (~30% of total transferred) and O-methyl and O-methyl-glucuronidated catechins were the predominant metabolites detected in the serosal fluid (~50%), suggesting these as the most bioavailable forms (Fig. 4). As with the other flavonoids tested in this model, there was a lower level of metabolism occurring in the ileum,

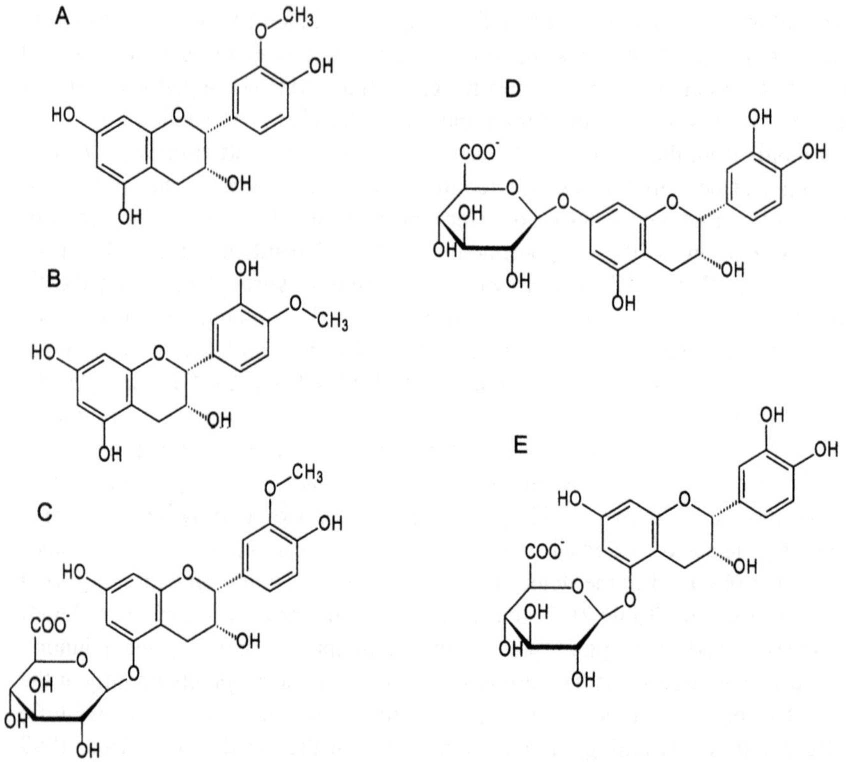

Figure 4 The structures of the main small intestinal metabolites of epicatechin produced in the isolated rat small intestine perfusion model. (A) 3'-O-methyl epicatechin; (B) 4'-O-methyl epicatechin; (C) 3'-O-methyl epicatechin-5-glucuronide; (D) epicatechin-5-glucuronide; (E) epicatechin-7-glucuronide.

although the total amounts of both catechin and epicatechin absorbed were much higher than in the jejunum (Tables 1 and 2). The greater susceptibility to methylation of flavanols than of other flavonoids in the jejunum may reside in the specificity of catechol-O-methyltransferase (COMT) for these compounds [65]. These data were confirmed in 2001 in a similar model, in which the rat jejunum and ileum were perfused with catechin [51]. In this study, catechin was absorbed into intestinal cells and metabolized extensively to a point where no native catechin could be detected in plasma from the mesenteric vein. Mesenteric plasma contained glucuronide conjugates of catechin and 3'-O-methyl catechin, indicating the intestinal origin of these conjugates and the large role the small intestine plays in the biotransformation of fiavanols during absorption [51,59]. Although most studies have identified flavanol metabolites as the main forms found entering the hepatic portal vein after absorption from the small intestine, the native flavanols are detected in small amounts [59]. Oral administration of the tea catechins, epicatechin, epigallocatechin, epicatechin gallate, and epigallocatechin gallate, to rats led to the detection of all four flavanols in the portal blood [66], clearly indicating that these flavonoids may be absorbed intact to a small degree. Studies have also shown that tannic acid and catechin may both interact with the small intestine, but only catechin appears able to traverse the gut [56]. This finding may be due to the binding of tannic acid and catechin by endogenous proteins in the intestinal lumen, limiting their absorption from the small intestine. It should also be noted that catechins have bactericidal properties and may play several roles in the digestive tract that may be linked to their protein-binding properties. In the small intestine, catechins inhibit α-amylase activity but do not affect lactic acid bacteria [67]. The inclusion of tea catechins in the diet for several weeks had the effect of decreasing putrefactive products and at the same time increasing organic acids by lowering pH [67].

Procyanidins have a high affinity for proteins, and their absorption through the gut barrier is most likely limited to lower oligomeric forms and to the metabolites formed by the colonic microflora. In 2001, perfusion of isolated small intestine with the procyanidin dimers B2 and B5 extracted from cocoa indicated that both forms of dimer are transferred to the serosal side of enterocytes but only to a very small extent (<1% of the total transferred flavanol-like compounds) [68]. Perfusion of dimer mainly resulted in detection of large amounts of unmetabolized/unconjugated epicatechin monomers on the serosal side ($\sim 95.8\%$). Low levels of O-methylated dimer were also detected ($\sim 3.2\%$), but no conjugates and metabolites of epicatechin, indicating that metabolism of monomer and dimer is limited during dimer cleavage/translocation. Experiments with normal Caco-2 cells and radiolabeled procyanidins suggested that dimer and trimer were transferred to the same extent as the epicatechin monomer, whereas oligomers with an average degree of polymerization of 7 were not [29].

In addition to flavan-3-ols, O-methylated derivatives of quercetin have been shown to be generated in the small intestine [50]. Quercetin is taken up into enterocytes and transferred to the plasma as glucuronidated, O-methylated, and sulfated derivatives with a large fraction of the absorbed quercetin re-excreted into the lumen as conjugated derivatives both directly and via the biliary duct [50]. This study indicates that sulfation of quercetin might occur in intestinal cells; however, because of the lack of precise identification of sulfated conjugates it remains unclear whether sulfation occurs in the small intestinal tract. It has been observed that flavonoids can inhibit the sulfation of resveratrol in the duodenum [69] and human cytosolic sulfotransferases show a high sulfating potential with flavonoids and isoflavones [70]. However, it is unclear to what extent these enzymes are present in the small intestine, and most studies indicate that the origin of most circulating sulfated flavonoids is the liver [26,51].

Another approach to obtain a better understanding of the bioavailability of flavonoids and their absorption and metabolism in the small intestine has used cultured human caco-2 cells (see Chap. 13).

C. Flavonoid Glycoside Processing

As there is little or no cleavage of dietary flavonoid glycosides in the mouth and stomach, the glycosidic forms of the flavonoids must enter the small intestine, where they are presented for absorption [1,29]. Because flavonoid glycosides are generally relatively polar in nature, their passive diffusion across the membranes of small intestinal brush border is limited. Many studies, however, have suggested that flavonoid glycosides are subject to the action of β-glucosidases before their absorption in the jejunum and ileum [40,52,71–79], and it is generally believed that the removal of the glycosidic moiety is necessary before absorption of the flavonoid can take place. The cleaved aglycone is thought then to undergo passive diffusion across the intestine brush border; however, the exact mechanism of uptake of these compounds is still unknown. It has been suggested that removal of the sugar and subsequent transport by proteins such as lactate phloridzin hydrolase [75] may occur in the small intestine; although this process may not occur with all flavonoid glycosides. The ability of cell-free extracts from human small intestine to deglycosylate various flavonoid glycosides has been investigated; it has been observed that quercetin-4′-glucoside, naringenin-7-glucoside, apigenin-7-glucoside, genistein-7-glucoside, and daidzein-7-glucoside are rapidly deglycosylated, whereas quercetin-3,4′-diglucoside, quercetin-3-glucoside, kaempferol-3-glucoside, quercetin-3-rhamnoglucoside, and naringenin-7-rhamnoglucoside remained unchanged [78]. In a similar study, the hydrolysis of quercetin glucosides, including quercetin-3-glucoside and rutin, by β-glucosidase isolated from the rat small intestine did occur, for the activity of jejunal β-glucosidase was highest for quercetin-4′-glucoside, whereas rutin was a poor substrate [79]. Furthermore, luteolin-7-glucoside, kaempferol-3-glucoside, and quercetin-3-glucoside are cleaved by rat jejunal or ileal mucosa, suggesting the presence of β-glucosidase in the enterocytes [52].

Most investigations and controversy have surrounded the absorption of quercetin glucosides in the small intestine [1,71,75]. Initial investigations pointed to the absorption of quercetin glucosides in the small intestine [17,24–26,80–83]; however, more recently these observations have been questioned. A number of studies reported the uptake of quercetin glucosides into the circulation. Quercetin glucosides were reported to be absorbed from onions fed to ileotomized volunteers [81], and investigations made since the late 1990s have postulated that flavonoid glucosides may be absorbed by the small intestine via the sodium-dependent glucose transporter (SGLT-1) [42,84–86]. However, a similar study in ileostomy patients, fed a meal containing high concentrations of both quercetin mono-and diglucosides, resulted in no detection of the these compounds in ileostomy fluid [74]. In contrast, the amounts of the aglycone quercetin were substantial, suggesting that both quercetin glycosides are efficiently hydrolyzed in the small intestine by β-glucosidases to quercetin [74]. Observations of the absorption of quercetin-3-glucoside and other gluco-sides of quercetin may have been confused by the coelution of these compounds with corresponding quercetin glucuronides on high-performance liquid chroma-tography (HPLC). With the introduction of new mass spectrometric techniques for the detection of flavonoids and their metabolites [87–89] it should be possible to solve beyond doubt whether quercetin-3-glucoside is absorbed intact in the small intestine or not. One argument against the uptake of intact quercetin glucosides is that the metabolic capacity of β-glucosidase in the small intestine, and of the liver, is too great for quercetin glucosides to escape deglycosylation [71]. In support of this, an analysis of human plasma using HPLC with coularray detection after oral administration of quercetin-3-glucoside or quercetin-4′-glucoside determined that no intact quercetin glucosides were present [90]. The major components in plasma were detectable by coularray detection to be quercetin glucuronides, as confirmed by the disappearance of the glucuronide peaks after treatment of the plasma β-glucuronidase.

The absorption of quercetin glycosides in the small intestine has also been investigated by using the Caco-2 cell model [86,91–93]. Initial observations suggested the facile absorption of quercetin through the human intestinal epithelium but did not support an active transport process for quercetin gluco-sides [91]. However, more recent investigations suggest that transport of one of the predominant dietary forms of quercetin, quercetin-4′-β-glucoside, across the apical membrane of enterocytes may involve both the apical multidrug resis-tance–associated protein 2 (MRP-2) [92] and/or the SGLT-1 [84–86], meaning that the transfer of quercetin glycosides in the small intestine might be possible. However, addition of plasma on the basolateral side significantly reduced the efflux of quercetin by 94%, and therefore the effect of plasma binding can result in an overestimation of basolateral to apical efflux and result in misleading net flux calculations in these types of experiments [93]. Both quercetin-3-glucoside (isoquercitrin) and quercetin-4′-glucoside (spiraeoside) significantly inhibit SGLT-1-mediated mucosal uptake of the glucose analog methyl-α-D-glucopy-ranoside (MDG), whereas the aglycone quercetin and quercetin-3-rhamnogluco-side (rutin) were ineffective [85]. In addition, the transport activity of SGLT-1 was markedly inhibited by green tea polyphenols [84] and was most pronounced with epicatechin gallate (ECG) and epigallocatechin gallate (EGCG). These studies suggest that quercetin glucosides may be capable of interacting with SGLT-1 in the mucosal epithelium and may therefore be absorbed by the small intestine in vivo. Whether or not they may also escape deglycosylation in the enterocytes and the liver is still to be addressed, although human feeding studies would suggest that this is so.

Further evidence that confounds observations of the inability of glycosides to cross the small intestine is derived from the many reports that the anthocyanidin glycosides (anthocyanins) are readily absorbed intact without initial cleavage of the sugar groups in the lumen of the small intestine [82,94–96]. Cyanidin-3-O-β-D-glucoside rapidly appeared in the plasma of rats after administerion; however, the cyanidin aglycone was not detected [95], although it was present in the jejunum. Furthermore, both cyanidin-3-glucoside and cyanidin-3,5-diglucoside were rapidly incorporated into the plasma of rats and humans after oral dosage [94], again indicating that anthocyanins may be absorbed from the digestive tract into the blood circulation system in mammals without structural alteration of the glycoside forms. Other glycosides such as the rhamnoglucoside of quercetin, rutin, are absorbed in the small intestine intact [52,97,98]. With an isolated rat small intestine model, about 10% of the administered rutin appeared on the vascular side, chiefly as free rutin (5.6%), but some rutin sulfate (2.5%) and glucuronide (2.0%) were also detected [97]. In a similar investigation rutin was also observed on the serosal side after perfusion of a rat jejunum and ileum perfusion model; however, metabolites of rutin were not detectable [52].

IV. COLONIC METABOLISM

Studies have suggested that the extent of absorption of dietary polyphenols in the small intestine is relatively small (10–20%) [52,58,59]. The implications of this low absorption in the small intestine are that the majority of ingested polyphenols, including those absorbed and conjugated in the enterocytes and/or the liver before transport back out into the lumen either directly or via the bile [50], reach the large intestine, where they encounter colonic microflora. The colon contains approximately 10^{12} microorganisms/cm^3, which have an enormous catalytic and hydrolytic potential, and this enzymatic degradation of flavonoids by the colonic microflora results in a huge array of new metabolites. For example, bacterial enzymes may catalyze many reactions, including hydrolysis, dehydroxylation, demethylation, ring cleavage, and decarboxylation, as well as rapid deconjugation [30]. Unlike human enzymes, the microflora catalyze the breakdown of the flavonoid backbone itself to simpler molecules such as phenolic acids. Specific metabolites have been observed in urine after consumption of a variety of phenolics. For example, the glycine conjugate of benzoic acid, hippuric acid, is primarily derived from plant phenolics and aromatic amino acids through the action of intestinal bacteria, and, consequently, the level of hippuric acid would be expected to increase in the urine of individuals consuming diets rich in flavanols or polyphenols in general. It must be noted, however, that hippuric acid could possibly derive from other sources such as quinic acid or, in quantitative terms, more importantly from the aromatic amino acids tryptophan, tyrosine, and phenylalanine, as well as from the use of benzoic acid as a food preservative. To date, most studies looking at the metabolism of flavonoids in the large intestine have used either flavanols or flavonols, and there are few data on the metabolism of other commonly consumed flavonoids and other polyphenols.

A. Flavanols

The 5,7,3',4'-hydroxylation pattern of flavan-3-ols is believed to enhance ring opening after hydrolysis [28,30], and metabolism of flavanols by enzymes of the microflora of the large intestine results in many metabolites: 3,4-dihydrophenyl-acetic acid, 3-hydroxyphenylacetic acid, homovanillic acid, and their conjugates derived from the B-ring [30] and phenolic acids from the C-ring. Flavanols, because of their structures (no C-4 carbonyl group), can also degrade to the specific metabolites phenylvalerolactones. Phenylpropionic acids (which may undergo further metabolism to benzoic acids) may also be the products of flavanol metabolism in animal studies, which demonstrate fission of the A-ring [30]. Only 3.1% of the ingested catechin was extractable from feces after feeding of rats, indicating that major absorption and/or degradation of catechin had occurred in the gastrointestinal tract [99].

Such metabolites of flavanols have been detected in human plasma and urine after a single ingestion of green tea [100], suggesting that that there may be significant metabolism by gut microflora in the colon. The two metabolites (-)-5-(3',4',5'-trihydroxyphenyl)-γ-valerolactone and (-)-5-(3',4'-dihydroxy-phenyl)-γ-valerolactone were identified in urine by both LC-mass spectrometry (LC-MS/MS) and nuclear magnetic resonance (NMR), appearing 7.5–13.5 h after ingestion (after a 3-h lag time), whereas EC and EGC peaked at 2 h. As well as their late excretion profiles, the amounts of metabolite excreted were 8- to 25-fold greater than those of epicatechin and epigallocatechin (EGC) excretion and accounted for 6–39% of the EC and EGC ingested. The late excretion and high levels of these metabolites would suggest that they are generated from the precursors epicatechin and EGC by the intestinal microorganisms. Before this human study, similar observations were made in rats fed with labeled catechin [101]. Here catechin glucuronides were observed in the bile after dosage of rats with the catechin, and m- and p-hydroxyphenylproprionic acid, δ-(3-hydroxyphenyl)-γ-valerolactone, and δ-(3,4-dihydroxyphenyl)-γ-valerolac-tone were identified as metabolites that arose through action of the colonic microflora [101].

In humans, studies on black tea consumption [38] suggest an association between polyphenol intake and excreted amounts of hippuric acid found. The formation of hippuric acid and hydroxyhippuric acids seems to be a possible central metabolic pathway for dietary flavonoids, in which the colon and the liver are active metabolic sites [36]. Other hydroxybenzoic acid glycine derivatives such as 4-hydroxyhippuric acid, vanilloylglycine, and isovanilloylglycine might also appear in reasonable amounts in urine after polyphenol consumption in general.

The metabolism of procyanidins by incubated human colonic microflora has been studied in vitro under anoxic conditions, using nonlabeled and ^{14}C-labeled purified proanthocyanidin polymers [102]. Interestingly, the oligomers were almost totally degraded after 48 h of incubation, and meta- or para-monohydroxylated-phenylacetic, phenylpropionic, and phenylvaleric acids were identified as metabolites, providing the first evidence that dietary flavan-3-ol polymers can be degraded to low-molecular-weight aromatic compounds in the body [102].

B. Flavonols

Quercetin-3-rhamnoglucoside and quercetin-3-rhamnoside may be hydrolysed by strains of colonic *Bacteroides distasonis*, *B. uniformis*, and *B. ovatus*, which may cleave the sugar by using α-rhamnosidase and β-glucosidase to liberate quercetin aglycone [103]. For example, a cell-free extract of *B. distasonis*, containing β-glucosidase, displayed an enzymatic activity of 1 μmol/10 min/10 mg of protein [103]. Other bacteria, such as *Enterococcus casseliflavus*, may utilize the sugar to yield formate, acetate, and lactate but do not further metabolize the aglycone [104]. *Eubacterium ramulus* occurs at numbers of approximately 10^8/g dry feces in humans and has been observed to degrade quercetin-3-glucoside [104], luteolin-7-glucoside, rutin, quercetin, kaempferol, luteolin, eriodictyol, naringenin, taxifolin, and phloretin [105] to phenolic acids. It may also hydrolyze kaempferol-3-sorphoroside-7-glucoside to kaempferol-3-sorphoroside and transform 3,4-dihydroxyphenylacetic acid, a product of anaerobic quercetin degradation [104], to nonaromatic fermentation products [105]. *E. ramulus* is capable of degrading the aromatic ring system of quercetin, producing the transient intermediate phloroglucinol. However, this bacterium was observed not to grow on phloroglucinol or quercetin aglycone itself and only to cleave the flavonoid ring system when glucose was present as a cosubstrate [104].

When quercetin-3-rhamnoside was incubated anaerobically with human intestinal bacteria, quercetin, 3,4-hihydroxyphenylacetic acid, and 4-hydroxybenzoic acid were produced as metabolites [106]. Analysis of urinary metabolites after orally administered rutin labeled with deuterium [(2′,5′,6′-2H]rutin led to the detection of 3-hydroxyphenylacetic acid, 3-methoxy-4-hydroxyphenylacetic acid, 3,4-dihydroxyphenylacetic acid, 3,4-dihydroxytoluene, and 3-(m-hydrox-yphenyl)-propionic acid as rutin metabolites. Unmetabolized rutin and quercetin were not present in the urine [102], suggesting the flavonol had been metabolized to phenolic acid metabolites by colonic microorganisms.

C. Other Studies

The flavonoid glycosides rutin, hesperidin, naringin, and poncirin are also metabolized to phenolic acids, via aglycones, by human intestinal microflora that produce α-rhamnosidase, exo-β-glucosidase, endo-β-glucosidase, and/or β-glu-curonidase enzymes [107]. In addition, baicalin, puerarin, and daidzin were transformed to their aglycones by the bacteria, producing β-glucuronidase, C-glycosidase, and β-glycosidase, respectively. β-Glucosidase (EC 3.2.1.21) has been purified from *Bacteroides* JY-6, a human intestinal anaerobic bacterium [108]. Protocatechuic acid has been detected in the plasma of rats after administration of cyanidin 3-*O*-β-D-glucoside, and it is proposed that this metabolite is produced by degradation of cyanidin by the microflora [95]. The metabolite was present in the plasma at a concentration that was approximately eightfold higher than that of cyanidin 3-*O*-β-D-glucoside, which had been absorbed intact in the small intestine.

V. CONCLUSIONS

Figure 5 summarizes the sites of biotransformation of dietary flavonoids in the body. It is clear that the gastrointestinal tract plays a very significant role in the metabolism and conjugation of these polyphenols before the liver is reached. In the jejunum and ileum of the small intestine there is efficient glucuronidation of nearly all flavonoids to differing extents by the action of UDP-glucuronosyl-transferase enzymes. In the case of catechol-containing B-ring flavonoids there is also extensive O-methylation by the action of COMT. Unabsorbed flavonoids, and those taken up, metabolized in the small intestine and liver, and transported back into the intestinal lumen, reach the large intestine, where they are further metabolized by the gut microflora to smaller phenolic acids. The extent to which these phenolic acids are absorbed in the colon is unknown; however, they are detected in the plasma and are often further conjugated and metabolized in the liver. Remaining compounds derived from flavonoid intake pass out in the feces.

It is the action of these flavonoid metabolites, in particular the O-methylated flavonoids deriving from small intestinal absorption, that is of great current interest. For example, the ability of 3′-*O*-methyl epicatechin and epicatechin glucuronides to protect against apoptotic cell death induced by

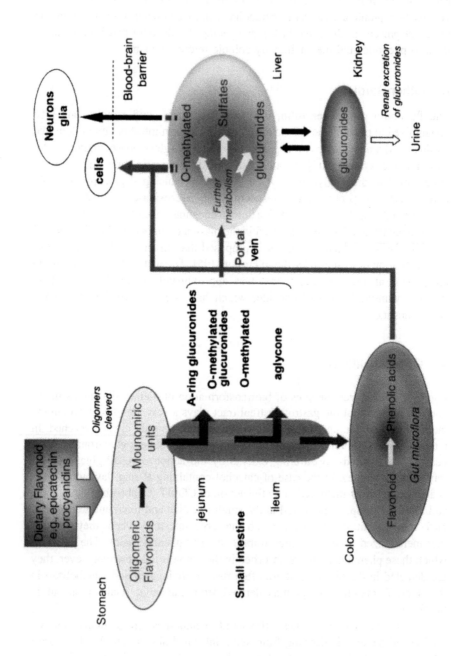

hydrogen peroxide or oxidized low-density lipoprotein (LDL) has been investigated [109–111]; it is discussed further in Chapter 9. It is also possible that flavonoid glucuronides might be cleaved by the action of β-glucourosidases located in human tissues such as the liver and neutrophils (Fig. 6). Various quercetin glucuronides are deconjugated by liver cell-free extracts and by pure recombinant human β-glucuronidase, indicating that the cleavage of glucuronides to free aglycones may occur in vivo [112]. Rat studies suggest that metabolic conversion of flavonoid does not influence the antioxidative ability of rat plasma, indicating that conjugated metabolites may participate in the antioxidant defense in blood plasma [55]. In terms of other possible modes of action, studies suggest that the metabolism of flavonoids by phase I and II enzymes in the small intestine may aid detoxification of potential carcinogens. For example, α-napthoflavone increases the activity of P450/3A in human jejunal and ileal microsomes [113], and chrysin causes an induction of intestinal UDP-glucuronosyltransferase (UGT1A1) in Caco-2-cells [114,115].

There is also a need to assess the role of the colonic microflora in the overall bioavailability and potential bioactivity of dietary flavonoids. The amount of absorption of colonic metabolites is unclear at this time, and there is growing interest in the potential effects of the phenolic acids and their derivatives as potentially beneficial agents. For example, the human intestinal bacteria metabolites of rutin and quercetin, 3,4-dihydroxyphenylacetic acid, and 4-hydroxyl-phenyl-acetic acid have been shown to possess more effective antiplatelet aggregation activity than rutin and quercetin [107]. Furthermore, 2,4,6-trihydroxybenzaldehyde and quercetin were more effective than rutin in their cytotoxicity against tumor cell lines. The effects the phenolics themselves have on the microflora are an emerging field, and it is possible that a flavonoid-induced change in the rich colonic bacterial population may have an influence on the overall health of the individual.

Over recent years we have gained greater knowledge of the bioavailable metabolites of dietary flavonoids, and it is now essential to evaluate fully the role of these conjugates and metabolites in disease prevention. It will be important to assess whether the observed metabolism aids entry into cells and/or renders them better or worse in providing protection against different stresses, such as oxidative or nitrative stress. New data in the field are already beginning to

---

Figure 5 Summary of the formation of metabolites and conjugates of flavonoids in humans. Cleavage of procyanidins may occur in the stomach in environments of low pH. All classes of flavonoids undergo extensive metabolism in the jejunum and ileum of the small intestine, and resulting metabolites enter the portal vein and undergo further metabolism in the liver. Colonic microflora degrade flavonoids into smaller phenolic acids, which may also be absorbed. The fate of most of these metabolites is renal excretion; however, the extent to which these compounds enter cells and tissues is unknown.

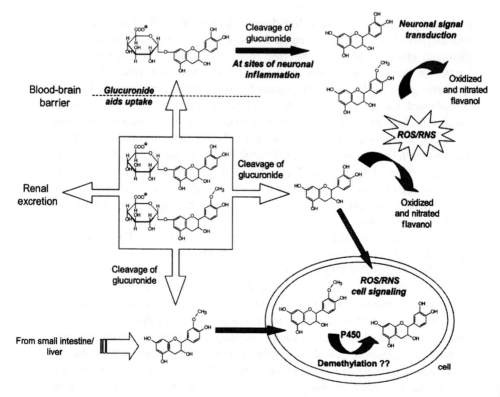

Figure 6 Possible fate and biological action of flavanol metabolites/conjugates in vivo.

suggest that flavonoids may act to protect cells by more complex mechanisms than was once thought [110]. Eventually it is hoped that these studies will allow specific dietary recommendations that will increase general health in the population to be made.

REFERENCES

1. Rice-Evans C, Spencer JPE, Schroeter H, Rechner AR. Bioavailability of flavonoids and potential bioactive forms in vivo. Drug Metabol Drug Interact 2000; 17:291–310.
2. Rice-Evans C. Flavonoid antioxidants. Curr Med Chem 2001; 8:797–807.
3. Rice-Evans CA, Miller NJ, Paganga G. Structure-antioxidant activity relationships of flavonoids and phenolic acids. Free Radic Biol Med 1996; 20:933–956.
4. Rice-Evans C. Plant polyphenols: free radical scavengers or chain-breaking antioxidants? Biochem Soc Symp 1995; 61:103–116.

5. Rice-Evans CA, Miller NJ. Antioxidant activities of flavonoids as bioactive components of food. Biochem Soc Trans 1996; 24:790–795.

6. Rice-Evans CA, Miller NJ, Bolwell PG, Bramley PM, Pridham JB. The relative antioxidant activities of plant-derived polyphenolic flavonoids. Free Radic Res 1995; 22:375–383.

7. Sekher PA, Chan TS, O'Brien PJ, Rice-Evans CA. Flavonoid B-ring chemistry and antioxidant activity: fast reaction kinetics. Biochem Biophys Res Commun 2001; 282:1161–1168.

8. Brown JE, Khodr H, Hider RC, Rice-Evans CA. Structural dependence of flavonoid interactions with Cu2+ ions: implications for their antioxidant properties. Biochem J 1998; 330 (pt 3):1173–1178.

9. Oldreive C, Zhao K, Paganga G, Halliwell B, Rice-Evans C. Inhibition of nitrous acid-dependent tyrosine nitration and DNA base deamination by flavonoids and other phenolic compounds. Chem Res Toxicol 1998; 11:1574–1579.

10. Pannala AS, Rice-Evans C. Rapid screening method for relative antioxidant activities of flavonoids and phenolics. Methods Enzymol 1901; 335:266–272.

11. Packer L, Rimbach G, Virgili F. Antioxidant activity and biologic properties of a procyanidin-rich extract from pine (*Pinus maritima*) bark, pycnogenol. Free Radic Biol Med 1999; 27:704–724.

12. Lairon D, Amiot MJ. Flavonoids in food and natural antioxidants in wine. Curr Opin Lipidol 1999; 10:23–28.

13. Gyorgy I, Foldiak G. Free-radical induced redox reactions of drugs—a pulse-radiolysis study. Magyar Kemiai Folyoirat 1992; 98:361–366.

14. deOliveira TT, Nagem TJ, daSilva MC, deMiranda LCG, Teixeira MA. Antioxidant action of the flavonoids derivatives. Pesquisa Agropecuaria Brasileira 1999; 34:879–883.

15. Bohm H, Boeing H, Hempel J, Raab B, Kroke A. Flavonols, flavones and anthocyanins as native antioxidants of food and their possible role in the prevention of chronic diseases. Z Ernahrungswissenschaft 1998; 37:147–163.

16. Hollman PC, Hertog MG, Katan MB. Role of dietary flavonoids in protection against cancer and coronary heart disease. Biochem Soc Trans 1996; 24:785–789.

17. Hollman PC, Katan MB. Absorption, metabolism and health effects of dietary flavonoids in man. Biomed Pharmacother 1997; 51:305–310.

18. Hollman PC, Feskens EJ, Katan MB. Tea flavonols in cardiovascular disease and cancer epidemiology. Proc Soc Exp Biol Med 1999; 220:198–202.

19. Arts IC, Hollman PC, Bueno dMH, Feskens EJ, Kromhout D. Dietary catechins and epithelial cancer incidence: the Zutphen elderly study. Int J Cancer 2001; 92: 298–302.

20. Arts IC, Hollman PC, Feskens EJ, Bueno dMH, Kromhout D. Catechin intake might explain the inverse relation between tea consumption and ischemic heart disease: the Zutphen Elderly Study. Am J Clin Nutr 2001; 74:227–232.

21. Hertog MG, Feskens EJ, Hollman PC, Katan MB, Kromhout D. Dietary antioxidant flavonoids and risk of coronary heart disease: the Zutphen Elderly Study. Lancet 1993; 342:1007–1011.

22. Hertog MG, Hollman PC. Potential health effects of the dietary flavonol quercetin. Eur J Clin Nutr 1996; 50:63–71.

23. Hollman PCH, Katan MB. Absorption, metabolism and health effects of dietary flavonoids in man. Biomed Pharmacother 1997; 51:305–310.

24. Hollman PC, Katan MB. Health effects and bioavailability of dietary flavonols. Free Radic Res 1999; 31 (suppl):S75–S80

25. Hollman PC, Katan MB. Dietary flavonoids: intake, health effects and bioavailability. Food Chem Toxicol 1999; 37:937–942.

26. Hollman PC, Katan MB. Bioavailability and health effects of dietary flavonols in man. Arch Toxicol Suppl 1998; 20:237–248.

27. Hollman PC. Bioavailability of flavonoids. Eur J Clin Nutr 1997; 51 (suppl 1): S66–S69.

28. Spencer JPE, Schroeter H, Rechner A, Rice-Evans C. Bioavailability of flavan-3-ols and procyanidins: gastrointestinal tract influences and their relevance to bioactive forms in vivo. Antiox Redox Signal 2001; 3:1023–1040.

29. Scalbert A, Williamson G. Dietary intake and bioavailability of polyphenols. J Nutr x2000; 130:2073S–2085S.

30. Scheline RR. Metabolism of oxygen heterocyclic compounds. In: CRC Handbook of mammalian metabolism of plant compounds. Boca Raton, FL: CRC Press, 1999: 243–295.

31. Okushio K, Suzuki M, Matsumoto N, Nanjo F, Hara Y. Methylation of tea catechins by rat liver homogenates. Bioscie Biotech Biochem 1999; 63:430–432.

32. Okushio K, Suzuki M, Matsumoto N, Nanjo F, Hara Y. Identification of (-)-epicatechin metabolites and their metabolic fate in the rat. Drug Metab Dispos 1999; 27:309–316.

33. Tsuchiya H, Sato M, Kato H, Okubo T, Juneja LR, Kim M. Simultaneous determination of catechins in human saliva by high-performance liquid chromatography. J Chromatogr B 1997; 703:253–258.

34. Yang CS, Lee MJ, Chen L. Human salivary tea catechin levels and catechin esterase activities: implication in human cancer prevention studies. Cancer Epidemiol Biomarkers Prev 1999; 8:83–89.

35. Laires A, Pacheco P, Rueff J. Mutagenicity of rutin and the glycosidic activity of cultured cell-free microbial preparations of human faeces and saliva. Food Chem Toxicol 1989; 27:437–443.

36. Macdonald IA, Mader JA, Bussard RG. The role of rutin and quercitrin in stimulating flavonol glycosidase activity by cultured cell-free microbial preparations of human feces and saliva. Mutat Res 1983; 122:95–102.

37. Parisis DM, Pritchard ET. Activation of rutin by human oral bacterial isolates to the carcinogen-mutagen quercetin. Arch Oral Biol 1983; 28:583–590.

38. de F, V, Mateus N. Structural features of procyanidin interactions with salivary proteins. J Agric Food Chem 2001; 49:940–945.

39. Nishioka H, Nishi K, Kyokane K. Human saliva inactivates mutagenicity of carcinogens. Mutat Res 1981; 85:323–333.

40. Spencer JPE, Chaudry F, Pannala AS, Srai SK, Debnam E, Rice-Evans C. Decomposition of cocoa procyanidins in the gastric milieu. Biochem Biophys Res Commun 2000; 272:236–241.

41. Porter IJ. Tannins. In: Harborne JB, ed. Plant Biochemistry. Vol. I. Plant Phenolics. London: Academic Press, 2002:389–418.

42. Gee JM, Dupont MS, Rhodes MJC, Johnson IT. Quercetin glucosides interact with the intestinal glucose transport pathway. Free Radic Biol Med 1998; 25: 19–25.

43. Bae EA, Han MJ, Kim DH. In vitro anti-Helicobacter pylori activity of some flavonoids and their metabolites. Planta Med 1999; 65:442–443.

44. Lin JH, Chiba M, Baillie TA. Is the role of the small intestine in first-pass metabolism overemphasized? Pharmacol Rev 1999; 51:135–158.

45. Higuchi WI, Ho NF, Park JY, Komiya I. Rate-limiting steps and factors in drug absorption. In: Prescott LF, Nimno WS, eds. Drug Absorption. New York: ADIS Press, 1981:35–60.

46. Ho NF, Park JY, Ni PF, Higuchi WI. Advancing quantitative and mechanistic approaches in interfacing gastrointestinal drug absorption studies in animals and humans. In: Crouthamel W, Sarapu AC, eds. Animal Models for Oral Drug Delivery: In Situ and In Vivo Approaches. Washington DC: American Pharmaceutics Association, 1983:27–106.

47. Yoshino K, Suzuki M, Sasaki K, Miyase T, Sano M. Formation of antioxidants from (-)-epigallocatechin gallate in mild alkaline fluids, such as authentic intestinal juice and mouse plasma. J Nutr Biochem 1999; 10:223–229.

48. Chen ZY, Zhu QY, Wong YF, Zhang ZS, Chung HY. Stabilizing effect of ascorbic acid on green tea catechins. J Agric Food Chem 1998; 46:2512–2516.

49. Cheng Z, Radominska-Pandya A, Tephly TR. Studies on the substrate specificity of human intestinal UDP-lucuronosyltransferases 1A8 and 1A10. Drug Metab Dispos 1999; 27:1165–1170.

50. Crespy V, Morand C, Manach C, Besson C, Demigne C, Remesy C. Part of quercetin absorbed in the small intestine is conjugated and further secreted in the intestinal lumen. Am J Physiol 1999; 277:G120–G126

51. Donovan JL, Crespy V, Manach C, Morand C, Besson C, Scalbert A, Remesy C. Catechin is metabolized by both the small intestine and liver of rats. J Nutr 2001; 131: 1753–1757.

52. Spencer JPE, Chowrimootoo G, Choudhury R, Debnam ES, Srai SK, Rice-Evans C. The small intestine can both absorb and glucuronidate luminal flavonoids. FEBS Lett 1999; 458:224–230.

53. Windmill KF, McKinnon RA, Zhu X, Gaedigk A, Grant DM, McManus ME. The role of xenobiotic metabolizing enzymes in arylamine toxicity and carcinogenesis: functional and localization studies. Mutat Res 1997; 376:153–160.

54. Franski R, Bednarek P, Siatkowska D, Wojtaszek P, Stobiecki M. Application of mass spectrometry to structural identification of flavonoid monoglycosides isolated from shoot of lupin (*Lupinus luteus L.*). Acta Biochim Pol 1999; 46:459–473.

55. Terao J. Dietary flavonoids as antioxidants in vivo: conjugated metabolites of (-)-epicatechin and quercetin participate in antioxidative defense in blood plasma. J Med Invest 1999; 46:159–168.

56. Carbonaro M, Grant G, Pusztai A. Evaluation of polyphenol bioavailability in rat small intestine. Eur J Nutr 2001; 40:84–90.

57. Fisher RB, Gardner ML. A kinetic approach to the study of absorption of solutes by isolated perfused small intestine. J Physiol 1974; 241:211–234.

58. Kuhnle G, Spencer JPE, Chowrimootoo G, Schroeter H, Debnam ES, Srai SKS,

Rice-Evans C, Hahn U. Resveratrol is absorbed in the small intestine as resveratrol glucuronide. Biochem Biophys Res Commun 2000; 272:212–217.

59. Kuhnle G, Spencer JPE, Schroeter H, Shenoy B, Debnam ES, Srai SK, Rice-Evans C, Hahn U. Epicatechin and catechin are O-methylated and glucuronidated in the small intestine. Biochem Biophys Res Commun 2000; 277:507–512.

60. Crevoisier C, Buri P, Boucherat J. The transport of three flavonoids across artificial and biological membranes. 5. Transport in situ across the small intestine of the rat. Pharm Acta Helv 1975; 50:231–236.

61. Choudhury R, Chowrimootoo G, Srai K, Debnam E, Rice-Evans CA. Interactions of the flavonoid naringenin in the gastrointestinal tract and the influence of glycosylation. Biochem Biophys Res Commun 1999; 265:410–415.

62. Shimoi K, Okada H, Furugori M, Goda T, Takase S, Suzuki M, Hara Y, Yamamoto H, Kinae N. Intestinal absorption of luteolin and luteolin 7-O-beta-glucoside in rats and humans. FEBS Letters 1998; 438:220–224.

63. Harada M, Kan Y, Naoki H, Fukui Y, Kageyama N, Nakai M. Identification of the major antioxidative metabolites in biological fluids of the rat with ingested (+)-catechin and (-)-epicatechin. Biosci Biotechnol Biochem 1999; 63: 973–977.

64. Boutin JA, Meunier F, Lambert PH, Hennig P, Bertin D, Serkiz B. In-vivo and in-vitro glucuronidation of the flavonoid diosmetin in rats. Drug Metab Dispos 1993; 21:1157–1166.

65. Mannisto PT, Kaakkola S. Catechol-O-methyltransferase (COMT): biochemistry, molecular biology, pharmacology, and clinical efficacy of the new selective COMT inhibitors. Pharmacol Rev 1999; 51:593–628.

66. Okushio K, Matsumoto N, Kohri T, Suzuki M, Nanjo F, Hara Y. Absorption of tea catechins into rat portal vein. Biol Pharm Bull 1996; 19:326–329.

67. Hara Y. Influence of tea catechins on the digestive tract. J Cell Biochem Suppl 1997; 27:52–58.

68. Spencer JPE, Schroeter H, Shenoy B, Srai SK, Debnam ES, Rice-Evans C. Epicatechin is the primary bioavailable form of the procyanidin dimers B2 and B5 after transfer across the small intestine. Biochem Biophys Res Commun 2001; 285:588–593.

69. de Santi C, Pietrabissa A, Spisni R, Mosca F, Pacifici GM. Sulphation of resveratrol, a natural compound present in wine, and its inhibition by natural flavonoids. Xenobiotica 2000; 30:857–866.

70. Pai TG, Suiko M, Sakakibara Y, Liu MC. Sulfation of flavonoids and other phenolic dietary compounds by the human cytosolic sulfotransferases. Biochem Biophys Res Commun 2001; 285:1175–1179.

71. Day AJ, Williamson G. Biomarkers for exposure to dietary flavonoids: a review of the current evidence for identification of quercetin glycosides in plasma. Br J Nutr 2001; 86 (suppl 1):105–110.

72. Hollman PCH, van Trijp JMP, Buysman MNCP, VanderGaag MS, Mengelers MJB, deVries JHM. Relative bioavailability of the antioxidant flavonoid quercetin from various foods in man. Febs Lett 1997; 418:152–156.

73. Hollman PC, Bijsman MN, van Gameren Y, Cnossen EP, de Vries JH, Katan MB. The sugar moiety is a major determinant of the absorption of dietary flavonoid glycosides in man. Free Radic Res 1999; 31:569–573.

74. Walle T, Otake Y, Walle UK, Wilson FA. Quercetin glucosides are completely hydrolyzed in ileostomy patients before absorption. J Nutr 2000; 130: 2658–2661.

75. Day AJ, Canada FJ, Diaz JC, Kroon PA, Mclauchlan R, Faulds CB. Dietary flavonoid and isoflavone glycosides are hydrolysed by the lactase site of lactase phlorizin hydrolase. FEBS Lett 2000; 468:166–170.

76. Gee JM, Dupont MS, Day AJ, Plumb GW, Williamson G, Johnson IT. Intestinal transport of quercetin glycosides in rats involves both deglycosylation and interaction with the hexose transport pathway. J Nutr 2000; 130: 2765–2771.

77. Morand C, Manach C, Crespy V, Remesy C. Respective bioavailability of quercetin aglycone and its glycosides in a rat model. Biofactors 2000; 12:169–174.

78. Day AJ, Dupont MS, Ridley S, Rhodes M, Rhodes MJ, Morgan MR. Deglycosylation of flavonoid and isoflavonoid glycosides by human small intestine and liver beta-glucosidase activity. FEBS Lett 1998; 436:71–75.

79. Ioku K, Pongpiriyadacha Y, Konishi Y, Takei Y, Nakatani N, Terao J. Beta-glucosidase activity in the rat small intestine toward quercetin monoglucosides. Biosci Biotech Biochem 1998; 62:1428–1431.

80. Hollman PCH, vanderGaag M, Mengelers MJB, van Trijp JMP, deVries JH, Katan MB. Absorption and disposition kinetics of the dietary antioxidant quercetin in man. Free Radic Biol Med 1996; 21:703–707.

81. Hollman PC, de Vries JH, van Leeuwen SD, Mengelers MJ, Katan MB. Absorption of dietary quercetin glycosides and quercetin in healthy ileostomy volunteers. Am J Clin Nutr 1995; 62:1276–1282.

82. Paganga G, RiceEvans CA. The identification of flavonoids as glycosides in human plasma. FEBS Lett 1997; 401:78–82.

83. Hollman PC, van Trijp JM, Buysman MN, van der Gaag MS, Mengelers MJ, de Vries JHM, Katan MB. Relative bioavailability of the antioxidant flavonoid quercetin from various foods in man. FEBS Lett 1997; 418:152–156.

84. Kobayashi Y, Suzuki M, Satsu H, Arai S, Hara Y, Suzuki K, Miyamoto Y, Shimizu M. Green tea polyphenols inhibit the sodium-dependent glucose transporter of intestinal epithelial cells by a competitive mechanism. J Agric Food Chem 2000; 48:5618–5623.

85. Ader P, Block M, Pietzsch S, Wolffram S. Interaction of quercetin glucosides with the intestinal sodium/glucose co-transporter (SGLT-1). Cancer Lett 2001; 162: 175–180.

86. Walgren RA, Lin JT, Kinne RK, Walle T. Cellular uptake of dietary flavonoid quercetin 4′-beta-glucoside by sodium-dependent glucose transporter SGLT1. J Pharmacol Exp Ther 2000; 294:837–843.

87. Lommen A, Godejohann M, Venema DP, Hollman PC, Spraul M. Application of directly coupled HPLC-NMR-MS to the identification and confirmation of quercetin glycosides and phloretin glycosides in apple peel. Anal Chem 2000; 72: 1793–1797.

88. Justesen U. Collision-induced fragmentation of deprotonated methoxylated flavonoids, obtained by electrospray ionization mass spectrometry. J Mass Spectrom 2001; 36:169–178.

89. Justesen U, Arrigoni E. Electrospray ionisation mass spectrometric study of degradation products of quercetin, quercetin-3-glucoside and quercetin-3-rhamno-

glucoside, produced by in vitro fermentation with human faecal flora. Rapid Commun Mass Spectrom 2001; 15:477–483.

90. Sesink AL, O'Leary KA, Hollman PC. Quercetin glucuronides but not glucosides are present in human plasma after consumption of quercetin-3-glucoside or quercetin-4′-glucoside. J Nutr 2001; 131:1938–1941.

91. Walgren RA, Walle UK, Walle T. Transport of quercetin and its glucosides across human intestinal epithelial Caco-2 cells. Biochem Pharmacol 1998; 55:1721–1727.

92. Walgren RA, Karnaky KJJ, Lindenmayer GE, Walle T. Efflux of dietary flavonoid quercetin 4′-beta-glucoside across human intestinal Caco-2 cell monolayers by apical multidrug resistance-associated protein-2. J Pharmacol Exp Ther 2000; 294: 830–836.

93. Walgren RA, Walle T. The influence of plasma binding on absorption/exsorption in the Caco-2 model of human intestinal absorption. J Pharm Pharmacol 1999; 51: 1037–1040.

94. Miyazawa T, Nakagawa K, Kudo M, Muraishi K, Someya K. Direct intestinal absorption of red fruit anthocyanins, cyanidin-3-glucoside and cyanidin-3,5-diglucoside, into rats and humans. J Agric Food Chem 1999; 47:1083–1091.

95. Tsuda T, Horio F, Osawa T. Absorption and metabolism of cyanidin 3-O-beta-D-glucoside in rats. FEBS Lett 1999; 449:179–182.

96. Lapidot T, Harel S, Granit R, Kanner J. Bioavailability of red wine anthocyanins as detected in human urine. J Agric Food Chem 1998; 46:4297–4302.

97. Andlauer W, Stumpf C, Furst P. Intestinal absorption of rutin in free and conjugated forms. Biochem Pharmacol 2001; 62:369–374.

98. Olthof MR, Hollman PC, Vree TB, Katan MB. Bioavailabilities of quercetin-3-glucoside and quercetin-4′-glucoside do not differ in humans. J Nutr 2000; 130: 1200–1203.

99. Bravo L, Abia R, Eastwood MA, Saura-Calixto F. Degradation of polyphenols (catechin and tannic acid) in the rat intestinal tract. Effect on colonic fermentation and faecal output. Br J Nutr 1994; 71:933–946.

100. Li C, Lee MJ, Sheng SQ, Meng XF, Prabhu S, Winnik B, Huang BM, Chung JY, Yan SQ, Ho CT, Yang CS. Structural identification of two metabolites of catechins and their kinetics in human urine and blood after tea ingestion. Chem Res Toxicol 2000; 13:177–184.

101. Das NP, Griffiths LA. Studies on flavonoid metabolism:metabolism of (+)-[14C] catechin in the rat and guinea pig. Biochem J 1969; 115:831–836.

102. Deprez S, Brezillon C, Rabot S, Philippe C, Mila I, Lapierre C, Scalbert A. Polymeric proanthocyanidins are catabolized by human colonic microflora into low-molecular-weight phenolic acids. J Nutr 2000; 130:2733–2738.

103. Bokkenheuser VD, Shackleton CH, Winter J. Hydrolysis of dietary flavonoid glycosides by strains of intestinal Bacteroides from humans. Biochem J 1987; 248: 953–956.

104. Schneider H, Schwiertz A, Collins MD, Blaut M. Anaerobic transformation of quercetin-3-glucoside by bacteria from the human intestinal tract. Arch Microbiol 1999; 171:81–91.

105. Schneider H, Blaut M. Anaerobic degradation of flavonoids by Eubacterium ramulus. Arch Microbiol 2000; 173:71–75.

106. Baba S, Furuta T, Fujioka M, Goromaru T. Studies on drug metabolism by use of isotopes XXVII: urinary metabolites of rutin in rats and the role of intestinal microflora in the metabolism of rutin. J Pharm Sci 1983; 72:1155–1158.

107. Kim DH, Jung EA, Sohng IS, Han JA, Kim TH, Han MJ. Intestinal bacterial metabolism of flavonoids and its relation to some biological activities. Arch Pharm Res 1998; 21:17–23.

108. Kim DH, Sohng IS, Kobashi K, Han MJ. Purification and characterization of beta-glucosidase from Bacteroides JY-6, a human intestinal bacterium. Biol Pharma Bull 1996; 19:1121–1125.

109. Spencer JPE, Schroeter H, Kuhnle G, Srai SKS, Tyrrell RM, Hahn U, Rice-Evans C. Epicatechin and its in vivo metabolite, 3′-O-methyl epicatechin, protect human fibroblasts from oxidative-stress-induced cell death involving caspase-3 activation. Biochem J 2001; 354:493–500.

110. Schroeter H, Spencer JPE, Rice-Evans C, Williams RJ. Flavonoids protect neurons from oxidized low-density-lipoprotein-induced apoptosis involving c-Jun N-terminal kinase (JNK), c-Jun and caspase-3. Biochem J 2001; 358:547–557.

111. Spencer JPE, Schroeter H, Crossthwaithe AJ, Kuhnle G, Williams RJ, Rice-Evans C. Contrasting influences of glucuronidation and O-methylation of epicatechin on hydrogen peroxide-induced cell death in neurons and fibroblasts. Free Radic Biol Med 2001; 31:1139–1146.

112. O'Leary KA, Day AJ, Needs PW, Sly WS, O'Brien NM, Williamson G. Flavonoid glucuronides are substrates for human liver beta-glucuronidase. FEBS Lett 2001; 503:103–106.

113. McKinnon RA, Burgess WM, Hall PM, Abdul-Aziz Z, McManus ME. Metabolism of food-derived heterocyclic amines in human and rabbit tissues by P4503A proteins in the presence of flavonoids. Cancer Res 1992; 52:2108s–2113s.

114. Galijatovic A, Otake Y, Walle UK, Walle T. Induction of UDP-glucuronosyl-transferase UGT1A1 by the flavonoid chrysin in Caco-2 cells—potential role in carcinogen bioinactivation. Pharm Res 2001; 18:374–379.

115. Walle T, Otake Y, Galijatovic A, Ritter JK, Walle UK. Induction of UDP-glucuronosyltransferase UGT1A1 by the flavonoid chrysin in the human hepatoma cell line hep G2. Drug Metab Dispos 2000; 28:1077–1082.

15
Absorption of Quercetin Glycosides

Andrea J. Day
University of Leeds
Leeds, England

Gary Williamson
Institute of Food Research
Norwich, England

I. INTRODUCTION

Understanding dietary flavonoid absorption is a fundamental requirement for the determination of the biological activity of these compounds in humans. The degree of absorption from the diet clearly affects the biological activities at various flavonoid-responsive sites within the body. Of the flavonoid subclasses, flavonols are ubiquitous in plant food. Quercetin is the major flavonol to be consumed; the estimated average Dutch intake of quercetin was calculated as 16 mg/day (73% of the flavonol intake) [1]. The main sources of quercetin in the diet include tea, onion, broccoli, apple, and red wine. Quercetin is rarely found in plants in a free form but is usually conjugated to sugar residues, and so most foods contain quercetin glycosides and *not* the aglycone. Processing and cooking of plant foods rarely cause deglycosylation [2], but treatments that release hydrolytic enzymes from the plant tissue [3] or fermentation with microorganisms can result in an increased level of aglycone. For example, in some types of red wine more than 50% of the quercetin exists in the free form [4]. The conjugation of quercetin and other flavonoids affects the mechanism by which the compound is absorbed by altering the basic physicochemical properties and thus the ability to enter cells, to interact with transporters, and to interact with cellular (lipo)proteins.

Dietary flavonoid glycosides, with the exception of anthocyanin glycosides [5], are deglycosylated during the process of absorption from the lumen of the

391

intestine into the systemic circulation. The site of deglycosylation and uptake through the epithelial cells of the intestine depends on the nature of the flavonoid, the nature of the attached sugar, and the position of attachment of the sugar on the flavonoid ring. Quercetin glucosides are deglycosylated in the small intestine by endogenous β-glucosidases, whereas quercetin rhamnoglucosides are not substrates for these enzymes and are only deglycosylated by microflora in the colon. This distinction is illustrated by the dramatic differences in pharmacokinetics between quercetin-3-glucoside and quercetin-3-rhamnoglucoside. Deglycosylation in the small intestine occurs either in the gut lumen by lactase phlorizin hydrolase (LPH) or after transport into epithelial cells [possibly by sodium-dependent glucose transporter 1 (SGLT-1) and other sugar transporters] by cytosolic β-glucosidase. The flavonoid aglycone generated by LPH diffuses into the epithelial cell as a result of its increased octanol/water partition coefficient. The intracellular aglycone, which results from either LPH/diffusion or SGLT-1/ cytosolic β-glucosidase, is then conjugated with glucuronic acid and pumped out of the cell by multidrug resistance–associated protein (MRP) or p-glycoprotein family transporters, either into the systemic circulation or back into the gut lumen. The apparent result of these metabolic processes is that the quercetin glucoside is efficiently absorbed by the small intestine but the serosal product in the hepatic portal vein is a quercetin glucuronide. The aim of this chapter is to describe the current available experimental evidence to support the preceding hypothesis and the consequences for health and disease.

II. STUDIES OF QUERCETIN UPTAKE BEFORE 1995

Considerable research on absorption of quercetin was carried out between 1960 and 1985 [6]. Many of these studies involved feeding rats high doses of either quercetin or quercetin-3-rhamnoglucoside (rutin), a major glycosylated form of quercetin present in plant foods. The work demonstrated that quercetin was absorbed to some extent in rats, but that rutin absorption was dependent on intestinal microflora activity; there was no significant absorption in germ-free or antibiotic-treated rats, suggesting that the glycoside conjugate required hydrolysis before the free quercetin could be absorbed. The high level of microbial metabolism also resulted in degradation of the aglycone ring structure. Although the phenolic acids produced from degradation can be absorbed, low levels of the flavonol aglycone in the urine would result. Several human studies were also carried out during this period, again feeding either quercetin or rutin as pure compounds rather than flavonols from food. For example, Gugler and associates [7] fed 4 g of quercetin aglycone (as solid) to human volunteers and detected no quercetin in the urine (detection limit 0.04 g; 1% of dose). The low yield in urine was also observed in more recent studies on quercetin absorption. Originally the

Table 1 Limitations of Approaches Used to Study Absorption of Flavonoid Glycosides

Study method	Information gained/advantages	Limitations
Human intervention	1. Pharmacokinetics parameters of uptake and excretion after feeding foods or purified compounds 2. Identification and quantification of plasma metabolites formed by humans	1. No mechanistic information 2. Inability to use inhibitors 3. Absorption from specific sections of intestinal tract and biliary excretion not easily measured
Animal intervention	1. Easier to conduct and control than human intervention studies 2. Information on identification and quantification of plasma metabolites in the animal 3. Effect of higher administered doses on absorption 4. Biliary excretion monitored easily	1. Nonhuman data not easily extrapolated to humans 2. Different distribution of microflora than human 3. Different profile of pharmacokinetic data than human 4. Different metabolites from human metabolites formed
Animal tissue (small intestine) ex vivo	1. Intact tissue study possible 2. Effect of inhibitors able to be studied to gain mechanistic information	1. Blood flow or diffusional sink potentially not present 2. Tissue not of human origin
Cell culture (e.g., Caco-2 cell line)	1. Immortalized cell lines that allow repeated studies with controlled conditions 2. Whole cell activity monitoring, allowing substrate interaction within compartments to be followed 3. Regulation of protein expression or use of stably transfected cell lines, and use of inhibitors to provide mechanistic information	1. Key proteins (e.g., enzymes and transporters) potentially not expressed or expressed to different levels than normal cells 2. No interaction with other food components
Enzyme studies	1. Substrate specificity and rates, along with inhibitor studies to facilitate study of detailed mechanisms	1. No account of accessibility and cellular compartmentalization of enzymes

observations were interpreted as meaning that quercetin was not absorbed effectively, but it is now known that—despite the evidence of low urinary yields—the quercetin in plasma can reach significant levels [up to ~7 μm (see late discussion)].

Hollman and colleagues [8] developed a new analytical method that allowed low levels of quercetin to be measured (detection limit of 0.15 ng/mL). Flavonols were derivatised post column with aluminum, resulting in fluorescent compounds that could be measured at a high sensitivity. Using ileostomized subjects who had had the colon surgically removed and so are believed not to have the experimental complication of microbial metabolism), Hollman and associates showed that quercetin was indeed bioavailable in humans [9]. In the study, subjects were fed either fried onion (a rich source of quercetin-4'-glucoside and quercetin-3,4'-glucoside), quercetin aglycone, or rutin supplements; urine and ileostomy effluent were collected. By calculating the amount of quercetin remaining in the ileostomy bag, an indirect measure of absorbed flavonol was obtained. From onions, 52% of the quercetin was absorbed, whereas from the quercetin aglycone or rutin supplement only 24% and 17% were absorbed, respectively. Thus, it was suggested not only that the type of attached sugar affected absorption, but that in the case of the glucoside the sugar actually promoted uptake of quercetin.

Many different approaches have been used to study bioavailability of quercetin glycosides. These include human and animal intervention studies, which monitor appearance of compounds in plasma and urine. With animals, particular sections of the gastrointestinal tract can be isolated or the bile duct can be cannulated, to allow additional information that would not necessarily be available from human studies to be generated. Enzyme and cell culture studies provide further evidence of the mechanical processes involved in absorption, metabolism, and distribution of flavonol glycosides. The properties of enzymes can also be studied by using the tools of molecular biology. Enzymes can be expressed at high levels, to a high degree of purity, or in such a way that interactions with other cellular components can be studied with less noise than is present in vivo. With each of these approaches, there are significant limitations of the information gained from individual methods (Table 1). It is only by combining data generated from complementary methods that we can start to understand the processes involved in bioavailability.

III. HUMAN INTERVENTION STUDIES

Studies investigating the absorption and disposition of flavonol glycosides from foods in human subjects entail measurement of quercetin in the plasma and allow 24-h recovery in urine to be monitored after ingestion of different foods. Resulting plasma concentrations of quercetin metabolites range from 0.1 μM to

7.6 μM in plasma, depending on the source and level of quercetin administered, typically with 0.1–1.0% of dose excreted in urine. Table 2 summarizes reported plasma and urine quercetin levels after consumption of quercetin-rich foods in human subjects. Some of the studies show the dramatic effect that the nature of the glycoside conjugate exerts on absorption of quercetin.

A study by Hollman and coworkers [10] showed that quercetin glucosides from onions were rapidly absorbed* across the small intestine (time to reach maximal plasma concentration, t_{max} < 0.7 h), whereas rutin exhibited a delayed absorption (t_{max} 9.3 h), typical of absorption in the colon, in other studies [15,18]*, pure quercetin-3-glucoside, quercetin-4′-glucoside, and rutin administered to volunteers displayed no significant difference in the rate (t_{max}, 0.6 h) or extent (maximal plasma concentration, C_{max}, 5 μM) of absorption of the two glucosides. However, rutin was absorbed much more slowly (t_{max} 6 h) and to a lesser extent than quercetin glucosides in terms of peak plasma concentrations (C_{max} 0.2 μM). These were critical experiments since they demonstrated conclusively that the nature of the sugar and not the position of attachment was the main factor in determining the manner of absorption of quercetin glycosides.

Graefe and coworkers [21] in 2001 conducted a similar human study comparing absorption from supplements with that from foods rich in the same type of quercetin conjugate. Peak plasma concentrations of quercetin from either onion or quercetin-4′-glucoside were similar (C_{max} 7.6, 7.0 μM, respectively), as were the times to reach maximal concentration (t_{max} 0.7 h for both). Even at twice the dose, the peak plasma concentration of quercetin from either buckwheat tea or purified rutin was significantly lower (C_{max} 2.0, 1.0 μM, respectively) and took a longer time to be reached (t_{max} 4.3, 7.0 h, respectively) than for quercetin from the other sources. A conclusion from the work is that the food matrix may influence absorption of the compounds to some extent, but that the most significant effect on absorption is the form of the conjugate attached to quercetin.

In a study comparing the absorption from dietary supplements of quercetin aglycone with rutin at various doses [16], rutin showed a clear maximal plasma concentration (of quercetin metabolites) at 6 h, with almost no quercetin in plasma at 4 h, indicative of absorption in the distal parts of the small intestine or in the colon. Late absorption was observed with all doses administered (16, 40, 100 mg rutin). In contrast, quercetin (8, 16, 50 mg aglycone) showed no clear maximal plasma concentration (t_{max} 1.9, 2.7, 4.9 h, respectively), but some absorption was observed from 30 min after ingestion, as is consistent with absorption along the length of the small and large intestine.

Comparing the available data provides evidence that quercetin appears in the plasma within a short period if present in the diet as the aglycone or a glucoside

---

* In the studies of Hollman and associates and Olthof and associates all samples were hydrolyzed and only the aglycone was measured.

Table 2 Typical Quercetin Concentrations in Plasma and Urine After Dietary Supplementation in Humans[a]

Food	Flavonol glycoside	N° of subject	Amount ingested (mg)[b]	C_{max}[c] (μM)	T_{max} (h)	% Excreted
Onion[10]	Quercetin glucosides	9	68	0.74	<0.7	1.39
Apple[10]	Quercetin glycosides	9	98	0.3	2.5	0.44
Supplement[10]	Rutin	9	100	0.3	9.3	0.35
Onion[11]	Quercetin glucosides	5	186	2.18	1.3–1.9[d]	0.8
Tea[12]	Rutin	15	49	0.1[e]	—	0.5
Onion[12]	Quercetin glucosides	15	13	0.07[e]	—	1.1
Onion[13]	Quercetin glycosides	24	114	1.48[f]	—	
Flavonol-rich meal[14]	Quercetin glycosides	10	87	0.37[g]	—	nd
Supplement[15]	Rutin	9	100	0.23	5.6	nd
Supplement[15]	Quercetin-4'-glucoside	9	100	3.2	<0.6	nd
Onion[16]	Quercetin supplement	16	8, 20, 50	0.14, 0.22, 0.29	1.9, 2.7, 4.9	nd
Supplement[16]	Rutin supplement	16	8, 20, 50	0.08, 0.16, 0.30	6.5, 7.4, 7.5	nd
Onion[17]	Quercetin glucosides	7	68–94	0.63[h]		nd
Supplement[18]	Quercetin-4'-glucoside	9	100	4.5	0.5	3.1
Supplement[18]	Quercetin-3-glucoside	9	100	5.0	0.6	3.6
Onion[19]	Quercetin glucosides	4	79	0.6[i]	nd	nd
Onion[20]	Quercetin glucosides	12	15.9	0.05[j]	nd	1.0
Tea[20]	Rutin	12	13.7	0.03[j]	nd	0.6
Red Wine[20]	Quercetin glycosides	12	14.2	0.03[j]	nd	0.8
Onion[21]	Quercetin glucosides	12	100	7.6	0.7	nd
Supplement[21]	Quercetin-4'-glucoside	12	100	7.0	0.7	nd
Buckwheat tea[21]	Rutin	12	200	2.0	4.3	nd
Supplement[21]	Rutin	12	200	1.0	7.0	nd

[a] C_{max} is maximal concentration of quercetin metabolites in plasma; T_{max} is time to reach maximal plasma concentration.
[b] Quercetin aglycone equivalent.
[c] Maximal concentration unless otherwise stated.
[d] Peak concentration for different compound analyzed.
[e] Daily portion split over three occasions for a 7-day intervention period. Blood was sampled approximately 4 h after first portion.
[f] Blood sampled 90 min after intake on day 7 of intervention.
[g] Blood sampled 3 h after meal.
[h] Daily portion split over three occasions for a 7-day intervention period. Blood sampled after 10 h of fasting.
[i] Average concentration 1.5 h after consumption of onion meal.
[j] Daily portion split over two to three occasions for a 4-day intervention period. Blood was sampled on two to three occasions on day 4 and flavonol levels were found to be consistent; nd, not determined.

conjugate rather than as a rhamnoglucoside (rutinoside) conjugate (Fig. 1). Quercetin from apples reaches a maximal plasma concentration intermediate to that of quercetin glucosides or rutin, probably reflecting that apples contain a mixture of glycoside conjugates [10,22].

Apart from the studies of Hollman and associates [9] and Erlund and colleagues [16], which measured uptake of quercetin from a supplement, there have been no other studies investigating the absorption of quercetin in a free form present within a food matrix. Quercetin is sparingly soluble in water, and studies with rats investigating the effect of solubility on uptake have shown that quercetin absorption is highly dependent on the solvent in which it is ingested [23]. Although free quercetin is rarely found in food, there is still no convincing evidence that quercetin glucosides are absorbed better than the aglycone in humans.

One study investigated the bioavailability of quercetin after consumption of red wine compared with consumption of either onions or tea over 3 days [20]. At the time the blood samples were taken, the plasma concentration after red wine consumption was between those observed after consumption of onions and tea. However, there was no difference in the 24-h urinary excretion after consumption of quercetin from red wine and onions (although for tea excretion

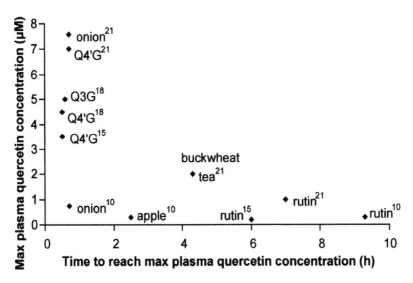

Figure 1 Summary of published data showing the time taken to reach maximal concentration of quercetin in plasma after consumption of various flavonol-rich foods or supplements. Q3G, quercetin-3-glucoside; Q4′G, quercetin-4′-gluco-side. Amount of quercetin glucoside consumed varies with study (see Table 2 for details).

was significantly lower since it contain quercetin predominantly in the form of rutin). Red wine has been shown to contain significant amounts of free quercetin [4], but in the human study described, the conjugation pattern of quercetin was not recorded. Therefore, it is not possible to conclude whether the similar excretion (and hence absorption) of quercetin from onions and red wine is due to the presence of similar quercetin conjugates or to comparable absorption of free quercetin from a soluble matrix.

There is now unequivocal evidence that some quercetin glycosides are absorbed from the upper parts of the gastrointestinal tract in humans. However, there are conflicting reports on whether quercetin glycosides are present in the plasma. Some researchers have reported the presence of quercetin glycosides in plasma or urine. Paganga and Rice-Evans [24] identified rutin along with several other unidentified flavonoid glycosides in human plasma. Aziz and associates [11] identified quercetin-4'-glucoside and 3'-methylquercetin-4'-glucoside, and Boyle and colleagues [25] identified quercetin-3-glucoside and 3'-methylquercetin-4'-glucoside in plasma after consumption of onions. All these reports utilized methods that were based on high-performance liquid chromatography (HPLC) under acidic conditions with identification based on retention time of flavonol glucoside standards after detection by diode array or fluorescence. Mauri and coworkers [26] identified rutin in plasma after consumption of tomato with detection by LC-mass spectrometry (LC-MS).

In contrast, Moon and associates [17], Walle and colleagues [27], and Day and coworkers [19] found no quercetin glucosides in plasma from subjects who had consumed an onion meal. The methodology used was HPLC with detection by diode array with pure standards as markers. These conditions would have allowed detection of quercetin glucosides if present. Wittig and associates [28] could not detect, using tandom LC-MS/MS analysis, quercetin glucosides in plasma after subjects consumed onions. Graefe and colleagues [21], using coularray detection, could not detect quercetin glucosides in the plasma or urine of subjects consuming onions or buckwheat tea or consuming quercetin-4'-glucoside or rutin as supplements. Likewise, Sesink and coworkers [29], also using coularray detection, could not detect quercetin-3-glucoside or quercetin-4'-glucoside in the plasma of subjects consuming these pure compounds. Erlund and associates [16], using electrochemical detection, could not detect rutin after supplementation. Furthermore, both Graefe and coworkers and Sesink and coworkers observed identical metabolic profiles in subjects after consumption of the different quercetin glycoside sources. In the study of Morand and coworkers [30], the same quercetin metabolic profile was also observed after rats were fed quercetin, rutin, or quercetin-3-glucoside. These results suggest that, post deglycosylation, quercetin follows the same metabolic pathway regardless of the form of quercetin glycosides administered and that plasma metabolites are quercetin (and methylquercetin) sulfates and glucuronides [19].

The apparently contradictory results—the presence or absence of quercetin glucosides in plasma—are probably explained by misidentification of quercetin glucosides since quercetin glucosides can have very similar retention times to those of the equivalent glucuronides under acidic conditions on HPLC and almost identical ultraviolet (UV) absorption spectra [31].

IV. ANIMAL INTERVENTION STUDIES

As described previously, many animal studies between 1950 and 1980 investigated the uptake and metabolism of quercetin and rutin in rats. More recently, a few animal intervention studies also considered absorption of other quercetin glycosides in order to gain more information about the site of absorption from the gastrointestinal tract. Manach and colleagues [32] investigated the uptake of quercetin and rutin administered to rats either as a single dose or after a 14-day dietary regimen. With the acute dose quercetin metabolites were shown to be present in plasma within 2 h of administration of quercetin, but, as with humans, quercetin metabolites formed from rutin were not found in plasma until 4 h after ingestion, suggesting a site of absorption at the more distal parts of the small intestine. The authors also showed that progressively more quercetin accumulated in the cecal contents of the rats fed rutin. With an extended feeding study of quercetin in the diet, plasma quercetin levels remained high, but absorption of quercetin became less efficient than with nonadapted rats. It is interesting to consider that most studies involving absorption of flavonols have used single meal (or dose) intervention and have not considered the effect of typical diet on the absorption of flavonols.

Choudhury and coworkers [33] measured the amounts of quercetin, quercetin-3-glucoside, or rutin in the urine after either an intravenous or an oral dose. After intravenous administration the authors detected a proportion of each of the compounds in the urine. However, after oral administration of the flavonol-free quercetin, quercetin metabolites or rutin could not be detected in the urine. Only a small percentage of quercetin-3-glucoside was detected in the urine after the glucoside was administered. In contrast, quercetin-3-glucoside was not detected in plasma of rats fed quercetin-3-glucoside [30]. The study compared absorption of quercetin, quercetin-3-glucoside, quercetin-3-rhamnoside, and rutin. Quercetin was absorbed at approximately 33% of the level of quercetin-3-glucoside when given at the same dose. At 4 h after administration, rutin absorption was 10% of quercetin-3-glucoside absorption, and quercetin-3-rhamnoside absorption was not detected. This suggests that, at least in rats, quercetin-3-glucoside is much more efficiently absorbed than the aglycone or other glycoside conjugates.

V. ENZYME STUDIES

It was assumed originally that since secreted digestive juices do not possess
β-glucosidase activity, and since quercetin β-glycosides resist hydrolysis by
stomach acid and other pancreatic enzymes [34], flavonoid glycosides would not
be hydrolysed until they reached the large intestine. In the colon, various
bacteria are able to release the aglycone but also further metabolize or degrade
the flavonoid ring structure. However, high levels of endogenous β-glucosidase
activity have been demonstrated in both human and rat small intestine epithelial
cells [35,36] and in human liver [35]. The activity is attributed to a broad-
specificity soluble cytosolic β-glucosidase [37] present in both tissues, and to
membrane-bound LPH [38] found on the brush border of the small intestine.
Only quercetin glucosides that are not conjugated through the 3-position of the
flavonol are substrates for the broad-specificity cytosolic β-glucosidase [39].
Rutin is not a substrate for either LPH or cytosolic β-glucosidase, and the
absence of endogenous hydrolytic enzymes explains why rutin is not absorbed in
the small intestine but passes to the colon, where it is deglycosylated by
microbial α-rhamnosidases and β-glucosidases.

Table 3 shows the substrate specificity of cytosolic β-glucosidase and LPH.
It should be noted that the flavonols quercetin-3-glucoside and quercetin-4'-
glucoside were substrates for the lactase domain of LPH and not the phlorizin
hydrolase domain, as may have been expected, given that phlorizin has a

Table 3 Specific Activities of Purified Endogenous Enzymes on Quercetin
Glycosides and on Lactose[a]

	Lactase phlorizin hydrolase[b]	Cytosolic β-glucosidase[c]	Substrates for human small intestine extract[d]	Absorption in small intestine
Quercetin-4'-glucoside	170	34	Yes	Yes
Quercetin-3-glucoside	137	0	Yes	Yes
Lactose	4	0	nd	Yes[e]
Rutin	0	0	No	No

[a] Values are expressed as catalytic efficiencies (V_{max}/K_m) at pH 7.4 and 37 °C; nd, not determined.
[b] Lactase phlorizin hydrolase (LPH) was purified from sheep small intestine [40].
[c] Recombinant human enzyme expressed in *Pichia pastoris* [39].
[d] Cell-free extracts of human small intestine were assessed for activity [35].
[e] Absorbed as glucose and galactose after hydrolysis.

flavonoid structure [38]. The specificity of the two human enzymes is the key factor in determining whether or not absorption occurs in the small intestine in humans. The rat small intestine differs from the human small intestine in its higher population of microbes. The relatively high content of microbes in the rat gut may possess sufficient activity to hydrolyse flavonoid glycosides. In contrast, the luminal contents of the human small intestine (which has only $\sim 10^4$ microbes/mL in the duodenum compared to $\sim 10^{12}$ microbes/g in the colon contents) probably do not exhibit significant activity on flavonoid glycosides. However, it cannot be excluded that the human small intestine lumen may contain endogenous proteolytically released LPH, cytosolic β-glucosidase from sloughed off epithelial cells, or some cytosolic β-glucosidase from the bile [27].

The demonstrated endogenous mammalian β-glucosidase activity has several implications for the absorption of quercetin glucosides. First, as LPH is present on the luminal side of the small intestine, any hydrolysis of quercetin glycosides would result in liberation of the aglycone that could readily diffuse across the epithelial cells as a result of increased hydrophobicity of the aglycone and release within the unstirred layer. Thus LPH provides a plausible explanation for absorption of quercetin in the small intestine and quercetin glycosides would not have to enter epithelial cells for this to occur. Second, if quercetin glycosides were transported across the intestinal wall, then the cytosolic β-glucosidase in both the small intestine and the liver would be active on the glycosidic bond of those flavonols not conjugated in the 3-position. Thus, if any flavonol-3-glycoside were absorbed intact, it would not be deglycosylated and would remain unchanged circulating in plasma until possible resecretion in the bile. Third, populations with polymorphisms in these enzymes would absorb and metabolize flavonols differently. LPH levels vary widely: 75% of the world's population show lactose maldigestion caused by reduced levels of LPH in adulthood. This may have implications for flavonol absorption that have not been explored.

VI. CELL CULTURE STUDIES

Caco-2 cells, derived from a human colon adenocarcinoma, differentiate spontaneously when grown on plates or filters into a continuous monolayer of cells presenting morphological and functional characteristics typical of normal ileal enterocytes [41–43] with a pseudo–brush border membrane. In the differentiated form, the monolayer mimics in some respects the microvilli of the small intestine epithelial cells. However, as for all immortalized cell lines, several key enzymes present in the human small intestine in vivo either are not expressed or are expressed to different extents between clones. The Caco-2 model has been used to study the intestinal absorption of flavonoids [44–49]. The evidence for transport of quercetin and quercetin glucosides across Caco-2 cells is confusing and is sometimes contradictory to the results from isolated small intestine (discussed later). Walgren and associates [44] demonstrated that although quercetin was absorbed efficiently across Caco-2 cells (by diffusion [47]), quercetin-4'-glucoside was not. In subsequent experiments, Walgren and colleagues showed that quercetin-4'-glucoside was pumped back out of Caco-2 cells by (MRP-2) (a multidrug-resistant protein present on the apical side of the intestinal epithelium [45]. By inhibiting MRP-2, the authors were able to demonstrate that quercetin-4'-glucoside could be transported by the SGLT-1 and confirmed the result in a nonintestinal cell line, Chinese hamster ovary cells, that had been stably transfected with SGLT-1 (G6C3 cells [46]). However, Caco-2 cells typically have very low levels of LPH [50], and so deglycosylation as a route for absorption would probably have been missing from these experiments.

In similar experiments, Murota and colleagues [49] demonstrated the absorption and metabolism of quercetin, quercetin-3-glucoside, quercetin-4'-glucoside, and quercetin-3,4'-glucoside by using Caco-2 cells. Absorption of quercetin was more efficient than that of any of the glucosides, although when compared to the other glucosides, quercetin-4'-glucoside was absorbed and metabolized to the greatest extent. Again, although not measured, the cells would have been deficient in LPH compared to that of normal small intestine, a difference that precludes quantitative comparison of the models. The authors suggested that on the basis of octanol-water partition coefficients of the quercetin conjugates the lipophilicity was the determining factor in absorption of the compound. Quercetin glucosides and quercetin conjugates (including methylated quercetin metabolites) were measured in the basolateral solutions. Despite the limitations, the results suggest that (1) quercetin glucosides can be transported across intestinal cells, (2) quercetin glucosides are deglycosylated during transport across intestinal cells, and (3) quercetin is metabolized by Caco-2 cells. Furthermore, as MRP-2 was not inhibited in these experiments, reduced transfer of compound to the basolateral side may have occurred as a result of greater apical efflux of the glucoside than would normally be expected. Figure 2 summarizes the observations from Caco-2 cells.

If the Caco-2 cells are good experimental models for the human and the rat, then the results with Caco-2 cells appear to contradict the results observed in the human and rat studies. Quercetin glucosides are absorbed efficiently in vivo. Despite the efflux effect of MRP-2 on net transfer of quercetin glucosides, and the fact that none of the experiments incorporated a basolateral sink (the effect of plasma protein binding [48,51,52]), a more efficient transfer of quercetin from quercetin glucosides would have been expected than that observed. As described, one possible explanation for a low percentage transfer in Caco-2 cells lies in the level of expression of the β-glucosidase, LPH. Although Caco-2 cells express LPH, the levels of the enzyme are significantly lower than those found in vivo

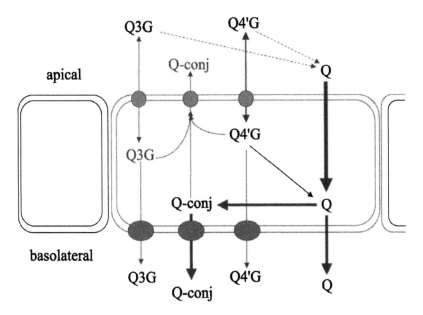

Figure 2 Schematic representation of data obtained after incubation of quercetin and quercetin glucosides with Caco-2 cells. Q, quercetin; Q3G, quercetin-3-glucoside; Q4'G, quercetin-4'-glucoside; Q-conj, quercetin conjugated metabolites; dotted arrows represent very low conversion to aglycone; thick arrows represent major pathway of uptake and metabolism. , MRP-2/SGLT-1 or similar transporters on the apical surface of the membrane; , unknown transporters on the basolateral surface of the enterocyte. (Data summarized from Ref. 49). Quercetin glucosides, metabolites, and free aglycone were measured in the basolateral solutions after incubation with Caco-2 cells. (From Refs. 45,46.) Quercetin-4'-glucoside was measured within the enterocyte, the MRP-2 transporter was shown to be involved in the efflux of the glucoside back into the apical solution, and β-glucosidase activity was demonstrated. (From Ref. 56.) A low level of apical deglycosylation was found when quercetin-3-glucoside and quercetin-4'-glucoside were incubated with Caco-2 cells. All researchers showed a preference of quercetin uptake over the glucoside uptake. MRP-2, multidrug resistance–associated protein 2; SGLT-1, sodium-dependent glucose transporter 1.

(Table 4). As an example of quercetin metabolism by LPH-free cells, HepG2 cells, derived from human liver, contain only cytosolic β-glucosidase and do not express LPH. HepG2 cells metabolize quercetin-4'-glucoside but do not metabolize quercetin-3-glucoside [55]. Because of the low LPH levels, Caco-2 cells also rapidly metabolize quercetin-4'-glucoside [46] but metabolize quercetin-3-glucoside very slowly if at all [56]. In marked contrast, rat small intestine metabolizes both quercetin-4'- and 3-glucosides [57].

Table 4 Relative Activities of Lactase Phlorizin Hydrolase in Cells and Tissues

	Typical specific activity on lactose (mU/mg protein)
Caco2 cells (undifferentiated)	0–0.12[a]
Caco2 cells (differentiated)	0.1 (50)
Caco2 cells clone PD7 (differentiated)	0.3 (50)
Human small intestine cell-free extract	20–80 (53)
Human small intestine cell free extract (from lactose-intolerant individuals)	2–10 (53)
Rat small intestine cell free extract	10–20 (54)

[a] Unpublished data.

VII. RAT INTESTINAL STUDIES

Everted-jejunal sacs have been used to study uptake of quercetin glycosides in vitro. Gee and associates [58] demonstrated, using efflux of [^{14}C]-galactose across rat everted-jejunal sacs, that quercetin-3- and 4'-glucoside, but not rutin, interacted with (but were not necessarily transported by) SGLT-1. Ader and colleagues [59] have also shown an interaction of quercetin-3- and 4'-glucoside, but not quercetin or rutin, with SGLT-1, by competitive Na$^+$-dependent mucosal uptake of methyl-α-D-glucopyranoside (a nonmetabolizable glucose analog). An interaction, however, does not show that transport has occurred since phenolic glucosides can interact by inhibiting sugar transport but at the same time are not transported across the membrane. For example, phlorizin, the classical inhibitor of SGLT-1, is not transported across the brush border because of the bulky nature of the aglycone [60]. There is considerable interspecies variation in properties of SGLT-1: the K_i for phlorizin is 10 μM for rat, 250 μM for human, and 750 μM for rabbit; p-nitrophenyl β-glucoside is transported by the rabbit form but inhibits the human and rat transporters [61], and these differences make comparisons of results from interspecies studies difficult.

Using a similar model, the rat perfused small intestine, Spencer and colleagues [62] found that a substantial proportion of quercetin-3-glucoside was transferred as the intact glucoside after a 90-min incubation, although some hydrolysis had occurred during absorption, as evident by the presence of some aglycone and of glucuronide metabolites. In contrast, Gee and coworkers [57] and Crespy and associates [63] did not detect quercetin-3-glucoside in experiments with everted-jejunum or in situ perfused rat intestine for a shorter period (using 15- to 30-min incubation, respectively). Transfer of quercetin had occurred, since significant concentrations of quercetin metabolites were found.

Rutin has been detected on the vascular side after incubation with perfused rat intestine [62,64], although quercetin aglycone was not detected.

VIII. COMPARISON OF EXPERIMENTAL MODELS

Quercetin was more efficiently transferred across Caco-2 cells than quercetin glucosides [49], but quercetin glucosides were more efficiently transferred across isolated rat small intestine than quercetin [57], and similar results were seen in vivo in rats [30]. The conflicting evidence may be a result of differences between the experimental models, either interspecies variation in transporter properties (human model compared to rat model), or variation in expression of proteins such as LPH (cell lines compared to normal tissue).

Quercetin-4'-glucoside can be transported by SGLT-1 into Caco-2 cells, and interactions between quercetin-3- and 4'-glucoside and SGLT-1 have been demonstrated in rat small intestine. LPH has activity toward flavonol glycosides, which may constitute the major role in the uptake of quercetin glycosides. Crespy and associates [63] provided some evidence for the involvement of LPH in the uptake of quercetin-3-glucoside. In the experiment, quercetin-3-glucoside was incubated with perfused rat intestine in the presence or absence of phlorizin (an inhibitor of SGLT-1). Quercetin-3-glucoside was hydrolysed by the lactase-active site and not the phlorizin hydrolase–active site of LPH [38], and so phlorizin should not affect quercetin-3-glucoside hydrolysis. The composition of flavonoids in the intestinal lumen contents was analyzed at the end of the 30-min incubation. The results showed that in the presence of phlorizin there was no significant effect on quercetin aglycone formed from quercetin-3-glucoside by the action of LPH, but also there was no difference in the total quercetin-3-glucoside absorbed (i.e., net absorption+ effluxed metabolites). However, there was a difference in the effluxed proportion of quercetin metabolites, probably due to competition of quercetin-3-glucoside and phlorizin for both conjugating enzymes and the efflux transporters such as MRP-2 [65]. Crespy and colleagues [63] showed that although twice the level of quercetin from quercetin-3-glucoside, compared to the level from the aglycone, was absorbed to the serosal side of the rat intestine (net absorption), the total absorption, including the metabolites that were secreted to the luminal side, was the same for both compounds. Evidently the glucoside does not help the transfer of quercetin into the enterocyte, but it does slow the rate of efflux, and so eventually a greater proportion of quercetin metabolites from quercetin-3-glucoside enters the bloodstream.

In order to demonstrate further the role of LPH in absorption of quercetin-3- and 4'-glucoside, N-(m-butyl)-deoxy-galactonojirimycin (NBDGM, an inhibitor of LPH), or phlorizin (an inhibitor of SGLT-1) was employed in the rat everted-jejunum model. The results demonstrate that by inhibiting LPH in a rat model, luminal deglycosylation of quercetin-3- and 4'-glucoside is significantly reduced. Consequently, transfer of quercetin-3- and 4'-glucoside to the vascular side of the tissue is lower (Fig. 3). In the presence of phlorizin, luminal hydrolysis is unaffected, quercetin-4'-glucoside transfer is reduced, but interestingly there is no effect on quercetin-3-glucoside. A plausible explanation of the data is that quercetin-3- and 4'-glucoside are absorbed across the intestine by different mechanisms: quercetin-3-glucoside is hydrolyzed by LPH, with SGLT-1 possibly only contributing to a minor extent; quercetin-4'-glucoside, in contrast, has two routes of passage into the enterocyte, luminal hydrolysis by LPH to quercetin followed by diffusion of the aglycone, or transport by SGLT-1 with subsequent deglycosylation by cytosolic β-glucosidase.

The preceding hypothesis makes understanding the differences observed in the Caco-2 studies easier. As a result of low expression of LPH in Caco-2 cells, SGLT-1 plays the dominant role in uptake. Hence, quercetin-4'-glucoside is absorbed more efficiently than quercetin-3-glucoside; however, overall absorption of the glucosides is still lower than in the rat model, in which both SGLT-1 and LPH are active. This does not exclude the possibility that higher expression of MRP-2 (due to the adenocarcinoma origin of the cells) contributes to the rapid efflux of the

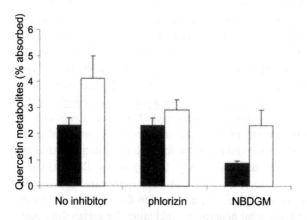

Figure 3 Effect of an SGLT-1 or lactase inhibitor on the absorption of quercetin-3-glucoside or quercetin-4'-glucoside across rat everted-jejunal sacs. Solid bars, quercetin-3-glucoside; open bars, quercetin-4'-glucoside; NBDGM, N-(m-butyl)-deoxy-galactonojirimycin. Quercetin glucosides (100 μM) were incubated with rat everted-jejunal sacs for 15 min as previously described [57] in the presence or absence of phlorizin (25 μM) as an SGLT-1 inhibitor or NBDGM (250 μM) as a lactase inhibitor. Total quercetin (as quercetin glucuronide metabolites) was measured in the serosal solution at the end of the incubation and is expressed as percentage absorbed. SGLT-1, sodium-dependent glucose transporter 1.

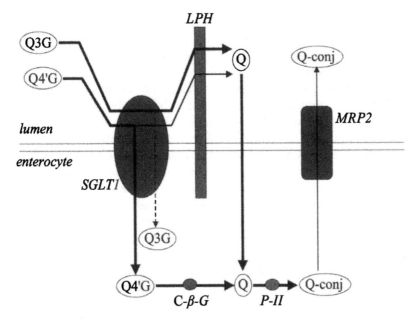

Figure 4 Potential mechanisms for quercetin glucoside absorption across the human small intestine. Q, quercetin; Q3G, quercetin-3-glucoside; Q4'G, quercetin-4'-glucoside; Q-conj, quercetin conjugated metabolites; LPH, lactase phlorizin hydrolase; SGLT-1, sodium-dependent glucose transporter; MRP-2, multidrug resistant protein; C-β-G, cytosolic β-glucosidase; P-II, phase 2 conjugating enzymes; dotted arrows, very low transport; thick arrows, major pathway of absorption and metabolism. We hypothesize that quercetin-4'-glucoside can be transported by SGLT-1, but quercetin-3-glucoside is a poor substrate. Both compounds are actively deglycosylated by LPH and uptake is significantly reduced if lactase is inhibited. Only quercetin-4'-glucoside can be deglycosylated by cytosolic β-glucosidase.

glucoside conjugate from the enterocyte. Figure 4 represents the proposed mechanism of uptake of quercetin glucosides in the human small intestine.

IX. CONCLUSIONS

The absorption of quercetin glycosides is complex. Flavonol-rich foods contain a mixture of many different types of conjugate, all of which may be absorbed in a different manner. Quercetin rhamnoglucosides such as rutin, commonly found in plant foods, appear only to be absorbed in the large intestine after hydrolysis by the colon microflora. Late absorption in the gastrointestinal tract leads to significantly lower overall absorption, probably as a result of degradation of the flavonoid ring structure. Quercetin glucosides, in contrast, are absorbed much earlier in the small intestine, leading to higher overall absorption of the compounds, and probably greater enterohepatic circulation. However, the quantitative contribution of each step to overall absorption in vivo remains unclear. LPH appears to play an important role in hydrolyzing the glucosidic bond, releasing the aglycone within the intestinal lumen for absorption by a diffusion mechanism. LPH has broad specificity, excepting certain sugar types (excluding rhamnosides) and positions of substitution on the flavonoid structure. SGLT-1, or another transporter, may also contribute to absorption of some compounds. Little, if any, glucoside conjugate appears in the plasma, probably because of efficient hydrolysis within the enterocyte or hepatocyte by cytosolic β-glucosidase. Further studies on the substrate specificity of SGLT-1 to transport the various flavonol glycoside moieties, particularly in populations with low LPH expression, are required. The ability of other flavonoid subclasses to interact with LPH and SGLT-1 must also be determined as early absorption leads to a greater bioavailablity of these bioactive compounds.

ACKNOWLEDGMENTS

We would like to thank Mike Morgan, Geoff Plumb, Paul Kroon, Jean-Guy Berrin, Karen O'leary, Susan DuPont, Jenny Gee, and Ian Johnson for valuable contributions. We would also like to thank the Biotechnology and Biological Sciences Research Council and the EU Framework V project Polybind QLK1-1999-00505 for contribution to funding in some of the experiments described.

REFERENCES

1. Hertog MGL, Hollman PCH, Katan MB, Kromhout D. Intake of potentially anticarcinogenic flavonoids and their determinants in adults in The Netherlands. Nutr Cancer 1993; 20:21–29.
2. Price KR, Bacon JR, Rhodes, MJC. Effect of storage and domestic processing on the content and composition of flavonol glucosides in onion (*Allium cepa*). J Agric Food Chem 1997; 45:938–942.
3. Price KR, Rhodes MJC. Analysis of the major flavonol glycosides present in four varieties of onion (*Alium cepa*) and changes in composition resulting from autolysis. J Sci Food Agric 1997; 74:331–339.
4. McDonald MS, Hughes M, Burns J, Lean MEJ, Matthews D, Crozier A. Survey of the free and conjugated myricetin and quercetin content of red wines of different geographical origins. J Agric Food Chem 1998; 46:368–375.

5. Miyazawa T, Nakagawa K, Kudo M, Muraishi K, Someya K. Direct intestinal absorption of red fruit anthocyanins, cyaniding-3-glucoside and cyaniding-3,5-diglucoside, into rats and humans. J Agric Food Chem 1999; 47:1083–1091.

6. Scheline RR. Handbook of Mammalian Metabolism of Plant Compounds. Boca Raton, FL: CRC Press, 1991:267–287.

7. Gugler R, Leschik M, Dengler MJ. Disposition of quercetin in man after single oral and intravenous doses. Eur J Clin Pharmacol 1975; 9:229–234.

8. Hollman PCH, van Trijp JMP, Buysman, MNCP. Fluorescence detection of flavonols in HPLC by postcolumn chelation with aluminium. Anal Chem 1996; 68:3511–3515.

9. Hollman PCH, de Vries JH, van Leeuwen SD, Mengelers MJ, Katan MB. Absorption of dietary quercetin glycosides and quercetin in healthy ileostomy volunteers. Am J Clin Nutr 1995; 62:1276–1282.

10. Hollman PCH, van Trijp JMP, Buysman MNCP, Gaag MS, Menglers MJB, deVries JHM, Katan MB. Relative bioavailablity of the antioxidant quercetin from various foods in man. FEBS Lett 1997; 418:152–156.

11. Aziz AA, Edwards CA, Lean MEJ, Crozier A. Absorption and excretion of conjugated flavonols, including quercetin 4'-O-β-glucoside and isorhamnetin 4'-O-β-glucoside, by human volunteers. Free Radic Res 1998; 29:257–269.

12. DeVries JHM, Hollman PCH, Meyboom S, Buysman, MNCP, Zock PL, van Staveren WA, Katan MB. Plasma concentrations and urinary excretion of the antioxidant flavonols quercetin and kaempferol as biomarkers for dietary intake. Am J Clin Nutr 1998; 68:60–65.

13. Janssen PLTMK, Mensink RP, Cox FJJ, Harryvan JL, Hovenier R, Hollman PCH, Katan MB. Effects of the flavonoids quercetin and apigenin on hemostasis in healthy volunteers: results from an in vitro and a dietary supplement study. Am J Clin Nutr 1998; 67:255–262.

14. Manach C, Morand C, Crespy V, Démigné C, Texier O, Régérat F, Rémésy C. Quercetin is recovered in human plasma as conjugated derivatives which retain antioxidant properties. FEBS Lett 1998; 426:331–336.

15. Hollman PCH, Buysman MNCP, van Gameren Y, Cnossen EPJ, deVries JHM, Katan MB. The sugar moiety is a major determinant of the absorption of dietary flavonoid glycosides in man. Free Radic Res 1999; 31:569–573.

16. Erlund I, Kosonen T, Alfthan G, Maenpaa J, Perttunen K, Kenraali J, Parantainen J, Aro A. Pharmacokinetics of quercetin from quercetin aglycone and rutin in healthy volunteers. Eur J Clin Pharmacol 2000; 56:545–553.

17. Moon J-H, Nakata R, Oshima S, Inakuma T, Terao J. Accumulation of quercetin conjugates in blood plasma after short-term ingestion of onion by women. Am J Physiol 2000; 279:R461–R467.

18. Olthof MR, Hollman PCH, Vree TB, Katan MB. Bioavailabilities of quercetin-3-glucoside and quercetin-4'-glucoside do not differ in humans. J Nutr 2000; 130:1200–1203.

19. Day AJ, Mellon F, Barron D, Sarrazin G, Morgan MRA, Williamson G. Human metabolism of dietary flavonoids: identification of plasma metabolites of quercetin. Free Radic Res 2001. In press.

20. de Vries JHM, Hollman PCH, van Amersfoort I, Olthof MR, Katan MB. Red wine is a poor source of bioavailable flavonols in men. J Nutr 2001; 131:745–748.

21. Graefe EU, Wittig J, Mueller S, Riethling AK, Uehleke B, Drewelow B, Pforte H, Jacobasch G, Derendorf H, Veit M. Pharmacokinetics and bioavailability of quercetin glycosides in humans. J Clin Pharmacol 2001; 41:492–499.

22. Price KR, Prosser T, Richetin AMF, Rhodes MJC. A comparison of the flavonol content and composition in dessert, cooking and cider-making apples, distribution within the fruit and effect of juicing. Food Chem 1999; 66:489–494.

23. Piskula MK, Terao J. Quercetin's solubility affects its accumulation in rat plasma after oral administration. J Agric Food Chem 1998; 46:4313–4317.

24. Paganga G, Rice-Evans CA. The identification of flavonoids as glycosides in human plasma. FEBS Lett 1997; 401:78–82.

25. Boyle SP, Dobson VL, Duthie SJ, Kyle JAM, Collins AR. Absorption and DNA protective effects of flavonoid glycosides from an onion meal. Eur J Nutr 2000; 39:213–223.

26. Mauri PL, Iemoli L, Gardana C, Riso P, Simonetti P, Porrini M, Pietta PG. Liquid chromatography electrospray ionization mass spectrometric characterization of flavonol glycosides in tomato extracts and human plasma. Rapid Commun Mass Spectrom 1999; 13:924–931.

27. Walle T, Otake Y, Walle UK, Wilson FA. Quercetin glucosides are completely hydrolyzed in ileostomy patients before absorption. J Nutr 2000; 130:2658–2661.

28. Wittig J, Herderich M, Graefe EU, Veit M. Identification of quercetin glucuronides in human plasma by high-performance liquid chromatography-tandem mass spectrometry. J Chromatogr B 2001; 753:237–243.

29. Sesink ALA, O'Leary KA, Hollman PCH. Quercetin glucuronides but not glucosides are present in human plasma after consumption of quercetin-3-glucoside or quercetin-4'-glucoside. J Nutr 2001; 131:1938–1941.

30. Morand C, Manach C, Crespy V, Remesy C. Quercetin 3-O-beta-glucoside is better absorbed than other quercetin forms and is not present in rat plasma. Free Radic Res 2000; 33:667–672.

31. Day AJ, Williamson G. Biomarkers for exposure to dietary flavonoids: a review of the current evidence for identification of quercetin glycosides in plasma. Br J Nutr 2001; 86:S105–S110.

32. Manach C, Morand C, Démigné C, Texier O, Régérat F, Rémésy C. Bioavailability of rutin and quercetin in rats. FEBS Lett 1997; 409:12–16.

33. Choudhury R, Srai SK, Debnam E and Rice-Evans CA. Urinary excretion of hydroxycinnamates and flavonoids after oral and intravenous administration. Free Radic Biol Med 1999; 27:278–286.

34. DuPont MS, Gee JM, Price KR, Johnson IT. The availability of flavonol glycosides for small intestinal transport. Gut 1999; 44:TH517.

35. Day AJ, DuPont MS, Ridley S, Rhodes M, Rhodes MJC, Morgan MRA, Williamson G. Deglycosylation of flavonoid and isoflavonoid glycosides by human small intestine and liver β-glucosidase activity. FEBS Lett 1998; 436:71–75.

36. Ioku K, Pongpiriyadacha Y, Konishi Y, Takei Y, Nakatani N, Terao J. β-

glucosidase activity in the rat small intestine towards quercetin monoglucosides. Biosci Biotech Biochem 1998; 62:1428–1431.

37. Lambert N, Kroon PA, Faulds CB, Plumb GW, McLauchlan WR, Day AJ, Williamson G. Purification of cytosolic beta-glucosidase from pig liver and its reactivity towards flavonoid glycosides. Biochim Biophys Acta 1998; 1435:110–116.

38. Day AJ, Cañada FJ, Díaz JC, Kroon PA, Mclauchlan R, Faulds CB, Plumb GW, Morgan MRA, Williamson G. Dietary flavonoid and isoflavone glycosides are hydrolysed by the lactase site of lactase phlorizin hydrolase. FEBS Lett 1998; 468:166–170.

39. Berrin J-G, McLauchlan WR, Juge N, Williamson G, Kroon PA. The human cbg1-1 gene encodes a β-glucosidase which hydrolyses a broad range of xenobiotic glycosides. Eur J Biochem. In press.

40. Rivera-Sagredo A, Cañada FJ, Nieto O, Jimenez-Barbero J, Martín-Lomas M. Substrate-specificity of small-intestinal lactase—assessment of the role of the substrate hydroxyl-groups. Eur J Biochem 1992; 209:415–422.

41. Hidalgo IJ, Raub TJ, Borchardt RT. Characterization of the human-colon carcinoma cell-line (Caco-2) as a model system for intestinal epithelial permeability. Gastroenterology 1989; 96:736–749.

42. Pinto M, Robine-Leon S, Appay MD. Enterocyte-like differentiation and polarization of the human colon carcinoma cell line Caco-2 in culture. Biol Cell 1983; 47:323–330.

43. Artursson P. Cell cultures as models for drug absorption across the intestinal mucosa. Crit Rev Ther Drug Carrier Syst 1991; 8:305–330.

44. Walgren RA, Walle UK, Walle T. Transport of quercetin and its glucosides across human intestinal epithelial Caco-2 cells. Biochem Pharmacol 1998; 55:1721–1727.

45. Walgren RA, Karnaky KJ, Lindenmayer GE, Walle T. Efflux of dietary flavonoid quercetin 4'-beta-glucoside across human intestinal Caco-2 cell monolayers by apical multidrug resistance-associated protein-2. J Pharmacol Exp Ther 2000; 294:830–836.

46. Walgren RA, Lin J-T, Kinne RK-H, Walle T. Cellular uptake of dietary flavonoid quercetin 4'-β-glucoside by the sodium dependent glucose transporter, SGLT1. J Pharmacol Exp Ther 2000; 294:837–843.

47. Kuo SM. Transepithelial transport and accumulation of flavone in human intestinal Caco-2 cells. Life Sci 1998; 63:2323–2331.

48. Walgren RA, Walle T. The influence of plasma binding on absorption/exsorption in the Caco-2 model of human intestinal absorption. J Pharm Pharmacol 1999; 51:1037–1040.

49. Murota K, Shimizu S, Chujo H, Moon JH, Terao J. Efficiency of absorption and metabolic conversion of quercetin and its glucosides in human intestinal cell line Caco-2. Arch Biochem Biophys 2000; 384:391–397.

50. Chantret I, Rodolosse A, Barbat A, Dussaulx E, Brotlaroche E, Zweibaum A, Rousset M. Differential expression of sucrase-isomaltase in clones isolated from early and late passages of the cell-line caco-2—evidence for glucose-dependent negative regulation. J Cell Sci 1994; 107:213–225.

51. Manach C, Morand C, Texier O, Favier ML, Agullo G, Demigné C, Régérat F,

Rémésy C. Quercetin metabolites in plasma of rats fed diets containing rutin or quercetin. J Nutr 1995; 125:1911–1922.

52. Boulton DW, Walle UK, Walle T. Fate of the flavonoid quercetin in human cell lines: chemical instability and metabolism. J Pharm Pharmacol 1999; 51:353–359.
53. Rossi M, Maiuri L, Fusco MI, Salvati VM, Fuccio A, Auricchio S, Mantei N, Zecca L, Gloor SM, Semenza G. Lactase persistence versus decline in human adults: multifactorial events are involved in down-regulation after weaning. Gastroenterology 1997; 112:1506–1514.
54. Goda T, Yasutake H, Tanaka T, Takase S. Lactase-phlorizin hydrolase and sucrase-isomaltase genes are expressed differently along the villus-crypt axis of rat jejunum. J Nutr 1999; 129:1107–1113.
55. Day AJ. Human Absorption and Metabolism of flavonoid glycosides Ph.D. dissertation, University of East Anglia, England, 2000.
56. Plumb GW, McLauchlan R, Kroon PA, Day AJ, Faulds CB, Gee JM, DuPont MS, Williamson G. Uptake and metabolism of quercetin glycosides in Caco-2 cell culture and everted rat gut sacs. Polyphenol Communications 2000, Freising-Weihenstephan Germany, Sept 11-15 2000:401–402.
57. Gee JM, DuPont MS, Day AJ, Plumb GW, Williamson G, Johnson IT. Intestinal transport of quercetin glycosides in the rat involves both deglycosylation and interaction with the hexose transport pathway. J Nutr 2000; 130:2765–2771.
58. Gee JM, DuPont MS, Rhodes MJC, Johnson IT. Quercetin glucosides interact with the intestinal glucose transport pathway. Free Radic Biol Med 1998; 25:19–25.
59. Ader P, Block M, Pietzsch S, Wolffram S. Interaction of quercetin glucosides with the intestinal sodium/glucose co-transporter (SGLT-1). Cancer Lett 2001; 162:175–180.
60. Lostao MP, Hirayama BA, Loo DDF, Wright EM. Phenylglucosides and the Na+/glucose cotransporter (SGLT1)- analysis of interactions. J Membr Biol 1994; 142:161–170.
61. Hirayama BA, Lostao MP, PanayotovaHeiermann M, Loo DDF, Turk E, Wright EM. Kinetic and specificity differences between rat, human, and rabbit Na+-glucose cotransporters (SGLT-1). Am J Physiol 1996; 33:G919–G926.
62. Spencer JPE, Chowrimootoo G, Choudhury R, Debnam ES, Srai SK, Rice-Evans C. The small intestine can both absorb and glucuronidate luminal flavonoids. FEBS Lett 1999; 458:224–230.
63. Crespy V, Morand C, Besson C, Manach C, Demigne C, Remesy C. Comparison of the intestinal absorption of quercetin, phloretin and their glucosides in rats. J Nutr 2001; 131:2109–2114.
64. Andlauer W, Stumpf C and Furst P Intestinal absorption of rutin in free and conjugated forms. Biochem Pharmacol 2001; 62: 369–374.
65. Mizuma T, Awazu S. Inhibitory effect of phloridzin and phloretin on glucuronidation of p-nitrophenol, acetaminophen and 1-naphthol: kinetic demonstration of the influence of glucuronidation metabolism on intestinal absorption in rats. Biochim Biophys Acta 1998; 1425:398–404.

16

Bioavailability of Flavanol Monomers

Jennifer L. Donovan
Medical University of South Carolina
Charleston, South Carolina, U.S.A.

Andrew L. Waterhouse
University of California, Davis
Davis, California, U.S.A.

I. INTRODUCTION

Flavanols are one of the most abundant classes of flavonoids, often referred to as *flavan-3-ols* or *catechins*. They are present as monomers, oligomers, and polymers and are often esterified with gallic acid [1]. The focus of this chapter are bioavailability and metabolism of the flavanol monomers, catechin, epicatechin, and the green tea flavanols epigallocatechin gallate, epigallocatechin, and epicatechin gallate (Fig. 1).

Flavanols are abundant components of many foods and beverages. Red wine, apples, tea, and chocolate are among the richest food sources [2–5]. The intake of the flavanol monomers was determined to be 50 mg/day in a Dutch cohort with tea, chocolate, apples, and pears as the main sources [6]. The dietary intake of flavanols is likely to vary greatly among individuals and cultures. Cultures that consume green tea and red wine by custom such as France or Japan would be expected to have higher intakes.

Numerous in vivo studies have indicated that flavanols have diverse biological activities including antioxidant activity [7–11] and anticancer properties [12–14]. Flavanols may inhibit platelet aggregation as well as other vascular and inflammatory processes that contribute to disease [15–17]. These effects are thought to be mediated by changes in eicosonoid synthesis after consumption of flavanol-containing foods [9,18]. Flavanol metabolites isolated from urine were shown to reduce monocyte adhesion to endothelial cells in the inflammatory

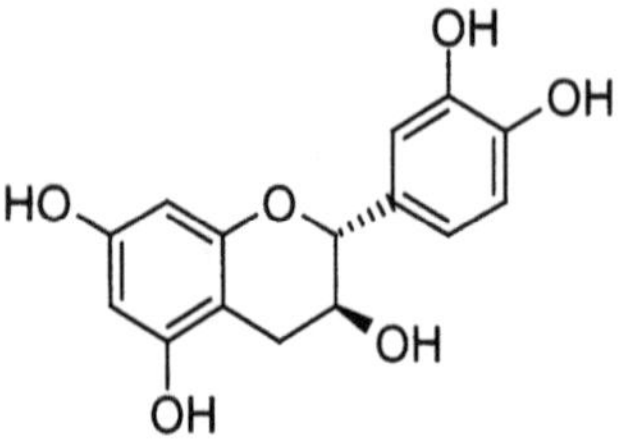

Catechin

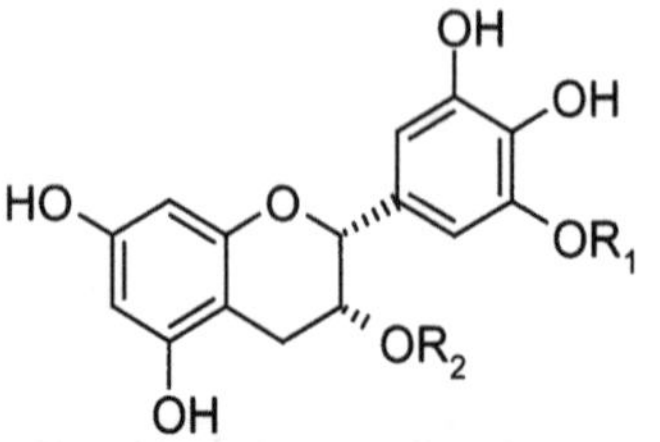

R₁=H, R₂=H Epicatechin
R₁=OH, R₂=H Epigallocatechin
R₁=H, R₂=Gallate; Epicatechin Gallate
R₁=OH, R₂=Gallate; Epigallocatechin Gallate

Figure 1   Chemical structures of monomeric flavanols.

process of atherosclerosis [7]. Feeding studies in several animal models have also demonstrated reduced progression of fatty streaks or atherosclerosis [19–21].

The study of flavanol absorption and metabolism has been carried out for some time, and there are a number of prior reviews on this subject [22–25]. Studies on the absorption and metabolism of monomeric flavanols are extensive, going back to the 1950s, though all early work was conducted on pharmacological doses of catechin [26–29]. The metabolic pathways of catechin elucidated in these studies continue to be the models for other flavonoids. They indicated that the potential pathways of metabolism are glucuronidation and sulfation as well as methylation, but the plasma and tissue levels of specific metabolites or their proportions are not reflective of those present after consumption of the same flavanols from foods [30].

Research in the last several years has focused on the bioavailability of flavanols when consumed in green tea, red wine, and chocolate. Plasma levels of flavanols and metabolites as well as some pharmacokinetic parameters have now been documented in humans after short-term feeding studies with these foods. Additionally, animal and cell culture models as well as in vitro experiments have been used to increase our understanding of the mechanisms of absorption, metabolism, and elimination that ultimately regulate the bioavailability of flavanols.

II. MECHANISMS REGULATING THE BIOAVAILABILITY OF FLAVANOLS

A. Absorption

Flavanols are an unusual class of flavonoids in that they are not present as glycosides in the diet so a sugar moiety does not play a role in the site or mechanism of absorption, as has been described for other flavonoids [31–33] (see Chap. 15). Catechin, epicatechin, epigallocatechin, and epigallocatechin gallate are able to be absorbed in their intact forms and have been identified as conjugated metabolites in plasma [8,30,34,35]. Epicatechin gallate may not be absorbed intact as most studies find negligible amounts in plasma and urine after tea consumption [34,36]. Epicatechin gallate has been identified in the rat portal vein as well as in humans after consumption of large doses of the pure compound [37,38].

1. Sources of Bioavailable Monomers

In addition to the common dietary sources of flavanols [39,40], there may be indirect routes to obtaining monomers or types of monomers. Studies indicate that although epigallocatechin gallate has been identified in vivo, cleavage of the gallate moiety from the flavanol ring system may also occur. Amelsvoort and colleagues [37] showed that 3% of epicatechin gallate and 5% of epigallocatechin gallate were in the hydrolyzed form in plasma after the administration of purified tea catechins to humans. Gallate esters that reach the large intestine certainly have contact with bacterial esterases, although subsequent absorption of the hydrolysis products has not been documented. A 1999 study reported cleavage of epigallocatechin gallate in the mouth. The esterase responsible was identified in saliva and is thought to be derived from human epithelial cells (as opposed to oral bacteria), although this hypothesis remains to be confirmed. The esterase was not reported to be present in human plasma or liver [41]. Nevertheless, it is possible that at least some epicatechin and epigallocatechin are produced from their respective gallate esters before absorption.

Proanthocyanidins have also been investigated as a potential source of monomeric flavanols. In vitro experiments indicate that proanthocyanidin dimers may be hydrolyzed into monomers by acid present in the stomach [42]. The degree of polymerization of proanthocyanidins has been reported to decrease in the rat intestine [43], and an energy-dependent cleavage of dimers was described in the rat small intestine [44]. Authors have also suggested that proanthocyanidins are cleaved by microflora in the large intestine although monomers have not been directly detected after incubation with intestinal bacteria [45–47]. Conversely, when rats were fed purified procyanidin B_3, a catechin dimer, no catechin or metabolites could be detected in plasma or urine [48]. The study also showed that plasma levels of catechin are not different when catechin is consumed alone or along with oligomeric procyanidins. These data show that in vivo, oligomeric procyanidins are not cleaved into bioavailable monomers at any point during the digestive process in the rat. Future experiments must be performed to determine whether proanthocyanidins are a source of bioavailable monomers in humans.

2. Mechanisms of Absorption

Most flavanol absorption is thought to occur in the small intestine. The mechanism of catechin and epicatechin absorption has been investigated by using intestinal perfusion models in the rat. Both catechin and epicatechin appear to be efficiently absorbed by the jejunum and ileum. Extensive metabolism also occurs within the small intestine, demonstrating that these flavanols are absorbed into intestinal cells, and not by the paracellular route [49,50].

The effect of dose on catechin absorption was also studied, using an intestinal perfusion [49]. In that study one-third of the catechin dose was absorbed at all concentrations ranging from 1 to 100 μM, suggesting that absorption of catechin by the small intestine is directly proportional to the dose over a fairly wide range. The data indicate that catechin enters intestinal epithelial cells by passive diffusion, a mechanism that is generally proportional to the dose. The process is relatively nonspecific but increases with the hydrophobicity of the molecule.

The mechanism of absorption of the gallate esters has not been studied directly. A 1999 study demonstrated that absorption of epigallocatechin gallate can begin to occur in the mouth [41]. The contrast between epigallocatechin gallate and epicatechin gallate is intriguing as the molecules differ by only a single hydroxyl group. Furthermore, the more hydrophilic molecule, epigallocatechin gallate, is absorbed, whereas epicatechin gallate appears not to be. Thus, passive diffusion is not the sole determinant of uptake—a more specific process must be involved. A plausible explanation is that a transporter with specificity for one of these gallates is involved in their uptake or presystemic elimination. Epicatechin gallate could also be completely hydrolyzed by an esterase before reaching the circulating blood.

The flavanols that enter epithelial cells likely undergo extensive metabolism within the small intestine (discussed later) before being delivered to the liver and the circulating blood and tissues.

B. Metabolism

As are most other flavonoids, the flavanols are metabolized by the phase II metabolic processes glucuronidation, sulfation, and methylation. Metabolism appears to be quite efficient as catechin and epicatechin are present exclusively as metabolites in vivo after doses in foods [30,34]. Free catechins are only present in plasma after large pharmaceutical doses are administered, suggesting the metabolic enzymes are saturated [26]. A fraction of the tea catechins epigallocatechingallate and epigallocatechin exist in their native forms after green tea consumption, the conjugated forms predominate [34]. The position and type of conjugate certainly are major determinants of biological activity [7] as well as their distribution and clearance in vivo. A general summary of major metabolites is shown in Table 1. The classes of metabolites and the positions of the specific conjugates are discussed later.

1. Methylated Metabolites

Methylation may occur on flavanols that contain ortho-hydroxy functional groups. The position of methylation of catechin and epicatechin has been identified as the 3'- in humans and several other species [30,48,51–55] although small amounts of 4'-*O*-methyl conjugates of catechin and epicatechin have been described in both human and rat urine [55,56].

Gallocatechins, on the other hand, appear to be methylated predominantly at the 4'-position. After a 1.5-g dose of purified epigallocatechin in humans, levels of 4'-*O*-methylated metabolites were four- to sixfold higher than levels of unmethylated forms in plasma [57]. Substitution at the 4'-position indicates different substrate specificity for the trihydroxylated functional group.

The gallic acid esters may be methylated on both the flavonoid ring and the gallic acid residue. Methylated metabolites of epigallocatechin gallate have been identified in bile after a 100-mg dose was administered to the rat. Methylation could occur at the 3'- or 4'-position of the flavonoid ring and at the 3- or 4-position of the gallic acid residue. It was also possible for a single metabolite to be methylated on both the flavonoid and the gallic acid residue [58].

2. Glucuronidated and Sulfated Metabolites

The majority of flavanols that exist in vivo are conjugated with glucuronic acid or sulfate. Mixed conjugates containing both glucuronide and sulfate have been described, and methylated metabolites are generally further conjugated with glucuronide or sulfate [30,52]. The addition of either of these groups results in metabolites that are negatively charged at physiological pH and are more polar, facilitating excretion in urine.

The major site of glucuronidation of catechin and epicatechin has been identified in rats as the 5-position. Okushio and associates [55] and Harada and colleagues [52] fed rats catechin or epicatechin, either purified or from red wine or green tea. The major metabolites in urine and bile were identified as 3'-*O*-methyl(epi)catechin 5-*O*-β-glucuronide and (epi)catechin 5-*O*-β-glucuronide. Both research groups purified the metabolites from rat urine and positively identified the positions of the methyl and glucuronide moieties by nuclear Overhauser effect (NOE) and heteronuclear multiple bond connectivity (HMBC) spectroscopy experiments. Figure 2 shows the structure of the major catechin and epicatechin metabolites in the rat. The position of glucuronidation of flavanols has not been studied in humans.

Table 1 A Summary of Metabolites for Each of the Monomeric Flavanols[a]

Metabolites	Catechin	Epicatechin	Epigallocatechin	Epicatechin gallate	Epigallocatechin gallate
Glucuronides Sulfates	Mainly mixed conjugates Glucuronide 5-position	Glucuronide sulfate Mixed conjugates Glucuronide 5-position	Mainly glucuronide	Not detected except after black tea consumption	Mainly sulfate
Methylated	20% of metabolites; 3'-position; also glucuronidated, sulfated	20–30% of metabolites; 3'-position also glucuronidated, sulfated	4'Position	Not detected	3' or 4'position; 3- or 4-position of gallic acid
Hydrolysis	Phenylvalerolactones, phenolic acids	Phenylvalerolactones, phenolic acids	Phenylvalerolactones, phenolic acids	Phenylvalerolactones, phenolic acids; possible hydrolysis of gallic acid residue	Phenylvalerolactones, phenolic acids; possible hydrolysis of gallic acid residue
Native	Detected only after consumption of foods that contain large doses	Detected only after consumption of foods that contain large doses	3–13% of dose	Not detected	12–28% of dose
Urinary excretion	3–10% of dose	5–7% of dose	2–4% of dose	Not detected except after black tea consumption	Not detected except after black tea consumption

[a] Metabolites have been identified and measured in human intervention studies [26–30], animal studies [34–37], and in vitro experiments [52–67].

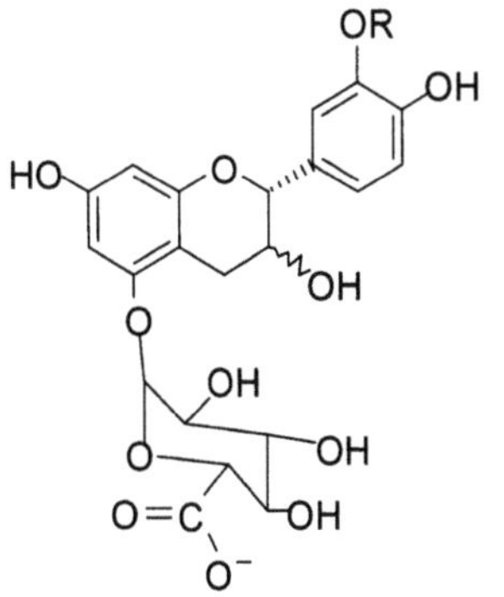

Figure 2 Glucuronide conjugates catechin and epicatechin identified in rats; R=H or CH_3. Glucuronidation occurs primarily at the 5-position. (From Refs. 52,55.)

There are no studies that have investigated the position of the sulfate moiety. Donovan and colleagues [30] reported that 3' -O-methylcatechin was not sulfated at all after red wine consumption, suggesting that conjugation with one precludes conjugation with the other. However, catechin metabolites that are both 3' -O-methylated and sulfated have been identified after larger doses in humans as well as in several other species [26,53]. Methylated epicatechin has also been identified as being predominantly in the sulfate form in humans [59].

3. Ring-Fission Metabolites

Low-molecular-weight metabolites form from fission of the heterocyclic ring and the A-ring and then degradation into phenolic acids and lactones. Metabolites of flavanol monomers are mainly hydroxylated derivatives of benzoic, phenylacetic, and phenylpropionic acids. Hippuric acid and its hydroxylated forms originate from the conjugation of glycine with benzoic acids. Phenylvalerolactones, derived from the B-ring and fragments of the C- and A-rings, have also been identified. One important aspect of this metabolic pathway is the deoxygenation of the B-ring, such that the 4' hydroxyl group from the B-ring is lost. A few studies have also shown that carbon dioxide is formed. Structures of some ring-fission metabolites are shown in Figure 3.

Early animal studies on rats, mice, and guinea pigs were used to identify the metabolites of catechin [60] and established a pathway for some of the steps illustrated in Fig. 3. Subsequently, a human clinical study that administered 4–6 of catechin per person showed that these ring-fission metabolites were absent in the urine for the first 12 h but predominated for the next 36 h [27]. They identified 3-hydroxyphenylpropionic as well as several hydroxylated phenylvalerolactones as the major metabolites of catechin and showed that that the

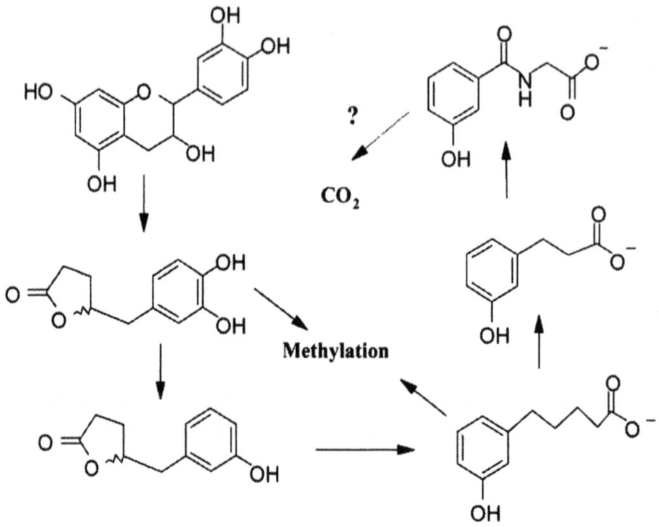

Figure 3 Structures of some ring-fission metabolites of monomeric flavanols. Phenolic acid metabolites may be conjugated with glycine to form hippuric acids before excretion in urine. (Pathway adapted from Refs. 60,65.)

majority of these metabolites exist as glucuronide or sulfate conjugates. One study by Kohri and coworkers [61] identified a specific glucuronic acid conjugate of a new valerolactone yielded from epigallocatechin. This product is shown in Figure 4.

A later study with a 2-g dose of catechin reported that the ring-fission metabolites accounted for only 10% of catechin excretion in urine over a 24-h period. Among the products excreted were 3-hydroxybenzoic, 3-hydroxy hippuric acid, and 3-hydroxyphenylpropionic acid [26]. However, no valerolactones were detected, and this study was carried out by using ^{14}C-labeled catechin. Much of the analysis used high-performance liquid chromatography (HPLC); some used thin-layer chromatography (TLC). Differences between these studies were initially attributed to the dose, although the effect of dose on ring-fission metabolites has never been studied directly. However, a number of studies have shown that the intestinal microflora have a profound effect on the production of ring-fission products. For instance, Das and Griffiths were able to suppress their production in the guinea pig completely after treatment with two antibiotics [62]. A later study with [14C]-catechin showed that antibiotics greatly suppressed blood levels and urinary elimination of metabolites when the ring fission was arrested [63], apparently by interrupting the production of the lactones from the lower gut. Nearly identical results were obtained in rats with epigallocatechin

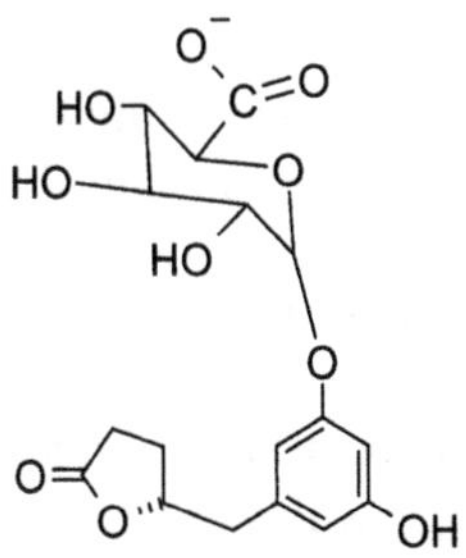

Figure 4 Glucuronide conjugate of a valerolactone. (Valerolactone originates from epigallocatechin identified in Ref. 61.)

[61]. A 2001 study with humans also showed that antibiotics drastically reduce the appearance of the valerolactones in urine after catechin consumption [64].

Meselhy and colleagues [65] incubated epicatechin gallate and other components of green tea with a human fecal or rat fecal suspension. The authors identified several new compounds, including the ring-opened and deoxygenated analog of the valerolactones. In particular they showed a progression of flavonoid ring A fission and then, after opening the lactone ring, loss of acetate. Another important reaction they reported was the conversion of the catechol group to a phenol group by deoxygenation of the phenolic ring.

Few studies have reported levels of ring-fission metabolites after doses in common foods. After green tea was administered to humans at high doses (400 mg total catechins), 4-hydroxy benzoic, 3,4-dihydihydroxybenzoic, and 3-methoxy 4-hydroxy benzoic as well as 3-methoxy 4-hydroxyhippuric acid were identified in plasma and urine, but it is not clear whether the valerolactones were investigated. The total amount of these metabolites accounted for 15% of the dose in urine [66]. Li and colleagues [67] identified a dihydroxylated and a tri-hydroxylated phenylvalerolactone in human plasma and urine after ingestion of a single dose of green tea. The metabolites were presumed to originate from epicatechin and epigallocatechin and were present at 8–25 times the concentration of their flavonoid precursors in urine. They accounted for 6–39% of the doses of epicatechin and epigallocatechin in urine [67].

So, it appears to be very likely that the difference between the earlier study by Das and associates [27] and the 1983 study by Hackett and colleagues [26] is due to different gut microflora in the subject populations. This hypothesis is strengthened by the results in 2000 of Li and coworkers [67], in that study the variation in the levels of epicatechin and epigallocatechin conjugates excreted in the urine varied fourfold between subjects, similarly to results of other reports, but the levels of the valerolactones varied by more than 10 times.

Finally, a large proportion of catechin, in particular the A-ring, appears to be totally catabolized by microbial action to carbon dioxide. Das and Griffiths [60] observed 18% of the radioactivity in catechin appear in this form in animal experiments, and Gott and Griffiths [63] observed that the proportion of flavonoid carbons released as carbon dioxide dropped from 45% to 6% (over a 72-h period) when the rats were treated with antibiotics.

The ring-fission process appears to be caused by microbial action, not mammalian metabolism. It involves breakdown of the A-ring, possibly releasing acetate or carbon dioxide. The remaining carbons of the C- and A-rings form a γ-lactone between the 3-hydroxy group and the newly formed carboxylic acid group at the end of the remaining 5-carbon chain. The lactone can be broken and suffer a loss of one or more "acetate" 2-carbon fragments. Other important reactions are the loss of one of the catechol oxygens as well as methylation. In urine all these substance are present mostly as polar conjugates.

4. Metabolism in Different Tissues

The phase II enzymes responsible for flavanol metabolism are widely distributed among tissues, including liver, lung, intestine, and kidney. The levels of the enzymes and their distribution among tissues determine the pattern of metabolites. Each of the reactions occurs within cells. Glucuronidation occurs on the luminal side of the endoplasmic reticulum by uridine-5′-diphosphate glucuronosyltransferases (UGTs), a large family of enzymes. Sulfation and methylation both occur in the cytosol by sulfotransferases (SULTs), and catechol-O-methyltransferases (COMTs). The specific isoforms involved have not been identified for flavanols. The UGT1A family is thought to be responsible for glucuronidation of other flavonoids [68]. SULT1A1 and SULT1A2 are implicated in the sulfation of phenol-type substrates and are also known as the *P-forms* or *thermostable* (*TS*) forms [69]. Future recognition of the specific isoforms involved in flavanol metabolism will aid in our understanding of interindividual variability and help predict interactions with other xenobiotics and drugs [70–72].

The metabolism of flavanols by specific tissues has only been investigated in rats. The small intestine is thought to be the major organ of glucuronidation of flavanols. Piskula and Terao [73] measured the activity of UGTs in preparations of rat liver, kidney, lung, intestine, and plasma, using epicatechin as a substrate. The small intestine had the highest capacity for glucuronidation and had approximately 10 times the activity of the liver. Perfusion studies in the rat also indicate that the small intestine is the most important organ of flavanol glucuronidation. Methylation may also occur in the small intestine but the process appears less efficient than glucuronidation [49,50,73]. Sulfation does not appear to occur significantly in the rat small intestine, although it should not be discounted in humans, as intestinal SULT activity is characteristically much higher in humans than in rats [74,75].

Donovan and coworkers [49] showed that glucuronides, and perhaps methylated glucuronides, produced in the small intestine are able to enter hepatocytes and become further metabolized therein. The mechanism of uptake into hepatocytes is still unknown, but uptake of other glucuronides by the liver has been previously documented [76,77].

Within the liver, SULT activity is predominant, but further methylation may also occur. Catechin glucuronides were subsequently sulfated, as well as methylated, in rat liver after in situ perfusion [49]. Piskula and Terao [73] also reported that SULT activity for epicatechin was present exclusively in liver. Within hepatocytes, it is also possible that catechin glucuronides are deglucuronidated and then reglucuronidated before reaching systemic circulation [49].

Further metabolism of flavanols may occur in peripheral tissues. Piskula and Terao [73] also showed that methylation was very active in the kidney. Some UGT activity, although significantly less than in the intestine, was also present in kidney, lung, and plasma. A schematic representation of the proposed mechanisms of flavanol metabolism is shown in Figure 5. Further

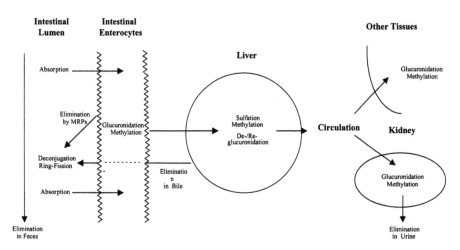

Figure 5 A schematic representation of the possible mechanisms of absorption, metabolism, and elimination of flavanols. Absorbed flavanols enter intestinal epithelial cells, where they are extensively glucuronidated and sometimes methylated. Some of these metabolites may pass through the liver without actually entering the hepatocytes. Other glucuronides are able to enter into hepatocytes and are possibly de-re-glucuronidated. In the cytosol of hepatocytes, metabolites may be methylated as well as sulfated before entry into circulation or elimination in bile.

methylation and possible glucuronidation may occur in the kidneys, as suggested by Piskula and Terao [73], although these metabolites are likely to be excreted in urine thereafter.

C. Elimination

Flavanols are quickly eliminated from plasma. The elimination half-lives of flavanols are far shorter than reported for the flavonols, which are around 24 h [78]. Half-lives in plasma after food doses range from 2 to 4 h for catechin and epigallocatechin [30,36]. Epicatechin has a slightly longer half-life (3–6 h) [36]. The half-lives indicate that most flavanols would be cleared from the body in 10–20 h. The half-life for epigallocatechingallate was reported to range from 5 to 7 h, and so it may exist slightly longer in vivo [36]. Few studies are able to detect appreciable levels in plasma 24 h after consumption [30,35,36,59].

Three major mechanisms of elimination of flavanols have been reported: elimination by the kidneys in urine, elimination by the liver in bile, and elimination by the small intestine by an active efflux transport mechanism. The final fate of most flavanols after elimination by the intestine or liver is likely to be through metabolism by microflora in the colon.

1. Elimination in Bile

Elimination in bile may be the most important mechanism of flavanol elimination. After an intravenous injection of ^{14}C-labeled catechin to rats, 33–44% of the dose was excreted in bile during the first 24 h [28]. The major biliary metabolites of catechin after oral administration of 10 or 100 mg of catechin to the rat were two glucuronide conjugates of 3' -O-methylcatechin, which accounted for 17% of the administered dose [54]. Both catechin and 3' -O-methylcatechin have also been reported to be present as glucuronide conjugates as well as mixed glucuronide/sulfate conjugates in rat bile after intestinal perfusion [49].

Elimination in bile has not been studied directly in humans; however, in general, larger, extensively conjugated metabolites are more likely to be eliminated in bile. Differences in the proportions of metabolites in plasma and urine also indicate that larger metabolites (i.e., glucuronide and mixed glucuronide/sulfate conjugates) are excreted in bile [79]. Mixed glucuronide and sulfate metabolites predominated in plasma after wine consumption but made up a smaller fraction of metabolites in urine [56]. After green tea consumption epigallocatechin was mainly in the glucuronide form in plasma but was mostly sulfated in urine [34]. Epigallocatechin gallate is present at high concentrations in plasma after green tea consumption but not detected in urine [34,80].

2. Elimination in Urine

After consumption of foods containing flavanols 1–10% of the dose is typically recovered in urine as forms containing the intact flavonoid ring (Table 1). The recovery of catechin ranged from 3% to 10% of the dose after red wine consumption [56]. Epicatechin and epigallocatechin recoveries range from 2% to 7% after green tea consumption [34,36]. One study reports 25–30% recovery of epicatechin after chocolate consumption, much of it in unconjugated form [59]. The gallate esters are not detected in urine after green tea but small amounts were detected after black tea consumption [34–36].

Excretion of intact flavanols in urine cannot be considered a reliable indicator of absorption as it may be only a minor route of excretion. The proportion excreted in urine may also vary with dose as well as the experimental conditions. One study with three different doses of green tea catechins reported that there was no correlation between the amount of epicatechin or epigallocatechin in urine and the areas under the curve (AUCs) in plasma [36].

The ring-fission products constitute a large proportion of the urinary metabolites in most, but not all studies [26,51,81]. The presence of these components appears to be dependent on microbial action in the lower intestine [62], so their elimination in the urine may be an indicator of gut microflora status.

Ethanol in red wine has been shown to enhance the elimination of catechin in urine [56]. Twenty percent more catechin was excreted after wine containing ethanol compared to dealcoholized wine. This study administered only 12.4 g of ethanol in 120 mL of red wine. The effect of higher amounts of ethanol may be more dramatic, although this possibility has not been studied. Ethanol was suggested to enhance the elimination of flavanols through a diuretic effect, which would not be specific to wine. Thus, a similar increase in urinary excretion may occur after the consumption of other alcoholic beverages along with flavanol-containing foods.

3. Elimination by the Small Intestine

Intestinal excretion is an important mechanism in the elimination of flavonoid conjugates [82–84]. This is likely mediated by the multidrug resistance–associated protein (MRP) pumps. MRPs actively export conjugated metabolites out of the small intestine back into the intestinal lumen and so prevent or reduce systemic circulation [85]. An experiment using cultured Caco-2 cells showed that two metabolites of epicatechin were excreted on the apical side of the cells. Their elimination has been attributed to MRP-2, as efflux was significantly reduced by a competitive MRP-2 inhibitor [86]. Conversely, intestinal perfusion experiments with catechin indicate that catechin metabolites are not substrates for these transport proteins [49].

D. Plasma Levels and Pharmacokinetics of Flavanols in Humans

Human intervention studies are ultimately necessary to determine the bioavailability of flavonoids from specific foods. A summary of recent studies including the doses, maximal plasma concentrations, and urinary excretion is shown in Table 2. There are large differences in the reported plasma levels among studies using seemingly similar food sources. The large interstudy variability cannot be explained by the doses used but may be explained by differences in experimental conditions and analytical methodology.

1. Analysis of Flavanols in Biological Samples

Early investigations of flavanols in biological samples relied on colorimetric measures of polyphenol content [87] or thin-layer chromatography to assess the presence of specific metabolites [60,62]. Because it is difficult to separate the polar conjugates chromatographically, they are typically hydrolyzed to release the free forms, which are subsequently analyzed. Acid had been used to hydrolyze the conjugates, but there is ample evidence of flavanol degradation by acid treatment, so enzymatic treatment is essential.

As a result of the low concentration of flavanols in blood fractions such as plasma when food level "doses" are consumed, there are only a few techniques that have sufficient sensitivity. The first of these is gas chromatography coupled to a mass spectrometer for detection [88]. This technique is quite sensitive and provides the assurance of accurate identification of the analyte when the molecular and major fragmentation ions are monitored. The drawbacks are that it is necessary to derivatize the flavanols to achieve sufficient volatility and the technique cannot be adapted for analysis of glucuronide and sulfate conjugates. This method has been used to quantify the plasma levels of catechin and its methylated derivatives after wine consumption, and a modified version is able to detect catechin, quercetin, and resveratrol simultaneously [89].

Because of the nonvolatile nature of flavanols or their metabolites, liquid chromatography is ideally suited for their separation. Multielectrode electrochemical detection is a sensitive and selective technique that has been used to measure flavanols and other flavonoids in biological samples [48,49,90]. The method has been adapted to measure specific conjugate forms of flavonols [33,91] and could be adapted for flavanol conjugates. Several other approaches seem to have adequate sensitivity, including fluorescence and [92] and chemiluminescence [93]. Both techniques are very selective and have little interference in biological samples. Chemiluminescence is dependent on the presence of the catechol functional group, and sensitivity is lost after some known metabolic conversions, such as methylation. Mass spectral detection has both sensitivity and the generality to allow for the measurement of nearly all

Table 2 A Summary of Human Clinical Studies Reporting Plasma Levels and Urinary Excretion After Consumption of Flavanols from Foods[a]

Flavanol	Dose	Source	Plasma Cmax (nmol/L)	Urine[b] (% of dose)	Reference
Catechin	35	Red wine	50–176	3–10	[30]
Epicatechin	164	Chocolate	700	NM	[97]
Epicatechin	137	Chocolate	257	NM	[8]
Epicatechin	35, 69, 104	Chocolate	133, 258, 355	NM	[8]
Epicatechin	220	Chocolate	4770	25–30	[59]
Epicatechin	111	Chocolate	36	NM	[9]
Epicatechin	32	Green tea	165–275	5–7	[34]
EGC	82		268–673	3–4	
EGCG	88		98–572	ND	
ECG	33		ND	ND	
Epicatechin	38, 75, 113	Reconstituted	189, 651, 655	4.7	[36]
EGC	102, 204, 306	green tea	484, 1660, 1797	2.6	
EGCG	110, 219, 329		254, 696, 685	ND	
ECG	33, 66, 99		ND	ND	
Epicatechin	7	Canned green tea	NM	6.1	[108]
EGC	24		NM	3.7	
EGCG	33		NM	ND	
ECG	6		NM	ND	
Catechin	13		NM	1.8	
GC	52		NM	0.4	
CG	6		NM	ND	
GCG	36		NM	ND	
Epicatechin	146	Black tea	174	2.6	[35]
EGC	62		145	3.7	
EGCG	67		20	0.14	
ECG	124		51	0.12	
EGCG	118	Green tea	5000	NM	[99]
EGCG	105	Green tea extract	135–303	NM	[107]
EGCG	225, 375, 525	Green tea extract	657, 4300, 4410	NM	[100]
EGC	8, 13, 18		35, 144, 255		

[a] Reported levels represent the total of free, glucuronidated, and sulfated metabolites, as well as the methylated metabolites if measured. Unno et al. [107] and Nakagawa et al. [100] report only free forms in plasma. EGC, epigallocatechin; EGCG, epigallocatechin gallate; CG, catechin gallate. When several doses are shown, plasma levels are shown in respective order.

[b] ND, not detected; NM, not measured.

metabolites, including unanticipated ones. It would thus appear that liquid chromatography-mass spectrometry (LC-MS) will be the most important tool in future studies of flavanol metabolism, and recent successes rely on this technology, including the identification and quantification of the new valerolactone metabolites from epigallocatechin [61,67].

Careful consideration must be given to the analytical methodology employed when flavanols are reported in biological samples. Recovery and stability experiments in plasma should be reported and hydrolysis techniques should be verified. Detection and quantitation must be performed with a method with some specificity for the analyte such as fluorescence, multi-electrode coulometry, or, ideally, mass spectrometry. Studies reporting very high levels of free forms without any hydrolysis of conjugates after normal doses, although potentially correct under certain conditions, must be regarded with some skepticism.

2. Pharmacological Doses

Earlier research on flavanol absorption and metabolism was performed on catechin as it was reported to be an effective treatment for viral hepatitis. These studies used doses that exceed the amounts in foods by a factor of at least 50 [26,27,51,94]. As a result, large amounts of native catechin reported to be present in plasma were likely due to the saturation of enzymatic pathways involved in metabolism.

Hackett and coworkers [26] showed that after ingestion of 2 g of purified ^{14}C-labeled catechin, plasma radioactivity peaked at a level corresponding to 40 μmol/L catechin. Most of the catechin was present as metabolites; 12.5% was in the native form. Half of the dose was excreted in urine, mainly as 3' -O-methylcatechin glucuronide, 3' -O-methylcatechin sulfate, and catechin glucuronide. Wermeille and colleagues [51] also showed that 60% of catechin was methylated at the 3' -position and that catechin and 3' -O-methylcatechin were present as sulfate and/or glucuronide conjugates in human urine after a 1-g dose.

3. Red Wine

Red wine is a rich source of flavanols as well as several other classes of flavonoids [95]. Plasma levels of catechin and metabolites were determined after a single serving of red wine (120 mL) containing 35 mg of catechin [30]. Maximal levels were achieved at 1 h and ranged from 50 to 176 nM, representing some variability among subjects. Catechin was present mainly as a mixed glucuronide/sulfate conjugate and to a much lesser extent a sulfate or glucuronide conjugate. Methylated metabolites accounted for 20% of the total, and 3-O-methylcatechin existed mostly as a glucuronide conjugate (Fig. 6). The

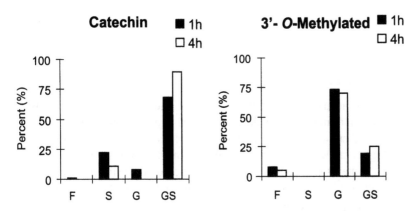

Figure 6 Proportions of the conjugate forms of catechin and 3'-O-methylcatechin 1 h after red wine consumption. (From Ref. 30.) Forms are free (F), sulfated (S), glucuronidated (G), or glucuronidated and sulfated (GS).

elimination half-life of all metabolites ranged from 2 and 4 h; sulfated metabolites were eliminated most rapidly. Recovery in urine ranged from 3% to 10% of the dose and was enhanced by presence of ethanol presence in wine [56].

4. Chocolate

Chocolate is a rich source of epicatechin monomer in addition to its high procyanidin content [96]. Several studies have fed chocolate samples and determined the total amount of glucuronide and sulfate conjugates in plasma [8–10,97]. Epicatechin levels in the experimental chocolate samples ranged from 35 to 164 mg. Average maximal levels of epicatechin in plasma in these studies ranged from 36 nM to 700 nM.

The specific forms of epicatechin metabolites as well as methylated forms were measured after chocolate and cocoa consumption. This study reported much higher maximal plasma levels (4.8 µM) than all other studies as well as a 25–30% recovery rate in urine. Epicatechin was present as a mixed sulfate/glucuronide, a sulfate conjugate, and a glucuronide conjugate. Methylated epicatechin accounted for up to 40% of the total amount of metabolites and was present as a sulfate or a mixed sulfate/glucuronide conjugate [59].

Multiple doses of chocolate were administered in a crossover study by Wang and coworkers [10]. Subjects consumed chocolate samples containing 27, 53, or 80 mg of epicatechin. The total amount of glucuronide and sulfate conjugates in plasma were reported. Maximal plasma levels and the area under the curve (AUC) increased proportionally after the three doses. The authors also suggest that food (bread) delays the absorption of epicatechin but does not significantly affect the total amount absorbed, as reflected by the AUCs.

5. Green and Black Tea

Tea is the main dietary source of the gallic acid esters of flavanols. Tea also contains significant amounts of gallocatechins (tri-hydroxylated B-ring). Green tea contains higher levels of flavanol monomers than black tea, which also contains higher-molecular-weight products of flavanol oxidation [98]. Although tea contains epicatechin monomer, it should be considered separately in terms of bioavailability, as gallic acid esters may be hydrolyzed to form epicatechin or epigallocatechin.

Flavanols from green tea have been studied most extensively of any flavanols (Table 2). Maximal plasma levels of epigallocatechin gallate, the major flavonoid in green tea, generally range from 100 to 600 nM after green tea consumption; one study reports 5 µM epigallocatechin gallate after similar doses [99]. Epicatechin generally ranges from 100 to 300 nM and epigallocatechin from 200 to 700 nM. As discussed earlier, epicatechin gallate is never detected in plasma after green tea consumption [34,36].

Only one study reported the different types of conjugate forms after tea consumption [34]. Epicatechin was exclusively in the conjugated form: two-thirds in the sulfate form and one-third in the glucuronide form. Epigallocatechin, conversely, was present mostly in the glucuronide form, followed by the sulfate. As much as 3–13% was present in the free form in plasma. Epigallocatechin gallate was present mainly in the sulfate form, and the free form accounted for 12–28%. Mixed conjugates and methylated metabolites were not identified in that study.

Three separate studies have investigated the effect of dose on plasma levels of tea catechins. One reported that at the highest level, no further increase in total plasma metabolites was observed [36]. Nakagawa and colleagues [100] reported a similar trend, although that study measured only the native forms in plasma. A third study, using a mixture of semipurified tea polyphenols, concluded the opposite and reported that AUCs at the highest doses were proportionally higher. The study also showed that the percentage of free EGCG increased with increasing dose [80].

A 2001 experiment administered black tea four times during a 6-h period [35]. Each cup contained approximately four times the amount of monomeric flavanols in a typical brew, and so subjects consumed the equivalent of 12 cups of tea. Maximal levels of epicatechin, epigallocatechin, and epigallocatechin gallate were achieved in 5–8 h. Levels of epigallocatechin gallate were markedly lower than has been observed after consumption of green tea, even considering the lower dose. Another difference between the two types of teas was that after consumption of black tea, epicatechin gallate was detected in plasma. Epicatechin gallate level was reported to increase linearly in plasma for 24 h after consumption. The reason for its appearance in plasma after consumption of only black tea as well as its curious kinetic properties remain unclear but may well be related to microbial action in the lower intestine. Small amounts of both gallate esters were also detected in urine. The specific conjugate forms of flavanols and the methylated forms were not studied. As expected, only small percentages of the initial doses were recovered in urine and less than 1% of the dose could be recovered in feces, indicating that gut microflora have metabolized a large percentage of the ingested dose.

Flavanols have also been measured in plasma after green and black tea consumption by using colorimetric detection after extraction and complexation with 4-dimethylaminocinnamaldehyde (DMACA) [101]. This method is specific for flavanols, but it is unknown whether there are interferences in biological samples. Additionally, the reaction does not occur if the A-ring is conjugated [102] and the authors did not attempt to hydrolyze the conjugate forms before to extraction. Nevertheless, the reported levels increased twice as much after green tea compared to black tea consumption. The addition of milk had little effect on plasma flavanols, in agreement with an earlier study using similar methodology [103]. The authors also suggested, on the basis of their assay, that approximately half of the catechins in plasma were associated with lipoproteins, mainly high-density lipoprotein (HDL) [101].

III. CONCLUSIONS

The flavanol monomers catechin, epicatechin, and epigallocatechin gallate are absorbed in humans and animals. Epicatechin gallate is generally not identified in vivo but may be present at low concentrations after black tea consumption. Catechin and epicatechin are not present in their native forms but as combinations of methylated, glucuronidated, and sulfated metabolites. Epigallocatechin and epigallocatechin gallate exist to some extent in their native forms but are also mainly present as conjugates. The total amount of metabolites is generally between 100 nM and 1 μM in plasma after consumption of foods very rich in flavanol content.

The conjugated forms of flavanols are eliminated rather quickly with reported half-lives ranging from 2 to7 h after consumption of wine, tea, or chocolate. This indicates that dosage must occur several times per day to maintain reasonable steady-state plasma concentrations throughout the day. The elimination of flavanols is expected to be nearly complete after an overnight fast.

The ring-fission products appear to persist for a much longer time, and thus if these components have important physiological roles, the effect may be persistent for about 24 h after consumption of foods high in flavanols. Much work remains to be done to characterize levels of these compounds in circulating blood and other tissues, but the presence of these substances may be responsible for the increase in the capacity of human plasma or LDL to resist oxidation when wine, tea, or chocolate is ingested [8,19,104–106], an increase that cannot be justified by the increased level of the intact flavonoids [66].

Mechanisms that regulate the bioavailability of flavanols have been further clarified in the last several years. The general pathways of metabolism—glucuronidation, sulfation, and methylation—have been identified, and the importance of the small intestine and the liver in metabolism has been recognized. Future identification of the isoforms of the enzymes involved in flavanol metabolism will be a key to predicting interactions with drugs and xenobiotics as well as understanding the interindividual variability in levels of flavanol metabolites.

The flavanol metabolites that exist in vivo need to be characterized further as they are the major determinants of biological activity. Significant amounts of ring-fission metabolites are likely present in vivo, but many of the specific structures and levels in humans are not known. Additional information is also needed on the positions of the glucuronide and sulfate moieties in human metabolism. Determining the identities of specific metabolites and their distribution within tissues and cells will be a key to accessing their biological activity, the major task at hand.

REFERENCES

1. Haslam E. Practical Polyphenolics: From Structure to Molecular Recognition and Physiological Action. Cambridge: Cambridge University Press, 1998:422.
2. de Pascual-Teresa S, Santos-Buelga C, Rivas-Gonzalo J. Quantitative analysis of flavan-3-ols in Spanish foodstuffs and beverages. J Agric Food Chem 2000; 48(11):5331–5337.
3. Guyot S, Marnet N, Laraba D, Sanoner P, Drilleau J-F. Reversed-phase HPLC following thiolysis for quantitative estimation and characterization of the four main classes of phenolic compounds in different tissue zones of a French cider apple variety (*malus domestica* var. Kermerrien). J Agric Food Chem 1998; 46(5):1698–1705.
4. Hammerstone JF, Lazarus SA, Mitchell AE, Rucker R, Schmitz HH. Identification of procyanidins in cocoa (*theobroma cacao*) and chocolate using high-performance liquid chromatography mass spectrometry. J Agric Food Chem 1999; 47(2):490–496.
5. Hammerstone JF, Lazarus SA, Schmitz HH. Proanthocyanidin content and variation in some commonly consumed foods. J Nutr 2000; 130 (suppl 8): 2086S–2092S.

6. Arts I, Hollman P, Feskens E, de Mesquita H, Kromhout D. Catechin intake and associated dietary and lifestyle factors in a representative sample of Dutch men and women. Eur J Clin Nutr 2001; 55(2):76–81.

7. Koga T, Meydani M. Effect of plasma metaboites of (+)-catechin and quercetin on monocyte adhesion to human aortic endothelial cells. Am J Clin Nutr 2001; 73(5):941–948.

8. Rein D, Lotito S, Holt RR, Keen CL, Schmitz HH, Fraga CG. Epicatechin in human plasma: in vivo determination and effect of chocolate consumption on plasma oxidation status. J Nutr 2000; 130 (suppl 8):2109–2114.

9. Wan Y, Vinson JA, Etherton TD, Proch J, Lazarus SA, Kris-Etherton PM. Effects of cocoa powder and dark chocolate on LDL oxidative susceptibility and prostaglandin concentrations in humans. Am J Clin Nutr 2001; 74(5):596–602.

10. Wang JF, Schramm DD, Holt RR, Ensunsa JL, Fraga CG, Schmitz HH, Keen CL. A dose-response effect from chocolate consumption on plasma epicatechin and oxidative damage. J Nutr 2000; 130 (suppl 8):2115S–2119S.

11. Whitehead TP, Robinson D, Allaway S, Syms J, Hale A. Effect of red wine ingestion on the antioxidant capacity of serum. Clin Chem 1995; 41(1):32–35.

12. Clifford AJ, Ebeler SE, Ebeler JD, Bills ND, Hinrichs SH, Teissedre PL, Waterhouse AL. Delayed tumor onset in transgenic mice fed an amino acid-based diet supplemented with red wine solids. Am J Clin Nutr 1996; 64(5):748–756.

13. Liao SS, Umekita Y, Guo JT, Kokontis JM, Hiipakka RA. Growth inhibition and regression of human prostate and breast tumors in athymic mice by tea epigallocatechin gallate. Cancer Lett 1995; 96(2):239–243.

14. Yin PZ, Zhao JY, Cheng SJ, Hara Y, Zhu QF, Liu ZG. Experimental studies of the inhibitory effects of green tea catechin on mice large intestinal cancers induced by 1,2-dimethylhydrazine. Cancer Lett 1994; 79(1):33–38.

15. Keevil J, Osman H, Reed J, Folts J. Grape juice, but not orange juice or grapefruit juice, inhibits human platelet aggregation. J Nutr 2000; 130(1):53–56.

16. Putter M, Grotemeyer KH, Wurthwein G, Araghi-Niknam M, Watson RR, Hosseini S, Rohdewald P. Inhibition of smoking-induced platelet aggregation by aspirin and pycnogenol. Thromb Res 1999; 95(4):155–161.

17. Rein D, Paglieroni TG, Pearson DA, Wun T, Schmitz HH, Gosselin R, Keen CL. Cocoa and wine polyphenols modulate platelet activation and function. J Nutr 2000; 130 (suppl 8):2120–2126.

18. Schramm DD, Wang JF, Holt RR, Ensunsa JL, Gonsalves JL, Lazarus SA, Schmitz HH, German JB, Keen CL. Chocolate procyanidins decrease the leukotriene-prostacyclin ratio in humans and human aortic endothelial cells. Am J Clin Nutr 2001; 73(1):36–40.

19. Hayek T, Fuhrman B, Vaya J, Rosenblat M, Belinky P, Coleman R, Elis A, Aviram M. Reduced progression of atherosclerosis in apolipoprotein E-deficient mice following consumption of red wine, or its polyphenols quercetin or catechin, is associated with reduced susceptibility of LDL to oxidation and aggregation. Arterioscler Thromb Vasc Biol 1997; 17(11):2744–2752.

20. Xu R, Yokoyama WH, Irving D, Rein D, Walzem R, German JB. Effect of dietary

catechin and vitamin E on aortic fatty streak development in hypercholesterolemic hamsters. Atherosclerosis 1998; 137(1):29–36.

21. Yamakoshi J, Kataoka S, Koga T, Ariga T. Proanthocyanidin-rich extract from grape seeds attenuates the developement of aortic atherosclerosis in cholesterol-fed rabbits. Atherosclerosis 1999; 142(1):139–149.

22. Hollman PC, Tijburg LB, Yang CS. Bioavailability of flavonoids from tea. Crit Rev Food Sci Nutr 1997; 37(8):719–738.

23. Scalbert A, Williamson G. Dietary intake and bioavailability of polyphenols. J Nutr 2000; 130 (suppl 8):2073S–2085S.

24. Hackett AM. The metabolism of flavonoid compounds in mammals In: Cody V, Middleton EJ, Harborne JB, eds. Plant Flavonoids in Biology and Medicine: Biochemical, Pharmacological, and Structure-Activity Relationships. New York: Alan R. Liss 1986:177–194.

25. Hollman PCH, Arts ICW. Flavonols, flavones and flavanols—nature, occurrence and dietary burden. J Sci Food Agric 2000; 80(7):1081–1093.

26. Hackett AM, Griffiths LA, Broillet A, Wermeille M. The metabolism and excretion of (+)-[^{14}C]cyanidol-3 in man following oral administration. Xenobiotica 1983; 13(5):279–286.

27. Das NP. Studies on flavonoid metabolism. Absorption and metabolism of (+)-catechin in man. Biochem Pharmacol 1971; 20(12):3435–3445.

28. Das NP, Sothy SP. Studies on flavonoid metabolism: biliary and urinary excretion of metabolites of (+)-[U-^{14}C]catechin. Biochem J 1971; 125(2):417–423.

29. Watanabe H. The chemical structure of the intermediate metabolites of catechin I-IV. Bull Agric Chem Soc Jpn 1959; 23(25):257–271.

30. Donovan JL, Bell JR, Kasim-Karakas S, German JB, Walzem RL, Hansen RJ, Waterhouse AL. Catechin is present as metabolites in human plasma after consumption of red wine. J Nutr 1999; 129(9):1662–1668.

31. Hollman PCH, Devries JHM, Vanleeuwen SD, Mengelers MJB, Katan MB. Absorption of dietary quercetin glycosides and quercetin in healthy ileostomy volunteers. Am J Clin Nutr 1995; 62(6):1276–1282.

32. Hollman PC, Bijsman MN, van Gameren Y, Cnossen EP, de Vries JH, Katan MB. The sugar moiety is a major determinant of the absorption of dietary flavonoid glycosides in man. Free Radic Res 1999; 31(6):569–573.

33. Morand C, Manach C, Crespy V, Rémésy C. Quercetin-3-O-β-glucoside is better absorbed than other quercetin forms and is not present in rat plasma. Free Radic Res 2000; 33(5):667–676.

34. Lee M-J, Wang Z-Y, Li H, Chen L, Sun Y, Gobbo S, Balentine DA, Yang CS. Analysis of plasma and urinary tea polyphenols in human subjects. Cancer Epidemiol Biomarkers Prev 1995; 4(4):393–399.

35. Warden B, Smith L, Beecher G, Balentine D, Clevidence B. Catechins are bioavailable in men and women drinking black tea throughout the day. J Nutr 2001; 131(6):1731–1737.

36. Yang C, Chen L, Lee M, Balentine D, Kuo M, Schantz S. Blood and urine levels of tea catechins after ingestion of different amounts of green tea by human volunteers. Cancer Epidemiol Biomarkers Prev 1998; 7(4):351–354.

37. Amelsvoort JMM, van het Hof KH, Mathot JNJJ, Mulder TPJ, Wiersma A, Tijburg LBM. Plasma concentrations of individual tea catechins after a single oral dose in humans. Xemobiotica 2001; 31(12):891–901.
38. Okushio K, Matsumoto N, Kohri T, Suzuki M, Nanjo F, Hara Y. Absorption of tea catechins into rat portal vein. Biol Pharm Bull 1996; 19(2):326–329.
39. Arts I, van de Putte B, Hollman P. Catechin contents of foods commonly consumed in the Netherlands. 1. Fruits, vegetables, staple foods, and processed foods. J Agric Food Chem 2000; 48(5):1746–1751.
40. Arts I, van de Putte B, Hollman P. Catechin contents of foods commonly consumed in the Netherlands. 2. Tea, wine, fruit juices, and chocolate milk. J Agric Food Chem 2000; 48(5):1752–1757.
41. Yang C, Lee M, Chen L. Human salivary tea catechin levels and catechin esterase activities: implication in human cancer prevention studies. Cancer Epidemiol Biomark Prev 1999; 8(1):83–89.
42. Spencer JP, Chaudry F, Pannala AS, Srai SK, Debnam E, Rice-Evans C. Decomposition of cocoa procyanidins in the gastric milieu. Biochem Biophys Res Commun 2000; 272(1):236–241.
43. Abia R, Fry SC. Degradation and metabolism of [14]c-labelled proanthocyanidins from carob (*ceratonia siliqua*) pods in the gastrointestinal tract of the rat. J Sci Food Agric 2001; 81(12):1156–1165.
44. Spencer JPE, Schroeter H, Shenoy B, Srai KS, Debnam ES, Rice-Evans C. Epicatechin is the primary bioavailable form of the procyanidin dimers B2 and B5 after transfer across the small intestine. Biochem Biophys Res Commun 2001; 285(3):558–593.
45. Déprez S, Brézillon C, Rabot S, Philippe C, Mila I, Lapierre C, Scalbert A. Polymeric proanthocyanidins are catabolized by a human colonic microflora into low molecular weight phenolic acids. J Nutr 2000; 130(11):2733–2738.
46. Groenewoud G, Hundt HKL. The microbial metabolism of condensed (+)-catechins by rat-caecal microflora. Xenobiotica 1986; 16(2):99–107.
47. Groenewoud G, Hundt HKL. The microbial metabolism of (+)-catechin to two novel diarylpropan-2-ol metabolites in vitro. Xenobiotica 1984; 14(9):711–717.
48. Donovan JL, Manach C, Rios L, Morand C, Scalbert A, Rémésy C. Procyanidins are not bioavailable in rats fed a single meal containing a grapeseed extract or the procyanidin dimer B3. Br J Nutr 2002; 87(4):299–306.
49. Donovan JL, Crespy V, Manach C, Morand C, Besson C, Scalbert A, Rémésy C. Catechin is metabolized by both the small intestine and liver in rats. J Nutr 2001; 131(6):1753–1757.
50. Kuhnle G, Spencer JPE, Schroeter H, Shenoy B, Debnam ES, Srai KS, Rice-Evans C, Hahn U. Epicatechin and catechin are *O*-methylated and glucuronidated in the small intestine. Biochem Biophys Res Commun 2000; 277(2):507–512.
51. Wermeille M, Turin E, Griffiths LA. Identification of the major urinary metabolites of (+)-catechin and 3-*O*-methyl-(+)-catechin in man. Eur J Drug Metab Pharmacokinet 1983; 8(1):77–84.
52. Harada M, Kan Y, Naoki H, Fukui Y, Kageyama N, Nakai M, Miki W, Kiso Y. Identification of the major antioxidative metabolites in biological fluids of the rat

with ingested (+)-catechin and (−)-epicatechin. Biosci Biotechnol Biochem 1999; 63(6):973–977.

53. Hackett AM, Shaw IC, Griffiths LA. 3′-O-methyl-(+)-catechin glucuronide and 3′-O-methyl-(+)-catechin sulphate: New urinary metabolites of (+)-catechin in the rat and the marmoset. Experimentia 1982; 38(5):538–540.

54. Shaw IC, Griffiths LA. Identification of the major biliary metabolite of (+)-catechin in the rat. Xenobiotica 1980; 10(112):905–911.

55. Okushio K, Suzuki M, Matsumoto N, Nanjo F, Hara Y. Identification of (−)-epicatechin metabolites and their metabolic fate in the rat. Drug Metab Dispos 1999; 27(2):309–316.

56. Donovan JL, Kasim-Karakas S, German JB, Waterhouse AL. Urinary excretion of catechin metabolites by human subjects after red wine consumption. Br J Nutr 2002; 87(1):31–37.

57. Meng X, Lee M, Li C, Sheng S, Zhu N, Sang S, Ho C, Yang C. Formation and identification of 4′-O-methyl-(−)-epigallocatechin in humans. Drug Metab Dispos 2001; 29(6):789–793.

58. Kida K, Suzuki M, Matsumoto N, Nanjo F, Hara Y. Identification of biliary metabolites of (−)-epigallocatechin gallate in rats. J Agric Food Chem 2000; 48(9):4151–4155.

59. Baba S, Osakabe N, Yasuda A, Natsume M, Takizawa T, Nakamura T, Terao J. Bioavailability of (−)-epicatechin upon intake of chocolate and cocoa in human volunteers. Free Radic Res 2000; 33(5):635–641.

60. Das NP, Griffiths LA. Studies on flavonoid metabolism: metabolism of (+)-[^{14}C]catechin in the rat and guinea pig. Biochem J 1969; 115(4):831–836.

61. Kohri T, Matsumoto N, Yamakama M, Suzuki M, Nanjo F, Hara Y, Oku N. Metabolic fate of (−)-[4-^{3}H] epigallocatechin gallate in rats after oral administration. J Agric Food Chem 2001; 49(8):4102–4112.

62. Das NP, Griffiths LA. Studies on flavonoid metabolism: metabolism of (+)-catechin in the guinea pig. Biochem J 1968; 110(3):449–456.

63. Gott D, Griffiths L. Effects of antibiotic pretreatments on the metabolism and excretion of [U14C](+)-catechin [(U14C](+)-cyanidanol-3) and its metabolite, 3′-O-methyl-(+)-catechin. Xenobiotica 1987; 17(4):423–434.

64. Li C, Meng X, Winnik B, Lee M, Lu H, Sheng S, Buckley B, Yang C. Analysis of urinary metabolites of tea catechins by liquid chromatography/electrospray ionization mass spectrometry. Chem Res Toxicol 2001; 14(6):702–707.

65. Meselhy MR, Nakamura N, Hattori M. Biotransformation of (−)-epicatechin 3-O-gallate by human intestinal bacteria. Chem Pharm Bull 1997; 45(5):888–893.

66. Pietta P, Simonetti P, Gardana C, Brusamolino A, Morazzoni P, Bombardelli E. Catechin metabolites after intake of green tea infusions. Biofactors 1998; 8(1–2):111–118.

67. Li C, Lee M, Sheng S, Meng X, Prabhu S, Winnik B, Huang B, Chung J, Yan S, Ho C, Yang C. Structural identification of two metabolites of catechins and their kinetics in human urine and blood after tea ingestion. Chem Res Toxicol 2000; 13(3):177–184.

68. Cheng Z, Radominska-Pandya A, Tephly TR. Studies on the substrate specificity

of human intestinal UDP-glucuronosyltransferases 1A8 and 1A10. Drug Metab Dispos 1999; 27(10):1165–1170.

69. Ghazali RA, Waring RH. The effects of flavonoids on human phenolsulpho-transferases: potential in drug metabolism and chemoprevention. Life Sci 1999; 65(16):1625–1632.

70. Galijatovic A, Walle UK, Walle T. Induction of UDP-glucuronosyltransferase by the flavonoids chrysin and quercetin in Caco-2 cells. Pharm Res 2000; 17(1): 21–26.

71. Eaton EA, Walle UK, Lewis AJ, Hudson T, Wilson AA, Walle T. Flavonoids, potent inhibitors of the human p-form phenolsulfotransferase—potential role in drug metabolism and chemoprevention. Drug Metab Dispos 1996; 24(2):232–237.

72. Raftogianis RB, Wood TC, Weinshilboum RM. Human phenol sulfotransferases *sult1A2* and *sult1A1*: genetic polymorphisms, allozyme properties, and human liver genotype-phenotype correlations. Biochem Pharmacol 1999; 58(4):605–616.

73. Piskula MK, Terao J. Accumulation of (−)-epicatechin metabolites in rat plasma after oral administration and distribution of conjugation enzymes in rat tissues. J Nutr 1998; 128(7):1172–1178.

74. Pacifici G, Franchi M, Bencini C, Repetti F, DiLascio N, Murano GB. Tissue distribution of drug-metabolizing enzymes in humans. Xenobiotica 1988; 18(7):849–856.

75. Dunn RTI, Klaassen CD. Tissue-specific expression of rat sulfotransferase messenger RNAs. Drug Metab Dispos 1998; 26(6):598–604.

76. Meier PJ, Eckhardt U, Schroeder A, Hagenbuch B, Steiger B. Substrate specificity of sinusoidal bile acid and organic anion uptake systems in rat and human liver. Hepatology 1997; 26(6):1667–1677.

77. Csala M, Banhegyi G, Braun L, Szirmai R, Burchell A, Burchell B, Benedetti A, Mandl J. β-glucuronidase latency in isolated murine hematocytes. Biochem Pharmacol, 2000; 59(7):801–805.

78. Hollman PCH, van Trijp JMP, Buysman MNCP, Gaag MSvd, Mengelers MJB, de Vries JHM, Katan MB. Relative bioavailability of the antioxidant flavonoid quercetin from various foods in man. FEBS Lett 1997; 418(1–2):152–156.

79. Rozman KK, Klaassen CD. Absorption, distribution, and excretion of toxicants. In: Klaassen CD, Amdur MO, Doull J, eds. Casarett and Doull's Toxicology. New York: McGraw-Hill, 1996:91–112.

80. Chow H-HS, Cai Y, Albert DS, Hakim I, Dorr R, Shahi F, Crowell J, Yang C, Hara Y. Phase I pharmacokinetic study of tea polyphenols following single-dose administration of epigallocatechin gallate and polyphenon E. Cancer Epidemiol Biomark Prev 2001; 10(1):53–58.

81. Griffiths LA. Studies on flavonoid metabolism: identification of the metabolites of (+)-catechin in rat urine. Biochem J 1964; 92(1):173–179.

82. Crespy V, Morand C, Manach C, Besson C, Demigne C, Remesy C. Part of quercetin absorbed in the small intestine is conjugated and further secreted in the intestinal lumen. Am J Physiol 1999; 277(pt 1):G120–G126.

83. Walgren RA, Karnaky KJ Jr, Lindenmayer GE, Walle T. Efflux of dietary flavonoid quercetin 4′-β-glucoside across human intestinal Caco-2 cell monolayers

by apical multidrug resistance-associated protein-2. J Pharmacol Exp Ther 2000; 294(3):830–836.

84. Walle UK, Galijatovic A, Walle T. Transport of the flavonoid chrysin and its conjugated metabolites by the human intestinal cell line Caco-2. Biochem Pharmacol 1999; 58(3):431–438.

85. Borst P, Evers R, Kool M, Wijnholds J. The multi-drug resistance protein family. Biochem Biophys Acta 1999; 1461(2):347–357.

86. Vaidyanathan JB, Walle T. Transport and metabolism of the tea flavonoid (−)-epicatechin by the human intestinal cell line Caco-2. Pharmaceut Res 2001; 18(10):1420–1425.

87. Bray HG, Thorpe WV. Analysis of phenolic compounds of interest in metabolism. Methods Biochem Anal 1954; 1(1):27–52.

88. Luthria DL, Jones AD, Donovan JL, Waterhouse AL. GC-MS determination of catechin and epicatechin levels in human plasma. J High Resol Chromatogr 1997; 20(11):621–623.

89. Soleas GJ, Yan J, Goldberg DM. Measurement of trans-resveratrol, (+)- catechin, and quercetin in rat and human blood and urine by gas chromatography with mass selective detection. Methods Enzymol 2001; 335(1):130–145.

90. Manach C, Texier O, Morand C, Crespy V, Regerat F, Demigne C, Remesy C. Comparison of the bioavailability of quercetin and catechin in rats. Free Radic Biol Med 1999; 27(11–12):1259–1266.

91. Sesink A, O'Leary K, Hollman P. Quercetin glucuronides but not glycosides are present in human plasma after consumption of quercetin-3-glucoside or quercetin-4'-glucoside. J Nutr 2001; 131(7):1938–1941.

92. Donovan JL, Luthria DL, Stremple P, Waterhouse AL. Analysis of (+)- catechin, (−)-epicatechin and their 3'- and 4'-O-methylated analogs: a comparison of sensitive methods. J Chromatogr B 1999; 726(1–2):277–283.

93. Nakagawa K, Miyazawa T. Chemiluminescence high-performance liquid chromatographic determination of tea catechin, (−)-epigallocatechin 3-gallate, at picomole levels in rat and human plasma. Anal Biochem 1997; 248(1):41–49.

94. Giles AR, Gumma A. Biopharmaceutical evaluation of cyanidanol tablets using pharmacokinetic techniques. Arzneim Forsch 1973; 23(1):98–100.

95. Ritchey JG, Waterhouse AL. A standard red wine: monomeric phenolic analysis of commercial cabernet sauvignon wines. Am J Enol Vitic 1999; 50(1):91–100.

96. Waterhouse AL, Shirley JR, Donovan JL. Antioxidants in chocolate. Lancet 1996; 348(9030):834.

97. Richelle M, Tavazzi I, Enslen M, Offord EA. Plasma kinetics in man of epicatechin from black chocolate. Eur J Clin Nutr 1999; 53(1):22–26.

98. Balentine DA. Manufacturing and chemistry of tea. In: Ho C-T, Lee CY and Huang M-T, eds. Phenolic Compounds in Foods and Their Effects on Health 1. Analysis, Occurrence and Chemistry. Washington, DC: American Chemical Society, 1992:102–117.

99. Maiani G, Serafini M, Salucci M, Azzini E, FerroLuzzi A. Application of a new high-performance liquid chromatographic method for measuring selected polyphenols in human plasma. J Chromatogr B 1997; 692(2):311–317.

100. Nakagawa K, Okuda S, Miyazawa T. Dose-dependent incorporation of tea catechins, (−)-epigallocatechin-3-gallate and (−)-epigallocatechin, into human plasma. Biosci Biotech Biochem 1997; 61(12):1981–1985.

101. van het Hof K, Kivits G, Weststrate J, Tijburg L. Bioavailability of catechins from tea: the effect of milk. Eur J Clin Nutr 1998; 52(5):356–359.

102. Treutter D. Chemical reaction detection of catechins and proanthocyanidins with 4-dimethylaminocinnamaldehyde. J Chromatogr A 1989; 467(1):185–193.

103. He YH, Kies C. Green and black tea consumption by humans: impact on polyphenol concentrations in feces, blood and urine. Plant Foods Hum Nutr 1994; 46(3):221–229.

104. Serafini M, Maiani G, Ferro-Luzzi A. Alcohol-free red wine enhances plasma antioxidant capacity in humans. J Nutr 1998; 128(6):1003–1007.

105. Serafini M, Ghiselli A, Ferro-Luzzi A. In vivo antioxidant effect of green and black tea in man. Eur J Clin Nutr 1996; 50(1):28–32.

106. Serafini M, Ghiselli A, Ferro-Luzzi A. Red wine, tea, and antioxidants. Lancet 1994; 344(8922):626.

107. Unno T, Kondo K, Itakura H, Takeo T. Analysis of (−)-epigallocatechin gallate in human serum obtained after ingesting green tea. Biosci Biotech Biochem 1996; 60(12):2066–2068.

108. Yang B, Arai K, Kusu FD. Determination of catechins in human urine subsequent to tea ingestion by high-performance liquid chromatography with electrochemical detection. Anal Biochem 2000; 238(1):77–82.

17

Absorption and Metabolism of Hydroxycinnamates

Andreas R. Rechner
King's College London
London, England

I. INTRODUCTION

The hydroxycinnamates are ubiquitous in the plant kingdom [1]. Derivatives of the major hydroxycinnamates, such as p-coumaric acid (4-hydroxycinnamic acid), caffeic acid (3,4-dihydroxycinnamic acid), ferulic acid (4-hydroxy-3-methoxycinnamic acid), and sinapic acid (4-hydroxy-3,5-dimethoxycinnamic acid), are found in fruits, vegetables, grains, and coffee [1–3]. They are also central compounds to the polyphenol biosynthetic pathway of plants. The shikimate pathway involves the metabolism of phenylalanine, which is converted to *trans*-cinnamic acid, followed by a hydroxylation at the 4-position of the aromatic ring forming 4-hydroxycinnamic acid or p-coumaric acid. Further hydroxylation at the 3-position yields caffeic acid, with subsequent O-methylation of the 3-hydroxy group, resulting in ferulic acid. In fruits and vegetables the hydroxycinnamates predominantly occur as esters with organic acids, such as quinic acid, tartaric acid, or malic acid, or with sugars, such as glucose, whereas glycosides are seldom found [1–3]. Oligomeric forms, especially of ferulic acid, are found in grains [2]. The most extensively occurring hydroxycinnamic acid derivatives are esters of caffeic acid with quinic acid, predominantly chlorogenic acid (IUPAC name: 5-caffeoyl quinic acid) [4], the main constituent in coffee, apple juice, artichoke, eggplant, and peach. The 3-caffeoyl quinic acid ester is present as a major component of cherry, plum, elderberry, and apricot, 4-caffeoyl quinic acid and various dicaffeoyl quinic acids are minor constituents of some fruits and vegetables [1–3].

441

The dietary intake of hydroxycinnamates is thought to range from 25 mg/day to 1 g/day (predominantly caffeic acid), comparing noncoffee drinkers who consume few fruits and vegetables with coffee drinkers who have a high intake of fruits and vegetables [3]. Caffeic acid conjugates usually dominate the daily dietary intake of hydroxycinnamates from the sources mentioned, but bran-containing food contains high levels of ferulic acid conjugates [3].

II. ABSORPTION AND METABOLISM

The absorption, metabolism, and elimination of hydroxycinnamates, either from dietary sources or in form of the pure compound, have been studied in humans and animals. The pioneering work of Booth and DeEds in the 1950s and 1960s in animals and humans largely contributed to the understanding of hydroxycinnamate metabolism. In oral administration studies in rats, rabbits, and humans, as well as in in vitro experiments with isolated rat or rabbit liver, a number of metabolic events have been identified. These include the cleavage of ester bonds, O-methylation via catechol O-methyl transferase, dehydroxylation, reduction of the residual double bond, β-oxidation, decarboxylation, and glycination [5–16]. In nature as well as in the diet derivatives of p-coumaric, caffeic, ferulic, and sinapic acids represent the most commonly occurring hydroxycinnamates. The following sections deal with the metabolic fate of the free acids and their numerous derivatives, which are the forms of their natural occurrence.

A. Caffeic Acid and Its Derivatives

The metabolites and conjugates of caffeic acid and caffeic derivatives identified in the different studies in humans and animals are summarized in Table 1. The majority of studies in humans and animals on absorption and metabolism of hydroxycinnamates have been undertaken with caffeic acid and its main natural conjugate, chlorogenic acid. The possession of a catechol moiety results in extensive metabolism of caffeic acid, which leads to the formation of a number of metabolites. The metabolic pathways of caffeic acid are schematically summarized in Figure 1 [6,16].

In early studies in humans and rats, Booth and associates [6] detected and identified by paper chromatography 10 metabolites in human urine after the ingestion of caffeic acid, namely, caffeic acid itself, ferulic acid, dihydrocaffeic acid, dihydroferulic acid, vanillic acid, m-coumaric acid, 3-(3-hydroxyphenyl)-propionic acid, 3-hydroxyhippuric acid, feruloylglycine, and vanilloylglycine. All identified metabolites appeared until 8 h post ingestion, except 3-hydroxyhippuric acid, which appeared between 8 and 24 h post ingestion [6]. Rates of conjugation with glucuronic acid were not investigated. Interestingly, the metabolism of caffeic acid in rats differed significantly, displaying 3-(3-hydroxyphenyl)-propionic acid as the major urinary metabolite. In contrast to humans, rats also metabolized caffeic acid to compounds possessing a 3-carbon side chain with only traces of 1-carbon side chain compounds. The ingestion of 1 g of chlorogenic acid in humans resulted in the presence of caffeic acid, 3-hydroxyhippuric acid, m-coumaric acid, and dihydroferulic acid in urine, whereas the ingestion of coffee (equivalent to approximately 2 g of chlorogenic acid) led only to the excretion of 3-hydroxyhippuric acid and m-coumaric acid. As for caffeic acid, a significant difference of chlorogenic acid metabolism was observed between humans and rats: 3-(3-hydroxyphenyl)-propionic acid was the major urinary metabolite in rats.

The urinary excretion of unchanged caffeic acid post β-glucuronidase treatment was also observed after the apllication of 1 g of pure caffeic acid in humans [17]. In addition, vanillic, ferulic, and isoferulic acids were also identified as key urinary elimination products after β-glucuronidase treatment. The majority of the identified metabolites were excreted in the first 4 h post ingestion, and the total recovery of caffeic acid and its identified metabolites after 48 h was approximately 10% of the ingested dose. The metabolic fate of the remaining caffeic acid could not be determined as no caffeic acid was detected in fecal samples during the post-ingestion period of 48 h. Caffeic acid was also identified as a metabolite of ferulic acid in addition to ferulic acid itself and two unidentified metabolites in urine post β-glucuronidase treatment. After ingestion of 1 L of coffee, only ferulic acid was identified in urine post β-glucuronidase treatment.

In a human study, in which 2 cups of coffee were ingested three times at 4-h intervals (149.7 mg hydroxycinnamates per cup equivalent to a total intake of 898.4 mg hydroxycinnamates relative to chlorogenic acid) after 1 day of a polyphenol-free diet, the methylated metabolites ferulic, isoferulic, and vanillic acids were major elimination products detected in urine post β-glucuronidase treatment [18]. No caffeic acid or chlorogenic acid was found. Unconjugated dihydroferulic acid was also identified, and 3-hydroxyhippuric acid (3-hydroxy-benzoylglycine) expressed a strong association between the amounts excreted and the ingestion of the coffee. The latter two compounds were excreted in much higher amounts than the conjugates ferulic, isoferulic, and vanillic acids but showed maximal excretion more than 12 h after the first coffee ingestion, indicating a prolonged metabolic route. The highest amounts of the conjugates of the other metabolites were excreted 1–3 h after any of the three coffee ingestions, implying fast absorption, metabolism, and elimination. A representative time profile of the amounts excreted of the identified metabolites is shown in Fig. 2 [18]. An association between the coffee ingestion and the amounts excreted of hippuric acid (benzoylglycine), a general aromatic metabolite in

Table 1 Nature of Urinary and Plasma Metabolites of Hydroxycinnamates in Humans and Rats

Reference	Humans		Rats	
	Urine	Plasma	Urine	Plasma
Caffeic acid[a]				
[6]	c–l	—	d,f,g,i,j,l	—
[17]	c,d,g,m	—	—	—
[23]	—	—	—	c,d
[21]	c	—	—	—
Chlorogenic acid and caffeic acid derivatives				
[6][a,b]	c,f,h,j	—	e,f,h–j	—
[17][a,b]	d	—	—	—
[24]	—	—	—	—
[20][b]	d,f,i,k (13)	—	—	—
[23][a]	—	—	—	c,d
[21][a]	c,n	—	—	—
[22][b]	c,d,n	c,d,n	—	—
[18][b]	d,f,g,j,m (13)	—	—	—
[19][b]	d,f,g,m	d,m	—	—
Ferulic acid				
[6][a]	—	—	d,f,g,i,k,l	—
[17][b]	c, d, g	—	—	—
[24][a]	—	—	d	—
[28][b]	d	—		
[29][b]	d	—	—	—
p-Coumaric acid				
[9][a]	—	—	p–t	—
[14][a]	—	—	p–t	—
Sinapic acid				
[33][a]	—	—	s, t, u–y	—
[34][a]	—	—	w	—
[35][a]	—	—	u, w	—

[a] Pure compound.
[b] Natural or nutritional source.
[c] Caffeic acid (3,4-dihydroxycinnamic acid).
[d] Ferulic acid (4-hydroxy-3-methoxycinnamic acid).
[e] Dihydrocaffeic acid (3-(3,4-dihydroxyphenyl)-propionic acid).
[f] Dihydroferulic acid (3-(4-hydroxy-3-methoxyphenyl)-propionic acid.
[g] Vanillic acid (4-hydroxy-3-methoxybenzoic acid).
[h] m-Coumaric acid (3-hydroxycinammic acid).

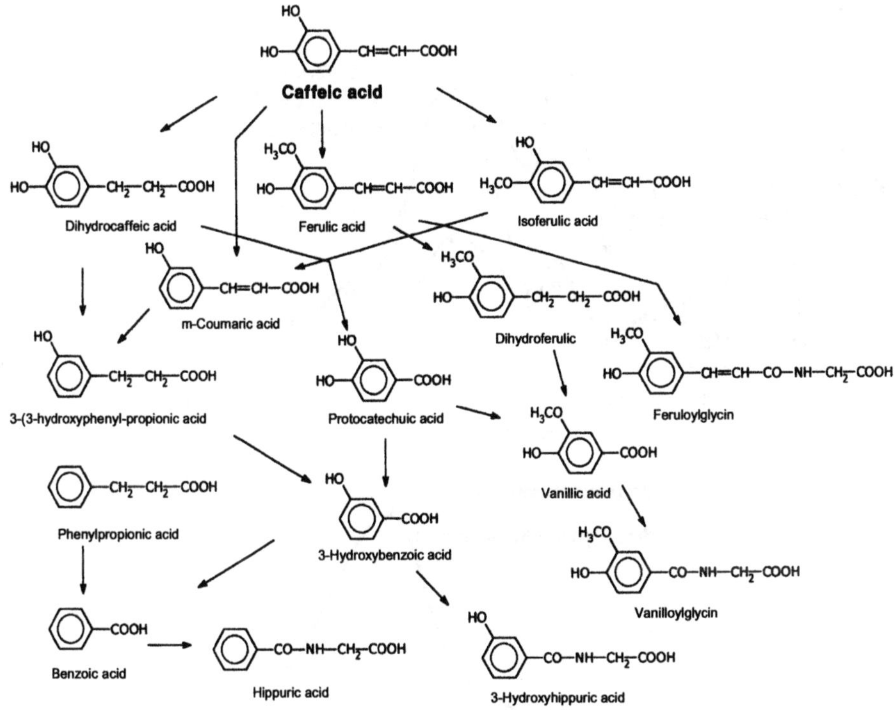

Figure 1   The metabolic pathways of caffeic acid. (From Refs. 6,16.)

Notes to Table 1:
[i] 3-(3-Hydroxyphenyl)-propionic acid.
[j] 3-Hydroxyhippuric acid (3-hydroxybenzoylglycine).
[k] Feruloylglycine.
[l] Vanilloylglycine.
[m] Isoferulic acid (3-hydroxy-4-methoxycinnamic acid).
[n] Chlorogenic acid (5-caffeoyl quinic acid).
[o] Hippuric acid (benzoylglycine).
[p] p-Coumaric acid (4-hydroxycinnamic acid).
[q] p-Coumaroylglycine.
[r] 4-Hydroxybenzoic acid.
[s] 3-(4-Hydroxyphenyl)-propionic acid.
[t] 4-Hydroxyhippuric acid (4-hydroxybenzoylglycine).
[u] Sinapic acid (4-hydroxy-3,5-dimethoxycinnamic acid).
[v] Dihydrosinapic acid (3-(4-hydroxy-3,5-dimethoxyphenyl)-propionic acid.
[w] 3-(3,5-Dihydroxyphenyl)-propionic acid.
[x] 3-Hydroxy-5-methoxycinnamic acid.
[y] 3-(3-Hydroxy-5-methoxyphenyl)-propionic acid.

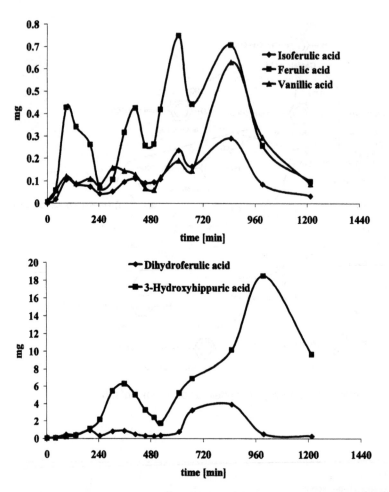

Figure 2 Time-amount profile of the excretion of the chlorogenic acid metabolites, ferulic, isoferulic, vanillic, dihydroferulic, and 3-hydroxyhippuric acid, in
single elimination samples after the consumption of coffee (three times in 4-h
intervals) for one volunteer. (From Ref. 18.)

urine, was also detected, leading to speculations about the metabolic fate of the
ingested hydroxycinnamates. Because of small amounts of ferulic acid derivatives in coffee, the identified metabolite could not be accounted for exclusively
by caffeic acid derivatives, apart from isoferulic acid, which most likely derives
from O-methylation of caffeic acid.

In another human study with a similar design, artichoke extract capsules
were administered three times in 4-h intervals (total hydroxycinnamate intake
123.9 mg relative to chlorogenic acid) [19]. In contrast to coffee, artichoke

contains only caffeic acid esters but no ferulic acid derivatives or other hydroxycinnamates. Investigation of urinary excretion of conjugates and metabolites post β-glucuronidase treatment revealed ferulic, isoferulic, vanillic, and dihydroferulic acids as metabolites of the ingested caffeic acid derivatives.

Many other studies have reported the appearance of methylated products as metabolites of caffeic acid derivatives. In particular, ferulic and dihydroferulic acids and additionally feruloylglycine were identified as urinary metabolites of caffeic acid derivatives from a crude extract of *Equisetum arvense* in a human study [20]. The rate of conjugation with glucuronic acid in this study was >90% for ferulic acid, 60–80% for dihydroferulic acid, and 16%   for feruloylglycine.

The absorption of caffeic acid and chlorogenic acid has been examined in healthy ileostomy (no colon) patients [21]. The extent of absorption was calculated as the amount ingested minus the amount excreted in ileostomy effluent. Using this calculation, complete absorption of caffeic acid and 33 ± 17% absorption of chlorogenic acid were observed, whereas only 11% of the ingested caffeic acid and traces of chlorogenic acid were detected in urine. No other metabolites were mentioned. Incubation with gastric juice, duodenal fluid, and ileostomy fluid had no effect on both compounds, which were completely recovered. The potential mechanism of degradation or metabolism of the compounds in the small intestine and their transfer to the circulation remain to be clarified.

Caffeic acid 42–96 nmol/L, entirely as glucuronide) and ferulic (not detectable–78 nmol/L, more than 50% as free acid) were detected in plasma of three volunteers post ingestion of 100 g of prunes after a 2-day low-hydroxycinnamate diet [22]. Furthermore, two of the three volunteers displayed *basal* plasma levels of caffeic acid (as glucuronide) and ferulic acid (more than 50% as free acid). Conjugated and free caffeic (0.146–0.496 μmol, 36–87% as free acid), ferulic acid (0.008–0.032 μmol), and chlorogenic acid (0.019–0.045 μmol as free acid) were identified post ingestion in the urine of the three volunteers, of whom two showed *basal* urinary levels of the compounds. Thus, there is considerable contradictory evidence for the potential absorption of chlorogenic acid per se.

In a study investigating the absorption of caffeic acid and chlorogenic acid in rats after oral administration, no absorption of chlorogenic acid and no metabolism in the small intestine were observed [23]. In contrast, caffeic acid was absorbed and detected in plasma post enzyme treatment accompanied by ferulic acid as a metabolite. Chlorogenic acid was only detected in urine of rats after intravenous administration (50 mg/kg) but also not after oral administration; that finding is consistent with the hypothesis that major hydrolysis to release free caffeic acid occurs in the colon [24]. In an in vitro model using isolated rat small intestine, chlorogenic acid was also not absorbed, but caffeic acid was found to be absorbed through the intestinal lumen [25]. The perfused caffeic acid was glucuronidated to a high extent (approximately 60%), most likely in the epithelium of the small intestine. Biliary excretion of glucuronidated caffeic acid into small intestine was found to be only a minor metabolic pathway in rats [26].

However, the administration of pure caffeic acid, especially in the high doses used in the studies, to humans or animals does not mirror the reality of the diet. It more or less represents an exposure to an unusual, unknown compound (i.e., in small intestine) in an overdose. This might result in a slightly distorted picture of caffeic acid metabolism, which does not start with caffeic acid itself but with caffeic acid derivative. It is clear that much remains to be clarified concerning the bioavailability of chlorogenic acid and caffeic acid deriving from chlorogenic acid, and the mechanisms and sites of action. Differences in study design and specifics of the volunteers studied, as well as the analytical system applied, especially the mode of detection and sample preparation, might be responsible for the present uncertainty.

The action of catechol O-methyl transferase might be a central metabolic event after the absorption of free caffeic acid or some of its metabolites with a still intact catechol moiety, such as dihydrocaffeic acid or protocatechuic acid. In most studies administering chlorogenic acid or preparations rich in caffeic acid derivatives (i.e., coffee), only O-methylated metabolites but no metabolites with an intact catechol group were detected in urine, supporting the central role of O-methylation of caffeic acid post absorption [6,17,18]. Studying the O-methylation of caffeic acid in vitro by using rat or rabbit liver slices or preparations of liver, both possible O-methylation products, ferulic and isoferulic acids, were formed, and a meta/para ratio of 2.8:1 was recorded [13]. In addition, the ability to reduce the residual double bond was also observed in vitro with rat or rabbit liver slices [10].

Crucial for the understanding of potential bioactivities of caffeic acid and its metabolites and derivatives is knowledge about their intracellular metabolism in the tissues after absorption and presence in the circulation. In vitro studies have demonstrated the ability of caffeic acid, chlorogenic acid, and dihydrocaffeic acid to form quinones, when oxidized with peroxidase/H_2O_2 or tyrosinase/O_2, which react with glutathione to form a number of conjugates [27]. The glutathione conjugates were also formed in the presence of reduced nicotinamide-adenine dinucleotide phosphate (NADPH) in isolated rat hepatocyte microsomes. This was prevented by a cytochrome P450 inhibitor, suggesting the involvement of cytochrome P450 in the formation of the glutathione conjugates. Furthermore, the cytotoxicity of three compounds to isolated rat hepatocyates was enhanced by H_2O_2 or cumene hydroperoxide–supported cytochrome P450 and was again prevented by a cytochrome P450 inhibitor. These results suggest that cytochrome P450 metabolically activates hydroxycinnamates and related compounds with a catechol moiety to form cytotoxic quinons or quinoid metabolites.

This cytotoxicity of caffeic acid, based on its catechol moiety and the interaction with cytochrome P450, might be the reason for the extensive O-methylation and glucuronidation of caffeic acid in humans observed in most studies investigating the absorption, metabolism, and elimination of dietary caffeic acid derivatives.

B. Ferulic Acid and Its Derivatives

The metabolites and conjugates of ferulic acid and ferulic derivatives identified in the different studies in humans and animals are summarized in Table 1. The metabolic fate of ferulic acid is qualitatively almost identical to that of caffeic acid (see Fig. 1). The same 3-hydroxyphenyl and 4-hydroxy-3-methoxyphenyl acids were detected in human urine after the ingestion of ferulic acid and caffeic acid [7]. When ferulic acid was fed to rats, 3-(3-hydroxyphenyl)-propionic acid, feruloylglycine, dihydroferulic acid, vanillic acid, and vanilloylglycine were identified in urine, also showing the close relationship to caffeic acid metabolism [6]. Also in rats, 3-(3-hydroxyphenyl)-propionic acid was reported as the major urinary metabolite of ingested ferulic acid, accompanied by smaller amounts of vanillic acid [12]. These early studies indicate that dehydroxylation, reduction of the side chain double bond, and β-oxidation are major metabolic events in ferulic acid metabolism. In humans caffeic acid was also detected alongside ferulic and vanillic acids and three unidentified metabolites in urine post β-glucuronidase treatment after the ingestion of 1 g of ferulic acid [17], suggesting the occurrence of demethylation. Conjugation with glucuronic acid was reported to be approximately 30–40% of the total amount excreted in urine after the consumption of a single bolus of tomatoes equivalent to an ingestion of 30.1 mg ferulic acid in the form of various derivatives [28]. The absorption rate of the ferulic acid from tomatoes, based on the free and conjugated ferulic acid excreted, varied between 11% and 25%, showing maximal excretion around 7 h post consumption. After the application of an extract from French maritime pine bark containing about 0.17% free ferulic acid and 0.47% ferulic acid in the form of ferulic acid glucoside, which totals 13.97 mg and 2.35 mg total ferulic acid ingested in two different administrations, in humans, an absorption rate of 36–43% based on total urinary excretion was observed [29]. The presence of free ferulic acid and the form of application (gelatin capsules) are suggested to cause the higher rate of absorption.

However, the O-methylation of the catechol group seems to influence the rate of absorption of ferulic acid compared to that of caffeic acid. In rats 10.5 ± 2.5% of an oral administered dose of 50 mg/kg was recovered in urine, approximately 50% as the free acid and the other 50% conjugated with glucuronic acid [24]. Intravenous administration resulted in similar recoveries of total ferulic acid in urine indicating extensive metabolism of ferulic acid in the circulation and tissues. In an in vitro model using isolated rat small intestine, the absorption of ferulic acid through the intestinal lumen was approximately four times higher than the absorption of caffeic acid [25]. The amount of conjugation with glucuronic acid was around 20%, much lower than the 60–70% observed for caffeic acid.

Ferulic acid was extensively excreted in the bile in the form of its glucuronide (30% of the dose) after intravenous or intradermal ingestion in rats [26]. In comparison, only small amounts of p-coumaric and caffeic acid (approximately 3% of the dose) were excreted in the bile. The biliary excretion of ferulic acid glucuronide might result in its absorption in the small intestine or, most likely, in its degradation in the colon. Dimers of ferulic acid, abundant structural components of plant cell walls, especially cereal brans, were released from the cell wall in the gastrointestinal tract of humans and rats by the colonic microflora expressing esterase activity [30]. In addition, diferulic acids were detected in rat plasma post β-glucuronidase/sulfatase treatment after the administration of diferulic acids in sunflower oil.

C. p-Coumaric Acid and Its Derivatives

The metabolites and conjugates of p-coumaric acid identified in the different studies in animals are summarized in Table 1. The metabolism of p-coumaric acid is relatively straightforward and is reported to involve conjugation with glucuronic acid or glycine in the liver or epithelium of small intestine, reduction of residual double bond to 3-(4-hydroxyphenyl)-propionic acid, and β-oxidation of the reduction product to 4-hydroxybenzoic acid. These products and the unconjugated p-coumaric acid have been detected and identified in urine of rats after the ingestion of p-coumaric acid [9,14]. The β-oxidation of p-coumaric acid occurs in the liver and is carried out by adenosine triphosphate–(ATP)-requiring enzymes localized in the mitochondria [31]. The reduction of the side chain double bond of p-coumaric acid was also identified as a metabolic action of colonic microorganisms [11]. Dehydroxylation or O-methylation of the phenolic hydroxyl group has not been described.

The absorption of p-coumaric acid was shown in the rat gut after the feeding of spinach cell walls with ^{14}C-labeled ferulic and p-coumaric acids to rats [32]. The two free hydroxycinnamic acids were absorbed after release from the cell wall in the cecum and colon, a finding that strongly suggested an involvement of colonic bacteria. In an in vitro study of the absorption of polyphenols using isolated small intestine from the rat, p-coumaric acid was absorbed to a higher extent than caffeic acid and to a lower extent than ferulic acid [25]. Glucuronidation of p-coumaric acid in the epithelium of the rat small intestine was not observed in this study. Biliary excretion of p-coumaric acid (as glucuronide) was shown to be only a minor metabolic pathway in rats [26]. The metabolism of p-coumaric acid is schematically summarized in Figure 3.

D. Sinapic Acid and Its Derivatives

The metabolites and conjugates of sinapic acid identified in the different studies in animals are summarized in Table 1. The metabolism of sinapic acid has been studied only in animals, rat and rabbit. The unchanged hydroxycinnamate, 3-(3,5-dihydroxyphenyl)-propionic acid, dihydrosinapic acid (3-(4-hydroxy-3, 5-dimethoxyphenyl)-propionic acid), 3-hydroxy-5-methoxycinnamic acid, and 3-(3-hydroxy-5-methoxyphenyl)-propionic acid were detected and identified as urinary metabolites in rats after an oral dose of 800 mg/kg [33,34]. Interestingly, the maximal excretion of the demethylated products occurred 2 or 3 days post ingestion, suggesting a prolonged metabolism. After the oral administration of 100 mg/kg to rats, only the unchanged sinapic acid and 3-(3,5-dihydroxyphenyl)-propionic acid were excreted, both partly conjugated [35]. Sinapic acid was excreted during 24 h post administration, whereas 3-(3,5-dihydroxyphenyl)-propionic acid excretion lasted 2 or 3 days. The 3-(3,5-dihydroxyphenyl)-propionic acid was shown to be the major urinary metabolites in rabbits fed with sinapic acid (200 mg/kg). The intestinal microflora of the rabbit was identified as the metabolic site of action for the demethylation. A more detailed study on the bacterial degradation revealed that sinapic acid was reduced to dihydrosinapic acid and then O-methylated and dehydroxylated, finally forming 3-(3,5-dihydroxyphenyl)-propionic acid [36].

The metabolic fate of sinapic acid in humans as well as its derivatives, which represent the naturally occurring form of sinapic acid, has not been studied yet and remains uncertain. But in a study on the metabolic fate of polyphenols from a polyphenol-rich diet, small amounts of sinapic acid were detected in urine post β-glucuronidase treatment during a 5-day feeding period [37]. This presence of sinapic acid (as glucuronide) in urine implies its bioavailability from the diet in humans and a metabolic fate of dietary sinapic acid derivatives similar to that of other hydroxycinnamic acid derivatives such as chlorogenic acid.

III. COLONIC METABOLISM

The colonic microflora are a central site of hydroxycinnamate metabolism in addition to the small intestine and liver. The majority of ingested hydroxycinnamates from dietary sources reach the colon, where they are exposed to the

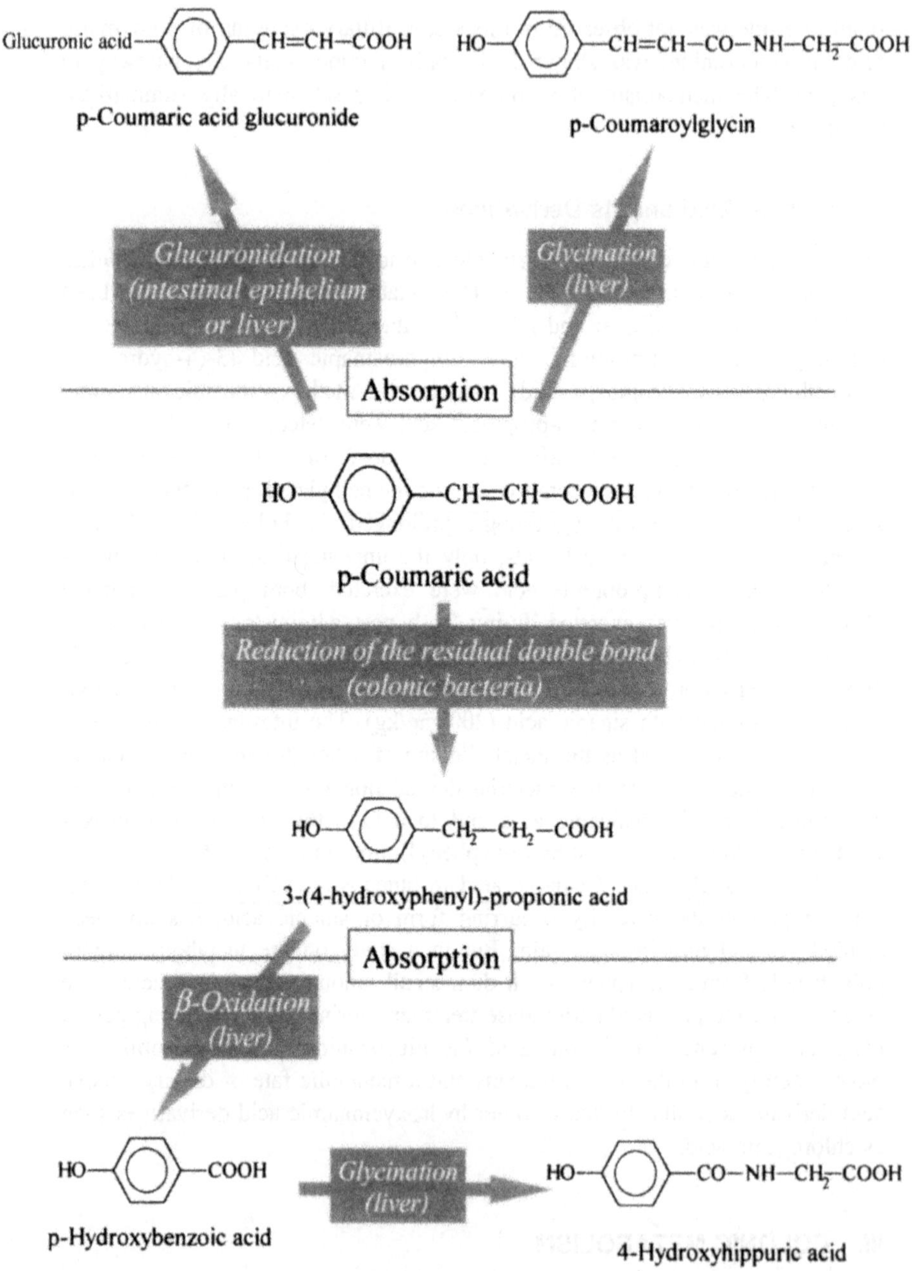

Figure 3 The metabolic pathways of p-coumaric acid.

colonic microflora and their metabolic properties. Many metabolic events have been suggested as actions of bacterial enzymes, including cleavage of the ester bond, dehydroxylation, demethoxylation, reduction of the residual double bond, and α- and β-oxidation [11]. In investigation of the possible site of cleavage of the ester bond between caffeic acid and quinic acid, the colon microflora were identified in an in vitro experiment as the only active extract, whereas extracts of small intestine epithelium, plasma, and liver expressed no esterase activity [38]. No cleavage of the ester bond of chlorogenic acids occurred at the acidic pH of the gastric lumen, studied by incubating the pure compound or extract rich in caffeic acid esters with simulated gastric juice [18,19,21]. Some human fecal microorganism expressing cinnamoyl esterase activity were isolated and genotypically characterized as *Escherichia coli* (three isolates) and *Bifidobacterium lactis* and *Lactobacillus gasseri* (two strains) [39]. The esterase activity of the characterized bacteria was essentially intracellular. Interestingly, the characterized bacteria did not express any other phenol-degrading activities.

Other phenol-degrading activities of the human colonic microflora were identified in an in vitro incubation trial with chlorogenic acid [40]. After the initial cleavage of the ester bond the reduction of the side chain double bond of caffeic and the dehydroxylation of the product dihydrocaffeic acid resulted in the formation of the endproduct 3-(3-hydroxyphenyl)-propionic acid. A delay of approximately 5 h from the start of the incubation until the cleavage of the ester bond occurred was observed, indicating a period of adaptation of the colonic microflora to the substrate. Further dehydroxylation or α- and β-oxidation of the identified endproduct was not observed, but protocatechuic acid (β-oxidation of dihydrocaffeic acid) and m-coumaric acid (dehydroxylation of caffeic acid) were detected as minor metabolites during the incubation. O-Methylation of the catechol moiety of chlorogenic, caffeic, or dihydrocaffeic acid also did not occur, pointing to other sites of action for this metabolic event. The most likely site for O-methylation as well as glucuronidation is the liver, but the epithelium of the colon might also be capable of this metabolic action. The epithelium of the small intestine was able to glucuronidate and O-methylate flavonoids and hydroxycinnamates in in vitro experiments with isolated rat small intestine [25,41].

The colon seems to be the essential site for the release of free hydroxycinnamic acids and their absorption. In a study of the uptake of ^{14}C-labeled hydroxycinnamates bound to spinach cell walls in rats, the foregut was localized as the site of absorption after the release of the labeled hydroxycinnamic acids form the cell wall by colonic microorganisms, when 25% of the ingested dose of label was found to be associated with body tissue after only 2 h [32]. On the basis of the results on absorption, metabolism, degradation, and elimination of chlorogenic acid in humans presented, its metabolic fate in

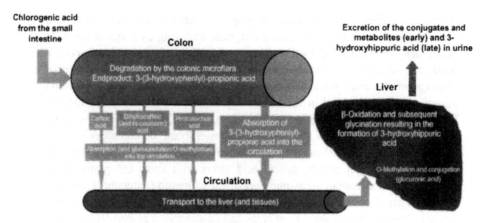

Figure 4   The metabolic fate of dietary chlorogenic acid.

humans is schematically summarized in Figure 4. However, infused into the rumen of the sheep, chlorogenic acid, caffeic acid, and ferulic acid underwent complete dehydroxylation and were mainly excreted as benzoic acid [15]. Small amounts of urinary cinnamic and 3-(3-hydroxyphenyl)-propionic acids were also detected. The complete dehydroxylation of free hydroxycinnamic acids would result in the formation of phenylpropionic acid, which is a proposed precursor of hippuric acid. An increase in hippuric acid excretion, a common urinary metabolite, was associated with a high dietary intake of polyphenols and/or hydroxycinnamates in humans and in rats [18,20,42,43]. Formed and absorbed in the colon, 3-phenylpropionic acid might contribute to the hippuric acid levels in plasma and urine after β-oxidation to benzoic acids and subsequent glycination in the liver. However, hippuric acid also derives from other dietary sources, such as aromatic amino acids, benzoates used as preservatives, and quinic acid, which together with caffeic acid forms chlorogenic acid [44,45].

IV.   CONCLUSIONS

In quantitative terms, colonic metabolism and the formed colonic metabolites are more important than the absorption and conjugation of the hydroxycinnamic acids. Whether these colonic metabolites express potential protective or preventive bioactivities in vivo remains to be studied. This also applies to the effect of the dietary hydroxycinnamates and their metabolites on the composition of the colonic microflora.

REFERENCES

1. Macheix JJ, Fleuriet A, Billot J. Fruit Phenolics. Boca Raton; FL: CRC Press, 1990.
2. Herrmann K. Occurrence and content of hydroxycinnamic and hydroxybenzoic acid compounds in foods. Crit Rev Food Sci Nutr 1989; 28:315–347.
3. Clifford MN. Chlorogenic acid and other cinnamates—nature, occurrence and dietary burden. J Sci Food Agric 1999; 79:362–372.
4. International Union of Pure and Applied Chemistry. Nomenclature of cyclitols. Biochem J 1976; 153:23–31.
5. DeEds F, Booth AN, Jones FT. Methylation of phenolic hydroxyl groups by rabbit. Fed Proc Fed Am Soc Exp Biol 1955, 14:189–205, 332.
6. Booth AN, Emerson OH, Jones FT, Deeds F. Urinary metabolites of caffeic and chlorogenic acids. J Biol Chem 1957; 229:51–59.
7. Shaw KNF, Trevarthen J. Exogenous sources of urinary phenol and indol acids. Nature 1958; 182:797.
8. Hill GA, Ratcliff J, Smith P. Urinary catechol ethers. Chem Ind 1959; 399:
9. Booth AN, Masri MS, Robbins DJ, Emerson OH, Jones FT, DeEds F. Urinary phenolic acid metabolites of tyrosine. J Biol Chem 1960; 235:2649.
10. Masri MS, Booth AN, DeEds F. O-Methylation in vitro of dihydroxyl- and trihydroxyl-phenolic compounds by liver slices. Biochim Biophys Acta 1962; 65:495.
11. Scheline RR. Metabolism of phenolic acids by the rat intestinal microflora. Acta Pharmacol Toxicol 1968; 26:189.
12. Teuchy H, Van Sumere CF. The metabolism of [1-^{14}C]phenylalanin, [3-^{14}C]cinnamic acid and [2-^{14}C]ferulic acid in the rat. Arch Int Physiol Biochim 1971, 79:589–597.
13. Creveling CR, Morris N, Shimizu H, Ong HH, Daly J. Catechol O-methyltransferase. IV. Factors affecting m-and p-mehylation of substituted catechols. Mol Pharmacol 1972; 8:398–409.
14. Griffiths LA, Smith GE, Metabolism of apigenin and related compounds in the rat. Biochem J 1972; 128:901–911.
15. Martin AK. The origin or urinary aromatic compounds excreted by ruminants. II. The metabolism of phenolic cinnamic acids to benzoic acid. Br J Nutr 1982; 47:155–164.
16. Scheline RR. Metabolism of acids, lactones, and esters. In: CRC Handbook of Mammalian Metabolism of Plant Compounds. Boca Raton; FL: CRC Press, 1991:167–175.
17. Jacobson EA, Newmark H, Baptista J, Bruce WR. A preliminary investigation of the metabolism of dietary phenolics in humans. Nutr Rep Int 1983; 28:1409–1417.
18. Rechner AR, Spencer JPE, Kuhnle G, Hahn U, Rice-Evans CA. Novel biomarkers of the bioavailability and metabolism of caffeic acid derivatives in humans. Free Radic Biol Med 2001; 30:1213–1222.
19. Rechner AR, Pannala AS, Rice-Evans CA. Caffeic acid derivatives in artichoke extract are metabolized to phenolic acids in vivo. Free Radic Res 2001; 35: 195–202.

20. Graefe EU, Veit M. Urinary metabolites of flavonoids and hydroxycinnamic acids in humans after application of a crude extract from Equisetum arvense. Phytomedicine 1999; 6:239–246.

21. Olthof MR, Hollman PCH, Katan MB. Chlorogenic acid and caffeic acid are absorbed in humans. J Nutr 2001; 131:66–71.

22. Cremin P, Kasim-Karakas S, Waterhouse AL. LC/ES-MS detection of hydroxycinnamates in human plasma and urine. J Agric Food Chem 2001; 49: 1747–1750

23. Azuma K, Ippuushi K, Nakayama M, Ito H, Higashio H, Terao J. Absorption of chlorogenic acid and caffeic acid in rats after oral administration. J Agric Food Chem 2000; 48:5496–5500.

24. Choudhury R, Srai K, Debnam E, Rice-Evans CA. Urinary excretion of hydroxycinnamates and flavonoids after oral and intravenous administration. Free Radic Biol Med 1999; 27:278–286.

25. Spencer JPE, Chowrimootoo G, Choudhury R, Debnam ES, Srai SK, Rice-Evans CA. The small intestine can both absorb and glucuronidate luminal flavonoids. FEBS Lett 1999; 458:224–230.

26. Westendorf J, Czok G. Studies on the pharmacokinetics of C-14 cinnamic-acid derivatives in rats. Z Ernährungswiss 1978; 17:26–36.

27. Moridani MY, Scobie H, Jamshidzadeh A, Salehi P, O'Brien PJ. Caffeic acid, chlorogenic acid, and dihydrocaffeic metabolism: glutathione conjugate formation. Drug Metab Dispos 2001; 29:1432–1439.

28. Bourne LC, Rice-Evans CA. Bioavailability of ferulic acid. Biochem Biophys Res Comm 1998; 253:222–227.

29. Grosse Düweler K, Rohdewald P. Urinary metabolites of French maritime pine bark extract in humans. Pharmazie 2000; 55:364–368.

30. Andreasen MF, Kroon PA, Williamson G, Garcia-Conesa MT. Intestinal release and uptake of phenolic antioxidant diferulic acids. Free Radic Biol Med 2001; 31:304–314.

31. Ranganathan S, Ramasarma T. Enzymic formation of p-hydroxybenzoate from p-hydroxycinnamate. Biochem J 1971; 122:487–493.

32. Buchanan CJ, Wallace G, Fry SC, Eastwood MA. In vivo release of 14C-labelled phenolic groups from intact dietary spinach cell walls during passage through the rat intestine. J Sci Food Chem 1996; 71:459–469.

33. Griffiths LA. Metabolism of sinapic acid and related compounds in the rat. Biochem J 1969; 113:603–609.

34. Griffiths LA. 3,5-dihydroxyphenylpropionic acid, a further metabolite of sinapic acid. Experientia 1970; 26:723–724.

35. Meyer T, Scheline RR. 3,4,5-trimethoxycinnamic acid and related compounds. I. Metabolism in the rat. Xenobiotica 1972; 2:391–398.

36. Meyer T, Scheline RR. 3,4,5-trimethoxycinnamic acid and related compounds. I. Metabolism in by the rat intestinal microflora. Xenobiotica 1972; 2:383–390.

37. Rechner AR, Kuhnle G, Bremner P, Hubbard GP, Moore KP, Rice-Evans CA. The metabolic fate of dietary polyphenols in humans. Free Rad Biol Med 2002; 33:220–235.

38. Plumb GW, Garcia-Conesa MT, Kroon PA, Rhodes M, Ridley S, Williamson G. Metabolism of chlorogenic acid by human plasma, liver, intestine and gut microflora. J Sci Food Agric 1999; 79:390–392.

39. Coteau D, McCartney AL, Gibson GR, Williamson G, Faulds CB. Isolation and characterization of human colonic bacteria able to hydrolyse chlorogenic acid. J Appl Microbiol 2001; 90:873–881.

40. Rechner AR, Watts WMP, Gibson GR, Rice-Evans CA. Colonic metabolism of dietary phenolics: implications for the nature of the bioactive component. Free Rad Biol Med 2001; 31:89S.

41. Donovan JL, Crespy V, Manach C, Morand C, Besson C, Scalbert A, Remesy C. Catechin is metabolized by both the small intestine and liver of rats. J Nutr 2001; 131:1753–1757.

42. Phibbs AN, Stewart J, Wright B, Wilson ID. Effect of diet on the urinary excretion of hippuric acid and other dietary-derived aromatics in rat: a complex interaction between diet, gut microflora and substrate specifity. Xenobiotica 1998; 28:527–537.

43. Clifford MN, Copeland EL, Bloxsidge JP, Mitchell LA. Hippuric acid as a major excretion product associated with black tea consumption. Xenobiotica 2000; 50: 317–326.

44. Chesson A, Provan GJ, Russell WR, Scobbie L, Richardson AJ, Stewart C. Hydroxycinnamic acids in the digestive tract of livestock and humans. J Sci Food Agric 1999; 79:373–378.

45. Indahl SR, Scheline RR. Quinic acid aromatization in the rat: urinary hippuric acid and catechol excretion following the singular or repeated administration of quinic acid. Xenobiotica 1973; 3:549–556.

Index